Process Systems Engineering

Edited by
Michael C. Georgiadis,
Julio R. Banga, and
Efstratios N. Pistikopoulos

Related Titles

Antonov, A.

Mathematica for Chemists and Chemical Engineers

The Program for Mathematical Methods in Chemistry

2011
ISBN: 978-3-527-32748-5

Haber, R., Bars, R., Schmitz, U.

Predictive Control in Process Engineering

From the Basics to the Applications

2008
ISBN: 978-3-527-31492-8

Buzzi-Ferraris, G., Manenti, F.

Fundamentals and Linear Algebra for the Chemical Engineer

Solving Numerical Problems

2010
ISBN: 978-3-527-32552-8

Buzzi-Ferraris, G., Manenti, F.

Interpolation and Regression Models for the Chemical Engineer

Solving Numerical Problems

2010
ISBN: 978-3-527-32652-5

Dimian, A. C., Bildea, C. S.

Chemical Process Design

Computer-Aided Case Studies

2008
ISBN: 978-3-527-31403-4

Engell, S. (ed.)

Logistic Optimization of Chemical Production Processes

2008
ISBN: 978-3-527-30830-9

Ingham, J., Dunn, I. J., Heinzle, E., Prenosil, J. E., Snape, J. B.

Chemical Engineering Dynamics

An Introduction to Modelling and Computer Simulation

2007
ISBN: 978-3-527-31678-6

Keil, F. J. (ed.)

Modeling of Process Intensification

2007
ISBN: 978-3-527-31143-9

Roffel, B., Betlem, B.

Process Dynamics and Control

Modeling for Control and Prediction

2006
E-Book
ISBN: 978-0-470-05877-0

Puigjaner, L., Heyen, G. (eds.)

Computer Aided Process and Product Engineering

2006
ISBN: 978-3-527-30804-0

Process Systems Engineering

Volume 7: Dynamic Process Modeling

Edited by
Michael C. Georgiadis, Julio R. Banga, and
Efstratios N. Pistikopoulos

WILEY-VCH Verlag GmbH & Co. KGaA

The Editors

Prof. Michael C. Georgiadis
Aristotle University of Thessaloniki
Department of Chemical Engineering
Thessaloniki 54124
Greece

and

Imperial College London
Dept. of Chemical Engineering
Centre for Process System Engineering
South Kensington Campus
London SW7 2AZ
United Kingdom

Dr. Julio R. Banga
IIM-CSIC
(Bio)Process Engin. Group
C/Eduardo Cabello 6
36208 Vigo
Spain

Prof. Efstratios N. Pistikopoulos
Imperial College London
Dept. of Chemical Engineering
Centre for Process System Engineering
South Kensington Campus
London SW7 2AZ
United Kingdom

All books published by **Wiley-VCH** are carefully produced. Nevertheless, authors, editors, and publisher do not warrant the information contained in these books, including this book, to be free of errors. Readers are advised to keep in mind that statements, data, illustrations, procedural details or other items may inadvertently be inaccurate.

Library of Congress Card No.:
applied for

British Library Cataloguing-in-Publication Data
A catalogue record for this book is available from the British Library.

Bibliographic information published by the Deutsche Nationalbibliothek
The Deutsche Nationalbibliothek lists this publication in the Deutsche Nationalbibliografie; detailed bibliographic data are available on the Internet at <http://dnb.d-nb.de>.

© 2011 WILEY-VCH Verlag GmbH & Co. KGaA, Boschstr. 12, 69469 Weinheim, Germany

All rights reserved (including those of translation into other languages). No part of this book may be reproduced in any form – by photoprinting, microfilm, or any other means – nor transmitted or translated into a machine language without written permission from the publishers. Registered names, trademarks, etc. used in this book, even when not specifically marked as such, are not to be considered unprotected by law.

Composition VTEX Typesetting, Vilnius

Printing and Binding betz-druck GmbH, Darmstadt

Cover Design Schulz Grafik-Design, Fußgönheim

Printed in the Federal Republic of Germany
Printed on acid-free paper

ISBN: 978-3-527-31696-0

Contents

Preface *XV*
List of Contributors *XXI*

Part I Chemical and Other Processing Systems 1

1 Dynamic Process Modeling: Combining Models and Experimental Data to Solve Industrial Problems *3*
  M. Matzopoulos 3
1.1   Introduction *3*
1.1.1   Mathematical Formulation *4*
1.1.2   Modeling Software *5*
1.2   Dynamic Process Modeling – Background and Basics *5*
1.2.1   Predictive Process Models *6*
1.2.2   Dynamic Process Modeling *6*
1.2.3   Key Considerations for Dynamic Process Models *7*
1.2.4   Modeling of Operating Procedures *9*
1.2.5   Key Modeling Concepts *10*
1.2.5.1   First-Principles Modeling *10*
1.2.5.2   Multiscale Modeling *10*
1.2.5.3   Equation-Based Modeling Tools *11*
1.2.5.4   Distributed Systems Modeling *12*
1.2.5.5   Multiple Activities from the Same Model *13*
1.2.5.6   Simulation vs. Modeling *13*
1.3   A Model-Based Engineering Approach *14*
1.3.1   High-Fidelity Predictive Models *14*
1.3.2   Model-Targeted Experimentation *16*
1.3.3   Constructing High-Fidelity Predictive Models – A Step-by-Step Approach *16*
1.3.4   Incorporating Hydrodynamics Using Hybrid Modeling Techniques *22*
1.3.5   Applying the High-Fidelity Predictive Model *22*

1.4	An Example: Multitubular Reactor Design	23
1.4.1	Multitubular Reactors – The Challenge	24
1.4.2	The Process	25
1.4.3	The Solution	25
1.4.4	Detailed Design Results	29
1.4.5	Discussion	30
1.5	Conclusions	31

2	**Dynamic Multiscale Modeling – An Application to Granulation Processes**	**35**
	G.D. Ingram and I.T. Cameron	35
2.1	Introduction	35
2.2	Granulation	36
2.2.1	The Operation and Its Significance	36
2.2.2	Equipment, Phenomena, and Mechanisms	37
2.2.3	The Need for and Challenges of Modeling Granulation	39
2.3	Multiscale Modeling of Process Systems	41
2.3.1	Characteristics of Multiscale Models	41
2.3.2	Approaches to Multiscale Modeling	43
2.4	Scales of Interest in Granulation	45
2.4.1	Overview	45
2.4.2	Primary Particle Scale	47
2.4.3	Granule Scale	48
2.4.4	Granule Bed Scale	48
2.4.5	Vessel Scale	49
2.4.6	Circuit Scale	50
2.5	Applications of Dynamic Multiscale Modeling to Granulation	52
2.5.1	Overview	52
2.5.2	Fault Diagnosis for Continuous Drum Granulation	55
2.5.3	Three-Dimensional Multiscale Modeling of Batch Drum Granulation	56
2.5.4	DEM-PBE Modeling of Batch High-Shear Granulation	58
2.5.5	DEM-PBE Modeling of Continuous Drum Granulation	59
2.6	Conclusions	61

3	**Modeling of Polymerization Processes**	**67**
	B.S. Amaro and E.N. Pistikopoulos	67
3.1	Introduction	67
3.2	Free-Radical Homopolymerization	68
3.2.1	Kinetic Modeling	68
3.2.2	Diffusion-Controlled Reactions	69
3.2.2.1	Fickian Description of Reactant Diffusion	71
3.2.2.2	Free-Volume Theory	72

3.2.2.3	Chain Length Dependent Rate Coefficients	73
3.2.2.4	Combination of the Free-Volume Theory and Chain Length Dependent Rate Coefficients	75
3.2.2.5	Fully Empirical Models	76
3.3	Free-Radical Multicomponent Polymerization	77
3.3.1	Overview	77
3.3.2	Pseudo-Homopolymerization Approximation	78
3.3.3	Polymer Composition	80
3.4	Modeling of Polymer Molecular Properties	80
3.4.1	Molecular Weight Distribution	80
3.5	A Practical Approach – SAN Bulk Polymerization	90
3.5.1	Model	90
3.5.1.1	Kinetic Diagram	90
3.5.1.2	Mass Balances	91
3.5.1.3	Diffusion Limitations	92
3.5.1.4	Pseudo-Homopolymerization Approximation	94
3.5.2	Illustrative Results	95
3.6	Conclusions	97

4 Modeling and Control of Proton Exchange Membrane Fuel Cells 105
C. Panos, K. Kouramas, M.C. Georgiadis and E.N. Pistikopoulos 105

4.1	Introduction	105
4.2	Literature Review	108
4.3	Motivation	109
4.3.1	Reactant Flow Management	112
4.3.2	Heat and Temperature Management	112
4.3.3	Water Management	113
4.4	PEM Fuel Cell Mathematical Model	113
4.4.1	Cathode	114
4.4.2	Anode	117
4.4.3	Anode Recirculation	119
4.4.4	Fuel Cell Outlet	120
4.4.5	Membrane Hydration Model	120
4.4.6	Electrochemistry	122
4.4.7	Thermodynamic Balance	123
4.4.8	Air Compressor and DC Motor Model	125
4.4.9	DC Motor	126
4.4.10	Cooling System	127
4.5	Reduced Order Model	128
4.6	Concluding Remarks	132

5 Modeling of Pressure Swing Adsorption Processes 137
E.S. Kikkinides, D. Nikolic and M.C. Georgiadis 137

5.1	Introduction	137

5.2	Model Formulation	*144*
5.2.1	Adsorbent Bed Models	*144*
5.2.2	Single-Bed Adsorber	*145*
5.2.3	Adsorption Layer Model	*146*
5.2.3.1	General Balance Equations	*146*
5.2.3.2	Mass Balance	*147*
5.2.3.3	Heat Balance	*147*
5.2.3.4	Momentum Balance	*148*
5.2.3.5	Equation of State	*148*
5.2.3.6	Thermophysical Properties	*148*
5.2.3.7	Axial Dispersion	*148*
5.2.3.8	Transport Properties	*149*
5.2.3.9	Boundary Conditions	*149*
5.2.4	Adsorbent Particle Model	*150*
5.2.4.1	General Mass Balance Equations	*150*
5.2.4.2	Local Equilibrium	*151*
5.2.4.3	Linear Driving Force (LDF)	*152*
5.2.4.4	Surface Diffusion	*152*
5.2.4.5	Pore Diffusion	*153*
5.2.4.6	Gas–Solid Phase Equilibrium Isotherms	*154*
5.2.5	Gas Valve Model	*157*
5.2.6	The Multibed PSA Model	*158*
5.2.7	The State Transition Network Approach	*158*
5.2.8	Numerical Solution	*162*
5.3	Case-Study Applications	*163*
5.3.1	Simulation Run I	*165*
5.3.2	Simulation Run II	*165*
5.3.3	Simulation Run III	*166*
5.4	Conclusions	*167*

6 **A Framework for the Modeling of Reactive Separations** *173*
E.Y. Kenig *173*

6.1	Introduction	*173*
6.2	Reactive Separations	*174*
6.3	Classification of Modeling Methods	*176*
6.4	Fluid-Dynamic Approach	*178*
6.5	Hydrodynamic Analogy Approach	*183*
6.6	Rate-Based Approach	*188*
6.7	Parameter Estimation and Virtual Experiments	*193*
6.8	Benefits of the Complementary Modeling	*196*
6.9	Concluding Remarks	*199*

7	**Efficient Reduced Order Dynamic Modeling of Complex Reactive and Multiphase Separation Processes Using Orthogonal Collocation on Finite Elements** *203*	
	P. Seferlis, T. Damartzis and N. Dalaouti *203*	
7.1	Introduction *203*	
7.2	NEQ/OCFE Model Formulation *205*	
7.2.1	Conventional and Reactive Absorption and Distillation *207*	
7.2.2	Multiphase Reactive Distillation *213*	
7.3	Adaptive NEQ/OCFE for Enhanced Performance *218*	
7.4	Dynamic Simulation Results *220*	
7.4.1	Reactive Absorption of NO_x *220*	
7.4.1.1	Process Description *220*	
7.4.1.2	Dynamic Simulation Results *223*	
7.4.2	Ethyl Acetate Production via Reactive Distillation *225*	
7.4.2.1	Process Description *225*	
7.4.2.2	Dynamic Simulation Results *227*	
7.4.3	Butyl Acetate Production via Reactive Multiphase Distillation *231*	
7.4.3.1	Process Description *231*	
7.4.3.2	Dynamic Simulation Results *232*	
7.5	Epilog *234*	
8	**Modeling of Crystallization Processes** *239*	
	A. Abbas, J. Romagnoli and D. Widenski *239*	
8.1	Introduction *239*	
8.2	Background *240*	
8.2.1	Crystallization Methods *241*	
8.2.1.1	Recrystallization Methods *241*	
8.2.2	Driving Force *242*	
8.3	Solubility Predictions *243*	
8.3.1	Empirical Approach *243*	
8.3.2	Correlative Thermodynamic *244*	
8.3.3	Predictive Thermodynamic *244*	
8.3.3.1	Jouyban–Acree Model *245*	
8.3.3.2	MOSCED Model *245*	
8.3.3.3	NRTL-SAC Model *246*	
8.3.3.4	UNIFAC Model *247*	
8.3.3.5	Solubility and Activity Coefficient Relationship *247*	
8.3.4	Solubility Examples *247*	
8.3.5	Solution Concentration Measurement Process Analytical Tools *250*	
8.4	Crystallization Mechanisms *251*	
8.4.1	Nucleation *251*	
8.4.1.1	Modeling Nucleation *252*	
8.4.2	Growth and Dissolution *254*	
8.4.3	Agglomeration and Aggregation *255*	

8.4.4 Attrition 255
8.5 Population, Mass, and Energy Balances 256
8.5.1 Population Balance 256
8.5.2 Solution Methods 257
8.5.2.1 Method of Moments 257
8.5.2.2 Discretization Method 258
8.5.3 Mass and Energy Balances 264
8.6 Crystal Characterization 264
8.6.1 Crystal Shape 264
8.6.2 Crystal Size 265
8.6.3 Crystal Distribution 265
8.6.4 Particle Measurement Process Analytical Tools 266
8.7 Solution Environment and Model Application 266
8.7.1 Simulation Environment 266
8.7.2 Experimental Design 267
8.7.3 Parameter Estimation 268
8.7.4 Validation 269
8.8 Optimization 270
8.8.1 Example 1: Antisolvent Feedrate Optimization 270
8.8.2 Example 2: Optimal Seeding in Cooling Crystallization 274
8.9 Future Outlook 276

9 Modeling Multistage Flash Desalination Process – Current Status and Future Development 287
I.M. Mujtaba 287
9.1 Introduction 287
9.2 Issues in MSF Desalination Process 289
9.3 State-of-the-Art in Steady-State Modeling of MSF Desalination Process 292
9.3.1 Scale Formation Modeling 299
9.3.1.1 Estimation of Dynamic Brine Heater Fouling Profile 301
9.3.1.2 Modeling the Effect of NCGs 301
9.3.1.3 Modeling of Environmental Impact 302
9.4 State-of-the-Art in Dynamic Modeling of MSF Desalination Process 303
9.5 Case Study 308
9.5.1 Steady-State Operation 308
9.5.2 Dynamic Operation 311
9.6 Future Challenges 312
9.6.1 Process Modeling 312
9.6.2 Steady-State and Dynamic Simulation 313
9.6.3 Tackling Environmental Issues 313
9.6.4 Process Optimization 314
9.7 Conclusions 315

Part II Biological, Bio-Processing and Biomedical Systems 319

10 Dynamic Models of Disease Progression: Toward a Multiscale Model of Systemic Inflammation in Humans 321
J.D. Scheff, P.T. Foteinou, S.E. Calvano, S.F. Lowry and I.P. Androulakis 321
10.1 Introduction 321
10.2 Background 322
10.2.1 *In-Silico* Modeling of Inflammation 323
10.2.2 Multiscale Models of Human Endotoxemia 325
10.2.3 Data Collection 327
10.3 Methods 328
10.3.1 Developing a Multilevel Human Inflammation Model 328
10.3.1.1 Identification of the Essential Transcriptional Responses 328
10.3.1.2 Modeling Inflammation at the Cellular Level 330
10.3.1.3 Modeling Inflammation at the Systemic Level 335
10.3.1.4 Modeling Neuroendocrine–Immune System Interactions 336
10.3.1.5 Modeling the Effect of Endotoxin Injury on Heart Rate Variability 338
10.4 Results 340
10.4.1 Transcriptional Analysis and Major Response Elements 340
10.4.2 Elements of a Multilevel Human Inflammation Model 343
10.4.3 Estimation of Relevant Model Parameters 345
10.4.4 Qualitative Assessment of the Model 347
10.4.4.1 Implications of Increased Insult 348
10.4.4.2 Modes of Dysregulation of the Inflammatory Response 349
10.4.4.3 The Emergence of Memory Effects 353
10.4.4.4 Evaluation of Stress Hormone Infusion in Modulating the Inflammatory Response 354
10.5 Conclusions 360

11 Dynamic Modeling and Simulation for Robust Control of Distributed Processes and Bioprocesses 369
A.A. Alonso, M.R. García and C. Vilas 369
11.1 Introduction 369
11.2 Model Reduction of DPS: Theoretical Background 372
11.2.1 Model Reduction in the Context of the Finite Element Method 374
11.2.1.1 Proper Orthogonal Decomposition 376
11.2.1.2 Laplacian Spectral Decomposition 377
11.3 Model Reduction in Identification of Bioprocesses 377
11.3.1 Illustrative Example: Production of Gluconic Acid in a Tubular Reactor 378
11.3.2 Observer Validation 379

11.4	Model Reduction in Control Applications	383
11.4.0.1	Model Equations	384
11.4.1	Robust Control of Tubular Reactors	386
11.4.1.1	Controller Synthesis	389
11.4.1.2	Robust Control with a Finite Number of Actuators	392
11.4.2	Real-Time Optimization: Multimodel Predictive Control	394
11.4.2.1	Optimization Problem	395
11.4.2.2	The Online Strategy	396
11.5	Conclusions	397

12 Model Development and Analysis of Mammalian Cell Culture Systems 403

A. Kiparissides, M. Koutinas, E.N. Pistikopoulos and A. Mantalaris 403

12.1	Introduction	403
12.2	Review of Mathematical Models of Mammalian Cell Culture Systems	406
12.3	Motivation	410
12.4	Dynamic Modeling of Biological Systems – An Illustrative Example	413
12.4.1	First Principles Model Derivation	415
12.4.2	Model Analysis	421
12.4.3	Design of Experiments and Model Validation	432
12.5	Concluding Remarks	435

13 Dynamic Model Building Using Optimal Identification Strategies, with Applications in Bioprocess Engineering 441

E. Balsa-Canto, J.R. Banga and M.R. García 441

13.1	Introduction	441
13.2	Parameter Estimation: Problem Formulation	443
13.2.1	Mathematical Model Formulation	444
13.2.2	Experimental Scheme and Experimental Data	444
13.2.3	Cost Function	445
13.2.4	Numerical Methods: Single Shooting vs. Multiple Shooting	446
13.3	Identifiability	447
13.4	Optimal Experimental Design	449
13.4.1	Numerical Methods: The Control Vector Parameterization Approach	450
13.5	Nonlinear Programming Solvers	450
13.6	Illustrative Examples	453
13.6.1	Modeling of the Microbial Growth	453
13.6.2	Modeling the Production of Gluconic Acid in a Fed-Batch Reactor	457
13.7	Overview	463

14	**Multiscale Modeling of Transport Phenomena in Plant-Based Foods** *469*	

Q.T. Ho, P. Verboven, B.E. Verlinden, E. Herremans and B.M. Nicolaï   *469*

14.1 Introduction   *469*
14.2 Length Scales of Biological Materials   *470*
14.3 Multiscale Modeling of Transport Phenomena   *472*
14.3.1 Mass Transport Fundamentals   *472*
14.3.2 Multiscale Transport Phenomena   *474*
14.3.2.1 Macroscale Approach   *474*
14.3.2.2 Microscale Approach   *474*
14.3.2.3 Kinetic Modeling   *475*
14.3.2.4 Multiscale Model   *476*
14.4 Numerical Solution   *476*
14.4.1 Geometrical Model   *476*
14.4.2 Discretization   *478*
14.5 Case Study: Application of Multiscale Gas Exchange in Fruit   *480*
14.5.1 Macroscale Model   *480*
14.5.2 Microscale Model   *482*
14.5.3 O_2 Transport Model   *482*
14.5.4 CO_2 Transport Model (Lumped CO_2 Transport Model)   *483*
14.6 Conclusions and Outlook   *485*

15	**Synthetic Biology: Dynamic Modeling and Construction of Cell Systems** *493*	

T.T. Marquez-Lago and M.A. Marchisio   *493*

15.1 Introduction   *493*
15.2 Constructing a Model with Parts   *494*
15.2.1 General Nomenclature   *494*
15.2.1.1 Parts and Devices   *494*
15.2.1.2 Common Signal Carriers   *496*
15.2.1.3 Pools and Fluxes   *497*
15.2.2 Part Models   *500*
15.2.2.1 Promoters   *500*
15.2.2.2 Ribosome-Binding Sites   *504*
15.2.2.3 Coding Regions   *508*
15.2.2.4 Noncoding DNA   *509*
15.2.2.5 Small RNA   *511*
15.2.2.6 Terminator   *511*
15.2.3 Introducing Parts and Fluxes into Deterministic Equations   *512*
15.3 Modeling Regimes and Simulation Techniques   *518*
15.3.1 Deterministic or Stochastic Modeling?   *519*
15.3.1.1 Deterministic Regime   *519*
15.3.1.2 Stochastic Regime   *520*

15.3.2 Stochastic Simulation Algorithms *522*
15.3.2.1 Exact Algorithms *522*
15.3.2.2 Coarse-Grained Methods *527*
15.4 Application *532*
15.4.1 The Repressilator *533*
15.5 Conclusions *541*

16 Identification of Physiological Models of Type 1 Diabetes Mellitus by Model-Based Design of Experiments *545*

F. Galvanin, M. Barolo, S. Macchietto and F. Bezzo *545*

16.1 Introduction *546*
16.1.1 Glucose Concentration Control Issues *547*
16.2 Introducing Physiological Models *548*
16.3 Identifying a Physiological Model: The Need for Experiment Design *548*
16.4 Standard Clinical Tests *550*
16.5 A Compartmental Model of Glucose Homeostasis *551*
16.6 Model Identifiability Issues *552*
16.6.1 A Discussion on the Identifiability of the Hovorka Model *554*
16.7 Design of Experiments Under Constraints for Physiological Models *556*
16.7.1 Design Procedure *558*
16.8 Design of Experimental Protocols *560*
16.8.1 Modified OGTT (mOGTT) *561*
16.8.1.1 Effect of the Number of Samples *562*
16.9 Dealing with Uncertainty *563*
16.9.1 Online Model-Based Redesign of Experiments *565*
16.9.2 Model-Based Design of Experiment with Backoff (MBDoE-B) *566*
16.9.2.1 Backoff Application *567*
16.9.3 Effect of a Structural Difference Between a Model and a Subject *569*
16.10 Conclusions *572*

Index *583*

Preface

The central role of process modeling in all aspects of process design and operation is now well recognized. Although most early models were steady state, more recently the emphasis has been on dynamic models that can be used for studying transient process behavior. Moreover, there is an increasing trend toward "high-fidelity" models, which can accurately predict the trajectories of the key variables that affect the process performance, safety, and economics. This demand for substantially higher model accuracy can often be traced to the need for extracting further gains in profitability out of processes that have already undergone incremental improvement over several decades. In other cases, it arises from the strict environmental and safety constraints and product specifications under which many processes currently operate [1].

Process modeling has always been an important component of process design, from the conceptual synthesis of the process flowsheet to the detailed design of specialized processing equipment such as advanced reaction and separation devices, and the design of their control systems. Recent years have witnessed the model-based approach being extended to the design of complex products, such as batteries, fuel cells, biomedical, biochemical, drug delivery systems, which can themselves be viewed as miniature plants produced in very large numbers. Inevitably, the modeling technology needed to fulfill the demands posed by such a diverse range of applications is very different from the standard steady-state flowsheeting packages that served the process industries so well in the past [2].

Volume 7 of this book series has attempted a review of some of the current trends in process modeling and its practical application during the past few years. It focuses on modeling frameworks for complex systems including chemical, biochemical, bio-processing, biological, and energy systems.

In Chapter 1, Mark Matzopoulos from Process Systems Enterprise Ltd summarizes his long experience in the area of process modeling. He presents a model-based engineering approach to the construction and application of detailed dynamic process models. He first summarizes the key concepts and consideration to be taken into account when building first principle of the dynamic modeling. Then he introduces a *model-based engineering* (MBE) approach, which involves engineering activities with the assistance of a mathematical model of the process under investigation. A step-by-step approach is nicely described for the construction of high-fidelity predictive models including estimation of model parameters from data, analysis of the experimental data, and design of experiments, if necessary. The applicability of the overall modeling approach is illustrated in a multitubular reactor design problem. A number of benefits using a high-fidelity predictive modeling approach are revealed.

In Chapter 2, Ingram and Cameron discuss a multiscale modeling approach for granulation processes. The industrial significance of granulation processes is first introduced and the multiscale nature of process systems, general characteristics of multiscale models, and the emerging practice of multiscale modeling are then presented in details. The relevant scales of observation for granulation processes are outlined and several examples of the modeling techniques used at each scale are provided. Key multiscale granulation models appearing in the literature are discussed and then a handful of them are reviewed in more detail.

In Chapter 3, Amaro and Pistikopoulos discuss a number of theoretical principals behind polymerization process modeling, applied to free-radical polymer reactions. Comprehensive kinetic schemes encompassing a large number of reactions that might occur during these processes were discussed in details. The specific modeling of polymer molecular properties was highlighted and exemplified for molecular weight distribution with different approaches regarding its representation. In conclusion, it was emphasized that modeling of polymerization processes is a powerful tool allowing researchers and companies to perform a broad variety of simulations, allowing for a number of model-based activities such as optimization and control.

Panos and coworkers in Chapter 4 present a detailed dynamic model for PEM fuel cell stack. The model has the great advantage of less computation time consuming while providing results consistent with the literature, and well oriented toward control. Then a reduced order state space model is designed for optimal control studies. Finally an explicit/multiparametric MPC controller has been developed to keep the controlled variables close to the set points while taking care of the physical constraints on the manipulated variables, namely the reactant and coolant mass flows. The controller finally selected shows good performance to resist the disturbances in the load.

In Chapter 5, Kikkinides and coworkers present a detailed modeling framework for pressure swing adsorption flowsheets. Several research challenges are identified and a generic modeling framework for the separation of gas mixtures using multibed PSA flowsheets is presented. The core of the framework represents a detailed adsorbent bed model relying on a coupled set of mixed algebraic and partial differential equations for mass, heat, and momentum balance at both bulk gas and particle level, equilibrium isotherm equations, and boundary conditions according to the operating steps. The adsorbent bed model provides the basis for building PSA flowsheets with all feasible interbed connectivities. Operating procedures are automatically generated, thus facilitating the development of complex PSA flowsheet for an arbitrary number of beds. Finally, a case study concerning the separation of hydrogen from steam-methane reforming of gas is used to illustrate the application and efficiency of the developed framework.

In Chapter 6, Kenig introduces a complementary modeling approach for the reactive separation process based on a reasonable and efficient combination of different approaches. He presented a classification of kinetics-based models based on the complexity of the process fluid dynamics. He concluded that for geometrically simple flows, the fluid dynamic approach (FDA) should be applied as it gives full information about the process in a purely theoretical manner. For very complex flow patterns, the rate-based approach (RBA) represents a good choice provided that the model parameters are determined properly. The hydrodynamic analogy approach serves as an intermediate between the FDA and RBA and is suitable for those processes in which a certain structure or order exists. Several case studies were used to highlight the use of all approaches.

Seferlis and coworkers in Chapter 7 discuss efficient *reduced order dynamic modeling* techniques of *complex reactive and multiphase separation processes*. They illustrated that combined *nonequilibrium* and *orthogonal collocation on finite elements* (NEQ/OCFE) models become quite attractive for real-time control applications of these processes. A novel optimization-based finite element partition algorithm is then presented to enhance the ability of the proposed models to control the approximation error along the column within reasonable levels despite the influence of exogenous disturbances responsible for the formation of steep fronts in the composition and temperature profiles. Case studies that involve reactive absorption, reactive distillation, and multiphase reactive distillation illustrated the strengths of the NEQ/OCFE techniques.

In Chapter 8, Abbas and coworkers discuss the modeling of crystallization processes. An overview of industrial crystallization, crystallization fundamentals, and mechanisms is first presented followed by detailed discussions on crystallization modeling, model solution techniques, and model analysis. Various model application areas are illustrated before finishing with two examples, namely, antisolvent crystallization and seeded cooling crystallization.

In Chapter 9, Mujtaba highlights the state-of-the-art and future challenges in modeling of multistage flash (MSF) desalination process. He presents how *computer-aided process engineering* modeling techniques and the practitioners of desalination can address sustainable freshwater issue of tomorrow's world via desalination. He also emphasizes that the exploitation of full economic benefit of replacing time-consuming and expensive experimental studies of MSF processes requires development of accurate mathematical models and model-based applications such as optimization and control. Several research challenges in this area are also introduced.

Recognizing the importance of a mechanistic systems approach to biological sciences, Androulakis in Chapter 10 discusses the potential role of systems-based approaches in the quest to better understand critical physiological responses. He demonstrated how quantitative models of inflammation can be used as minimal representations of biological reality to formulate and test hypotheses, reconcile observations, and guide future experimental design. He also demonstrated the possibility of the generalization of this framework in a wide range of disease progression models. It was emphasized that it is important to realize that *in silico* models will never replace either biological or clinical research. They could, however, rationalize the decision-making process by establishing the range of validity and predictability of intervention strategies, thus enabling the use of systems biology in translational research.

In Chapter 11, Alonso and coworkers consider the dynamic modeling of distributed (bio)processes, that is, those described by partial differential equations. Their contribution considers aspects that are particularly relevant for robust control, with emphasis on model reduction techniques for convection–diffusion–reaction processes. These techniques are illustrated with examples, including a bioreactor for the production of gluconic acid and the control of a tubular reactor.

Paving the way toward a "closed-loop" holistic framework for bioprocess automation, Kiparissides and coworkers in Chapter 12 cover the development of dynamical models of biological systems. They introduced and explained in a step-by-step fashion a biological model development framework. The scientific concerns, challenges, and "real-life" problems associated with each step of the framework were clearly highlighted. Adapting a "real-life" example from their previous work, the logical and systematic evolution of a model were presented from the conception to validation as it flows through the various steps of the model development framework. The key conclusion of this contribution is that by utilizing a systematic way of organizing available information, one can avoid conducting experiments for the sake of experimentation and develop models with an *a priori* set aim.

Balsa-Canto and coworkers in Chapter 13 consider optimal identification strategies and their application in bioprocess engineering. Dynamic model building is presented as an iterative loop with three key topics: parameter estimation (model calibration), identifiability analysis, and optimal experimental design. These authors highlight the need of checking identifiability and using global optimization techniques for proper parameter estimation in nonlinear dynamic models. Further, the use optimal experimental design to increase the identifiability is motivated. These techniques are illustrated with two examples – one related with dynamic modeling of microbial growth, and another with dynamic modeling of the production of gluconic acid in a fed-batch bioreactor.

In Chapter 14, Nicolaï and coworkers consider the multiscale dynamic modeling of transport phenomena in foods, with emphasis in plant-based foods. These authors show how the multiscale paradigm combines micro- and macro-scale models through homogenization and localization, and which numerical methods should be used to solve the resulting model. The use of this approach is exemplified with a case study considering application of multiscale gas exchange in fruit.

Marquez-Lago and Marchisio in Chapter 15 consider dynamic modeling in synthetic biology, that is, the engineering of novel biological functions and systems. These authors adopt a detailed modeling methodology based on the concept of composable parts. Further, they pay special attention to the selection of a correct simulation regime, highlighting the problems of using deterministic approaches, and discussing alternative stochastic simulation methods. These topics and techniques are illustrated considering a synthetic oscillator made of three genes.

In Chapter 16, Bezzo and coworkers reviewed the role of the optimal model-based design of experiments techniques with reference to the problem of individual parameter identification for complex physiological models of glucose homeostasis. It was emphasized that the parameter identification problem is a tradeoff between several issues: acquisition of a high information content from a clinical test, compliance to a number of constraints in the system inputs and outputs, practical applicability of the test. It was showed that model-based design of experiments does allow designing effective and safe clinical tests, where the administration of carbohydrates (i.e., glucose) and possibly insulin is exploited to provide dynamic excitation to the body system, and a proper schedule of blood samples is used to collect the information generated during the test.

This collection represents a set of stand-alone works that captures recent research trends in the development and application of modeling frameworks techniques of various process and biosystems. We hope that by the end of the book, the reader will have developed a commanding comprehension of the main aspects of dynamic process modeling, the ability to critically access the key characteristics and elements related to the construction and application of detailed models and the capacity to implement the new technology in practice.

We are extremely grateful to the authors for their outstanding contributions and for their patience, which have led to a final product that far exceeded our expectations.

References

1 PANTELIDES, C. C., URBAN, Z. E., Process modeling technology: a critical review of recent developments, in: *Proceedings of the 6th International Conference on Foundations of Computer-Aided Process Design*, Floudas, C. A., Agrawal, R. (eds.), CACHE Publications, **2004**, pp. 69–82.

2 PANTELIDES, C. C., New challenges and opportunities on process modelling, Presented in the European Symposium on Computer-Aideds Process Engineering – 11.

Michael C. Georgiadis
Julio R. Banga
Efstratios N. Pistikopoulos
January 2010

List of Contributors

Ali Abbas
University of Sydney
School of Chemical and Biomolecular Engineering
Sydney NSW 2006
Australia

Antonio A. Alonso
Spanish Council for Scientific Research
Instituto de Investigaciones Marinas
(C.S.I.C.)
(Bio)Process Engineering Group
C/Eduardo Cabello 6
36208 Vigo
Spain

Bruno S. Amaro
Imperial College London
Center for Process Systems Engineering, Department of Chemical Engineering
Roderic Hill Building, South Kensington Campus
London SW7 2AZ
UK

Ioannis P. Androulakis
Rutgers University
Biomedical Engineering
Piscatawaym, NJ 08854
USA
and

Chemical and Biochemical Engineering
Piscataway, NJ 08854
USA

Eva Balsa-Canto
Spanish Council for Scientific Research
Instituto de Investigaciones Marinas
(C.S.I.C.)
(Bio)Process Engineering Group
C/Eduardo Cabello 6
36208 Vigo
Spain

Julio R. Banga
Spanish Council for Scientific Research
Instituto de Investigaciones Marinas
(C.S.I.C.)
(Bio)Process Engineering Group
C/Eduardo Cabello 6
36208 Vigo
Spain

Massimiliano Barolo
Università di Padova
CAPE-Lab – Computer-Aided Process Engineering Laboratory, Dipartimento di Principi e Impianti di Ingegneria Chimica
via Marzolo 9
35131 Padova PD
Italy

Process Systems Engineering: Vol. 7 Dynamic Process Modeling
Edited by Michael C. Georgiadis, Julio R. Banga, and Efstratios N. Pistikopoulos
Copyright © 2011 WILEY-VCH Verlag GmbH & Co. KGaA, Weinheim
ISBN: 978-3-527-31696-0

List of Contributors

Fabrizio Bezzo
Università di Padova
CAPE-Lab – Computer-Aided Process
Engineering Laboratory
Dipartimento di Principi e Impianti di
Ingegneria Chimica
via Marzolo 9
35131 Padova PD
Italy

Steve E. Calvano
Department of Surgery
UMDNJ-Robert Wood Johnson Medical
School
New Brunswick, NJ 08901
USA

Ian T. Cameron
The University of Queensland
School of Chemical Engineering
Brisbane, Queensland 4072
Australia

Natassa Dalaouti
Testing Research & Standards Center
Public Power Corporation S.A.
Leontariou Str. 9
15351 Kantza-Pallinis/Attiki
Greece

Theodoros Damartzis
Chemical Process Engineering
Research Institute
Centre for Research and Technology –
Hellas
6th km Charilaou-Thermi Road P.O.
Box 60361
57001 Thermi-Thessaloniki
Greece

Panagiota T. Foteinou
Rutgers University
Biomedical Engineering
Piscataway, NJ 08854
USA

Federico Galvanin
Università di Padova
CAPE-Lab – Computer-Aided Process
Engineering Laboratory, Dipartimento
di Principi e Impianti di Ingegneria
Chimica
via Marzolo 9
35131 Padova PD
Italy

Miriam R. García
National University of Ireland
Maynooth
Hamilton Institute
Maynooth
Ireland

and

Spanish Council for Scientific Research
Instituto de Investigaciones Marinas
(C.S.I.C.)
(Bio)Process Engineering Group
C/Eduardo Cabello 6
36208 Vigo
Spain

Michael C. Georgiadis
Aristotle University of Thessaloniki
Department of Chemical Engineering
54124 Thessaloniki
Greece

and

Imperial College London
Centre for Process Systems Engineering
Department of Chemical Engineering
South Kensington Campus
London SW7 2AZ
UK

Els Herremans
Katholieke Universiteit Leuven
Flanders Center of Postharvest
Technology, BIOSYST-MeBioS
Willem de Croylaan 42
3001 Leuven
Belgium

Quang Tri Ho
Katholieke Universiteit Leuven
Flanders Center of Postharvest
Technology, BIOSYST-MeBioS
Willem de Croylaan 42
3001 Leuven
Belgium

Gordon D. Ingram
Curtin University of Technology
Department of Chemical Engineering
Perth WA
Australia

Eugeny Y. Kenig
University of Paderborn
Chair of Fluid Process Engineering
Pohlweg 55
33098 Paderborn
Germany

and

Gubkin Russian State University of Oil
and Gas
Chair of Thermodynamics and Heat
Engines
Leninsky Prospect 65
Moscow 119991
Russia

Eustathios S. Kikkinides
University of Western Macedonia
Department of Mechanical Engineering
Sialvera & Bakola Street
50100 Kozani
Greece

Alexandros Kiparissides
Imperial College London
Centre for Process Systems Engineering
Department of Chemical Engineering
South Kensington Campus
London SW7 2AZ
UK

Kostas Kouramas
Imperial College London
Centre for Process Systems Engineering
Department of Chemical Engineering
Roderic Hill Building
South Kensington Campus
London SW7 2AZ
UK

Michalis Koutinas
Imperial College London
Centre for Process Systems Engineering
Department of Chemical Engineering
South Kensington Campus
London SW7 2AZ
UK

Stephen F. Lowry
Department of Surgery
UMDNJ-Robert Wood Johnson Medical
School
New Brunswick, NJ 08901
USA

Sandro Macchietto
Imperial College London
Department of Chemical Engineering
South Kensington Campus
London SW7 2AZ
UK

Athanasios Mantalaris
Imperial College London
Centre for Process Systems Engineering
Department of Chemical Engineering
South Kensington Campus
London SW7 2AZ
UK

Mario Andrea Marchisio
Department of Biosystems Science and
Engineering
Swiss Federal Institute of Technology,
Computational Systems Biology Group
ETH Zentrum CAB J 71.5
8092 Zurich
Switzerland

Tatiana T. Marquez-Lago
Department of Biosystems Science and Engineering
Swiss Federal Institute of Technology, Computational Systems Biology Group
ETH Zentrum CAB J 71.5
8092 Zurich
Switzerland

Mark Matzopoulos
Process Systems Enterprise Ltd
6th Floor East, 26–28 Hammersmith Grove
London W6 7AH
UK

Iqbal M. Mujtaba
University of Bradford
School of Engineering Design and Technology
West Yorkshire BD7 1DP
UK

Bart M. Nicolaï
Katholieke Universiteit Leuven
Flanders Center of Postharvest Technology, BIOSYST-MeBioS
Willem de Croylaan 42
3001 Leuven
Belgium

Dragan Nikolic
University of Western Macedonia
Department of Mechanical Engineering
Sialvera & Bakola Street
50100 Kozani
Greece

Christos Panos
Imperial College London
Centre for Process Systems Engineering, Department of Chemical Engineering
Roderic Hill Building, South Kensington Campus
London SW7 2AZ
UK

Efstratios N. Pistikopoulos
Imperial College London
Center for Process Systems Engineering
Department of Chemical Engineering
South Kensington Campus
London SW7 2AZ
UK

Jose Romagnoli
Louisiana State University
Chemical Engineering Department
Baton Rouge, LA 70803
USA

Jeremy D. Scheff
Rutgers University
Biomedical Engineering
Piscataway, NJ 08854
USA

Panos Seferlis
Aristotle University of Thessaloniki
Department of Mechanical Engineering
P.O. Box 484
54124 Thessaloniki
Greece

and

Chemical Process Engineering Research Institute
Centre for Research and Technology – Hellas
6th km Charilaou-Thermi Road
P.O. Box 60361
57001 Thermi-Thessaloniki
Greece

Pieter Verboven
Katholieke Universiteit Leuven
Flanders Center of Postharvest Technology, BIOSYST-MeBioS
Willem de Croylaan 42
3001 Leuven
Belgium

Bert E. Verlinden
Katholieke Universiteit Leuven
Flanders Center of Postharvest
Technology, BIOSYST-MeBioS
Willem de Croylaan 42
3001 Leuven
Belgium

Carlos Vilas
Spanish Council for Scientific Research
Instituto de Investigaciones Marinas
(C.S.I.C.)
(Bio)Process Engineering Group
C/Eduardo Cabello 6
36208 Vigo
Spain

David Widenski
Louisiana State University
Chemical Engineering Department
Baton Rouge, LA 70803
USA

Part I
Chemical and Other Processing Systems

1
Dynamic Process Modeling: Combining Models and Experimental Data to Solve Industrial Problems

Mark Matzopoulos

Keywords

steady-state modeling, first-principles modeling, dynamic modeling, high fidelity modeling, Fischer–Tropsch reaction, model-based engineering (MBE), parameter estimation, computation fluid dynamic (CFD) model

1.1
Introduction

The operation of all processes varies over time. Sometimes this is an inherent part of the process design, as in batch processes or pressure swing adsorption; sometimes it is as a result of external disturbances, for example a change in feedstock or product grade, equipment failure, or regular operational changes such as the diurnal load variation of a power plant or the summer–winter operation of a refinery.

Many processes are intended to run and indeed appear to run at steady state, and for practical purposes of design they can be modeled as such. However for inherently dynamic processes, or when considering transient aspects of otherwise "steady-state" processes – for example, start-up, shutdown, depressurization, and so on – steady-state or pseudo-steady-state analysis is simply not applicable, or is at best a gross simplification. Such cases require dynamic process models.

By modeling the time-varying aspects of operation we are not simply "adding another dimension" to steady-state models. There are many factors that, when considered as a whole, represent a step change beyond the philosophy and approach taken in traditional steady-state modeling.

There are of course many different levels of fidelity possible when creating dynamic models, but this chapter uses "dynamic process models" as a shorthand for high-fidelity predictive process models (which generally include dynamics) that typically combine chemical engineering first principles theory with observed, "real-life" data. It looks at how these can be systematically applied to complex problems within industry to provide value in a whole range of process systems.

At the heart of the approaches described here is a family of methodologies that use models to perform many different tasks: not only to simulate and optimize process performance but also to estimate the parameters used in the equations on which those simulations and optimizations are based; to analyze experimental data and even to design the experiments themselves. The methodologies also make it possible to move from relatively little experimental data collected in small-scale equipment to a high-fidelity predictive model of a complex and large-scale industrial unit or process.

As these methodologies are underpinned by a mathematical model, or series of mathematical models, they are known as model-based methodologies – in particular *model-based engineering* and *model-based innovation* and their supporting techniques of model-based data analysis, model-targeted experimentation and model-based experiment design. This chapter describes how these techniques can be systematically applied to create models that are capable of predicting behavior over a wide range of conditions and scales of operation with unprecedented accuracy.

1.1.1
Mathematical Formulation

Process systems are often most readily specified in terms of a mixed set of integral, partial differential and algebraic equations, or IPDAEs, and this is the chosen representation of most established software designed specifically for process systems engineering.

Traditionally, IPDAE systems are reduced into a mixed set of ordinary differential and algebraic equations, or DAEs. While ordinary differential equations (ODEs) typically arise from the conservation of fundamental quantities of nature, algebraic equations (AEs) result from processes where accumulation is not important or transients are considered to occur instantaneously, as well as by introduction of auxiliary relationships among variables.

Discontinuities in the governing phenomena of physical, chemical, and biological systems frequently arise from thermodynamic, transport, and flow transitions as well as structural changes in the control configuration. While some of these transitions are intrinsically reversible and symmetric discontinuities, others are reversible and asymmetric or irreversible discontinuities. State-of-the-art languages incorporate powerful mechanisms to declare any type of transitions (for example, if-then-else and case constructs) which can be nested to an arbitrary depth.

A second class of discontinuities arises from external actions imposed by the surroundings on the system at a particular point in time, typically discrete manipulations and disturbances such as operating procedure or imposed failures.

The ability to incorporate complex sets of discontinuities describing a series of states in the formulation of a DAE system leads to the concept of state-task networks (STNs) [1]. Many, though not all, modern custom-modeling tools are capable of solving complex problems involving arbitrarily complex combinations of DAE and STN systems.

In this chapter we are not concerned with the mathematics of dynamic modeling; this is well covered in the literature. The focus is rather on the practical and systematic application of the high-fidelity process models, using now well-established principles of model-based engineering, to add value to industrial operations.

1.1.2
Modeling Software

Custom-modeling tools based on general-purpose, high-level, equation-oriented, declarative modeling languages are becoming the standard tools of the trade for the development and application of high-fidelity predictive models, particularly when full control over the scope and detail of the process model is required [2]. Such tools can support many of the activities essential for the effective dynamic process modeling that is the subject of this volume.

Established commercial examples of general-purpose custom-modeling platforms for integrated simulation and optimization (including in some cases model validation and design activities) include Aspen Technology's Aspen Custom Modeler®, Process Systems Enterprise's gPROMS® and Dassault Systèmes's Dymola®; open source initiatives include EMSO (Environment for Modeling, Simulation and Optimization) and Open Modelica.

1.2
Dynamic Process Modeling – Background and Basics

The state-of-the-art in process modeling has moved well beyond the "material and energy accounting" exercises of steady-state flowsheeting applications (which nevertheless still represent a large proportion of process modeling activity) to become a central platform for capture and deployment of companies' Intellectual Property (IP).

A systematic modeling activity using high-fidelity predictive models provides a set of tools that enables companies to adapt – rapidly if necessary – to changing market drivers and gain competitive advantage. Models, if properly constructed, embody valuable corporate knowledge in such a way that it can be rapidly accessed and deployed to generate immediate commercial value. For example, during an economic downturn the models originally used to optimize equipment design can be rapidly redeployed with a different objective: reducing operating cost.

For example, a model may be used to capture the definitive set of reaction kinetic relationships that may represent many years' worth of experimentation. This reaction network model can then be used as part of a reactor model to redesign the reaction process, or to adapt an operating process to improve its economics under new market conditions. The knowledge developed during the redesign or adaptation – for example, regarding the performance of alternative reactor internal configurations or catalyst formulations – is often valuable information in itself. This new information can be captured in the model for use in future analysis, or to form the basis of patent applications in support of licensed process development, and so on. It is possible in this way to enter a virtuous cycle where each use of the model generates information that increases its usefulness in the future; every subsequent activity increases the return on initial modeling investment.

1.2.1
Predictive Process Models

In order to describe modeling approach and practice, it is worth first defining what we mean by a "process model." For the purposes of this chapter a process model is considered to be *a set of mathematical relationships and data that describe the behavior of a process system, capable of being implemented within a process modeling environment in such a way that they can be used to generate further knowledge about that system through simulation, optimization, and other related activities.*

To be more specific, a chemical engineering model – whether written in a programming language such as Fortran or C or a modern modeling language such as gPROMS language or Modelica – is typically a collection of physics, chemistry, engineering, operating and economic knowledge in equation form, coupled in some way with empirically determined data.

To be useful in design and operational analysis, a model should have a *predictive capability* so that given, for example, a new set of feed or operating conditions or design parameters, it calculates accurate revised rates for the output values of interest.

At its best, a model is capable of predicting actual behavior with accuracy over a wide range of conditions without refitting the model parameters, making it possible to explore the design or operational space consistently and comprehensively and be confident in the results.

In order to be predictive, a model needs to take into account all the phenomena that significantly affect the output values of interest, to the required degree of accuracy. For example, a model that includes a very detailed reaction kinetic representation but ignores diffusion effects for a reaction that is severely diffusion-limited will not provide very useful results. Conversely there is little point in including detailed intrapore diffusion effects in a catalytic reaction model when it is known that reactions are fast and occur largely on the catalyst surface; the unnecessary detail may increase calculation times and reduce model robustness for very little gain in accuracy. Generally, current practice is to construct "fit-for-purpose" models wherever the modeling software environment allows this; in many cases these can be "parameterized" in order easily to include or exclude phenomena.

1.2.2
Dynamic Process Modeling

Most process engineers are familiar with steady-state simulation modeling from some stage of their careers, typically in the form of widely available process flowsheeting packages such as Aspen Technology's AspenPlus™ and Hysys®, and Invensys Process Systems' PRO/II®.

The key ostensible difference between the steady-state models of these packages and dynamic process models is the ability to take into account variation over time. However, as can be seen below the difference is not simply the addition of a time dimension; dynamic modeling often brings a whole different approach that results in dynamic models being a much truer-to-life representation of the process in many other respects.

While steady-state analysis is mainly used for process flowsheet design, usually to determine mass and energy balances and approximate equipment sizes, or perhaps stream properties, the ability of dynamic models to model transient behavior opens up a whole new world of application. Typical applications of dynamic models are as follows:

- analysis of transient behavior, including performance during start-up, shutdown, and load change;
- regulatory (i.e., PID) control scheme analysis and design;
- design of optimal operating procedures – for example, to optimize transition between product grades;
- design of batch processes;
- design of inherently dynamic continuous processes – for example, pressure swing adsorption;
- fitting data from nonsteady-state operations – for example, dynamic experiments, which contain much more information than steady-state experiments, or estimation of process parameters from transient plant data;
- safety analysis – for example, the determination of peak pressures on compressor trip;
- inventory accounting and reconciliation of plant data;
- online or offline parameter re-estimation to determine key operating parameters such as fouling or deactivation constants;
- online soft-sensing;
- operator training.

1.2.3 Key Considerations for Dynamic Process Models

There are of course many different types of dynamic model, from simple transfer-function models often used for control system design, through operator training simulation models and "design simulators" to fully fledged process models containing a high-fidelity representation of physics and chemistry and capable of being solved in a number of different ways. Here we are more interested in the latter end of this range.

In a dynamic model the simple assumptions of a steady-state model – for example, that material always flows from an upstream unit where the pressure is higher to a lower pressure downstream pressure unit – may no longer be valid. Transients in the system may cause the "downstream" pressure to become higher than the "upstream" pressure, causing flow reversal; the model has to allow for this possibility.

This requirement for a "pressure-driven" approach is just one example. There are many others: the following are some of the considerations that need to be taken into account in a reliable dynamic process model:

- *Material transport.* Material cannot simply be assumed to flow from one unit to another conveniently in "streams." The device in which it is transported (typically a pipe) itself needs to be treated as a piece of equipment and modeled. Flow – and the direction of flow – is determined by pressure gradients and resistance considerations, which may be different depending on the direction of flow, as well as flow regime.
- *Flow reversal.* As mentioned above, it is necessary to provide appropriate mechanisms for reverse flow into units where there is material accumulation. It is also necessary to provide for the fact that the intrinsic properties (for example, specific enthalpy) of the stream change when flow is reversed to correspond to the properties of the material leaving the downstream, rather than the upstream, unit. This requires proper definition of streams at the downstream boundary of the flowsheet to include reverse-flow "inlet" properties.
- *Equipment geometry.* The physical geometry of equipment may have a significant effect on the dynamics, and needs to be taken into account. For example, the time dynamics of a distillation column depend on the tray area, weir height, bubble cap vapor flow area, and so on. Not only do the spatial relationships need to be represented in the model, but also a whole additional category of data – typically taken from design specification sheets – is required.
- *Process control and control devices.* In steady-state modeling, as the term implies, state variables (typically those representing accumulation of mass and energy, or their close relatives such as level, temperature, and pressure) can simply be set to the required values as required. In dynamic models, as in real life, these values vary as a result of external influences – for example, pressure or level is affected by the changing flowrates in and out of a vessel – and cannot be set. As on the physical plant, the states need to be maintained at their "set point" values through the use of controllers and associated measurement and control devices. These devices need to be represented fully in the dynamic model.
- *Phase disappearance.* During transient behavior phases may appear and disappear. For example, a flash drum that would normally operate in a vapor–liquid two-phase region may generate liquid only if the pressure rises above the bubble-point pressure or the temperature drops below the bubble-point temperature. The transition between phase regimes is very challenging numerically and the solution engine needs to be able to handle the calculation robustly and efficiently.
- *Equilibrium.* In general, steady-state simulators assume equilibrium between phases, for example, in distillation calculations. During transient situations operation may depart significantly from equilibrium and mass transfer limitations become important. The assumption of equilibrium between phases can no longer be taken for granted, and rate-based mass transfer modeling is required (note that rate-based considerations can be equally important in a steady-state representation).

There are other areas where representation tends to be (or in many cases, needs to be) much more accurate in dynamic models than steady-state models – for example, in representation of reaction kinetics.

The proper inclusion of the phenomena listed above results in a significant enhancement in fidelity of representation and range of applicability over and above that afforded by the simple addition of accumulation terms.

The inclusion of these phenomena not only requires greater modeling effort; it also requires an appreciation of the structural consistency issues resulting from inclusion of many closely interrelated effects. When specifying the set or assigned variable values for a dynamic simulation, for example, it is generally necessary to specify quantities that would be fixed or considered constant in real-life operation, such as equipment dimensions, controller set points, and upstream and downstream boundary conditions. Also, in some cases the "simplifying assumptions" used in steady-state models (for example, constant holdups on distillation trays) lead to a high-index dynamic problem [3]; resolving this may take considerable additional modeling effort.

Choosing a consistent set of initial conditions, and providing meaningful numerical values for these variables, is not a trivial undertaking for those coming from a steady-state modeling paradigm. Many modelers struggle with this task which, though not complex, nevertheless requires attention, knowledge, and additional modeling effort.

The remainder of this chapter takes "*dynamic process models*" as a shorthand for *high-fidelity predictive models* that embody many or all of the representations described above.

1.2.4
Modeling of Operating Procedures

A significant and important advantage of dynamic process modeling is the ability to model – and hence optimize – operating procedures, including production recipes, simultaneously with equipment and process.

The ability to optimize operating procedures is of course an essential requirement in the design of batch process systems. However, it also has significant application to continuous processes.

For example, it is possible to model process start-up procedure as a series of tasks ("start pump P102, continue until level in vessel L101 is 70%, open steam valve CV113, ...," etc.), each of which may involve executing a set of defined subtasks. Given a suitably detailed start-up procedure and a model that is capable of providing accurate results (and executing robustly) over the range of operating conditions encountered – which may include significant nonequilibrium operation – it is possible not only to simulate the plant behavior during start-up but also to *optimize* the start-up sequence.

The objective function in such cases is usually economic – for example, "start-up to on-spec production in the minimum time/at the lowest cost." Constraints are typically rate limitations or material limits – for example, "subject to a maximum steam rate of x kg/s," "subject to the temperature in vessel L141 never exceeding y degrees C," etc.

Other applications are in the design and optimization of feedstock or product grade change procedures for complex processes. Many hours of off-spec production can often be eliminated by using dynamic optimization to determine the optimal process variable trajectories to maximize a profit objective function.

1.2.5
Key Modeling Concepts

There are a number of key terms and concepts that frequently arise in any discussion on high-fidelity predictive modeling that are worth touching on briefly.

1.2.5.1 First-Principles Modeling

For the purposes of this discussion, the term first-principles modeling is intended to reflect current generally accepted chemical engineering first-principles representation of natural phenomena derived from the conservation of mass, energy, and momentum. More specifically, this implies material and energy balances, hydraulic relationships, multicomponent diffusion, reaction kinetic relationships, material properties as provided by typical physical property packages, and so on, all applied to distinct physical components or pseudocomponents – rather than first-principles representation at a submolecular level. While there are emerging techniques that will allow incorporation of molecular-level representation within process models, allowing simultaneous design of molecule and process [4], for the purposes of this chapter we assume the existence of predefined fixed molecules with known properties.

1.2.5.2 Multiscale Modeling

In many chemical engineering problems important phenomena – i.e., those that have a significant effect on the behavior of a process – may occur at a number of different spatial and temporal scales.

For example, the size of a slurry-bed reactor for Fischer–Tropsch gas-to-liquid conversion (typically tens of meters in height) is strongly influenced by the micron-scale diffusion of reactants and product within catalyst particle pores. In order to successfully optimize the design of such a reactor, the model needs to represent key relationships and phenomena at both extremes as well as all the intermediate scales. If such a model is to be placed within a plant flowsheet, yet another scale needs to be taken into account.

In some cases – for example, rate-based distillation – well-defined theoretical models exist [5] that take into account the relevant phenomena, from film diffusion to whole column operation. It is relatively straightforward to implement these within a modeling package.

In others, the modeler needs to apply significant ingenuity to represent all the required scales within a computationally tractable solution. An example is a multitubular reactor model, where a full representation needs to couple an accurate shell-side fluid model with highly detailed models of the catalyst-filled tubes within the shell. The shell-side fluid model alone may require discretization (meshing) running to hundreds of thousands or even millions of cells. Tube models require at least a 2D representation (a third spatial dimension may be required to account for a distribution along the catalyst pore depth) in which the full reaction-with-diffusion model is solved at each grid point. There may be 20 000 such tubes in a multitubular reactor, each experiencing a different external temperature profile because of nonuniformity in the shell-side fluid flows at any given shell radius.

In such a case a full representation of all scales would result in hundreds of millions of equations to be solved. This is obviously not a viable approach, and the model needs to be reduced in size significantly and in such a way that the accuracy is preserved.

As software becomes more powerful, the challenge of representing different scales within a single model (or hierarchy of models) decreases; however, brute force is not necessarily the answer. There are numerous techniques and methodologies for reducing the computational effort required to deal with different scales, from the stepwise approach to parameter estimation at different scales described in Section 1.3.3 to the multizonal hybrid modeling approaches that couple the computational fluid dynamic (CFD) representation of hydrodynamics with phenomenological models of processes such as crystallization [6], to the interpolation techniques used to effect model reduction in fuel cell stacks. The example in Section 1.4 outlines a tractable approach for high-fidelity multitubular reactor modeling that is now applied routinely to provide accurate detailed design information.

1.2.5.3 Equation-Based Modeling Tools

Past practice tended to combine models – the relationships describing a process – with their mathematical solution, often with data thrown into the mix in the form of hard-coded values or empirical functions with limited range of applicability.

In many cases "black box" models, they were written in programming languages such as Fortran, DELPHI, or C, with users unable to modify or even see the underlying relationships. New unit models needed to be coded from scratch, including all their internal numerical solution code, which sometimes took months of effort. Such models were typically inflexible, limited in scope and predictive capability, capable of being solved in a single direction only, difficult to maintain, and often lacking in robustness.

Modern "equation-based" or "equation-oriented" tools separate the engineering aspects from the mathematical solution using fourth-generation languages or graphical representations; good modeling practice ensures that model and data are separated too. This means that a complex unit can be now described in tens of lines of physics and chemistry relationships rather than thousands of lines of code required in the past, with significant implications for development time, quality assurance, ease of understanding and maintenance and re-usability – and of course lifecycle cost. This fact alone has removed significant hurdle from the development and application of models.

The removal of the mathematics means that it is possible to extend the fidelity and range of application of models easily. For example, a lumped parameter

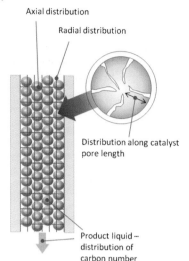

Fig. 1.1 A fixed-bed reactor model can involve a four-dimensional distribution – for example, the four spatial and one property distribution shown here.

pipeline or heat exchanger model can be turned into a much more accurate distributed system model with a few days' effort rather than months of programming.

1.2.5.4 Distributed Systems Modeling

The more sophisticated equation-based custom-modeling environments make it possible to model distributions in size, space, or any other relevant quantity easily.

Consider a fixed-bed Fischer–Tropsch tubular reaction process converting syngas into longer chain hydrocarbons (Fig. 1.1). Axial and radial distributions are used to represent the distribution of concentration, temperature, and other quantities along and across the tube. A further spatial distribution is used to representing changing conditions along the length of the catalyst pores (essential for this severely diffusion-limited reaction). Finally the bulk of the product from the Fischer–Tropsch reaction – longer chain hydrocarbons with various numbers of carbon atoms – is modeled as a distribution of carbon numbers.

Without any one of these distributions the model would not be able to represent the system behavior accurately.

Sophisticated tools go further, allowing for nonuniform grids. This can be important in reducing system size while maintaining accuracy. An example is the modeling of a catalyst tube in a highly exothermic reaction, where much of the action takes place close to the beginning of the active catalyst bed. In this area of steep gradients, it is desirable to have many discretization grid points; however, in the areas of relatively low reaction further down the tube having many points would simply result in unnecessary calculation. Some tools go further, providing adaptive grids that are capable of modeling moving-front reactions to high accuracy with a relatively small number of discretization points.

1.2.5.5 Multiple Activities from the Same Model

A key advantage of equation-based systems is that it is possible to perform a range of engineering activities with the same model. In other words, the same set of underlying equations representing the process can be solved in different ways to provide different results of interest to the engineer or researcher. Typical examples are:

- *Steady-state simulation*: for example, given a set of inlet conditions and unit parameters, calculate the product stream values.
- *Steady-state design*: for example, given inlet and outlet values, solve the "inverse problem" – i.e., calculate equipment parameters such as reactor diameter.
- *Steady-state optimization*: vary a selected set of optimization variables (for example, unit parameters such as column diameter, feed tray location, or specified process variables such as cooling water temperature) to find the optimal value of a (typically economic) objective function, subject to constraints.
- *Dynamic simulation*: given a consistent set of boundary conditions (for example, upstream compositions and flowrates or pressure and downstream pressure, unit parameters and initial state values, show how the process behaves over time. Dynamic simulations typically include the control infrastructure as an integral part of the simulation. The process is typically subject to various disturbances; the purpose of the simulation is usually to gauge the effect of these disturbances on operation.
- *Dynamic optimization*: determine the values and/or trajectories of optimization variables that maximize or minimize a (typically economic) objective function, subject to constraints. Dynamic optimization is used, for example, to determine the optimal trajectories for key process variables during feedstock or product grade change operations in order to minimize off-spec production. It effectively removes the need to perform repeated trial-and-error dynamic simulations.
- *Parameter estimation*: fit model parameters to data using formal optimization techniques applied to a model of the experiment, and provide confidence information on the fitted parameters.

There are many other ways in which model can be used, for example, for sensitivity analysis, model-based experiment design, model-based data analysis, generation of linearized models for use in other contexts (for example, in control design or within model-based predictive controllers) and so on.

1.2.5.6 Simulation vs. Modeling

It is useful at this point to distinguish between dynamic simulators or simulation tools and dynamic modeling tools. The former generally perform a single function only (in fact they are often limited by their architecture to doing this): the dynamic simulation activity described above. Frequently dynamic simulators cannot solve for steady state directly, but need to iterate to it. Operator training simulators are a typical example of dynamic simulators.

Fully featured dynamic process modeling tools can perform dynamic simulation as one of the set of possible model-based activities. They treat the model simply as a representation of the process in equation form, without prior assumption regarding how these equations will be solved.

1.3
A Model-Based Engineering Approach

As the term implies, a model-based engineering (MBE) approach involves performing engineering activities with the assistance of a mathematical model of the process under investigation.

MBE applies high-fidelity predictive models in combination with observed (laboratory, pilot, or plant) data to the engineering process in order to explore the design space as fully and effectively as possible and support design and operating decisions. Typical application is in new process development or optimization of existing plants.

Closely related is model-based innovation, where a similar approach is used at R&D level to bridge the gap between experimental programs and engineering design, and capture corporate knowledge in usable form, typically for early-stage process or product development.

Key objectives of both approaches are to:

- accelerate process or product innovation, by providing fast-track methods to explore the design space while reducing the need for physical testing;
- minimize (or effectively manage) technology risk by allowing full analysis of design and operational alternatives, and identifying areas of poor data accuracy;
- integrate R&D experimentation and engineering design in order to maximize effectiveness of both activities and save cost and time.

At the heart of the model-based engineering approach are high-fidelity predictive models that typically incorporate many of the attributes described in the previous section.

1.3.1
High-Fidelity Predictive Models

In order to create a high-fidelity predictive model it is necessary to combine theoretical knowledge and empirical information derived from real-world measurement (Fig. 1.2). Current best practice is to:

- fix (in equation form) everything that can reliably be known from theory. Typically this involves writing equations for heat and material balance relationships, reaction kinetics, hydraulic relationships, geometry, etc., many of which are based on well-understood chemical engineering theory or can be found in model libraries or literature;

1.3 A Model-Based Engineering Approach

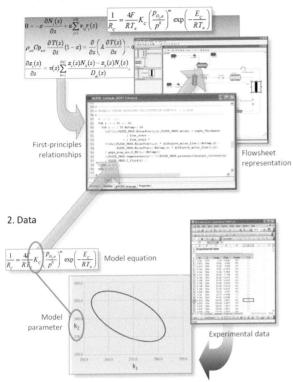

Fig. 1.2 The constituents of a high-fidelity predictive model: first-principles theoretical relationships and real-world (laboratory, pilot or operating) data.

- then estimate the empirical constants using the information already contained within the model in conjunction with experimental data.

The key objective is to arrive at a model that is accurately predictive over a range of scales and operating conditions. In order to do this it is necessary to include a representation of all the phenomena that will come into play at the scales and conditions for which the model will be executed, and to determine accurate *scale-invariant* model parameters, where is possible.

It is the combination of the first-principles theory and suitably refined data that provides the uniquely predictive capability of such models over a wide range of conditions.

This approach brings the best of both worlds, combining as it does well-known theory with actual observed values. But how does is it implemented in practice? This is explored in the subsequent sections.

1.3.2
Model-Targeted Experimentation

Model-based engineering requires a different approach to experimentation from that traditionally employed.

Traditionally experimentation is carried out in such a way that the results of the experiment can be used to optimize the process. For example, experiments are designed to maximize production of the main product rather than impurities; the knowledge gained from the experiment is used directly to attempt to maximize the production on the real plant.

Model-targeted experimentation instead focuses on experiments that *maximize the accuracy of the model*. The model, not the experimental data, is then used to optimize the design.

Model-targeted experimentation is just a likely to require an experiment that maximizes production of an *impurity* rather than the main product, in order to better characterize the kinetics of the side reaction. With accurate characterization of all key reactions it is possible accurately to predict the effect on product purity, for example, of minor changes in operating temperature.

1.3.3
Constructing High-Fidelity Predictive Models – A Step-by-Step Approach

Typically predictive models are constructed in a modular fashion in a number of steps, each of which may involve validation against laboratory or pilot plant experimental data. The validation steps will require a model of the process in which the experimental data is being collected – *not* a model of the full-scale equipment at this stage – that includes a description of the relevant phenomena occurring within that process.

For example, consider the case where reaction data are being collected using a small-scale semibatch reactor with a fixed charge of catalyst, for the purposes of estimating reaction kinetic parameter values to be used in a reactor design. The model used for parameter estimation needs to provide a high-fidelity representation of the semibatch experimental process, as well as the rate equations proposed for the expected set of reactions and any diffusion, heat transfer or other relevant effects. If the model does not suitably represent what is occurring in the experiment, the mismatch will be manifest as uncertainty in the estimated parameter values; this uncertainty will propagate through the design process and eventually surface as a risk for the successful operation of the final design.

The procedure below provides a condensed step-by-step approach to constructing a validated model of a full-scale process plant (Fig. 1.4). It assumes access to equation-based modeling software where it is possible to program first principles models and combine these in some hierarchical form in order to represent multiple scales, as well as into flowsheets in order to represent the broader process. It also assumes software with a parameter estimation capability.

The key steps in constructing a validated first-principles model with a good predictive capability over a range of scales are as follows:

Step 1 – Construct the first-principles model

This is probably perceived as the most daunting step for people not familiar with the art. However it need not be as onerous as it sounds.

If starting from scratch, many of the relationships that need to be included have been known for generations: conservation of mass, conservation of momentum, conservation of energy and pressure-flow relationships are standard tools of the trade for chemical engineers. Thermophysical properties are widely available in the form of physprop packages, either commercially or in-house. Maxwell–Stefan multicomponent diffusion techniques, though complicated, are well-established and can provide a highly accurate representation of mass transfer with minimal empirical (property) data. Most researchers dealing with reaction have a fair idea of the reaction set and a reasonable intuition regarding which reactions are predominant. Molecular modeling research can provide guidance on surface reactions. If necessary, a reaction set and rate equations can be proposed then verified (or otherwise) using *model-discrimination* techniques based on parameter estimation against experimental data [7].

Of course it is not necessary to go to extremes of modeling every time. The challenge to the modeler often lies in selecting the appropriate level of fidelity for the requirements and availability of data – if a reaction is rate-limited rather than diffusion-limited it is not necessary to include diffusion relationships. Restricting the modeled relationships strictly to those required to obtain an accurate answer helps make the resulting mathematical problem more robust and faster to solve, but it does restrict generality.

Having said that, most modelers do not start from scratch; there is an increasing body of first-principles models of many types of process available from university research, in equation form in the literature, or as commercially supplied libraries. Many first-principles models can be found in equation form with a careful literature search.

An interesting tendency is that as models increasingly are constructed from faithful representations of fundamental phenomena such as multicomponent mass transfer and reaction kinetics, the core models become more universally applicable to diverse processes. For example, the core models used to represent the reaction and mass transfer phenomena in a reactive distillation provide essential components for a bubble column reactor model; similarly the multi-dimensional fixed-bed catalytic reactor models that take into account intrapore diffusion into catalyst particles provide most of a model framework required for a high-fidelity model of a pressure-swing adsorber. Organizations investing in constructing universally applicable first-principles models – as opposed to "shortcut" or empirical representations – are likely to reap the fruit of this investment many times over in the future.

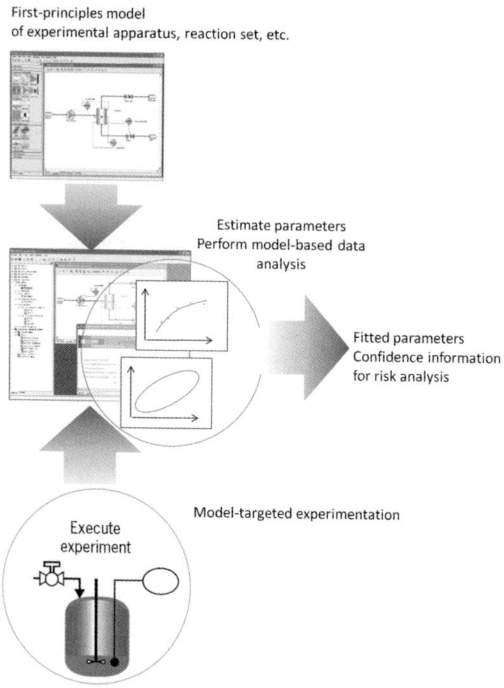

Fig. 1.3 Fitting experimental data to a model of the experiment in order to generate model parameter values.

Step 2 – Estimate the model parameters from data

Having constructed a first principles model, it will often be necessary – or desirable – to estimate some parameters from data. This is almost always the case when dealing with reaction, for example, where reaction kinetic parameters – virtually impossible to predict accurately based on theory – need to be inferred from experimental data. Other typical parameters are heat transfer coefficients, crystal growth kinetic parameters, and so on. Initial parameter values are usually taken from literature or corporate information sources as a starting point.

Parameter estimation techniques use models of the experimental procedure within an optimization framework to fit parameter information to experimental data (Fig. 1.3). With modern techniques it is possible to estimate simultaneously large numbers of the parameters occurring in complex nonlinear models – involving tens or hundreds of thousands of equations – using measurements from any number of dynamic and/or steady-state experiments. Maximum likelihood techniques allow instrument and other errors inherent in practical experimentation to be taken directly into account, or indeed to be estimated simultaneously with the model parameter values.

When dealing with a large number of parameters, parameters at different scales can be dealt with in sequential fashion rather than simultaneously. For example it is common practice to estimate reaction kinetics from small-scale (of the order of centimeters) experiments in isothermal conditions, then maintain these parameters constant when fitting the results of subsequent larger scale experiments to measure "equipment" parameters such as catalytic bed heat transfer coefficients.

Parameter estimation effectively "tunes" the model parameters to reflect the observed ("real-life") behavior as closely as possible. If performed correctly, with data gathered under suitably defined and controlled conditions, estimation can generate sets of parameter that are virtually scale invariant. This has enormous benefits for predictive modeling used in design scale-up. It means that, for instance, the reaction kinetic or crystal growth parameters determined from experimental apparatus at a scale of cubic centimeters can accurately predict performance in equipment several orders of magnitude larger, providing that effects that become significant at larger scales – such as hydrodynamic (e.g., mixing) effects – are suitably accounted for in the model.

Parameter estimation can be simpler or more challenging than it sounds, depending in the quantity and quality of data available. It is a process that often requires a significant amount of engineering judgment.

Step 3 – Analyze the experimental data

Parameter estimation has an important additional benefit: model-based data analysis. The estimation process produces quantitative measures of the degree of confidence associated with each estimated parameter value, as well as estimates of the error behavior of the measurement instruments. Confidence information is typically presented in the form of confidence ellipsoids.

Often this analysis exposes inadequacies in the data – for example certain reaction kinetic constants in which there is a poor degree of confidence, or which are closely correlated with other parameters. In some cases parameters can exhibit poor confidence intervals even if the values predicted using the parameters appears to fit the experimental data well. Using model-based data analysis it is usually easy to identify the parameters of most concern and devise additional experiments that will significantly enhance the accuracy of – and confidence in – subsequent designs.

Poor data can be linked to design and operational risk in a much more formal manner. Using techniques currently under development [8], global sensitivity analysis coupled with advanced optimization methods can be used to map the uncertainty in various parameters – in the form of probability distributions – onto the probability distributions for plant key performance indicators (KPIs). This translates the uncertainty inherent in, say, a reaction kinetic parameter, directly into a quantitative risk for the success of the plant performance. For example if the real parameter value turns out to be in a certain region within the confidence ellipsoid the designed plant may not make a profit, or may end up operating in an unsafe region.

The ability to relate KPIs such as profit directly to parameter uncertainty through the medium of a high-fidelity predictive model means that often-scarce research funding can be directed into the study of the parameters that have the greatest effect on these KPIs. This makes it possible to justify R&D spending based on a quantitative ranking.

Step 4 – Design additional experiments, if necessary

In some cases it will be obvious which additional experiments need to be carried out, and how they need to be carried out. For more complex situations, in particular cases where experiments are expensive or time consuming, there are benefits in applying more formal experiment design techniques.

A major development in the last few years has been the emergence of model-based techniques for the design of experiments [7]. In contrast to the usual statistically based techniques (e.g., factorial design), *model-based experiment design* takes advantage of the significant amount of information that is already available in the form of the mathematical model. This is used to design experiments which yield the maximum amount of parameter information, thereby minimizing the uncertainty in any parameters estimated from the results of these experiments.

This optimization-based technique is applicable to the design of both steady-state and dynamic experiments, and can be applied sequentially taking account of any experiments that have already been performed. Figure 1.4 shows how model-based experiment design fits within the MBE procedure.

Typical decision variables determined by model-based experiment design include the optimal conditions under which the new experiment is to be conducted (e.g., the temperature profile to be followed over the duration of the experiment), the optimal initial conditions (e.g., initial charges and temperature) and the optimal times at which measurements should be taken (e.g., sampling times for off-line analysis).

The overall effect is that the required accuracy in the estimated parameter values may be achieved using the minimum number of experiments. In addition to maximizing the improvement in parameter accuracy per experiment, the approach typically results in significantly reduced experimentation time and cost.

Steps 2 to 4 are repeated until parameter values are within acceptable accuracy. Determining "acceptable accuracy" in this context generally means evaluating the effects of possible values of parameters on key design criteria (for example, the conversion of a key component in a reactor) using the model, as part of a formal or informal risk assessment of the effect on plant KPIs. Generally, if any combination of the potential parameter values results in unacceptable predicted operation (for example, subspecification products or uneconomic operation) – even if other combinations demonstrate acceptable operation – the parameter values need to be refined by going around the experimentation cycle again.

Steps 1 to 4 may also be repeated to develop submodels at different scales, covering different phenomena. Typically the parameters fitted during earlier analyses are held constant (i.e., considered to be fixed), providing in effect a "model reduction" for the parameter estimation problem by reducing the number of parameters to be fitted at each step and allowing the experiments at each step to focus on the phenomena of interest. The procedure at each stage is the same: build a model that contains the key phenomena, conduct experiments that generate the required

1.3 A Model-Based Engineering Approach

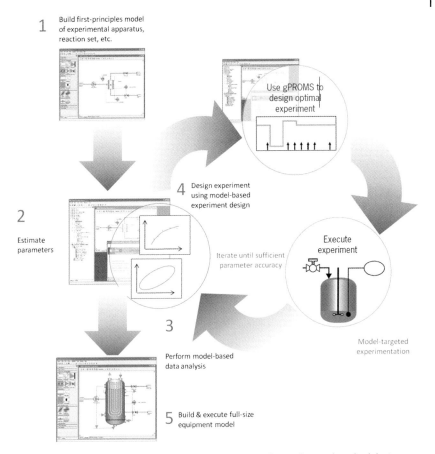

Fig. 1.4 Model-based engineering – a schematic overview of procedure and methodologies.

data as unambiguously as possible, estimate the relevant parameters, then "fix" this model and parameters in subsequent investigations at different scales.

Step 5 – Build the process model

Once the parameter confidence analysis indicates an acceptable margin of uncertainty in the parameters, the submodels can then be used in construction of the full equipment model. If necessary, this can be validated against operating or test run data to determine, for example, overall heat transfer or flow coefficients. It should not be necessary, and indeed would be counterproductive, to estimate parameters already fixed from the earlier laboratory or pilot validation at this stage.

Building the full model typically involves assembling the relevant submodels of phenomena at different scales, defining appropriate boundary conditions (flows, pressures, stream and external wall temperatures, etc.) and if necessary including ancillary equipment such as heat exchangers, pumps and compressors, recycles, and so on within a flowsheet representation.

1.3.4
Incorporating Hydrodynamics Using Hybrid Modeling Techniques

When scaling up, or simply constructing models of larger-scale equipment, the assumption of perfect mixing may not hold. In such cases the model needs to take into account fluid dynamics in addition to the accurate representation of diffusion, reaction and other "process" phenomena.

The IPDAE software modeling systems used for the process chemistry representation are not ideally suited for performing detailed hydrodynamic modeling of fluid behavior in irregular geometries. An approach that is increasingly being applied is to couple a computation fluid dynamic (CFD) model of the fluid dynamics to the process model in order to provide hydrodynamic information [6].

Such "hybrid modeling" applications fall into two broad classes: those where the CFD and process models model different aspects of the *same* volumes, and those where they model different volumes, sharing information at surface boundaries.

Such linking of CFD models with models of process and chemistry is a well-established practice used for design of crystallizers, slurry-bed reactors, multitubular reactors, fuel cells, and many other items of equipment. In most cases the CFD model provides initial input values to the process model; these may be updated periodically based on information generated by the process model and provided back to the CFD model but in general intensive iteration between the two is (fortunately) not required.

The combination makes available the best features of both modeling approaches. An example is crystallizer modeling, where each model represents different aspects of the same physical volume, usually a stirred vessel. The process model is divided into a number of well-mixed zones (typically between 5 and 50), each of which is mapped onto groups of cells in the CFD model. The CFD model calculates the velocity field, in particular the flows in and out of the zone, and may also provide turbulent dissipation energy information for the crystallization calculation. The process model performs the detailed crystallization population balance calculation for the zone, as well as calculating the kinetics of nucleation, growth, attrition and agglomeration and any other relevant phenomena, based on the information received from the CFD model.

1.3.5
Applying the High-Fidelity Predictive Model

If Steps 1 to 4 have been followed diligently and Step 5 model has been constructed with due care by taking into account all the relevant phenomena, the resulting model should be capable of predictive behavior over a *wide range of scales and conditions*.

The model can now be used to optimize many different aspects of design and operation. It is possible to, for example:

- accurately determine optimal designs and predict performance of complex items of equipment such as reactors or crystallizers [9];
- perform accurate scale-up taking into account all relevant phenomena, backed up by experimental data where necessary;
- optimize operation to – for example – minimize costs or maximize product throughput, subject to quality constraints;
- optimize batch process recipes and operating policy in order to reduce batch times and maximize product quality;
- perform detailed design taking into account both hydrodynamic and chemistry effects [10];
- support capital investment decisions based on accurate quantitative analysis;
- rank catalyst alternatives and select the optimal catalyst type based on their performance within a (modeled) reactor;
- optimize reactor operating conditions for maximum catalyst life [11];
- determine optimal trajectories for feedstock or load or product grade change, in order to minimize off-spec operation [12];
- troubleshoot poor operation;
- perform online monitoring, data reconciliation and yield accounting;
- generate linearized models for model-predictive control within the automation framework.

1.4
An Example: Multitubular Reactor Design

The example below describes the application of high-fidelity predictive modeling and hybrid modeling in the development of a new high-performance multitubular reactor by a Korean chemical company, as the core unit of a new licensed process [10].

The example has been chosen because it illustrates many different aspects of the model-based engineering approach described above, in particular:

- the use of complex first-principles models using spatial distributions, detailed reaction and diffusion modeling;
- modeling at multiple scales – from catalyst pore to full-scale equipment – using hierarchical models;
- parameter estimation at different scales;
- the combination of CFD hydrodynamic modeling with detailed catalytic reaction models using a hybrid modeling approach;
- the ability to perform detailed design of equipment geometry taking all relevant effects into account;
- a stepwise approach to a complex modeling problem.

It also represents a now-standard approach to modeling of equipment that until a few years ago was considered far too complex to contemplate.

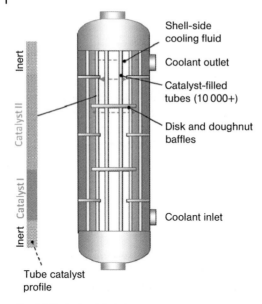

Fig. 1.5 Typical multitubular reactor.

1.4.1
Multitubular Reactors – The Challenge

Multitubular reactors (Fig. 1.5) typically comprise tens of thousands of catalyst-filled tubes within a cylindrical metal shell in which cooling fluid – for example, molten salt – circulates. They are very complex units that are not only difficult to design and operate, they are extremely difficult to model.

Each tube experiences a different external temperature profile down its length, which affects the rate of the exothermic catalytic reaction inside the tube. This in turn affects the local temperature of the cooling medium.

The consequences of temperature maldistribution can be significant. Tubes that experience different temperature profiles produce different concentrations of product and impurities. The interaction between cooling medium and the complex exothermic reaction inside the tubes can give rise to "hot spots" – areas of localized high temperature that can result in degradation and eventual burnout of catalyst in the tubes in that area, exacerbating nonuniformity of product over time, reducing reactor overall performance and reducing the time between catalyst changes. The need to avoid such areas of high temperature usually means that the reactor as a whole is run cooler than it need be, reducing overall conversion and affecting profitability.

Another complicating factor is that the mixture of inert to active catalyst – or even grades of catalyst – may differ down the length of tubes, in an attempt to minimize the formation of hotspots.

The challenge is to design the reactor internal geometry – for example, the tube and baffle spacing – and the profile of different catalyst formulations along the

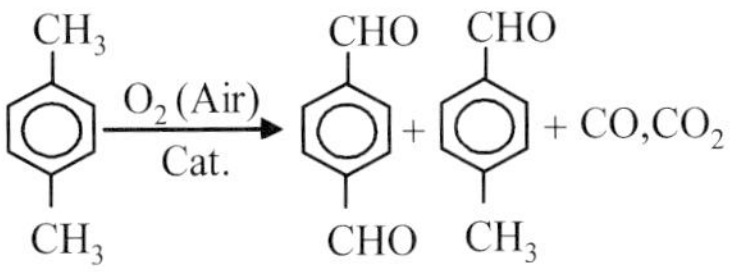

Fig. 1.6 The new process for selective oxidation of p-xylene to TPAL and PTAL.

length of the tubes, in such a way that all of the tubes in a particular part of the reactor experience the same external temperature at normal operating conditions.

Getting it right results in near-uniform conversion of reactants into products across all the tubes and reduction in the potential for hot spots. This means better higher overall quality of product and potentially substantial increase in catalyst life – significantly raising the profitability of the plant.

1.4.2
The Process

The company had developed a new catalyst for the environmentally friendly production (Fig. 1.6) of terephthaldehyde (TPAL), a promising intermediate for various kinds of polymer such as liquid crystal, electron conductive polymer and specialty polymer fibers among other things.

The existing TPAL process (Fig. 1.7) involved production of environmentally undesirable Cl_2 and HCl; their removal was a major factor in the high price of TPAL. Other catalytic routes gave too low a yield. The new catalyst addressed all these problems; all that was required was to develop a viable process. Because of the exothermic nature of the reaction, a multitubular reactor configuration was proposed.

In order to ensure the viability of the process economics, it was necessary to design a reactor that would produce uniform product from tubes at different points in the radial cross section and had a long catalyst cycle.

Design variables to be considered included reactor diameter, baffle window size, baffle span, inner and outer tube limit, coolant flowrate and temperature, tube size, tube arrangement and pitch, and so on (Fig. 1.10 below).

1.4.3
The Solution

In order to gauge the effect of relatively small changes to the geometry it was necessary to construct a high fidelity model of the unit with an accurate representation of catalytic reaction kinetics, reactant and product diffusion to and from the catalyst, heat transfer within the catalyst bed and between bed and tube wall and onward into the cooling medium, and cooling side hydrodynamics and heat transfer.

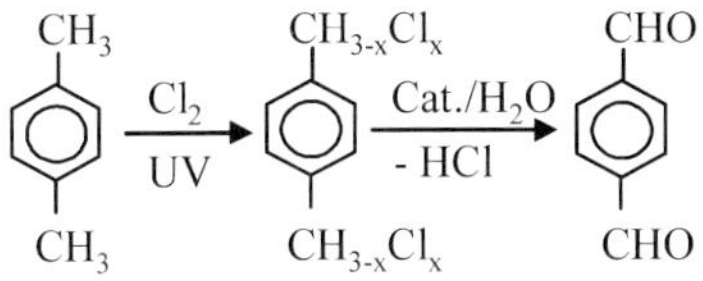

Fig. 1.7 The existing commercial process for TPAL production, which generates environmentally undesirable Cl_2 and HCl.

Constructing a model of the required fidelity required the following steps:

1. Build an accurate model of the catalytic reaction.
2. Develop an understanding of and characterize the fixed-bed performance.
3. Construct and validate a single tube model with the appropriate inert and catalyst loading.
4. Implement the detailed tube model within a full model of the reactor, incorporating the shell-side hydrodynamics.
5. Perform the detailed equipment design by investigating different physical configurations and select the optimal configuration.

In this case, the detailed design of the reactor itself was the key objective, and ancillary process equipment was not considered.

Step 1 – Build an accurate model of the catalytic reaction

First the reaction set model was constructed. This was done by postulating a reaction set (the reactions were relatively well known) and rate equations, using initial reaction kinetic constants from literature.

The reaction set model was implemented within a detailed catalyst model, which included phenomena such as diffusion from the bulk phase to the catalyst surface and into catalyst pores. The models were then implemented within a model of the experimental apparatus. The model parameters were then fitted to existing data from small-scale laboratory experimental apparatus.

The level of information in the experimental measurements allowed rate constants (i.e., the activation energy and pre-exponential factor in the Arrhenius equation) and adsorption equilibrium constants (heat of adsorption and pre-exponential factor in the van 't Hoff equation) to be fitted to a high degree of accuracy.

Examination of parameter estimation confidence intervals showed that the fit was suitably accurate not to warrant any further experimentation. The accuracy of fit also implied that the proposed reaction set was an appropriate one.

Step 2 – Characterize the fixed-bed performance

The reaction set model determined in Step 1 was then deployed in a two-dimensional catalyst-filled tube model which was used to determine the bed heat transfer characteristics against data from a single-tube pilot plant. The pilot plant setup is shown schematically in Fig. 1.8.

The following procedure was used:

First a model of the single-tube experiment was constructed from:

- the reaction set model constructed in Step 1 above, whose kinetic parameters were now considered to be fixed;
- existing library models for catalyst pellets (as used in Step 1); and
- 2D models of catalyst-filled tube (including inert sections)

as shown in Fig. 1.9.

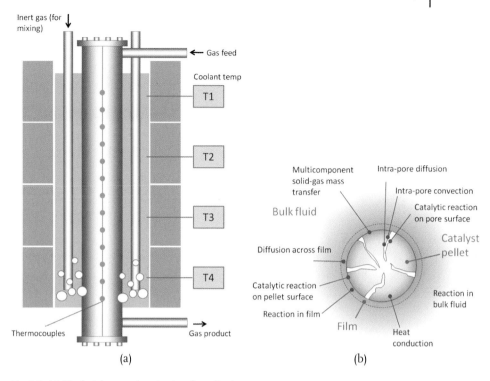

Fig. 1.8 (a) Single-tube experiment setup for collecting accurate temperature profile and concentration information prior to parameter estimation; (b) phenomena modeled in catalytic reaction.

Numerous pilot plant tests were then carried out at different feed compositions, flowrates and temperatures. The collected data were used to estimate the intrabed and bed-to-wall heat transfer coefficients.

Step 3 – Construct a reactor tube model

At this point an accurate model of the catalytic reaction and heat transfer occurring within a single tube had been established. A model of a "typical" reactor tube

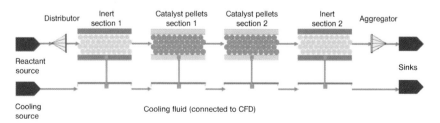

Fig. 1.9 Single-tube model showing sections of different catalyst activity. For the final design model, cooling sections were linked to the CFD model.

could now be constructed using a combination of inert and catalyst-filled sections to represent the catalyst loading.

Step 4 – Build the full reactor model

The next step was to combine tube models with a model of the shell side cooling fluid.

In many cases a reasonable approach would have been to represent the shell side using a network of linked "well-mixed" compartment (zone) models.

However in this case, because the detailed design of the complex baffle and tube bank geometry was the objective, it was necessary to provide a model that was capable of accurately representing the effects of small changes to geometry on the hydrodynamics. This was best done using computational fluid dynamics (CFD) model of the shell side, which was already available from earlier flow distribution studies.

The CFD and tube models were combined using a proprietary "hybrid modeling" interface. Rather than attempting to model the thousands of tubes in the reactor individually, the interface uses a user-specified number of representative tubes, each of which approximates a number of actual tubes within the reactor, interpolating values where necessary.

Step 5 – Perform the detailed equipment design

A number of cases were studied using the hybrid model.

First, the effect of catalyst layer height and the feed rate was studied to determine optimal height of the reactor.

Having established the height, the model was used to check whether there was any advantage in using a multistage shell structure. It was found that a single stage was sufficient for the required performance.

Then the detailed geometry of the three-dimensional tube bank and baffle structure was determined by investigating many alternatives.

Design variables considered (Fig. 1.10) included:

- *Reactor diameter.* As the MTR is effectively a shell-and-tube heat exchanger, the shell can be any diameter, depending on the number of tubes required and their pitch and spacing. However, it is essential that the geometric design provides suitable flow paths to achieve the required cooling in as uniform and controllable as possible a manner.
- *Baffle window size and baffle span.* These affect the fluid flow, and thus the pressure drop and temperature distribution across the shell fluid space.
- *Inner and outer tube limit.* As with the baffle spacing, these attributes can affect cooling patterns significantly.
- *Coolant flowrate and temperature.* These are critical quantities, as cooling has a major effect on the conversion achieved and catalyst life. Too much cooling, and side reactions are favored, resulting in poor conversion; too little and there is a danger of hotspot formation resulting in reduced catalyst life and possible runaway.

1.4 An Example: Multitubular Reactor Design

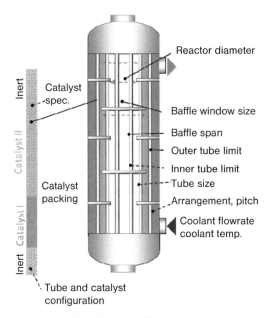

Fig. 1.10 The high-fidelity model allows optimization of many aspects of reactor geometry.

- *Catalyst packing and catalyst specification.* The type of catalyst used and the way it is packed with inert are extremely important operational decisions that can only be answered by running a high-accuracy model. Even extensive pilot plant testing cannot provide adequate information. The model tubes could be "packed" with different level catalyst.
- *Tube size.* This affects the bed diameter and the number of tubes it is possible to have within a given shell diameter, and hence the throughput.
- *Tube arrangement and pitch.* These naturally affect the flow characteristics of the shell-side fluid, and thus the pressure drops, heat transfer coefficients and cooling effectiveness.

The model provided the ability to investigate all of these alternatives in a way that had never been possible previously, and to fine-tune the design to provide the desired reaction characteristics.

Results from a "poor" case and an "optimal case" are presented here based on the reactor length and shell configuration determined in the first two studies. Values in results are normalized to preserve confidentiality.

1.4.4
Detailed Design Results

Some of the very small changes turned out to be the most important. For example, changing the reactor diameter did not have a very large effect. However small changes in baffle placement and design made a big difference.

Fig. 1.11 Tube center temperature distribution through the shell for poor (a) and optimal (b) design.

Figure 1.11 illustrates the effects of the above adjustments on the key design objective – the temperature distribution along the center of each catalyst-filled tube, over the entire reactor shell. The poor case has marked temperature gradients (color contours) across the tube bundle for tubes in the same axial position.

As a result of the considered adjustments to the internal geometry, the optimal case shows virtually uniform radial temperature profiles. This means that the reactants in all tubes are subject to the same or very similar external conditions at any cross section of the reactor, with no discrepancy of performance arising from the radial position of the tube within the tube bundle. It also means that the reactions occurring within the tubes – and hence conversion – are very similar for all the tubes across the bundle.

Figure 1.12 shows similar information: the tube center temperature profiles for selected tubes at different radial positions. The variation across the tubes is considerable in the original, up to 40 °C at certain axial positions.

Figure 1.13 shows the corresponding conversions for the same tubes. In the optimal design, all tubes have a very similar conversion. Not only is overall conversion better, but the different tubes perform in a similar way, meaning that catalyst deactivation is more uniform. This results in better overall performance and longer catalyst life.

1.4.5
Discussion

The results shown here are the effects of very subtle changes, which can only be represented accurately using very high fidelity predictive modeling at all scales of the process, from catalytic reaction to shell-side fluid modeling.

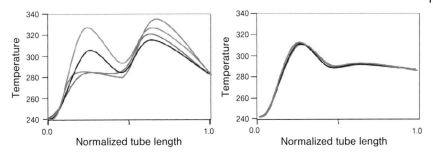

Fig. 1.12 Tube center axial temperature profiles for selected tubes at different radial positions for poor and optimal design.

The ability to perform this type of analysis gave the company involved a chance to perform detailed design themselves, saving millions of dollars on the engineering contract.

The design exercise here resulted in a reactor capable of being brought to market as a commercial process. It is just one of many similar types of reactor modeled by various chemical companies, some far more exothermic where the effects of achieving uniform production across the tube bundle radius are far more valuable.

1.5
Conclusions

It can be seen that it is possible to tackle enormously complicated and challenging modeling tasks with today's process modeling technology.

Not only are the tools available in which to describe and solve the models, but there is a set of well-proven methodologies that provide a logical step-by-step approach to achieving high-fidelity predictive models capable of an accurate representation over a wide range of operating conditions and scales. The procedures outlined above are all readily achievable and are being applied in many different areas of the process industries.

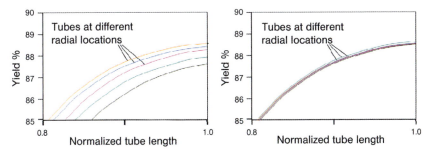

Fig. 1.13 Axial conversion profiles for selected tubes at different radial positions for poor and optimal design.

A good model can be used to explore a much wider design space than is possible via experimentation or construction of pilot plants, and in a much shorter time. This leads to "better designs faster," with shorter times-to-market (whether the "design" is an entire new process or simply an operational improvement on an existing plant). And as has been described above, the model-based engineering approach can generate information that is simply not available via other techniques, which can be used to significantly reduce technology risk.

Pilot plant testing can be reduced to the point where its sole purposes are to provide data for determining key parameters and to validate that the model is capable of predicting different states of operation accurately.

What this means in broad terms for the process organization is that it is now possible to:

- *Design process, equipment and operations* to an unprecedented level of accuracy. This includes the ability to apply formal optimization techniques to determine optimal solutions directly rather than rely on trial-and-error simulation.
- Simply by applying the procedures above to derive and apply accurate models it is possible to bring about percentage improvements in "already optimized" processes, translating to significant profits for products in competitive markets.
- *Represent very complex processes* – for example, crystallization and polymerization, or the complex physics, chemistry and electrochemistry of fuel cells – in a way that was not possible in the past. This facilitates and accelerates the design of new processes, enables reliable scale-up, and provides a quantitative basis for managing the risk inherent in any innovation.
- *Integrate R&D experimentation and engineering*, using models as a medium to capture, generate and transfer knowledge. By promoting parallel rather than sequential working, the effectiveness and speed of both activities is increased.

The ability to create models with a wide predictive capability that can be used in a number of contexts within the organization means that it is possible to recover any investment in developing models many times over. For example, models are increasingly embedded in customized interfaces and supplied to end users such as operations or purchasing personnel, who benefit from the use of model's power in providing advice for complex decisions without having to know anything about the underlying technology.

To assist this process, there is a growing body of high-fidelity models available commercially, and in equation form from university research and literature, as the fundamentals of complex processes become more and more reliably characterized.

It is evident that high-fidelity predictive modeling can provide significant value, and that to realize this value requires investment in modeling (as well as a certain amount of experimentation). The key remaining challenge is to effect a change in the way that the application of modeling is perceived by process industry management and those who allocate resources, as well as in some cases the technical personnel engaged in modeling, who need to broaden their perceptions of what can be achieved.

References

1 KONDILI, E., PANTELIDES, C. C., SARGENT, R. W. H., A general algorithm for short-term scheduling of batch operations – I. MILP formulation, *Computers and Chemical Engineering* 17(2) (**1993**), pp. 211–227.

2 FOSS, B. A., LOHMANN, B., MARQUARDT, W., A field study of the industrial modeling process, *Journal of Process Control* 8 (**1998**), pp. 325–338.

3 PANTELIDES, C. C., SARGENT, R. W. H., VASSILIADIS, V. S., Optimal control of multistage systems described by high-index differential-algebraic equations, in: *Computational Optimal Control*, Bulirsch, R., Kraft, D. (eds.), *Intl. Ser. Numer. Math.*, vol. 115, Birkhäuser Publishers, Basel, **1994**, pp. 177–191.

4 KESKES, E., ADJIMAN, C. S., GALINDO, A., JACKSON, G., Integrating advanced thermodynamics and process and solvent design for gas, in: *16th European Symposium on Computer Aided Process Engineering and 9th International Symposium on Process Systems Engineering*, Marquardt, W., Pantelides, C. (eds.), **2006**.

5 SCHNEIDER, R., SANDER, F., GORAK, A., Dynamic simulation of industrial reactive absorption processes, *Chemical Engineering and Processing* 42 (**2003**), pp. 955–964.

6 LIBERIS, L., URBAN, Z., Hybrid gPROMS-CFD modelling of an industrial scale crystalliser with rigorous crystal nucleation and growth kinetics and a full population balance, in: *Proceedings, Chemputers 1999, Dusseldorf, Germany, October 21–October 22, Session 12 – Process Simulation: On the Cutting Edge*.

7 ASPREY, S. P., MACCHIETTO, S., Statistical tools for optimal dynamic model building, *Computers and Chemical Engineering* 24 (**2000**), pp. 1261–1267.

8 RODRIGUEZ-FERNANDEZ, M., KUCHERENKO, S., PANTELIDES, C. C., SHAH, N., Optimal experimental design based on global sensitivity analysis, in: *17th European Symposium on Computer Aided Process Engineering – ESCAPE17*, Plesu, V., Agachi, P. S. (eds.).

9 BERMINGHAM, S. K., NEUMANN, A. M., KRAMER, H. J. M., VERHEIJEN, P. J. T., VAN ROSMALEN, G. M., GRIEVINK, J., A design procedure and predictive models for solution crystallisation processes, *AIChE Symposium Series* 323 (**2000**), pp. 250–264, presented at the 5th International Conference on Foundations of Computer-Aided Process Design, Colorado, USA, 18–23 July 1999.

10 SHIN, S. B., HAN, S. P., LEE, W. J., IM, Y. H., CHAE, J. H., LEE, D. I., LEE, W. H., URBAN, Z., Optimize terephthaldehyde reactor operations, *Hydrocarbon Processing (International edition)* 86(4) (**2007**), pp. 83–90.

11 BAUMLER, C., URBAN, Z., MATZOPOULOS, M., *Hydrocarbon Processing (International edition)* 86(6) (**2007**), pp. 71–78.

12 ROLANDI, P. A., ROMAGNOLI, J. A., A framework for on-line full optimizing control of chemical processes, in: *Proceedings of the ESCAPE15*, Elsevier, **2005**, pp. 1315–1320.

2
Dynamic Multiscale Modeling – An Application to Granulation Processes

Gordon D. Ingram and Ian T. Cameron

Keywords

granulation, multiscale modeling, granulation circuits, particle size distribution (PSD), multiscale model, computer-aided process modeling (CAPM) tools, population balance equation (PBE)

2.1
Introduction

Granulators take a fine powder feed, typically a hundred micrometers in size, and combine it with a binder to form millimeter-sized agglomerates. The process vessels have dimensions of meters and a commercial-scale granulation plant may have a circuit length one hundred meters long. Time scales range from microseconds for the duration of particle collisions to hours for the plant to recover from a process upset. Large-scale granulation circuits, like other particulate processes, are challenging to design and operate. High recycle ratios, flow surging, and off-specification production are not uncommon. The development of accurate, fit-for-purpose, dynamic models is a step toward addressing these practical problems. Powder-scale processes of surface wetting, agglomeration, attrition, and so forth directly influence the behavior of the circuit, but we simply are not able – and certainly do not wish – to model the whole circuit at the level of individual particles. We need efficient and accurate *dynamic* and *multiscale* models to bridge the gap in time and length scales. Granulation processes offer up an abundance of practical, theoretical, and numerical challenges that are worthy of investigation.

This chapter is organized as follows. Section 2.2 introduces granulation: its significance, rate processes, and typical equipment, and it presents the case why dynamic multiscale modeling is beneficial for this process. The multiscale nature of process systems, general characteristics of multiscale models, and the emerging practice of multiscale modeling are discussed in Section 2.3. In Section 2.4, we outline the relevant scales of observation for granulation processes and provide examples of the modeling techniques used at each scale. Section 2.5 summarizes key multiscale granulation models appearing in the literature and then reviews a handful of them in more detail. Our conclusions on the progress toward the multiscale modeling of granulation systems and comments on some open issues are presented in Section 2.6.

2.2
Granulation

In this section, we consider the importance of granulation in the production of a wide range of commodity products and consumer goods. Granules, and more generally particulate products, are ubiquitous in their geographical spread and in their use by society.

2.2.1
The Operation and Its Significance

Wherever we are or whatever we might be doing, we are confronted by a world of particles. Those particles can be naturally occurring materials such as mineral products or they can be manufactured particles, made through the agency of granulation and similar processes. These particulate materials may be used directly by society as consumer products, or they may be feedstocks for agriculture and industry. In the former case, a modern supermarket soon reveals products of granulation processes: washing powders, pharmaceuticals, slow release garden fertilizers, personal care products, and toothpaste additives are just a few of the myriad products that are the result of granulation processes.

Not only are consumer products the result of granulation processes, but so too are large-scale commodity products. Here we encounter a wide range of minerals products, industrial fertilizers, such as mono- and di-ammonium phosphates (MAP, DAP), granulated insulation and construction materials, slags, cements, and globally traded commodities that are granulated for transportation reasons.

Globally, the value of such commodities and consumer products is in the trillions of dollars. Hence, the granulation processes behind such products are vital industrial operations. Deep physicochemical insights are needed to devise better, fit-for-purpose products, process designs, control systems, and optimization methodologies. The process industry is continually seeking highly efficient solutions for the manufacture of granulated products.

Only in the last 20 years has there been a distinct move from empiricism in design to a more fundamental approach based on a deeper understanding of the basic physics and chemistry, as well as the representation of the underlying phenomena in the form of models. These models can be adapted and adopted into simulation systems for synthesizing and analyzing the design, behavior, and performance of these complex systems.

As well as consideration of the basic physics and chemistry, we need to take account of the many forms of process equipment in which granulation is conducted. These range from the low-shear drum granulators for powders with a liquid binder, through to high-shear mixer-granulators, as well as fluidized beds, extrusion systems, and dry granulation processes [1]. In each of these process types, there exist substantial challenges in understanding the basic phenomena underlying the granulation processes. Herein lie many academic and industrial challenges.

However, the last decade has seen a significant deepening of the knowledge around the physics and chemistry of granulation processes, which range from nucleation and growth, through to layering and agglomeration, and potentially to breakage phenomena [2]. This fundamental knowledge, together with consideration of the processing routes, encompasses a wide range of fundamental length and time scales over which phenomena occur. Extending our understanding of these systems and exploiting that understanding is a classic multiscale challenge.

At the same time, the growth in the practice of modeling such systems using a variety of computational approaches has led to a very wide range of models within many of the individual scales of consideration. Thus, model-centric approaches and the use of multiscale models are a growth area where it is becoming increasingly feasible to address product design, process design, control, and optimization for granulation systems, which is now a routine practice in other process areas such as petrochemicals and biochemical systems.

2.2.2
Equipment, Phenomena, and Mechanisms

Granulation is a vital industrial process with significant economic benefits as seen in Section 2.2.1. The practice of granulation takes place in a wide variety of equipment, each employing different basic mechanisms to generate particulate products of a wide range of physical and chemical characteristics. As well as the granulation device itself, there are often numerous other equipment items necessary to fulfill product quality and yield requirements. These can include:

- feed preparation and prereaction systems;
- filters and cyclones for particle separation purposes;
- tanks and feed systems for binder storage and application;
- heat transfer, fans, and feed systems for air supply to fluid beds;
- particulate dryers for product treatment;
- transfer systems for handling powders and granules;
- a variety of screens for intermediate and product separation purposes;
- crushers for reducing oversize materials.

Figure 2.1 shows a pilot-scale, continuous drum granulation circuit, typical of circuits for the granulation of powder feeds using a liquid binder. This figure shows both the major equipment in the circuit and an integrated monitoring and control system for process analysis purposes. The granulator is the heart of the process, but the associated equipment is vital for viable production of quality product.

A typical range of granulation devices, from high-shear granulators to fluidized bed and drum devices, is shown in Fig. 2.2. Other options such as pan granulation also exist. They demonstrate a variety of ways for contacting powder with a liquid to promote granule formation. These devices also possess quite distinct process

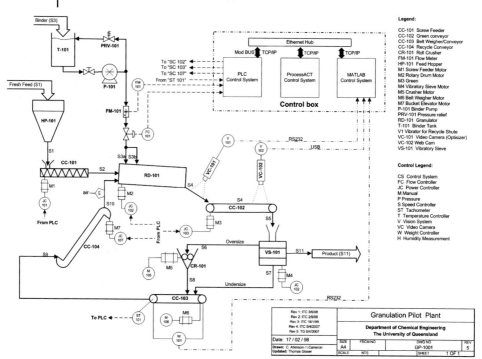

Fig. 2.1 Drum granulation circuit (pilot scale).

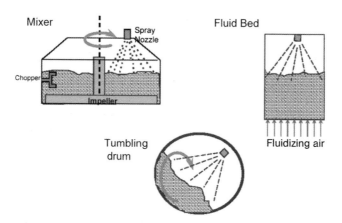

Fig. 2.2 Typical granulation technologies.

behaviors that are utilized for the purposes of small- or large-scale granule production.

Of key importance within the granulator are several mechanisms that occur due to initial contact of the fluid with the powder and then the subsequent behavior of the wetted powder. These include:

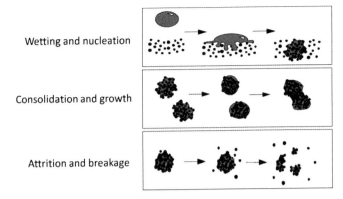

Fig. 2.3 Basic mechanisms in granulation.

- wetting and nucleation, where liquid contacts the powder in a variety of ways and with varying outcomes;
- consolidation and growth of particles, via a range of compaction phenomena and then subsequent growth;
- attrition and granule breakage, through impacts with other particles, the vessel walls, and any internals within the granulator.

Such phenomena are seen in Fig. 2.3. These mechanisms are fundamental to the modeling and simulation of granulation systems. Different phenomena happen at different times and locations within the granulator and around the circuit. The characteristic times of the various phenomena also vary within the granulator and around the circuit.

In some systems, the initial contact of liquid and powder does not immediately lead to growth, but an induction time is observed before growth takes place. This phenomenon is traceable to powder properties, specific process parameters, and the interaction of liquid with the powder surface.

It has become clear that these processes present significant research and development challenges to academe and to industry. In most cases, a multiscale approach is needed to deal with this multifaceted, dynamic behavior.

2.2.3
The Need for and Challenges of Modeling Granulation

The behavior of granulation circuits is complex and they can often be difficult to operate in a stable manner in the face of various disturbances to the system. Uncontrolled granule growth, surging, and cyclic behavior have been observed in industrial operations.

Figure 2.4 illustrates the principal circuit interactions that are of importance in granulation. Central to the interaction diagram is the population balance, which tracks dynamic changes in the particle size distribution (PSD) within the granulator and indeed in all solids streams in the granulation plant. In some cases,

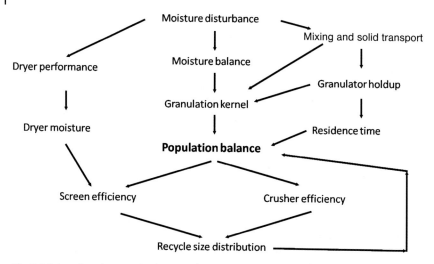

Fig. 2.4 Interactions in a continuous granulation circuit.

for example MAP-DAP production, other factors such as chemical reaction, liquid phase properties, and crystallization add to the complexity of the system [3].

Figure 2.4 also shows that there are numerous particulate properties, including moisture and size distributions, solids mixing, segregation, and transport behavior, which, when combined with process equipment performance factors, determine the final product attributes.

Much of the design and operational difficulties arise from the performance of the granulator itself, which is in turn a function of the process parameters and the underlying mechanisms operating within the granulator. As such, the dynamics of these systems become an important consideration in actual production, leading to a growing interest in considering micro-, meso-, and macroscale dynamics and the impacts they have on system design, control, and optimization.

Modeling in various forms provides one approach to help understand and address the current challenges in granulation systems. The challenges related to the effective modeling of these systems may be considered as follows:

- Microscale models are needed that effectively:
  o capture particle–binder and particle–particle interactions;
  o represent particle breakage mechanisms;
  o account for powder–liquid properties and also the equipment design and operational parameters.
- Mesoscale challenges, where the dominant issues include:
  o granule property predictions in time and space;
  o use of single and multidimensional population balances to capture process dynamics for multiple internal properties, such as size, liquid content, and porosity;
  o particle transport models that capture key operational regimes representing flow patterns and segregation phenomena;

    o integration of the microscale and mesoscale phenomena through forms of multiscale integration frameworks.
- Macroscale issues that include:
    o overall circuit dynamics that incorporate major equipment dynamics;
    o ability to consider control system structures and their effectiveness in disturbance rejection of state-driving applications;
    o optimization of steady-state behavior as well as optimal state-driving applications;
    o addressing adequately the area of plant and process failures for efficient diagnostic system design and implementation.

Along with these modeling challenges are the associated roles of simulation systems that must afford efficient, robust solution of multiscale models in an acceptable time. They should also facilitate the configuration of circuit redesigns and the implementation and testing of control system structures.

These challenges are being met through our current and growing understanding of the important granulation phenomena. This process understanding is being captured in a variety of useful models that are beginning to be linked into a multiscale approach to this important industrial application.

2.3
Multiscale Modeling of Process Systems

This section reinforces the notion that process engineering systems, including granulation systems, have a multiscale character. We discuss the typical features of multiscale process models and an approach to their construction.

2.3.1
Characteristics of Multiscale Models

Process systems are multiscale. They begin at the scales of chemical bond formation where characteristic times and distances are around 10^{-16} s and 10^{-13} m, and extend up to the scale of the environment in which the process and allied industries operate, say 10^9 s and 10^6 m [4–7]. Process vessels and plants fall somewhere between these extremes, perhaps 10^0–10^5 s and 10^{-2}–10^3 m, and it is at these scales that we exert a direct influence on process design and operations. The saleable products that are produced invariably have specifications on their chemical composition (10^{-9} m) and also likely on their physical structure (10^{-6}–10^{-2} m). This is certainly true for particulate products, where particle size and pore size distributions, morphology, mechanical strength, flowability, dissolution rate, and similar properties may be important for their end-use function. Models that bridge these scale gaps are needed to assist in process and product design, optimization, process control, fault diagnosis, and so on.

A multiscale model is a composite mathematical model that combines two or more constituent models that describe phenomena at different scales [8]. Single-scale models may be combined in many different ways. Coupled with the diversity of model types available at different scales, this makes multiscale modeling a fascinating and challenging endeavor. Multiscale models have been developed in a wide range of disciplines, for example, materials science, biology, medicine, ecology, and climatology, as well as in engineering [9, 10].

Some characteristics of multiscale models arise from the nature of the single-scale models, essentially the physics, chemistry, and engineering principles involved, but also whether they are:

- discrete, continuous, or hybrid forms;
- static or dynamic;
- deterministic or stochastic;
- computationally easy or demanding.

Other characteristics result from the scale-linking process that is the essence of multiscale modeling. Even given the wide range of applications developed, it is possible to make some general comments on the characteristics of multiscale models:

- most current models link only two or three scales;
- they are often required to span a wide range of characteristic length and time scales, and may have inputs and outputs at different levels;
- they may be multiphysics and multidisciplinary, meaning that different principles lie behind the equations at the various scales, and their development may require the cooperation of people with different backgrounds;
- specialized techniques have been developed for specific problems, for example, linking discrete and continuum models in materials science.

There are many open issues associated with developing and using multiscale models, which we mention only briefly:

- model formulation, which is discussed in Section 2.3.2;
- software integration, in the case where different modeling software or hardware platforms are involved;
- model solution, with an emphasis on efficiency;
- verification and validation of multiscale models, which is challenging given the range of time and length scales typically involved.

The issues, characteristics, and challenges of multiscale modeling are discussed further in [11, 12].

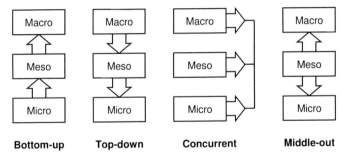

Fig. 2.5 Options for multiscale model construction in terms of the order of scale consideration for a three-scale problem.

2.3.2
Approaches to Multiscale Modeling

Multiscale modeling of process systems is a specific case of process modeling, for which an established general methodology is available [13]. There are, however, some additional matters to consider when developing multiscale models [8, 12]. Four tasks have been identified that supplement the general modeling procedure:

- identifying and selecting the relevant scales of interest for a particular modeling goal;
- deciding on the order of model construction in terms of scale;
- developing, or selecting and refining, models at each scale of interest;
- linking the individual models into a coherent multiscale model.

Most often, insight and experience are used to identify the scales of interest in a process, perhaps aided by the growing collection of scale-based hierarchical diagrams and length-time scale maps appearing in the literature. See for example [4, 14–16]. It is a matter of identifying coherent patterns, in terms of objects and behavior, from background noise as one's scale of observation increases from the atomic level upward. Multiscale analysis of data, for example, by wavelets, may assist in scale identification [17, 18]. As part of the iterative model building process, scales may be added, omitted, or modified to remedy some inadequacy noted in the evolving model.

Once a list of scales, perhaps tentative, has been established, consideration should be given to the order of detailed model construction. Four options have been used in the literature [8, 12] as illustrated in Fig. 2.5:

- bottom–up modeling, or model composition, refers to starting at the smallest scale of interest and applying first principles techniques to aggregate objects and behavior to larger and larger scales;
- top–down modeling, which is also known as iterative deepening or gradual model enrichment, begins at the largest scale and adds smaller scale models until model closure and accuracy requirements are met;

- concurrent modeling involves developing the models at all relevant scales simultaneously, and then linking them together;
- middle-out modeling begins at whichever scale is best characterized and then moves "outward" to larger and smaller scales as needed.

The top–down approach tends to be used in engineering applications. Further discussion may be found in [12].

If an existing model at some scale is suitable, it can be used, but if one is not, then the general model building procedure can be applied. The available modeling techniques span the range of scales of interest: computational chemistry, molecular simulation, a variety of mesoscale techniques, computational fluid dynamics (CFD), discrete element modeling (DEM), finite element modeling, unit operation modeling, flowsheeting, environmental simulation, supply chain and enterprise modeling. Selecting among competing modeling approaches at a given scale, for example, different computational chemistry or fluid dynamics methods, is aided by comparative information, such as [19, 20]. When considering the use of existing models, we should take account of the advice in [21]:

> "It should not be taken for granted that techniques (e.g. for CFD or molecular modelling) that have evolved over long periods of time for 'stand-alone' use automatically possess the properties that are necessary for their integration within wider computational schemes. Indeed, our experience has been that the contrary is often true and that significant theoretical and algorithmic effort may have to be invested in ensuring that these additional requirements are met."

Linking scales together lies at the heart of multiscale modeling. There have been several attempts to classify the way that the various scales of interest are linked to form a multiscale model, including [8, 21–25]. Such classifications are helpful because they can be used to analyze existing multiscale models and also because they set out the broad alternatives to consider in the construction of new multiscale models. The model classification scheme of [8, 12] uses the spatial relationship of the control or balance volumes of fine (micro) scale and coarse (macro) scale model pairs to distinguish between five broad multiscale integration frameworks. These include the "decoupling" serial and simultaneous frameworks, and the "interactive" embedded, multidomain and parallel frameworks. Figure 2.6 depicts these frameworks and provides an informal description of each one.

Inspired by general systems theory, [25] developed an ontology for multiscale models. Starting from the basic concepts of general systems, they define the notions of scale and interscale relationships, and differentiate between aggregation laws, disaggregation laws, and mereological connection laws. Consequently they classify multiscale models as scale-collecting, scale-connecting, or scale-integrating. This may be the first fundamental deductive approach to the conceptualization of multiscale models. It was demonstrated how several existing classification schemes for multiscale models, including the frameworks shown in Fig. 2.6, could be described by the new ontology.

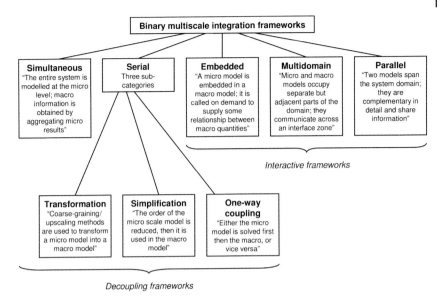

Fig. 2.6 One classification of multiscale models, adapted from [12].

It should be emphasized that all these classification schemes provide only broad guidance on the ways of linking individual scales into a multiscale model for any particular application. Doubtless, however, such work will help in developing multiscale extensions to existing computer-aided process modeling (CAPM) tools [25–27].

2.4
Scales of Interest in Granulation

This section is concerned with the range of modeling techniques used for granulation processes at particular scales of observation. Multiscale models that link these single-scale approaches are considered in Section 2.5.

2.4.1
Overview

Many authors have commented on the relevant time and length scales for particulate systems, along with their associated objects, mechanisms, and interconnections, for example [28–32]. Figure 2.7 shows a selection of these: different scales in gas–solid processes with an emphasis on fluidized bed systems, which are sometimes used in granulation [33]; particle production, especially in the context of high-shear granulation [34]; general granulation processes [35]; and our own view of the scales relevant to drum granulation [11, 36]. Even within the relatively narrow field of granulation modeling, there are different interpretations of the scales of interest.

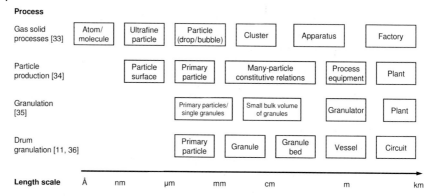

Fig. 2.7 Scales of interest in particle processing and granulation in particular.

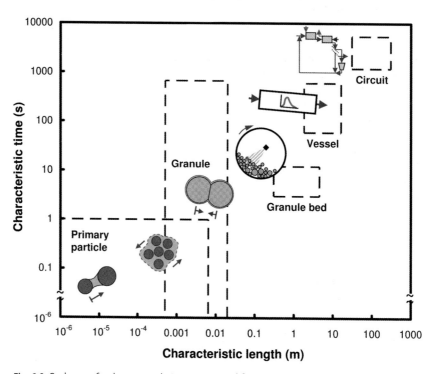

Fig. 2.8 Scale map for drum granulation [11], revised from [36].

A scale map for drum granulation is given in Fig. 2.8. Although scale maps show only order of magnitude ranges of characteristic times and lengths, some justification for the boundaries in this scale map is provided in [11, 36]. Sometimes a clean separation of time and length scales does not exist. This is illustrated by the very wide range of characteristic times exhibited by granules: microseconds for the duration of collisions, but spanning up to several minutes for granule densification. This potential overlap of scales also applies more generally [37].

Sections 2.4.2–2.4.6 consider each scale in turn, describing the relevant objects and phenomena involved, and summarizing the modeling techniques used, including typical input and output variables.

2.4.2
Primary Particle Scale

This scale is concerned with the primary particles, or powder feed, and the detailed interactions with the liquid binder and gas phases. It also includes whole granules that are resolved down to the scale of the primary particles. The processes of interest at this level include surface wetting, liquid bridge dynamics, and granule microstructure evolution. The modeling inputs at this scale may include the physical properties of the binder and powder; numbers and sizes of all the particles or binder drops considered; shape information for each particle; and information on the environment, for example, impact velocities, shear rates, and the local particle–binder–air concentrations. The potential outputs of interest include the granule nucleation rate, rates of layered growth and attrition, and effective granule properties. These granule-level properties can include:

- microstructural – solid–liquid–gas volume fractions, granule shape, height of asperities, binder-layer thickness, fraction of granule surface that is covered with binder;
- mechanical – yield stress, Young's modulus, granule–granule and granule–wall friction coefficients;
- transport – thermal conductivity, permeability, dissolution rate.

A variety of modeling approaches are used at this level, for instance:

- Granule structure formation was simulated using a ballistic deposition algorithm and volume of fill (VoF) technique for individual powder particles and binder droplets, which also included equilibrium liquid spreading and binder solidification. Microstructural and transport properties were deduced from the simulations [38–40].
- The nucleation of granules was studied on the assumption that the powder forms a porous continuum which is in contact with a binder droplet and that binder flow may result from either capillary or effective diffusion mechanisms. Simple expressions for nuclei growth kinetics were derived for both planar and spherical geometries [41].
- The strength of static and dynamic liquid bridges between particles has been extensively modeled using the methods of fluid mechanics [42, 43]. One application was a study of binder transfer between particles, for later use in DEM [44].
- Granule agglomeration has been studied via DEM, in which the motion of every particle in the colliding granules was resolved, for example [45].

2.4.3
Granule Scale

Granules are semipermanent agglomerations of primary particles, binder in liquid or solid form, and entrapped air. At this scale, we are concerned with the interactions of a small number of granules with each other and with the internals of the processing vessel, and also with the primary powder and binder when they are treated as continua. Several important granule-level phenomena may be identified: growth and agglomeration, attrition and breakage, consolidation or densification, and drying and rewetting, as mentioned in Section 2.2.2. Modeling inputs include granule size, moisture content, and porosity; granule mechanical properties; binder-layer thickness and coverage; and the environment experienced by a granule, such as the collision velocity with other granules. The outputs of modeling calculations may include quantities like the probability of coalescence or breakage, the fragment size distribution in breakage, the consolidation rate, and so on.

Modeling approaches for agglomeration include:

- Energy balance methods, such as [46], which compared the initial kinetic energy of colliding granules with the energy dissipation by viscous dissipation in the binder layer and elastic–plastic deformation of the granule matrix.
- Finite element analysis (FEA) to model granule collision parameters, such as the coefficient of restitution and contact area [47].

Granulation rate processes and their modeling are reviewed in [48, 49].

2.4.4
Granule Bed Scale

The granule bed level is not recognized by all researchers in this area as a separate scale (Fig. 2.7), although the modeling techniques considered are certainly used extensively. This scale attempts to describe the interactions of a large assembly of granules with the vessel internals, without necessarily representing the whole process vessel. Phenomena of interest include impact velocity distributions, solids transport and flow patterns, granule mixing and segregation, and the distribution of binder and powder. Model inputs include granule size, moisture and porosity distributions; granule mechanical properties; models of granule–granule interactions, regarding coalescence, for example; binder spray rates and spray pattern; and granulator design and operation details, for example, the design of the impeller in a high-shear mixer or superficial velocity in a fluidized bed. The potential model outputs at this scale are the rates of granulation processes (nucleation, growth, agglomeration, . . .) as influenced by local vessel design and operating conditions; and spatial distributions of granules, powder, and binder.

The preeminent modeling technique at this scale is DEM, in which the equations of linear and angular momentum are solved for every granule moving in an assembly. Comprehensive reviews of the theory and applications of discrete element methods are available in [50, 51]. DEM is classified into "hard-sphere" and

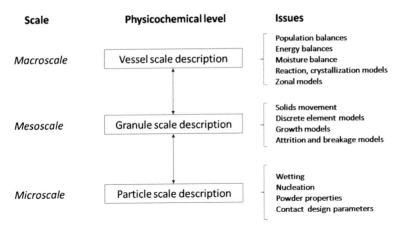

Fig. 2.9 Vessel–scale interactions with smaller scale considerations.

"soft-sphere" methods. The latter allows small virtual overlaps of the particles and tends to be used for dense-phase processes where there may be lasting particle contacts. For some applications, notably fluidized bed granulation, granule motion is strongly affected by the surrounding gas velocity field, requiring a coupled solution of the DEM equations with CFD. It should be noted that in granulation research, DEM has been applied both at the primary particle level and at the granule level.

Other modeling techniques potentially appropriate at this scale include Eulerian–Eulerian (two-fluid) CFD, discrete bubble models, Monte Carlo methods, convection–diffusion analogies, and the kinetic theory of granular flow (KTGF) [20, 52].

2.4.5
Vessel Scale

At the vessel scale, we are concerned with tracking the gross behavior of the PSD, along with other important indicators such as granule porosity and liquid content distribution. These are clearly important aspects of dynamic performance and are linked to smaller scale phenomena, for example, the success of collisions between particles inside the vessel. Vessel scale behavior is currently addressed through population balance models (PBMs) of various forms. An extensive discussion on this topic is given in [53–55].

Not only do the physical phenomena need consideration, but in many cases, due to complex concurrent processes such as reaction and crystallization, we also need to account for energy and moisture balances [3]. There are strong interacting effects among the variables involved. In this instance, we are incorporating particle and granule scale submodels into the vessel dynamics. Figure 2.9 shows the interactions and scale issues that exist between the vessel-scale description and smaller scales. Interactions such as these are captured in some of the multiscale models reviewed in Section 2.5.

The general population balance equation (PBE) that operates at the vessel scale involves tracking the evolution of a density function of particle class and location inside the vessel. It can be expressed in natural language form as

$$\left\{ \begin{array}{c} \text{density function change} \\ \text{in class, location, and time} \end{array} \right\}$$

$$= \left\{ \begin{array}{c} \text{dispersion in} \\ \text{through boundary} \end{array} \right\} - \left\{ \begin{array}{c} \text{dispersion out} \\ \text{through boundary} \end{array} \right\}$$

$$+ \left\{ \begin{array}{c} \text{flow in} \\ \text{through boundary} \end{array} \right\} - \left\{ \begin{array}{c} \text{flow out} \\ \text{through boundary} \end{array} \right\}$$

$$+ \left\{ \begin{array}{c} \text{growth in} \\ \text{from lower classes} \end{array} \right\} - \left\{ \begin{array}{c} \text{growth out} \\ \text{from current class} \end{array} \right\}$$

$$+ \left\{ \begin{array}{c} \text{birth due to} \\ \text{coalescence} \end{array} \right\} - \left\{ \begin{array}{c} \text{death due to} \\ \text{coalescence} \end{array} \right\}$$

$$+ \left\{ \begin{array}{c} \text{breakup in} \\ \text{from upper classes} \end{array} \right\} - \left\{ \begin{array}{c} \text{breakup out} \\ \text{from current class} \end{array} \right\}$$

Particle class refers primarily to size, but also to moisture and porosity – and possibly other intrinsic particle properties in specialized applications. Detailed discussions on the specific application of the general PBE and the development of the individual terms on the right-hand side are provided by [53–55].

A range of discretization and solution methods are applied to PBMs. We need to consider reduced order models for some applications, such as process control. A number of well-established discretization methods exist, such as those developed by [56] and [57], and more recently the multitiered approaches for multidimensional PBEs [58]. These are regularly used and simple to implement.

Other techniques based around methods of weighted residuals, moments, and wavelet approaches have found use in PBE solution, but standard discretization techniques dominate current practice.

Vessel scale dynamics are crucial to understand the behavior of the overall production system. This reinforces the importance of coupling population balances with models that capture smaller scale phenomena.

2.4.6
Circuit Scale

The aim at the circuit level of description is related to assessing a range of process performance issues. From a systems perspective, these can include:

- The effects of structural changes in the circuit equipment, ranging from new types of oversize crushers to improved double-decked screens and granulator internals, as well as the implementation of different operational modes.
- The effects on performance of changes in powder feed properties, ranging from different PSDs to the use of special powder additives for marketing advantages. There can also be changes in liquid binder conditions such as the degree of preneutralization in MAP-DAP production.
- Assessing dynamic behavior of the plant in the face of disturbances, including such issues as partial or full blockages of screens and other equipment, changes to the scrubbing water used for granulator off-gases, or changes in reaction or crystallization processes.
- Assessing the performance of new control schemes, from simple PID through cascade control to forms of model predictive control (MPC).
- State driving strategies related to product changes and to the start-up and shutdown of the plant.
- The behavior of the plant when a range of plant and process failures might occur. This addresses diagnostic and operator guidance systems (OGSs).

All these applications require a full description of the plant to some appropriate level of fidelity. This full description can be done through general process modeling and simulation systems, such as gPROMS [59], Aspen Custom Modeler [60], or Daesim Dynamics [61]. Other specific particle-based simulation tools also exist, for example, SolidSim [62]. Whatever the system used, it would ideally have the following functionality:

- capability to represent all relevant models at various scales of consideration (micro-, meso-, macro-, and higher scales);
- ability to incorporate alternate model descriptions;
- easy assembly of flowsheet equipment units into the integrated flowsheet;
- ability to track all dynamics and display the results informatively. This generally means spatial and temporal displays;
- allow the incorporation of a range of control systems to represent realistic plant behaviors;
- ability to induce faults and failures for testing design robustness and flexibility issues;
- ability to connect the simulation system to other software systems via external interfaces to help facilitate control, training, and diagnostic applications.

Figure 2.10 shows a commercial fertilizer granulation flowsheet that may be used in dynamic simulation studies. These types of flowsheets are complex, with multiple parallel pathways and significant recycle flows. They include a wide range of physical and chemical phenomena that necessitate the tracking of a significant range of particle sizes. Typically, some 20 to 25 particle size ranges are used in such studies. There are a growing number of studies that look at such flowsheets, but few where plant dynamic data have been used to validate the simulated circuit behavior [3]. Obtaining such data is a nontrivial exercise. It requires special sampling

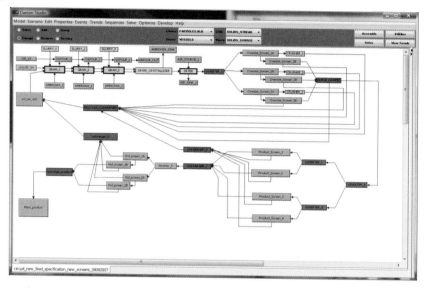

Fig. 2.10 Commercial granulation flowsheet simulation.

approaches, careful monitoring of the plant, and the ability to inject controlled disturbances into the granulator. This can be problematic for large-scale commercial operations where management policies try to avoid plant instabilities at all costs.

There is much to be done in developing reliable and easy-to-use modeling and simulation systems for performing analysis of commercial granulation circuits. Compared to the well-established tools in the petroleum and petrochemical sectors, total-plant tools for the dynamic simulation of commercial granulation processes are in their infancy. There is significant scope for further important developments in this area.

2.5
Applications of Dynamic Multiscale Modeling to Granulation

Section 2.4 explored the modeling options for granulation systems from the particle to the circuit scale, more or less in isolation. We now look into multiscale models that explicitly attempt to link together two or more models that represent different scales of observation.

2.5.1
Overview

Several examples of multiscale modeling of granulation systems are summarized in Table 2.1. A few noteworthy single-scale approaches that contain an element of multiscale thinking are also included. Having recognized that differences exist in the understanding of scale in particle systems, as demonstrated in Fig. 2.7, this

2.5 Applications of Dynamic Multiscale Modeling to Granulation

Table 2.1 Summary of selected applications of granulator modeling, mostly of multiscale type.

Granulator type	Primary particle	Granule	Granule bed	Vessel	Circuit	Comment	Reference
Batch high-shear		✓	✓	✓		Soft-sphere DEM, coalescence and PBE modeling	[63, 64]
Batch fluidized bed		✓	✓	✓		Hard-sphere DEM/CFD, KTGF, coalescence and PBE modeling	[65, 66]
Batch fluidized bed	✓	✓				Hard-sphere DEM/CFD with surface adhesion force and drying	[67]
Batch fluidized bed		✓	✓	✓		Hard-sphere DEM/CFD, KTGF-inspired kernel, PBE modeling	[68–70]
Batch fluidized bed		✓	✓	✓		Hard-sphere DEM and PBE modeling	[71]
Batch fluidized bed (Wurster type)	✓	✓	✓			VoF-derived granule property, two-fluid CFD, KTGF-inspired kernel, PBE modeling	[72]
Batch spout-fluidized bed		✓				Hard-sphere DEM/CFD with growth by layering only	[73]
Batch drum	✓					Agglomerates modeled at resolution of primary particles	[74]
Batch drum	✓	✓	✓	✓		Multidimensional PBE with mechanistic coalescence and nucleation modeling	[75–77]
Continuous drum		✓	✓	✓		Soft-sphere DEM, coalescence and PBE modeling	[11]
Continuous drum			✓	✓	✓	Fault diagnosis and loss prevention focus	[78, 79]

(continued on next page)

Table 2.1 (Continued.)

Granulator type	Scales considered					Comment	Reference
	Primary particle	Granule	Granule bed	Vessel	Circuit		
Batch and continuous drum				✓		Multiform PBE modeling for optimal control	[80]
Any low-shear type	✓	✓				Nuclei formation from individual (nonspherical) powder particles, VoF method	[38, 40]
Any	✓	✓				Continuum-based nucleation rate modeling	[41]
Any			✓	✓	✓	Hierarchical granulation modeling library	[31, 81]

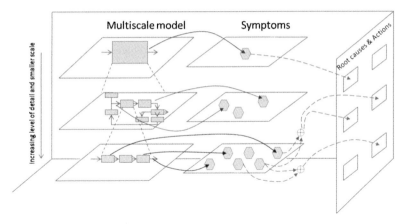

Fig. 2.11 Diagnostic multiscale structure.

table provides just one interpretation of the multiscale nature of the works cited. Almost all the models are dynamic on at least one of the scales considered.

Sections 2.5.2–2.5.5 examine several of the multiscale models in more detail. The purpose of the model and scales involved are described. The method used for linking the single-scale models is identified. A key aim of the following discussions is to highlight any modifications to the single-scale models needed to make them amenable to multiscale modeling. The main conclusions from each study are also summarized.

2.5.2
Fault Diagnosis for Continuous Drum Granulation

Fault diagnosis continues to be a vital part of industrial operations. It sits within the corporate risk management framework. In addressing fault diagnosis, models of the system play an important role. The approaches utilize both quantitative and qualitative models to provide fault detection, isolation, and corrective actions. Like many other issues already raised in this chapter, the models used can be classified in terms of the scale that they represent.

Early work on granulation diagnostics was done by Norsk Hydro [82] utilizing an expert systems framework. Other work combined Gensym Corporation's G2 real-time expert system with qualitative models [83]. More recently, Németh et al. [78, 79, 84] used qualitative multiscale models to address the issues of loss prevention in industrial granulation processes. Their study used a combination of two hazard identification methods producing qualitative descriptions of fault scenarios, likely causes and consequences. It combined both failure mode effects analysis (FMEA) and hazard and operability studies (HAZOPs) into a structured knowledge base that became the model for use by a multiagent diagnostic system.

Figure 2.11 shows the multiscale structure of the knowledge model used for diagnostic applications. Here the whole plant is viewed hierarchically from the overall input–output model, to the flowsheet level where the focus is on individual equipment, and then down again to the lowest level of mechanisms and operational factors within individual operations. Symptoms can be associated with each level in the hierarchical view, and this approach captures issues around time and length scales of the phenomena involved.

In generating the knowledge base, formal qualitative methods were used based on function-driven (HAZOP) and component-driven (FMEA) analyses. These were captured in tabular form, suitable for agent-based interrogation.

Using the qualitative information in terms of cause–deviation–consequence triplets from HAZOP or the failure mode–failure mode cause–effects triplet of FMEA provides rich knowledge about the hazard structure and causal links in a granulation system. This knowledge can be cast into a logic structure that is interrogated. Part of the knowledge representation capturing symptoms (deviations) is shown in Fig. 2.12, where key symptoms are shown as hexagons and causes – consequences are connected via AND and OR gates.

As well as the qualitative models, a full quantitative model was also incorporated for predictive purposes. Combining the multiscale knowledge bases generated from qualitative methods, quantitative predictive models, and a multiagent architecture or real-time expert system can lead to the development of an OGS. Such a system is helpful in providing timely advice to operators in the case of abnormal condition management. This study into granulation systems can lead to a general framework suitable across many application areas, including operator training purposes.

2.5.3
Three-Dimensional Multiscale Modeling of Batch Drum Granulation

It is well known that size is not all that matters in regard to granulation. Iveson [85] clearly noted that when it comes to a realistic representation of granule formation, other granule properties are vitally important, including liquid or binder content, porosity, and composition distributions. The purely one-dimensional PBE based on size alone is too simple a picture of reality for tracking granule growth dynamics. There is a need to improve our understanding of the time evolution of these granule properties within a multidimensional and multiscale framework.

Recent work by numerous research groups has concentrated on the modeling and validation of multidimensional PBMs. These have focused on size, porosity, and liquid binder content. Ramachandran *et al.* [75] carried out experimental studies in batch drum granulation to measure how the evolution of the three-dimensional particle size – binder content – porosity distribution was influenced by powder properties, binder properties, and processing conditions. They also sketched out the structure of a conceptual multiscale granulation model and its potential use in a feedback control system. Similar multidimensional experimental work for high-shear granulation has been reported by [86].

In further work, Poon *et al.* [76, 77] developed and validated a three-dimensional PBM that incorporated mechanistic submodels of nucleation and aggregation phenomena along with an empirical consolidation submodel. This approach tracked

2.5 Applications of Dynamic Multiscale Modeling to Granulation

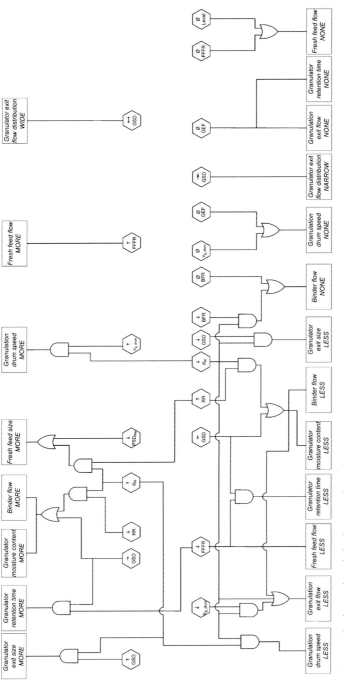

Fig. 2.12 Qualitative knowledge base as a logic diagram.

solid, liquid, and gas volume in granules over time, thus tackling the broader prediction of properties beyond size. The numerical solution approach was a two-tiered method [58, 87]. This study provides an excellent starting point for embedding such micro–meso scale phenomena into complete circuit simulations. Further experimental work will be required at both laboratory and plant level to further refine and adequately validate such a model.

2.5.4
DEM-PBE Modeling of Batch High-Shear Granulation

Multiscale modeling of a high-shear mixer-granulator was reported in [63, 64, 88]. Modeling elements at three different scales were combined: (i) DEM modeling to capture the bulk behavior of the granule assembly, (ii) coalescence modeling for colliding pairs of granules, and (iii) PBM to describe the vessel scale. The primary aim of this work was to develop a mechanistic model that links the physical properties of the colliding granules and processing conditions to the consequent rate of agglomeration in the vessel.

A three-dimensional, soft-sphere DEM model was developed for a $\pi/32$ radian vertical slice through a high-shear mixer. The base of the mixer was rotated to simulate a flat disk rotor and periodic boundary conditions were set up for the two vertical planar surfaces of the slice. An interesting finding in a preliminary study [88] was that the KTGF, which has been very successful in explaining velocity distributions and collision rates in fluidized bed granulators [68], was not so successful for a high-shear granulator containing particles with a low coefficient of restitution, such as wet granules.

Subsequent DEM work [63, 64] used the granule coalescence model of [46] to estimate the coefficient of restitution for each colliding granule pair based on the granule mechanical properties and impact velocity. Each granule was characterized not only by its size, but also by a porosity and pore saturation. No agglomeration was implemented in the DEM formulation. Several DEM-coalescence simulations were performed and analyzed to yield a correlation for the agglomeration kernel, which characterizes the rate of granule agglomeration. The kernel was correlated as a linear function of average granule porosity, pore saturation and rotor speed, and a more complex function of the colliding granule diameters.

A dynamic three-dimensional PBE model that tracked the size, porosity, and pore saturation distributions was implemented. The PBE model was solved using a constant-number Monte Carlo method with rare event simulation. The functional form of the correlation for the agglomeration kernel deduced in [64] was retained, but the correlation coefficients, and a factor to allow for tangential granule impacts, were regressed against a set of experimental data.

From a multiscale point of view, this model uses parallel integration (Fig. 2.6) to link the DEM and coalescence models. DEM provides the properties and impact speed of colliding granules to the coalescence model, which in turn estimates the coefficient of restitution used by the DEM model. These results were then reduced to a correlation for the agglomeration kernel as a function granule size and properties, and the kernel was then used in a PBE model, which is an example of serial integration by simplification (Fig. 2.6).

2.5.5
DEM-PBE Modeling of Continuous Drum Granulation

A model similar to that in Section 2.5.4, but for a rotary drum granulator, was developed in [11, 88]. Three different scales and model types were combined:

- coalescence modeling at the granule scale;
- DEM at the granule bed scale;
- dynamic PBE modeling at the vessel scale.

The model was constructed for the purpose of exploring the techniques of dynamic multiscale modeling.

To predict granule coalescence, the model of [46] was used, similar to the study reported in Section 2.5.4. Inputs to this coalescence model include granule sizes, mechanical properties, and binder layer thickness, as well as the impact velocity of the granules. The model of [46] assumes perfectly normal impacts along the line joining the granule centers, and no rotational motion. It predicts whether or not a given collision will result in coalescence or rebound of the granules involved.

At the granule bed level, the DEM model used was a soft-sphere type and it was assumed that air drag was negligible. The granules were modeled as spheres, but their motion was constrained to the plane of the drum cross-section, making the motion essentially two-dimensional. No nucleation, growth, consolidation, agglomeration, or breakage was implemented in the DEM formulation, only granule motion. The equipment and processing conditions chosen approximated those of the University of Queensland granulation pilot plant (Fig. 2.1). A series of simulations were run for different PSDs and levels of fill, and statistics were collected on the distributions of normal and tangential impact velocities, and the rate of granule–granule collisions. Figure 2.13 shows some typical results. All simulations suggested that there was a significant tangential component to the granule impact velocity. Recent work by Freireich *et al.* [90] addresses the importance of microscale particle collision issues in drum granulation.

In the light of the DEM results, the output from the coalescence model was further treated:

- The normal impact velocity distribution (Fig. 2.13(b)) was taken into account when applying the coalescence model, since only a fraction of granules in the drum will have velocities that will result in coalescence.
- Granule collisions having a high tangential component of the impact velocity were assumed not to result in coalescence.
- Newly formed granules that are judged to be fragile, that is, having only a small amount of plastic deformation of the granule matrix as predicted by the coalescence model, were assumed to break up immediately into their original granules.

Details of the implementation of these modifications are reported in [11].

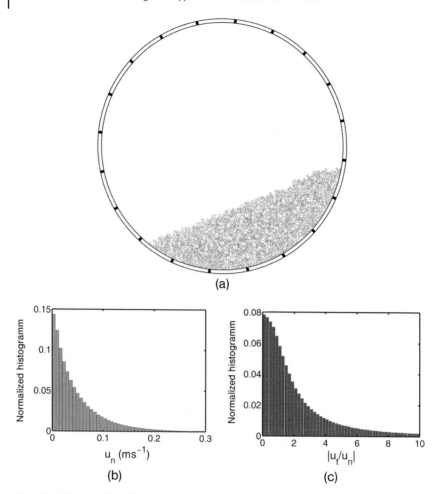

Fig. 2.13 DEM modeling of granules in a rotating drum: (a) snapshot of granule bed, (b) normal collision velocity distribution, and (c) distribution of ratio of tangential to normal collision velocities [11].

Dynamic vessel-scale behavior was modeled using the discretized population balance of [56]. Agglomeration, as characterized by the agglomeration kernel, was the only granulation rate process considered; however, accumulation and axial solids transport terms were included. The granulator drum was divided into three well-mixed axial regions and it was assumed that the granule flowrate through the drum was uniform. Sensitivity analyses confirmed the high sensitivity of the product granule size distribution to the agglomeration kernel.

The structure of the overall multiscale model and the flow of information between its parts are illustrated in Fig. 2.14. The multiscale nature of the model may be described as follows. The serial (simplification) integration framework (Fig. 2.6) was used to combine the PBE and DEM models. The DEM model results were pa-

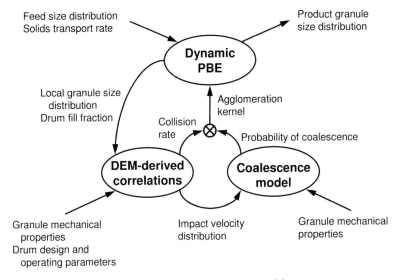

Fig. 2.14 Information flows in the multiscale granulation model.

rameterized (simplified) so that the mean collision rate, and distribution of normal and tangential impact velocities could be predicted from the granule size distribution. The modified coalescence model was linked using the embedded framework (Fig. 2.6) with the vessel-scale PBE model, in the sense that it was called as needed by the PBE model to provide coalescence results.

Notably, this model and its high-shear cousin in Section 2.5.4 both rely implicitly on the separation of time scales between granule-scale and vessel-scale phenomena. This allows the transient processes of granule collision and coalescence to be reduced to a pseudo-steady state (algebraic) agglomeration kernel.

The multiscale model results with an analysis of the model's properties, and alternative options for combining coalescence, DEM, and PBMs, are presented in [11]. One of the main outcomes of this granulation modeling exercise was the realization that current coalescence models should not be applied in a multiscale model without modifications such as the three previously mentioned.

2.6
Conclusions

Tremendous progress has been made in our understanding and representation of granulation phenomena across micro-, meso-, and macroscales. There is a growing body of knowledge around most aspects of the fundamental physics of particle–particle, particle–liquid, and particle–equipment interactions. Those micro- and mesoscale phenomena at the powder and granule level have been developed into various regime maps that show how zones of behavior are related to fundamental parameters of the system.

Significant progress has also been made in the area of multidimensional PBMs, which are able to track through space and time a range of particle-related properties: size, porosity, liquid content, composition, and others. Computational techniques for handling the solutions of such complex systems are becoming robust and efficient. The dynamics of solids movement have been addressed through discrete element and CFD techniques. However, these remain computationally intensive, and for most applications it is not yet possible to make a fast assessment of solids transport and segregation effects by these methods. There are opportunities to develop further zonal models, which are approximate model forms that represent subregions of processing vessels, for use in simulation tools.

Obtaining reliable data from production-scale operations still remains a major challenge. It is nevertheless vital for conducting convincing validations of many of the models used in current granulation simulations.

An area for possible future work is the development of computer-aided multiscale process modeling tools that can easily generate or capture single-scale models, and seamlessly link them into a multiscale model. Insights from the classification and ontology development for multiscale models are yet to be fully exploited [8, 25]. Particulate processes, and especially granulation, offer industrially important and challenging problems that would make ideal test cases to drive forward the development of multiscale CAPM tools.

References

1 SALMAN, A. D., HOUNSLOW, M. J., SEVILLE, J. P. K., *Granulation*, in: *Handbook of Powder Technology*, vol. 11, Elsevier, Amsterdam, **2007**.

2 ENNIS, B. J., in: *Handbook of Pharmaceutical Granulation Technology*, 2nd edn., Taylor & Francis Group, Boca Raton, FL, **2005**, p. 7.

3 BALLIU, N., CAMERON, I. T., *Powder Technology* 179 (**2007**), p. 12.

4 WADLEY, H. N. G., ZHOU, X., JOHNSON, R. A., NEUROCK, M., *Progress in Materials Science* 46 (**2001**), p. 329.

5 VILLERMAUX, J., in: *Proceedings of the 5th World Congress of Chemical Engineering*, American Institute of Chemical Engineers, New York, **1996**, p. 16.

6 GROSSMANN, I. E., WESTERBERG, A. W., *AIChE Journal* 46 (**2000**), p. 1700.

7 CHARPENTIER, J.-C., *Chemical Engineering Science* 57 (**2002**), p. 4667.

8 INGRAM, G. D., CAMERON, I. T., HANGOS, K. M., *Chemical Engineering Science* 59 (**2004**), p. 2171.

9 GLIMM, J., SHARP, D. H., *SIAM News* 30 (**1997**), p. 4.

10 LI, J., ZHANG, J., GE, W., LIU, X., *Chemical Engineering Science* 59 (**2004**), p. 1687.

11 INGRAM, G. D., Multiscale modelling and analysis of process systems, PhD Thesis, University of Queensland, Australia, **2005**.

12 CAMERON, I. T., INGRAM, G. D., HANGOS, K. M., in: *Computer Aided Process and Product Engineering*, vol. 1, Wiley-VCH, Weinheim, **2006**, p. 189.

13 HANGOS, K. M., CAMERON, I. T., *Process Modelling and Model Analysis*, Academic Press, London, **2001**.

14 ALKIRE, R., VERHOFF, M., *Chemical Engineering Science* 49 (**1994**), p. 4085.

15 WEI, J., *Chemical Engineering Science* 59 (**2004**), p. 1641.

16 Saez-Rodriguez, J., Kremling, A., Gilles, E. D., *Computers and Chemical Engineering* 29 (**2005**), p. 619.

17 Ren, J., Li, J., in: *Fluidization IX: Proceedings of the Ninth Engineering Foundation Conference on Fluidization*, Engineering Foundation, New York, **1998**, p. 629.

18 Ren, J., Mao, Q., Li, J., Lin, W., *Chemical Engineering Science* 56 (**2001**), p. 981.

19 Irikura, K. K., Frurip, D. J., *ACS Symposium Series* 677 (**1998**), p. 2.

20 van Sint Annaland, M., Deen, N. G., Kuipers, J. A. M., *Granulation*, in: *Handbook of Powder Technology*, vol. 11, Elsevier, Amsterdam, **2007**, p. 1071.

21 Pantelides, C. C., in: *European Symposium on Computer Aided Process Engineering (ESCAPE-11)*, Elsevier, Amsterdam, **2001**, p. 15.

22 Maroudas, D., *AIChE Journal* 46 (**2000**), p. 878.

23 Li, J., Kwauk, M., *Chemical Engineering Science* 58 (**2003**), p. 521.

24 Bianco, N., Immanuel, C. D., in: *Model Reduction and Coarse-Graining Approaches for Multiscale Phenomena*, Springer, Berlin, **2006**, p. 443.

25 Yang, A., Marquardt, W., *Computers and Chemical Engineering* 33 (**2009**), p. 822.

26 Morales-Rodríguez, R., Gani, R., in: *European Symposium on Computer Aided Process Engineering (ESCAPE-17)*, Elsevier, Amsterdam, **2007**, p. 207.

27 Yang, A., Zhao, Y., in: *10th International Symposium on Process Systems Engineering (PSE2009)*, Elsevier, Amsterdam, **2009**, p. 189.

28 Li, J., *Powder Technology* 111 (**2000**), p. 50.

29 Rielly, C. D., Marquis, A. J., *Chemical Engineering Science* 56 (**2001**), p. 2475.

30 Jefferson, B., Jarvis, P., Sharp, E., Wilson, S., Parsons, S. A., *Water Science and Technology* 50 (**2004**), p. 47.

31 Balliu, N., Cameron, I., Newell, R., in: *European Symposium on Computer Aided Process Engineering (ESCAPE-12)*, Elsevier, Amsterdam, **2002**, p. 427.

32 Bell, T. A., *Powder Technology* 150 (**2005**), p. 60.

33 Li, J., Kwauk, M., *Industrial and Engineering Chemistry Research* 40 (**2001**), p. 4227.

34 Mort, P. R., *Powder Technology* 150 (**2005**), p. 86.

35 Ennis, B. J., Litster, J. D., in: *Perry's Chemical Engineers' Handbook*, 8th edn., McGraw-Hill, New York, **2008**, p. 21–110.

36 Ingram, G. D., Cameron, I. T., *Developments in Chemical Engineering and Mineral Processing* 12 (**2004**), p. 293.

37 Ng, K. M., Li, J., Kwauk, M., *AIChE Journal* 51 (**2005**), p. 2620.

38 Stepánek, F., Ansari, M. A., *Chemical Engineering Science* 60 (**2005**), p. 4019.

39 Stepánek, F., *Granulation*, in: *Handbook of Powder Technology*, vol. 11, Elsevier, Amsterdam, **2007**, p. 1353.

40 Stepánek, F., Rajniak, P., Mancinelli, C., Chern, R. T., Ramachandran, R., *Powder Technology* 189 (**2009**), p. 376.

41 Hounslow, M. J., Oullion, M., Reynolds, G. K., *Powder Technology* 189 (**2009**), p. 177.

42 Simons, S. J. R., *Granulation*, in: *Handbook of Powder Technology*, vol. 11, Elsevier, Amsterdam, **2007**, p. 1257.

43 Willett, C. D., Johnson, S. A., Adams, M. J., Seville, J. P. K., *Granulation*, in: *Handbook of Powder Technology*, vol. 11, Elsevier, Amsterdam, **2007**, p. 1317.

44 Shi, D., McCarthy, J. J., *Powder Technology* 184 (**2008**), p. 64.

45 Lian, G., Thornton, C., Adams, M. J., *Chemical Engineering Science* 53 (**1998**), p. 3381.

46 Liu, L. X., Litster, J. D., Iveson, S. M., Ennis, B. J., *AIChE Journal* 46 (**2000**), p. 529.

47 Adams, M. J., Lawrence, C. J., Urso, M. E. D., Rance, J., *Powder Technology* 140 (**2004**), p. 268.

48 Iveson, S. M., Litster, J. D., Hapgood, K., Ennis, B. J., *Powder Technology* 117 (**2001**), p. 3.

49 Hapgood, K. P., Iveson, S. M., Litster, J. D., Liu, L. X., *Granulation*, in: *Handbook of Powder Technology*, vol. 11, Elsevier, Amsterdam, **2007**, p. 897.

50 Zhu, H. P., Zhou, Z. Y., Yang, R. Y., Yu, A. B., *Chemical Engineering Science* 62 (**2007**), p. 3378.

51 Zhu, H. P., Zhou, Z. Y., Yang, R. Y., Yu, A. B., *Chemical Engineering Science* 63 (**2008**), p. 5728.

52 Ottino, J. M., Khakhar, D. V., *Annual Review of Fluid Mechanics* 32 (**2000**), p. 55.

53 Abberger, T., *Granulation*, in: *Handbook of Powder Technology*, vol. 11, Elsevier, Amsterdam, **2007**, p. 1109.

54 Cameron, I. T., Wang, F. Y., in: *Handbook of Pharmaceutical Granulation Technology*, 2nd edn., Taylor & Francis Group, Boca Raton, FL, USA, **2005**, p. 555.

55 Ramkrishna, D., *Population Balances: Theory and Applications to Particulate Systems in Engineering*, Academic Press, San Diego, USA, **2000**.

56 Hounslow, M. J., Ryall, R. L., Marshall, V. R., *AIChE Journal* 34 (**1988**), p. 1821.

57 Kumar, S., Ramkrishna, D., *Chemical Engineering Science* 51 (**1996**), p. 1311.

58 Immanuel, C. D., Doyle III, F. J., *Powder Technology* 156 (**2005**), p. 213.

59 PSE, gPROMS Advanced Process Modelling, http://www.psenterprise.com/gproms/index.html (accessed: December 7, 2009), **2009**.

60 Aspentech, Aspen Custom Modeler, http://www.aspentech.com/products/aspen-custom-modeler.cfm (accessed: December 7, 2009), **2009**.

61 Daesim Technologies, Daesim Dynamics, http://www.daesim.com/software/daesim/index.html (accessed: December 7, 2009), **2009**.

62 SolidSim Engineering, SolidSim – A Novel Software Tool for the Flowsheet Simulation of Solids Processes, http://www.solidsim.com/www/ (accessed: December 7, 2009), **2009**.

63 Gantt, J. A., Gatzke, E. P., *AIChE Journal* 52 (**2006**), p. 3067.

64 Gantt, J. A., Cameron, I. T., Litster, J. D., Gatzke, E. P., *Powder Technology* 170 (**2006**), p. 53.

65 Goldschmidt, M. J. V., Hydrodynamic modelling of fluidised bed spray granulation, PhD Thesis, University of Twente, The Netherlands, **2001**.

66 Goldschmidt, M. J. V., Weijers, G. G. C., Boerefijn, R., Kuipers, J. A. M., *Powder Technology* 138 (**2003**), p. 39.

67 Kafui, D. K., Thornton, C., *Powder Technology* 184 (**2008**), p. 177.

68 Tan, H. S., Goldschmidt, M. J. V., Boerefijn, R., Hounslow, M. J., Salman, A. D., Kuipers, J. A. M., *Powder Technology* 142 (**2004**), p. 103.

69 Tan, H. S., Salman, A. D., Hounslow, M. J., *Chemical Engineering Science* 60 (**2005**), p. 3847.

70 Tan, H. S., Salman, A. D., Hounslow, M. J., *Chemical Engineering Science* 61 (**2006**), p. 3930.

71 Rao, N. N., Simulations for modelling of population balance equations of particulate processes using the discrete particle model (DPM), PhD Thesis, Otto-von-Guericke Universität, Magdeburg, Germany, **2009**.

72 Rajniak, P., Stepanek, F., Dhanasekharan, K., Fan, R., Mancinelli, C., Chern, R. T., *Powder Technology* 189 (**2009**), p. 190.

73 Link, J. M., Godlieb, W., Deen, N. G., Kuipers, J. A. M., *Chemical Engineering Science* 62 (**2007**), p. 195.

74 Mishra, B. K., Thornton, C., Bhimji, D., *Minerals Engineering* 15 (**2002**), p. 27.

75 Ramachandran, R., Poon, J. M.-H., Sanders, C. F. W., Glaser, T., Immanuel, C. D., Doyle III, F. J., Litster, J. D., Stepanek, F., Wang, F.-Y., Cameron, I. T., *Powder Technology* 188 (**2008**), p. 89.

76 Poon, J. M.-H., Immanuel, C. D., Doyle III F. J., Litster, J. D., *Chemical Engineering Science* 63 (**2008**), p. 1315.

77 Poon, J. M.-H., Ramachandran, R., Sanders, C. F. W., Glaser, T., Immanuel, C. D., Doyle III F. J., Litster, J. D., Stepanek, F., Wang, F.-Y., Cameron, I. T., *Chemical Engineering Science* 64 (**2009**), p. 775.

78 Németh, E., Cameron, I. T., Hangos, K. M., *Computers and Chemical Engineering* 29 (**2005**), p. 783.

79 Németh, E., Lakner, R., Hangos, K. M., Cameron, I. T., *Information Sciences* 177 (**2007**), p. 1916.

80 Wang, F. Y., Cameron, I. T., *Powder Technology* 179 (**2007**), p. 2.

81 Balliu, N. E., An object oriented approach to the modelling and dynamics of granulation circuits, PhD Thesis, University of Queensland, Australia, **2005**.

82 Saelid, S., Mjaavatten, A., Fjalestad, K., *Computers and Chemical Engineering* 16 (**1992**), p. S97.

83 Schelbach, D., Development of granulation circuit diagnostics for Gensym's G2 intelligent control system, BE Thesis, University of Queensland, Australia, **2000**.

84 NÉMETH, E., LAKNER, R., HANGOS, K. M., CAMERON, I. T., *Innovations in Applied Artificial Intelligence* 3533 (**2005**), p. 367.

85 IVESON, S. M., *Powder Technology* 124 (**2002**), p. 219.

86 LE, P. K., AVONTUUR, P., HOUNSLOW, M. J., SALMAN, A. D., *Powder Technology* 189 (**2009**), p. 149.

87 IMMANUEL, C. D., DOYLE, III, F. J., *Chemical Engineering Science* 58 (**2003**), p. 3681.

88 GANTT, J. A., GATZKE, E. P., *Industrial and Engineering Chemistry Research* 45 (**2006**), p. 6721.

89 INGRAM, G. D., CAMERON, I. T., in: *European Symposium on Computer Aided Process Engineering (ESCAPE-15)*, Elsevier, Amsterdam, **2005**, p. 481.

90 FREIREICH, B., LITSTER, J., WASSGREN, C., *Chemical Engineering Science* 64 (**2009**), p. 3407.

3
Modeling of Polymerization Processes
Bruno S. Amaro and Efstratios N. Pistikopoulos

Keywords
free-radical polymerization, diffusion-controlled phenomena, free-volume theory, geometric mean model (GMM), diffusion mean model (DMM), three-stage polymerization model (TSPM)

3.1
Introduction

In a growing, increasingly demanding and competitive market, polymer industries face different challenges in order to offer the best product at the lowest cost possible and come up with innovative solutions that meet clients needs. Many of the variables that directly affect relevant product quality indices are difficult (sometimes impossible) to measure along the process, within a reasonable time or cost. This often leads to poor product quality monitoring and control. The customer usually specifies values for end-use properties that have to be met by the producer. This is achieved by relating the nonmolecular parameters (e.g. tensile strength, impact strength, color, crack resistance, thermal stability, melt index, density, etc.) to fundamental polymer properties such as the molecular weight distribution (MWD), copolymer composition and its distribution, particle size distribution (PSD), etc. The latter variables, particularly MWD, PSD, and copolymer composition, are more easily measurable than the former variables, leading to inferential control of the former through feedback on the latter. Modeling of polymerization processes is therefore of great industrial importance, since polymer manufacturers find a better understanding of their existing polymerization reactions and process behaviors, enabling the design of more efficient technology and the development of new products [1].

Various chemical and physical phenomena occurring in a polymer reactor can be divided in three levels of modeling:

1. microscale chemical kinetic modeling;
2. mesoscale physical/transport modeling;
3. macroscale dynamic reactor modeling.

The first and lowest level includes the polymer reactions related to the kinetic mechanism and the most powerful approach for modeling this phenomena is the detailed species balance method. At the mesoscale stage, interphase transfers of heat and mass, interphase equilibrium, micromixing, polymer size distribution and particle morphology play important roles and further influence the polymer properties. The macroscale level deals with models describing the macromixing phenomena in the reactor [2].

This chapter will focus on the first of the three previous levels, considering the particular case of free-radical polymerization for both single monomer and multi-monomer systems.

3.2
Free-Radical Homopolymerization

3.2.1
Kinetic Modeling

For a linear addition polymerization, the equations must take into account initiation or catalytic site activation, chain propagation, chain termination, and chain transfer reactions.

The initiation reaction is the attack of a monomer molecule by a primary radical originating from the initiator. This process involves two reactions: decomposition of the initiator to form primary radicals and the actual initiation reaction.

The propagation reaction is repeated hundreds to many thousands of times for each chain formed. It is generally assumed that the rate constant of this reaction remains the same regardless of the length of the chain to which the radical site is attached.

Growing radicals can terminate with each other in two different ways. In the case of disproportionation, a hydrogen atom is transferred from one chain to the other leading to the formation of two molecules of "dead" polymer, one bearing a double bond at chain end. If the termination occurs by combination, a homopolar bond is formed by pairing the single electrons of the free-radical sites of two chains.

It can happen that a free-radical site reacts with a molecule of solvent, or initiator, or monomer, or even with polymer itself. The result of such reactions is that the growing site at chain end is removed. However, another radical arises from the molecule which has reacted, and if it is able to add monomer, it gives rise to another macromolecule. The total number of active radical sites is not modified by the occurrence of these transfer reactions; if the radicals formed upon transfer are active (with respect to the monomer), the kinetic consequences of transfer processes are of minor importance. But the occurrence of transfer processes strongly influences the molecular weight of the polymer formed. Instead of building one polymer molecule (or 1/2 molecule, in case of termination by combination) a given radical site can produce several polymer molecules during its lifetime, if transfer is involved. Primary radicals can also get involved in transfer reactions, while transfer to the initiator is usually very small and for this reason it is not considered in the kinetic models used in this work [3].

A typical kinetic scheme is shown below. It includes initiation (3.2), propagation (3.4), chain transfer to monomer (3.5), and to solvent (3.6) and termination of both by disproportionation (3.7) and combination (3.8).

Initiation:

$$I \xrightarrow{k_d} 2R^\bullet \qquad (3.1)$$

$$R^\bullet + M \xrightarrow{k_I} P_1 \qquad (3.2)$$

Propagation:

$$P_1 + M \xrightarrow{k_p} P_2 \qquad (3.3)$$

$$P_n + M \xrightarrow{k_p} P_{n+1} \qquad (3.4)$$

Chain transfer to monomer:

$$P_n + M \xrightarrow{k_{fm}} D_n + P_1 \qquad (3.5)$$

Chain transfer to solvent:

$$P_n + S \xrightarrow{k_{fs}} D_n + P_1 \qquad (3.6)$$

Termination by disproportionation:

$$P_n + P_m \xrightarrow{k_{td}} D_n + D_m \qquad (3.7)$$

Termination by combination:

$$P_n + P_m \xrightarrow{k_{tc}} D_{n+m} \qquad (3.8)$$

3.2.2
Diffusion-Controlled Reactions

In free-radical polymerizations, as the monomer conversion grows toward completion, the viscosity of the reaction medium increases significantly, causing a decrease in the mobility of the molecules and hence diffusion limitations. These will affect termination reactions, usually labeled as Trommsdorff or gel effect [4, 5], propagation reactions and initiation reaction, also known as glass and cage effect, respectively. These phenomena are depicted in Fig. 3.1 and will be described next.

Since termination in free-radical polymerization involves the reaction of two macroradicals it is now generally considered as always being diffusion controlled

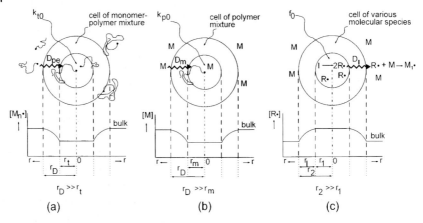

Fig. 3.1 Schematic diagram illustrating the coordinate system used in describing (a) radical termination process (gel effect) for k_t calculation, (b) propagation process (glass effect) for k_p calculation, and (c) primary radical diffusion process (cage effect) for f calculation [6].

from the beginning of polymerization and also in solvents with very low viscosity [7]. Accordingly, termination rate coefficient values not only depend on temperature and pressure, as rate coefficients usually do, but also on other parameters that can influence the diffusive motion of the polymeric radicals being terminated. These parameters include polymer weight fraction, solvent viscosity, interactions between polymer, monomer, and solvent, chain lengths of the terminating macroradicals, chain flexibility, dynamics of entanglements and MWD of the surrounding matrix polymer, through which a radical chain end must diffuse to encounter another radical chain end [8, 9]. However such big number of parameters that can even be interrelated, becomes difficult to be used in the mathematical description of macroradicals mobility. Several theoretical approaches can be found, when trying to model the termination rate constant, but still no single all-encompassing model is fulfilled [8].

Propagation involves the reaction of small monomer molecules and only one large radical, thus propagation is much less hindered during the reaction and k_p remains relatively unaffected until very high conversions. Around 80–90% of monomer conversion of this reaction rate tends asymptotically to zero and the reaction almost stops, "freezing" before the monomer is fully consumed. This phenomenon occurs in polymerizations taking place at temperatures below polymer glass transition temperatures and is associated to glass effect. From that high conversion point onward, the mobility of monomer molecules is constrained and propagation reaction is assumed to be diffusion controlled, therefore decreasing with conversion. At this stage, because k_t is now determined by reaction–diffusion mechanism, and is proportional both to propagation rate coefficient and monomer concentration, also reinforces the observed rapid drop. At the limiting conversion the glass transition temperature of the monomer–polymer mixture becomes equal to the reaction temperature [79].

In the past, the reaction rate reduction observed at high conversions was exclusively attributed to diffusion-controlled propagation. However, comparative experiments on MMA bulk and emulsion polymerization revealed that the initiation reaction can also be ruled by diffusion mechanisms [10]. Given that the initiator decomposition rate constant, k_d, is unlikely to be affected by the viscosity of reaction medium, the initiator efficiency, f_I is the parameter assumed to strongly depend on diffusion-controlled phenomena. Further experimental results show that initiator efficiency can dramatically change with monomer conversion [11, 12, 80]. The term "cage effect" was originally used in the definition of initiator efficiency. This empirical parameter was introduced in order to account for all side reactions that could take place between primary radicals formed from the initiator decomposition inside an "ideal" cage before they escape and react with monomer molecules (see Fig. 3.1(c)). In consequence of these facts, any modeling approach for f_I should include the effect of both the initiation reaction kinetics and the diffusion phenomena taking place at high conversions.

Throughout the years, several models have been developed to include these effects in polymerization simulations. These methodologies, in particular the ones extensively used over the last decade, can be divided, according to [79], into five categories differentiated by their theoretical background, as it is elucidated next.

3.2.2.1 Fickian Description of Reactant Diffusion

In this modeling approach, diffusion effects are viewed as an integral part of the termination, initiation and propagation reactions from the beginning until the end of polymerization. This eliminates the need for using critical break-points denoting the onset of diffusion effects and the associated segmentation of the model in different parts. The approach expressed next was the basis for the model originally developed by [13–15] and extended by [16, 17], known as CSS-AK model. The equations also account for considerations and suggestions proposed by [18] and [19].

Considering a reaction between to spherical particles, A and B, the effective (observed) rate coefficient, k_{eff}, is expressed through Eq. (3.9), where two terms, one kinetically controlled (k_0), the other diffusion controlled ($4\pi N_A r_{AB} D_{AB}$) are accounted for. The symbol k_0 represents the intrinsic reaction rate constant, r_{AB} denotes the spherical radius within which all diffusive motions have been completed and the two molecules can react, and D_{AB} is the mutual diffusion coefficient of the reactive molecules (usually taken as the sum of the two self-diffusion coefficients). Equation (3.9) is regarded as the basis for any model on diffusion control of rate coefficients:

$$\frac{1}{k_{eff}} = \frac{1}{k_0} + \frac{1}{4\pi N_A r_{AB} D_{AB}} \tag{3.9}$$

If, according to this methodology, the diffusion molecule B is a monomer (Fig. 3.1), k_{eff} symbolizes the effective propagation rate coefficient, expressed by

Eq. (3.10), where r_p is the radius of interaction for propagation and D_{MP} the mutual diffusion coefficient between a macroradical and a monomer molecule:

$$\frac{1}{k_{p,\text{eff}}} = \frac{1}{k_{p,0}} + \frac{1}{4\pi N_A r_p D_{MP}} \tag{3.10}$$

On the other hand, if molecule B is a macroradical, then k_{eff} refers to the effective termination rate coefficient and it can be calculated with Eq. (3.11). The radius of interaction for termination is denoted by r_t and $D_{p,xy}$ represents the mutual diffusion coefficient between two polymerizing chains with degrees of polymerization x and y, respectively:

$$\frac{1}{k_{t,\text{eff}}^{\text{diff}}} = \frac{1}{k_{t,0}} + \frac{1}{4\pi N_A r_t D_{p,xy}} \tag{3.11}$$

A similar strategy is used to describe the effect of diffusion-controlled phenomena on initiation reactions. The evolution of initiator efficiency, f_I, is described in Eq. (3.12) as depending on the diffusion coefficient of the primary initiator radicals D_I, where r_1 and r_2 are the radii of the two concentric spheres in which diffusion is assumed to occur, k_{I_0} is the intrinsic chain initiation rate constant and f_{I_0} the initial initiator efficiency. In this equation, the effect of possible loss of primary radicals through the recombination reactions is considered in f_{I_0} while, as D_I decreases at high conversions, the overall f_I diminishes, as well:

$$\frac{1}{f_I} = \frac{1}{f_{I_0}} + \frac{r_2^3}{3r_1} \frac{k_{I_0}[M]}{f_{I0}} \frac{1}{D_I} \tag{3.12}$$

3.2.2.2 Free-Volume Theory

The semiempirical model based on the free-volume theory is originally developed by the group of Hamielec and later extended and used by the groups of Hamielec, Penlidis, Kiparissides and Vivaldo-Lima [20–32]. The methodology is based on the calculation of a parameter K (3.13) that, when reaching its critical value K_{cr} (3.14), determines the beginning of the diffusion-controlled phenomena on the termination reaction [27]:

$$K = M_w^m \exp\left(\frac{A}{V_f}\right) \tag{3.13}$$

$$K_{\text{cr}} = A_{\text{cr}} \exp\left(\frac{E_{\text{cr}}}{RT}\right) \tag{3.14}$$

In the previous equations, M_w is the polymer cumulative weight-average molecular weight, V_f the total free-volume, and m, A, A_{cr}, and E_{cr} are the parameters dependent upon the monomer type. In the regime where K lies below K_{cr} the termination rate coefficient is ruled by segmental diffusion, characterized by a gradual increase in that coefficient, shown by Eq. (3.15), where k_{t0} is the chemically controlled rate constant, δ_c is a segmental diffusion parameter, C_p the polymer concentration, and MW the monomer molecular weight:

$$k_{t,\text{seg}} = k_{t0}(1 + \delta_c C_p MW) \tag{3.15}$$

A regime transition occurs when K_{cr} is reached, defining the onset of the gel effect region. The critical values of V_f and M_w ($V_{f,\text{cr}}$ and $M_{w,\text{cr}}$) also occur at this stage and the termination rate coefficient now decreases according to Eq. (3.16), with n and A being monomer-dependent adjustable parameters:

$$k_T = k_{t0}\left(\frac{M_{w,\text{cr}}}{M_w}\right)^n \exp\left[-A\left(\frac{1}{V_f} - \frac{1}{V_{f,\text{cr}}}\right)\right] \tag{3.16}$$

An overall termination rate coefficient, k_t, is defined by the summation of three contributions, as seen in Eq. (3.17), with $k_{t,\text{rd}}$ denoting the reaction diffusion [33, 24]:

$$k_t = k_{t,\text{seg}} + k_T + k_{t,\text{rd}} \tag{3.17}$$

Propagation rate coefficient is considered to be similarly affected by a diffusion-controlled mechanism, and a decrease is observed when the free-volume is under a critical value of the monomer free-volume, $V_{f,\text{crM}}$, mathematically observed in Eq. (3.18). Here, k_{p0} represents the chemically controlled propagation rate constant and B is a monomer-specific parameter:

$$k_p = k_{p0} \exp\left[-B\left(\frac{1}{V_f} - \frac{1}{V_{f,\text{crM}}}\right)\right] \tag{3.18}$$

Initiator efficiency is also assumed to diffusion controlled toward the end of the reaction, as initiator radicals get increasingly hindered from moving out of their cage due to the presence of larger molecules. When a critical initiator free-volume is reached, $V_{f,\text{crEff}}$, its efficiency is expected to drop dramatically, following Eq. (3.19), where f_0 is the initial initiator efficiency and C is a constant:

$$f = f_0 \exp\left[-C\left(\frac{1}{V_f} - \frac{1}{V_{f,\text{crEff}}}\right)\right] \tag{3.19}$$

3.2.2.3 Chain Length Dependent Rate Coefficients

According to this modeling approach one single rate coefficient, k_t, is not adequate to describe all the termination interactions occurring at any instant during a polymerization reaction. Since macroradicals of a variety of sizes are present, termination is better described in terms of $k_t^{i,j}$, the rate coefficient for termination between radicals of degree of polymerization i and j, respectively [7, 8]. Therefore, the termination rate coefficient should depend on the sizes of the two terminating chains. Three different types of models are more commonly used to describe the variation of termination rate with i and j. The first one is known as geometric mean model (GMM) and according to it, the evolution of $k_t^{i,j}$ is described by Eq. (3.20) [7, 34].

In this equation e quantifies the strength of the chain length dependence of termination. If, for instance, $e = 0$ then $k_t^{i,j} = k_t^{1,1}$ which means that termination would be chain length independent. Though GMM has no physical basis, it has been suggested to best approximate the functional diffusion process [35]:

$$k_t^{i,j} = k_t^{1,1}(\sqrt{ij})^{-e} = k_t^{1,1}(ij)^{-e/2} \tag{3.20}$$

The harmonic mean model (HMM) has been shown to be the functional form expected for $k_t^{i,j}$ (3.21) if chain-end encounter upon coil overlap is the rate determining step for termination [7]:

$$k_t^{i,j} = k_t^{1,1}\left(\frac{2ii}{i+j}\right)^{-e} \tag{3.21}$$

The third methodology, "diffusion mean" model (DMM), is of the functional form expected if translational diffusion is rate determining [35]. The DMM follows from the long-time limit of the Smoluchowski equation for a diffusion-controlled rate coefficient [7, 36] and, it is known to provide a reasonable description of the termination kinetics of small radicals, as in Eq. (3.22) [35]:

$$k_t^{i,j} = 0.5 k_t^{1,1}\left(i^{-e} + j^{-e}\right) \tag{3.22}$$

Regarding these three approaches, it has been shown that, for $e < 1$, the DMM value lies between the other two (3.23) [7, 34],

$$k_t^{i,j}(\text{GMM}) \leqslant k_t^{i,j}(\text{DMM}) \leqslant k_t^{i,j}(\text{HMM}) \tag{3.23}$$

Inclusion of chain length dependent rate coefficients in any free-radical polymerization kinetic model requires the solution of an infinite set of differential equations describing the evolution of the concentration of every radical and "dead" polymer with chain length i. Otherwise, the full MWD is characterized by solving only for the first few moments. Therefore a significantly higher computational effort is required and consequently, according to [79], it should be used only if needed.

Opposite to termination, the propagation rate coefficient is usually assumed to be chain length independent [37, 38]. However, according to [39, 40] the Langmuir-type expression represented in Eq. (3.24) would be appropriate to describe the k_p dependence on the chain length i. This equation can assume an exponential form, becoming Eq. (3.25). In the previous equations, A, B, k_p^0, k_p^∞, and k are constants:

$$k_p^i = k_p^0 - \frac{A}{B+i} i \tag{3.24}$$

$$k_p^i = (k_p^0 - k_p^\infty)\exp[-ki] + k_p^\infty \tag{3.25}$$

References [37, 38] propose a chain length averaged propagation rate coefficient, $\langle k_p \rangle$ defined by Eq. (3.26), in which k_p^i is the rate coefficient of an i-meric radical

$[R_i]$ adding to a monomer molecule [37, 38]:

$$\langle k_p \rangle = \sum_{i=1}^{\infty} k_p^i \frac{[R_i]}{[R]} \tag{3.26}$$

Based on the available experimental and theoretical data, a functional form for k_p^i is then proposed in Eqs. (3.27) and (3.28), where k_p denotes the long chain propagation rate coefficient and C_1 is the factor by which k_p^1 exceeds k_p. Similarly to what happens to the "half-life" of the first-order kinetics, $i_{1/2}$ is a measure that determines the chain length dependence of k_p^i:

$$k_p^i = k_p \left\{ 1 + C_1 \exp\left[\frac{-\ln 2}{i_{1/2}} (i-1) \right] \right\} \tag{3.27}$$

$$C_1 = \frac{k_p^1 - k_p}{k_p} \tag{3.28}$$

Since chain transfer to monomer reaction competes with propagation involving the same reactants, a similar chain length dependence to that for propagation is expected to be observed [7]. To quantify this dependence, [41] suggest the use of Eq. (3.29):

$$\frac{k_{tr,0}}{k_{tr}} = \frac{k_{tp,0}}{k_p} = 1 + \frac{k_{p,0}}{k_{p,\text{diff}}} \tag{3.29}$$

3.2.2.4 Combination of the Free-Volume Theory and Chain Length Dependent Rate Coefficients

Reference [42] proposes Eq. (3.30) for the estimation of k_t over the full monomer conversion, where the effects of segmental, translational, and reaction diffusion are accounted through rate coefficients $k_{t,\text{SD}}$, $k_{t,\text{TD}}$, and $k_{t,\text{RD}}$, respectively,

$$\frac{1}{\langle k_t \rangle} = \frac{1}{k_{t,\text{TD}}} + \frac{1}{k_{t,\text{SD}}} + k_{t,\text{RD}} \tag{3.30}$$

Based on the idea that initiation follows as a result of primary radical fragments diffusing away from each other, same authors [43] also suggest Eq. (3.31) to describe the variation of initiator efficiency with conversion, with D_{term} considered constant and D_I expressed by the generalized free-volume theory of Vrentas and Duda:

$$\frac{1}{f} = 1 + \frac{D_{\text{term}}}{D_I} \tag{3.31}$$

For the calculation of diffusion-controlled termination rate coefficient, between two radicals of chain lengths n and m, [44, 45] suggest Eq. (3.32). The expression is obtained by using an additive model and assuming that k_t is both chain length dependent and conversion dependent:

$$k_t(n, m) = 0.5 k_t(1,1)\bigl(\omega(n) + \omega(m)\bigr) + k_{tp} \qquad (3.32)$$

The chain length dependence of termination rate constant in the nonentangled region is achieved through $\omega(n)$, which is calculated with Eq. (3.33), where β and γ are constants and n_e represents the critical chain length for which entanglements start to appear:

$$\omega(n) = \begin{cases} n^{-\beta}; & n \leq n_e \\ n_e^{-(\beta-\gamma)} n^{-\gamma}; & n > n_e \end{cases} \qquad (3.33)$$

The reaction diffusion-controlled termination rate constant, k_{tp} is given by Eqs. (3.34) and (3.35). $[M]_0$ represents the initial monomer concentration, ζ is a parameter that accounts for flexibility limitations of the free-radical chain and α is the size of the monomer unit:

$$k_{tp} = k_p C'_{RD}(1 - X) \qquad (3.34)$$

$$C'_{RD} = \frac{8\pi}{3} \zeta [M]_0 \alpha^3 n_e^{1/2} N_A \qquad (3.35)$$

Considering the termination rate coefficient as chain length dependent implies that the balance equation for the time evolution of the concentration of free radicals of chain length n will be a partial differential equation, which requires more complex mathematical routines to be solved.

3.2.2.5 Fully Empirical Models

Having a motivation for online optimizing control applications, [46–48] use empirical models described in Eqs. (3.36) and (3.37) to quantify the effect of conversion on termination and propagation rate constants, respectively. Same correlations have been used by [49–52] with all A_i and B_i, functions of temperature, set as empirical constants determined from fitting to experimental data:

$$k_t = k_{t,0} \exp\bigl(A_1 + A_2 X + A_3 X^2 + A_4 X^3\bigr) \qquad (3.36)$$

$$k_p = k_{p,0} \exp\bigl(B_1 + B_2 X + B_3 X^2 + B_4 X^3\bigr) \qquad (3.37)$$

A three-stage polymerization model (TSPM) has been suggested by [53, 54] for treating kinetic data available from literature. Bulk free-radical polymerization is divided into three different stages (low conversion, gel effect, and glass effect) and classical free-radical equations are applied in each of them. Reference [79] states that, even considering the good performance of the model when fitting experimental data, it should be considered as a good data-fitting procedure rather than a predictive method.

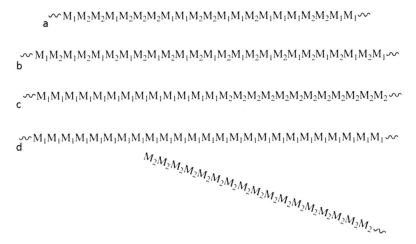

Fig. 3.2 Schematic diagram illustrating different copolymer structures: (a) statistical or random copolymer, (b) alternating copolymer, (c) block copolymer, and (d) graft copolymer (adapted from [56]).

For models earlier than the ones just presented, the reviews of [9, 33, 18, 55, 36] are recommended.

3.3
Free-Radical Multicomponent Polymerization

3.3.1
Overview

Chain polymerizations can be carried out with mixtures of two monomers to form polymeric products with two different structures in the polymer chain. The product of this particular type of chain polymerization is referred to as a copolymer, which is not an alloy of two homopolymers but contains units of both monomers incorporated into each molecule.

This type of copolymerization is relevant since much of the knowledge regarding reactivities of monomers, free radicals, carbocations, and carbanions in chain polymerization comes from copolymerization studies. The behavior of monomers in copolymerization reactions is especially useful for studying the effect of chemical structure on reactivity. From an industrial point of view, copolymerization enables companies to tailor-make polymer products with specifically desired properties, representing a major evolution compared to polymerization of a single monomer, which is relatively limited as to the number of different products that are possible to obtain.

Depending on the distribution of the monomer units in the copolymer molecules, these can be separated into different types. A statistical copolymer (Fig. 3.2(a)) has a distribution of the two monomer units along the copolymer chain that follows a specific statistical law, such as Bernoullian (zero-order Markov) or first- or second-order Markov. Copolymers formed via Bernoullian processes have the two monomer units distributed randomly and are referred to as random copolymers. A copolymer containing the two monomer units in equimolar amounts, in a regular alternating distribution, as shown in Fig. 3.2(b), is therefore named as alternating copolymer.

Block and graft copolymers differ from the previously presented copolymer structures in that there are long sequences of each monomer in the macromolecular chain. A block copolymer is a linear copolymer with one or more long uninterrupted sequences of each polymeric species while a graft copolymer is a branched copolymer with a backbone of one monomer to which are attached one or more side chains of another monomer, as depicted in Figs. 3.2(c) and (d), respectively [56].

Also the polymerization of three different monomer units, commonly denominated terpolymerization, has become increasingly important from the commercial perspective. In most of its uses the terpolymer has two of the monomers present in major amounts to obtain the gross properties desired, with the third monomer in a minor amount for modification of a special property [56].

Properties exhibited by this particular type of polymer units are determined by several factors, such as monomer type, molecular weight and architecture, sequence length, and polymer composition.

3.3.2
Pseudo-Homopolymerization Approximation

The "pseudo-homopolymerization approach" is presented by [57] for multi-monomer emulsion polymerization. By generalizing the Smith–Ewart theory [58], a system of polymer particle population balances is developed for describing the kinetics of emulsion polymerization processes involving any number of monomer species. Each population is characterized by the number of active radicals of each type inside each particle, and described through a size distribution function. The authors propose an approximation procedure that reduces the original system to that typical of homopolymerization processes. This way, each population is characterized through the overall number of radicals, without a significant loss of accuracy. This approximation is based on the observation that two types of transitions between particles of different populations are possible. The first one involves particles with different overall number of active radicals, and is due to radical entry or exit or termination mechanisms. The second one is related to propagation reactions and involves particle populations characterized by the same overall number of active radicals, but of different type. Since in polymerization systems the transitions of the latter type occur in a faster fashion, these are assumed to happen at any time instant under quasi-steady-state conditions, with respect to all transitions of the first type. The reliability of such approach is partially tested by [57], by comparing the model simulations with polymer composition vs. monomer conversion experimental data for the ternary system acrylonitrile–styrene–methyl methacrylate.

Despite the name differences, the methodology previously presented is similar to the "pseudo-kinetic rate constant method," developed by [59] for the calculation of molecular weight in linear chains and copolymer chains with long branches [60]. Considering long chain approximation, the propagation kinetic rate for a copolymerization system can be described by Eq. (3.38), where $[R_i^\bullet]$ is the concentration of live polymer chains, with an ending monomer unit of type i:

$$R_p = k_{p11}[R_1^\bullet][M_1] + k_{p21}[R_2^\bullet][M_1] + k_{p12}[R_1^\bullet][M_2] + k_{p22}[R_2^\bullet][M_2] \tag{3.38}$$

By introducing two new variables, f_i and $\Phi_i^\bullet$, representing the fractions of monomer i and live polymer chains ending with a type i unit, respectively, Eq. (3.38) is rearranged to give Eq. (3.39). The pseudo-kinetic rate constant is denoted by $\overline{k_p}$ and defined by (3.40), while $[M]$ and $[R^\bullet]$ are described by (3.41) and (3.42), respectively:

$$R_p = \overline{k_p}[M][R^\bullet] \tag{3.39}$$

$$\overline{k_p} = \sum_{i=1}^{2} \sum_{j=1}^{2} k_{pij} \Phi_i^\bullet f_j \tag{3.40}$$

$$[M] = \sum_{i=1}^{2} [M_i] \tag{3.41}$$

$$[R^\bullet] = \sum_{i=1}^{2} [R_i^\bullet] \tag{3.42}$$

Same principles can be applied to all other kinetic rate constants, with due modifications, which will be shown in Section 3.5, where a practical example is used to fully describe the kinetic modeling steps.

For the calculation of the pseudo-kinetic rate constants to be performed it is necessary to have a way of computing the polymeric radical fraction, $\Phi_i^\bullet$. For binary copolymerizations this can be achieved through Eqs. (3.43) and (3.44), resulting from the application of the steady-state hypothesis to $[R_i^\bullet]$, followed by a few simplifications:

$$[\Phi_1^\bullet] = \frac{k_{p21} f_1}{k_{p21} f_1 + k_{p12} f_2} \tag{3.43}$$

$$[\Phi_2^\bullet] = \frac{k_{p12} f_2}{k_{p21} f_1 + k_{p12} f_2} \tag{3.44}$$

3.3.3
Polymer Composition

Apart from some particular cases of initial monomer compositions (azeotropic mixtures), the copolymer composition of a typical free-radical copolymerization often exhibits fluctuations along the polymerization process. This phenomenon results both from different reactivities of the monomers and the growing polymer radicals.

For the calculation of final polymer composition for binary polymerization systems, [61] and [62] suggest Eq. (3.45), a ratio between concentrations of both monomers, where r_{ij} denote the different reactivity ratios for the two monomers i and j. The ratios can be defined by (3.46)

$$d[M_1] : d[M_2] = [M_1]\left\{[M_1] + \frac{[M_2]}{r_{12}}\right\}\left\{1 + \frac{r_{12}}{r_{21}}\right\} :$$

$$[M_2]\left\{\frac{[M_1]}{r_{21}} + [M_2]\right\}\left\{\frac{r_{21}}{r_{12}} + 1\right\} \tag{3.45}$$

$$r_{ij} = \frac{k_{pii}}{k_{pij}} \tag{3.46}$$

Equation (3.45) can be used directly for copolymer composition estimation or, it can be simplified to Eq. (3.47). If the common terms from this expression are divided out, (3.48) is obtained, which is recognized as the well-established and widely used "copolymer equation" for two component systems:

$$\frac{d[M_1]}{d[M_2]} = \frac{\frac{[M_1]}{r_{12}r_{21}}\{r_{12}[M_1] + [M_2]\}\{r_{21} + r_{12}\}}{\frac{[M_2]}{r_{12}r_{21}}\{[M_1] + r_{21}[M_2]\}\{r_{21} + r_{12}\}} \tag{3.47}$$

$$\frac{d[M_1]}{d[M_2]} = \frac{[M_1]\{r_{12}[M_1] + [M_2]\}}{[M_2]\{[M_1] + r_{21}[M_2]\}} \tag{3.48}$$

Later, [63] has also confirmed the previous expressions, as an extension of their work on the refinement for the established Alfrey–Goldfinger terpolymer equation. The product of that work will be used in the following chapter, as a starting point for the derivation of the terpolymer composition equation.

3.4
Modeling of Polymer Molecular Properties

3.4.1
Molecular Weight Distribution

Mathematical modeling can be used to characterize particular polymer molecular properties that depend both on the macroscopic mixing of different volume elements and molecular scale mixing [2].

Table 3.1 Effect of MWD of polyolefins on end-use properties [1].

Property	As MWD broadens, property
Stiffness	Decreases slightly
Impact strength	Decreases
Low temperature brittleness	Decreases
Softening point	Increases
Melt strength	Increases
Gloss	Decreases
Shrinkage	Decreases

The MWD is regarded as one of the most important polymer architectural parameters to be controlled in any industrial polymerization process since it is directly related to some important polymer properties. Table 3.1 shows the qualitative effects of MWD on key end-use properties of polymers, polyolefins in this particular case [1].

It is then useful to have a model that comprises an effective methodology for calculation of MWD, bringing improvements to both operation and control of processes as well as in the end-product properties of polymeric materials.

The MWD of a polymer is a record of the kinetic history of the reactions which occurred during its formation [64, 65]. Therefore, it is an important source of information about the kinetic processes that have taken place in polymerization systems. When speaking of the molecular weight of a polymer, it means something different from that which is applied to small-sized compounds. Polymers differ from these small-sized compounds in that they are polydisperse or heterogeneous in molecular weight [2].

The polydispersity index (PI) is a measure of the width of MWDs, theoretically very important. However in the past it alone has been used to characterize the polymer MWD. This is not correct since when considering real polymers, the importance of PI is partly lost. [66], study the origin and solutions for this problem, which are described through theoretical distribution functions.

Experimentally, MWD is determined either by gel-permeation chromatography (GPC) or by size exclusion chromatography (SEC). The PI is calculated as a ratio of the molecular weight averages, defined in Eq. (3.49). A higher polydispersity index is expected to imply a wider distribution but the reverse it is not always verified, as it is shown in this chapter:

$$\mathrm{PI} = \frac{\overline{M}_w}{\overline{M}_n} \qquad (3.49)$$

Indeed they present a few examples of simulated distributions having the same polydispersity index and various widths. One of them is depicted in Fig. 3.3, where a two-parameter Schulz distribution is used to compute the different MWDs. For comparison of various MWDs, a preprocessing of the "raw" distribution data is suggested in order to normalize it. The authors conclude that it is completely unac-

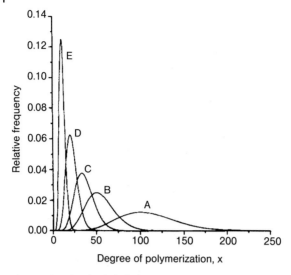

Fig. 3.3 Simulated Schulz distribution functions with identical polydispersity index PI = 1.1 [66].

ceptable to draw conclusions on the absolute width of MWDs of polymers uniquely on the basis of polydispersity index, which is contrary to the widely accepted belief at that moment.

A novel method for the calculation of weight chain length distribution (WCLD) in free-radical polymerization is presented by [67]. A standard kinetic model is considered, accounting for initiation, propagation, chain transfer to monomer and to solvent, and termination by disproportionation. As stated before, the limitation of using the molecular weight moment equations is that only molecular weight averages are calculated and a complete MWD is not obtainable. To override this problem, Crowley and Choi present a different methodology, based on the definition of a function, $f(m, n)$ that represents the weight fraction of polymer within a certain chain length range. The final form of this function is Eq. (3.50) and has to be solved simultaneously with the kinetic modeling equations, together with Eq. (3.51). The latter is the differential equation for the first moment of the "dead" polymer chains:

$$\frac{df(m,n)}{dt} = \left[\left\{ \frac{m(1-\alpha)+\alpha}{\alpha} \right\} \alpha^{m-1} - \left\{ \frac{(n+1)(1-\alpha)+\alpha}{\alpha} \right\} \alpha^n \right.$$

$$\left. - (2-\alpha) f(m,n) \right] \frac{k_p MP}{\lambda_1} \tag{3.50}$$

$$\frac{d\lambda_1}{dt} = (2-\alpha) k_p MP \tag{3.51}$$

To illustrate the proposed computational method, a batch free-radical solution polymerization of methyl methacrylate (MMA) is carried out. Figure 3.4 represents

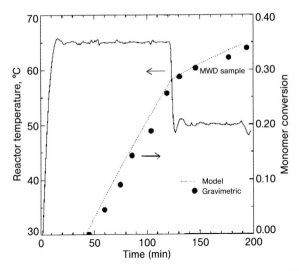

Fig. 3.4 Reactor temperature and monomer conversion profiles [67].

the experimental reactor temperature and monomer conversions measured off-line using a gravimetric method.

The monomer conversion curve predicted by the kinetic model is also shown in Fig. 3.4. The observed induction period is due to inhibitors in the monomer; therefore, in Crowley and Choi model simulation, the end of the induction period was set as the initial reaction time. The model predictions for MWD are compared graphically with experimental measurements obtained from GPC and the results are plotted in Fig. 3.5.

This is a simple method for the calculation of WCLD that makes use of the kinetic rate equations, molecular weight moment equations and the function that defines the weight fraction of polymer in a finite chain length interval. It is also important to note that the proposed method is based on several important assumptions including quasi-steady-state approximation for live polymers and chain length independent termination, and these may not be always applicable to certain free-radical polymerization processes.

This methodology proposed by [67] is later used by [68], extending it to a thermal polymerization of styrene in which termination by combination is the primary mode of chain termination. The MWD is calculated by using a modified function that represents the weight fraction of polymers of chain length above a given length n (3.52) and since $f(2, \infty) = 1$, the weight fraction of a chain length interval (m, n) is now computed as shown in Eq. (3.53). The authors apply the method to compute MWDs for dynamic models of batch polymerization and state that it is still possible to calculate the chain length distribution in continuous styrene polymerization processes:

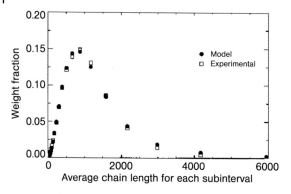

Fig. 3.5 Calculated and experimentally measured MWDs [67].

$$f(n+1, \infty) \equiv \frac{\sum_{i=n+1}^{\infty} i M_i}{\sum_{i=2}^{\infty} i M_i}$$

$$= \frac{\text{weight of polymer with chain lengths from } n+1 \text{ to } \infty}{\text{total weight of polymer}}$$

(3.52)

$$f(m, n) = \{1 - f(n+1, \infty)\} - \{1 - f(m, \infty)\}$$

$$= f(m, \infty) - f(n+1, \infty) \tag{3.53}$$

More recently, [69] proposes a systematic method for calculating the MWD moments in free-radical polymerization for the case of chain length dependent kinetics. The method is based on the moment technique and the central problem, according to the author, is to obtain the moments of the termination rate distribution (TRD). The termination rate of radicals of length n is defined in Eq. (3.54), and is used to later derive the equations of moments for the "live" and "dead" polymer moments, Eqs. (3.55) and (3.56), respectively. k_{tf} is the reaction rate for the transfer to monomer reaction, $k_{t,\text{av}}$ is the average termination coefficient and Z_j is defined by $\sum_{n=1}^{\infty} n^j k_{t,n,n} R_n^\bullet / k_{t,\text{av}} \lambda_j$:

$$R_{t,n} = R_n^\bullet \sum_{m=1}^{\infty} k_{t,n,m} R_m^\bullet \tag{3.54}$$

$$\frac{d\lambda_j}{dt} = 2 f k_d I_2 + k_{tf} M (\lambda_0 - \lambda_j) + k_p M \sum_{i=1}^{j} \binom{j}{i} \lambda_{j-i}$$

$$- k_{t,\text{av}} \lambda_0 \lambda_j (1 + Z_j)/2 - \frac{\lambda_j}{V} \frac{dV}{dt} \tag{3.55}$$

3.4 Modeling of Polymer Molecular Properties

$$\frac{d\mu_j}{dt} = k_{tf}M\lambda_j + \frac{1}{2}\frac{k_{td}}{k_t}k_{t,av}\lambda_0\lambda_j(1+Z_j)$$

$$+ \frac{1}{4}\frac{k_{tc}}{k_t}k_{t,av}\sum_{i=0}^{j}\binom{j}{i}\lambda_i\lambda_{j-i}(Z_i+Z_{j-i}) - \frac{\mu_j}{V}\frac{dV}{dt} \quad (3.56)$$

These moments can be solved by evaluating the moments of $k_{t,n,n}R_n^\bullet$. The methodology used in this work for evaluating these moments is the reconstruction of an approximate chain length distribution of radicals using orthogonal Laguerre polynomials [70] and the final expression is Eq. (3.57). Here k_{tp} is the coefficient of residual termination rate, k_{tvf} is the radical size term, r is an arbitrarily chosen value for convenience of the calculations, ρ is give by n/r and $R(\rho)$ is a continuous distribution that replaces the discrete one:

$$\sum_{n=1}^{\infty} n^j k_{t,n,n} R_n^\bullet$$

$$\approx k_{tp}\lambda_j + k_{tvf}r^{j+1}\left\{\int_0^{\rho_c} \rho^j R(\rho)\,d\rho + \int_{\rho_c}^{\infty} \rho^j \left(\frac{\rho}{\rho_c}\right)^{-2.4} R(\rho)\,d\rho\right\} \quad (3.57)$$

With this equation, the mathematical development to obtain the moments in a free-radical polymerization with chain length dependent termination is complete. The equations are solved numerically. The numerical character of this method proposed by [69] enables the easy inclusion of other diffusional effects such as cage effect and glass effect.

The novelty of the method presented is the strategy followed that includes the development of suitable equations of radical and polymer moments, the reconstruction of the radical distribution by means of a well-known procedure and the evaluation of the moments of the TRD.

Despite being an innovative strategy, MWD is not directly calculated (it is reconstructed) which may lead to the loss of significant information about the shape of the distribution.

For the calculation of MWD for methyl methacrylate bulk homopolymerization in a batch reactor, the method of direct integration is applied by [71]. This integration method is particularly effective in the treatment of critical points and allows a simultaneous solution of the initiator and monomer mass balances along with the macromolecular species mass balances. In this publication the authors present only the methodology to compute MWD for the case of chain length dependent termination. According to the authors, this calculation includes several steps:

(i) assume a value for the "live" radical total concentration λ_0;
(ii) assume a value for the summation S_0, defined in Eq. (3.58);
(iii) compute the entire MWD for "live" radicals by directly solving Eq. (3.60);
(iv) calculate the summation S_0;

(v) take this value for S_0 and repeat steps (iii)–(iv) until the calculated value and the assumed one for S_0 agree within a tolerance;
(vi) given the MWD, calculate the "live" radical total concentration λ_0, using Eq. (3.59);
(vii) assume this value for λ_0 and repeat steps (ii)–(vi) until the calculated value and the assumed one for λ_0 agree within a tolerance;
(viii) calculate the discrete summations transformed to integrals by using the 7-point Newton–Cotes integration rule, where "infinity" is replaced by a very large number;
(ix) calculate MWD for the "dead" polymer by solving macromolecular mass balances Eq. (3.61) with a varying step fourth-order Runge–Kutta method [72, 73]:

$$S_0 = \sum_{m=0}^{\infty} C 4\pi N_A r_t c^{-7/4} (m + N_c)^{-2} [R_m] \tag{3.58}$$

$$\lambda_0 = \sum_{n=0}^{\infty} R_n \tag{3.59}$$

$$\left(k_i R_0 [M] + k_{tm}[M]\lambda_0\right)\delta(n-1) - k_{tm}[M]R_m$$

$$- k_p[M] \frac{([R_n] - [R_{n-2\eta_{step}}])}{\eta_{step}}$$

$$+ \frac{1}{2} k_p[M] \frac{([R_n] - 2[R_{n-\eta_{step}}] + [R_{n-\eta_{step}}])}{\eta_{step}^2}$$

$$- [R_n] C 4\pi N_A r_t c^{-7/4} (n + N_c)^{-2} \lambda_0 + S_0 = 0 \tag{3.60}$$

$$r_{P_n} = k_{tm}[M][R_n]\lambda_0 + \frac{1}{2} \sum_{r=1}^{n-1} k_{tc,n-r,r}[R_r][R_{n-r}]$$

$$+ [R_n] \sum_{m=0}^{\infty} k_{td,n,m}[R_m] \tag{3.61}$$

The parameters and variables that appear in the previous expressions are the proportionality constant C, the Avogadro number N_A, the polymer concentration c, the critical chain length N_c, the Kronecker's delta δ, and the chain length step used in the discretization η_{step}. The "dead" polymer normalized weight molecular weight distribution (WMWD) can be directly calculated by using the methodology previously described.

In the study of [74] two numerical methods, namely the orthogonal collocation on finite elements and the fixed pivot technique, are employed to calculate the MWD in free-radical batch suspension polymerization. In the orthogonal collocation on

3.4 Modeling of Polymer Molecular Properties

finite elements (OCFE) formulation, the entire chain length domain is initially divided into a number of finite elements, N_e, and for each of them Lagrange polynomials are used as approximations for the variation of the state variables. These polynomials are expressed in terms of the values of the corresponding state variables, at the collocation points. The key assumption in OCFE methodology is that the model equations are satisfied only at these selected collocation points. Dynamic molar mass balances for "live" and "dead" polymer chains are written and must be satisfied at the selected collocation points when the functions $R_P(n)$ and $R_D(n)$ are orthogonal to the respective weighting functions, $W_{ij}^P(n)$ and $W_{ij}^D(n)$, over each finite element. The concentrations of both types of polymer chains are approximated over each element j by polynomial expressions:

$$[P(n)] = \sum_{i=0}^{n_c+1}[P(n_{ij})]\phi_{ij}(n) \tag{3.62}$$

$$[D(n)] = \sum_{i=0}^{n_c}[D(n_{ij})]\phi_{ij}(n) \tag{3.63}$$

where $[P(n)]$ and $[D(n)]$ are the unknown values of "live" and "dead" polymer chains concentrations at the i collocation point of element j and ϕ_{ij} are the Lagrange basis functions. Considering the orthogonality of the basis functions, together with the standard finite element formulation, it is possible to derive the discretized balance equations for the previously mentioned polymer chains:

$$\frac{d[P(n_{i,j})]}{dt} = \left\{k_I[PR^\bullet][M] + k_f[M]\sum_{n=1}^{N_f}[P(n)]\right\}\delta(n_{i,j}-1)$$

$$+ k_p\{[P(n_{i,j}-1)] - [P(n_{i,j})]\}[M] - k_f[M][P(n_{i,j})]$$

$$- (k_{t_c} + k_{t_d})[P(n_{i,j})]\sum_{n=1}^{N_f}[P(n)] - \frac{[P(n_{i,j})]}{V}\frac{dV}{dt} \tag{3.64}$$

$$\frac{d[D(n_{i,j})]}{dt}$$

$$= k_f[M][P(n_{i,j})]$$

$$+ \left(k_{t_d}[P(n_{i,j})]\sum_{n=1}^{N_f}[P(n)] + \frac{1}{2}k_{t_c}\right)\sum_{n=1}^{n_{i,j}-1}\{[P(n)][P(n_{i,j})-n)]\}$$

$$- \frac{[D(n_{i,j})]}{V}\frac{dV}{dt} \tag{3.65}$$

The resulting system of continuous–discrete balance equations is then integrated in time to calculate the dynamic evolution of the "live" and "dead" polymer concentrations at the selected collocation points. Accordingly, the WCLD of the "dead" polymer chains can be reconstructed using the following approximation, where MW is the molecular weight of a monomer unit,

$$WCLD = n_{i,j}\left[D(n_{i,j})\right]MW, \quad i = 0, \ldots, n_c, \; j = 1, \ldots, N_e \quad (3.66)$$

Mass balance equations for "live" and "dead" polymer chains are also solved using the fixed pivot technique (FPT) of [75]. Assuming constant concentrations in the discrete chain length domain (ζ_{j-1} to ζ_j) it is possible to define lumped polymer chain concentrations, $\tilde{P}_j$ and $\tilde{D}_j$, corresponding to the j element. Following, total chain length domain (1 to N_f) is divided into N_e elements, while $\tilde{P}_j$ and $\tilde{D}_j$ are assigned to the grid point n_j. From the application of FPT to population balance equations results a system of discrete differential equations, (3.67) and (3.68), where $\delta(x)$ is the Kronecker's delta function and $j = 1, 2, \ldots, N_e$:

$$\frac{d[\tilde{P}_j]}{dt} = 2fk_d[I]V\delta(j-1) + k_f[M]\sum_{k=1}^{N_e}\tilde{P}_k\delta(j-1)$$

$$+ k_p[M]\sum_{k=1}^{j}\tilde{P}_k B_{j,k} - k_p[M]\tilde{P}_j - k_f[M]\tilde{P}_j$$

$$- k_{t_d}\tilde{P}_j\sum_{k=1}^{N_e}\tilde{P}_k - k_{t_c}\tilde{P}_j\sum_{k=1}^{N_e}\tilde{P}_k - \frac{\tilde{P}_j}{V}\frac{dV}{dt} \quad (3.67)$$

$$\frac{d[\tilde{D}_j]}{dt} = k_f[M]\tilde{P}_j + k_{t_d}\tilde{P}_j\sum_{k=1}^{N_e}\tilde{P}_k + k_{t_c}\sum_{k=1}^{j}\sum_{m=k}^{j}A_{j,k,m}\tilde{P}_m\tilde{P}_k$$

$$- \frac{\tilde{D}_j}{V}\frac{dV}{dt} \quad (3.68)$$

According to this methodology, the initial infinite system of differential equations is now reduced to a finite system of discrete–continuous differential equations. This system is integrated in time to compute the dynamic evolution of $\tilde{P}_j$ and $\tilde{D}_j$. The WCLD is again reconstructed, this time using the approximation in (3.69), whose chain length domain is discretized through a logarithmic rule:

$$WCLD_j = \frac{(n_j\tilde{D}_j MW)}{(\zeta_{j-1} - \zeta_j)} \quad (3.69)$$

In Fig. 3.6 the authors compare both model approaches with MWD points experimentally obtained.

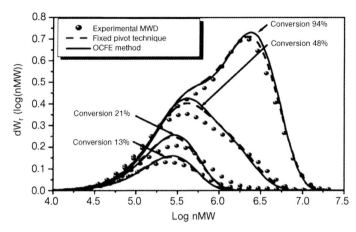

Fig. 3.6 Comparison of model predictions (by FPT and OCFE method) and experimental measurements on MWD at different monomer conversions [74].

For the representation of complete MWD, [76] proposes a method based on differentiation of the cumulative MWD, where the accumulated concentrations, evaluated at a finite number of chain lengths, are considered components in a reaction medium. The key concept for this approach is the fact that differential MWD can be obtained through differentiation of the cumulative distribution. The latter is achieved by computing the accumulated concentrations, UA, at a finite number of chain lengths in a finite domain N:

$$UA_m = \sum_{p=1}^{m} U_p, \quad m \in \Psi \tag{3.70}$$

$$\Psi = \{a_1, a_2, \ldots, a_{nr}\}, \quad a_{nr} = N \tag{3.71}$$

$$M = mMW \tag{3.72}$$

$$W_m = \frac{\sum_{p=1}^{m} pMWU_p}{\sum_{p=1}^{N} pMWU_p} = \frac{UA_{1,m}}{UA_{1,N}} \tag{3.73}$$

In these equations, U_p represents the concentration of "dead" polymer chains with p monomer units, a_i the chain lengths at which the accumulated concentrations are calculated, nr the number of bins in the domain N, MW and M are, respectively, the monomer and the polymer molecular weights and W_m is the normalized polymer mass fraction with chain length up to m. In order to get a smooth cumulative distribution (Fig. 3.7(a)) the authors apply an interpolation based in cubic splines. Differentiation of each of the cubic polynomials fitted between each pair of points will, therefore, generate the desired MWD (Fig. 3.7(b)). Different levels of detail can be obtained by changing the number of bins, nr, and the selected

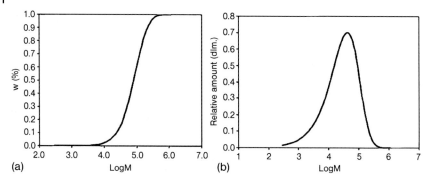

Fig. 3.7 (a) Cumulative and (b) differential MWDs [76].

chain lengths in Ψ. Hence, for the implementation of this method it only required to know how to compute the accumulated concentrations, treated as species and, consequently, having an associated mass balance.

3.5
A Practical Approach – SAN Bulk Polymerization

3.5.1
Model

3.5.1.1 Kinetic Diagram
For the demonstration of the modeling features presented above, the bulk copolymerization of styrene–acrylonitrile is selected. The kinetic scheme used in this practical demonstration of modeling capabilities is shown below, including initiation (3.75), propagation (3.76), chain transfer to monomer (3.77), to solvent (3.78), to modifier (3.79), and to polymer (3.80) and, termination both by disproportionation (3.81) and combination (3.82).

Initiation:

$$I \xrightarrow{k_d} 2R^\bullet \tag{3.74}$$

$$R^\bullet + M_j \xrightarrow{k_{Ij}} P_1^j, \quad j = 1, 2 \tag{3.75}$$

Propagation:

$$P_n^i + M_j \xrightarrow{k_{pij}} P_{n+1}^j, \quad i, j = 1, 2 \tag{3.76}$$

Chain transfer to monomer:

$$P_n^i + M_j \xrightarrow{k_{fmij}} D_n + P_1^j, \quad i, j = 1, 2 \tag{3.77}$$

Chain transfer to solvent:

$$P_n^i + S \xrightarrow{k_{fsi}} D_n + P_1^i, \quad i = 1, 2 \tag{3.78}$$

Chain transfer to modifier:

$$P_n^i + X \xrightarrow{k_{fxi}} D_n + P_1^i, \quad i = 1, 2 \tag{3.79}$$

Chain transfer to polymer:

$$P_n^i + D_m \xrightarrow{k_{fpij}} D_n + P_m^j, \quad i, j = 1, 2 \tag{3.80}$$

Termination by disproportionation:

$$P_n^i + P_m^j \xrightarrow{k_{tdij}} D_n + D_m, \quad i, j = 1, 2 \tag{3.81}$$

Termination by combination:

$$P_n^i + P_m^j \xrightarrow{k_{tcij}} D_{n+m}, \quad i, j = 1, 2 \tag{3.82}$$

3.5.1.2 Mass Balances

For all the species in this polymerization system the mathematical equations representing the mass balances are presented next. Initiator and both monomers can either be added in the beginning of the reaction or, in a semibatch fashion by setting a feed profile for these variables.

Initiator:

$$\frac{d(IV)}{dt} = F_I^{in} I^{in} - k_d I V \tag{3.83}$$

Monomers:

$$\frac{d(M_1 V)}{dt} = F_{M_1}^{in} \frac{\rho_{M_1}}{MW_{M_1}}$$

$$- \left[(k_{p11} + k_{fm11}) p_1 + (k_{p21} + k_{fm21}) p_2 \right] M_1 P V$$

$$- k_{I1} R^\bullet M_1 V \tag{3.84}$$

$$\frac{d(M_2 V)}{dt} = F_{M_2}^{in} \frac{\rho_{M_2}}{MW_{M_2}}$$

$$- \left[(k_{p12} + k_{fm12}) p_1 + (k_{p22} + k_{fm22}) p_2 \right] M_2 P V$$

$$- k_{I2} R^\bullet M_2 V \tag{3.85}$$

Solvent:

$$\frac{1}{V}\frac{d(SV)}{dt} = -k_{fs}SP \tag{3.86}$$

Modifier:

$$\frac{1}{V}\frac{d(XV)}{dt} = -k_{fx}XP \tag{3.87}$$

Radicals:

$$\frac{1}{V}\frac{d(R^\bullet V)}{dt} = 2f_I k_d I - k_{I1} R^\bullet M_1 - k_{I2} R^\bullet M_2 \tag{3.88}$$

Live polymer:

$$\frac{1}{V}\frac{d(PV)}{dt} = k_{I1} R^\bullet M_1 + k_{I2} R^\bullet M_2 - (k_{td} + k_{tc})P^2 \tag{3.89}$$

3.5.1.3 Diffusion Limitations

Similarly to what happens in free-radical homopolymerization the gradual build up of macromolecules causes a significant increase in the viscosity of the reacting mixture. Therefore, initiation, propagation and termination kinetic steps are affected by diffusion limitations that are accounted for in the present model, following the work of [16] and [77]. As mentioned above, a generalized mathematical framework is developed by these authors in order to describe diffusion-controlled reactions in free-radical copolymerization systems. The initiator efficiency as well as the propagation and termination rate constants are expressed in terms of a reaction-limited term and a diffusion-limited one. For the estimation of the reactive species diffusion coefficients the generalized free-volume theory of [78] for ternary systems is applied. The key equations resulting from the work of the authors, and used in the model described in this section, are presented next. This is a general approach that facilitates its application to different systems, provided that all the necessary parameters are available, or can be calculated.

Glass effect:

$$\frac{1}{f_I} = \frac{1}{f_{I_0}} + \frac{(k_{I1_0} M_1 + k_{I2_0} M_2) r_{I2}^3}{(3 r_{I1} f_{I_0} D_I)} \tag{3.90}$$

$$r_{I1} = \frac{1}{2}\left[\left(\frac{6 V^*_{M_1} MW_{M_1}}{\pi N_A}\right)^{1/3} + \left(\frac{6 V^*_{M_2} MW_{M_2}}{\pi N_A}\right)^{1/3}\right] \tag{3.91}$$

$$r_{I2} = 2 R_{H_0} \tag{3.92}$$

$$D_I = D_{I_0} \exp\left[-\frac{\gamma_I}{V_{FH}}\left(\omega_{M_1} V_{M_1}^* \frac{\xi_{IP}}{\xi_{M_1 P}} + \omega_{M_2} V_{M_2}^* \frac{\xi_{IP}}{\xi_{M_2 P}} + \omega_{P_1} V_{P_1}^* \xi_{IP}\right.\right.$$

$$\left.\left. + \omega_{P_2} V_{P_2}^* \xi_{IP}\right)\right] \tag{3.93}$$

$$D_{I_0} = D_{M_0} \tag{3.94}$$

$$k_{I i_0} = \epsilon_{i_0} k_{p i i_0}, \quad i = 1, 2 \tag{3.95}$$

$$\xi_{iP} = V_i^* \frac{MW_i}{V_{Pj}}, \quad i = I, M_1, M_2 \tag{3.96}$$

$$V_{Pj} = [0.6224 T_{gp} - 86.95] T_{gp} \geq 295 \text{ K} \tag{3.97}$$

$$V_{Pj} = [0.0925 T_{gp} + 69.47] T_{gp} < 295 \text{ K} \tag{3.98}$$

$$T_{gp} = F_{p1} T_{gp_1} + F_{p2} T_{gp_2} + \left(\frac{R_T}{100}\right)(T_{gp_{12}} - \overline{T}_{gp}) \tag{3.99}$$

$$\overline{T}_{gp} = \frac{1}{2}(T_{gp_1} + T_{gp_2}) \tag{3.100}$$

$$R_T = \frac{400 F_{p1} F_{p2}}{\{1 + [1 + 4 F_{p1} F_{p2}(r_1 r_2 - 1)]^{1/2}\}} \tag{3.101}$$

$$V_{FH} = \left[\alpha_{M_{1_0}} + \alpha_{M_1}(T - T_{gM_1})\right] V_{M_1}^* \omega_{M_1}$$

$$+ \left[\alpha_{M_{2_0}} + \alpha_{M_2}(T - T_{gM_2})\right] V_{M_2}^* \omega_{M_2}$$

$$+ \left[\alpha_{P_0} + \alpha_P(T - T_{gp})\right](V_{P_1}^* \omega_{P_1} + V_{P_2}^* \omega_{P_2}) \tag{3.102}$$

Cage effect:

$$\frac{1}{k_{pii}} = \frac{1}{k_{pii_0}} + \frac{1}{4\pi D_{M_i} r_{mi} N_A}, \quad i = 1, 2 \tag{3.103}$$

$$r_{mi} = r_{ti} \tag{3.104}$$

$$D_{M_1} = D_{M0_1} \exp\{-(\gamma_M / V_{FH})[\omega_{M_1} V_{M_1}^* + \omega_{M_2} V_{M_2}^*(\xi_{M_1 P}/\xi_{M_2 P})$$

$$+ \omega_{P_1} V_{P_1}^* \xi_{M_1 P} + \omega_{P_2} V_{P_2}^* \xi_{M_1 P}]\} \tag{3.105}$$

$$D_{M_2} = D_{M0_2} \exp\{-(\gamma_M / V_{FH})[\omega_{M_1} V_{M_1}^*(\xi_{M_2 P}/\xi_{M_1 P}) + \omega_{M_2} V_{M_2}^*$$

$$+ \omega_{P_1} V_{P_1}^* \xi_{M_2 P} + \omega_{P_2} V_{P_2}^* \xi_{M_2 P}]\} \tag{3.106}$$

Gel effect:

$$k_{tii} = k_{tii}^d + k_{tii}^{res}, \quad i = 1, 2 \tag{3.107}$$

$$k_{tii}^{res} = A_{pi} k_{pi} M_i \tag{3.108}$$

$$A_{pi} = \pi \delta_i^3 j_{ci} N_A \tag{3.109}$$

$$\frac{1}{k_{tii}^d} = \frac{1}{k_{tii0}} + \frac{1}{4\pi D_{Pe} r_{ti} N_A} \tag{3.110}$$

$$r_{ti} = \frac{\{\ln[\frac{1000\tau_i^3}{N_A P \pi^{3/2}}]\}^{1/2}}{\tau_i} \tag{3.111}$$

$$\tau_i = \left[\frac{3}{(2 j_{ci} \delta_i^2)}\right]^{1/2} \tag{3.112}$$

$$\frac{1}{j_{ci}} = \frac{1}{j_{ci0}} + \frac{2\phi_P}{X_{ci0}} \tag{3.113}$$

$$D_{Pe} = D_P F_{seg} \tag{3.114}$$

$$D_P = \left(\frac{D_{P0}}{M_w^2}\right) \exp[(-\gamma_P/V_{FH})(\omega_{M_1} V_{M_1}^* / \xi_{M_1 P} + \omega_{M_2} V_{M_2}^* / \xi_{M_2 P}$$
$$+ \omega_{P_1} V_{P_1}^* + \omega_{P_2} V_{P_2}^*)] \tag{3.115}$$

$$D_P' = \frac{k_B T}{6\pi \eta_{\text{mix}} R_H} \tag{3.116}$$

$$\ln \eta_{\text{mix}} = f_{m1} \ln \eta_1 + f_{m2} \ln \eta_2 + f_{m1} f_{m2} G_{12} \tag{3.117}$$

$$R_H = \left(\frac{3[\eta] M_w}{10\pi N_A}\right)^{1/3} \tag{3.118}$$

$$F_{seg} = \frac{r_e^3 [\pi r_e + 6\sqrt{2} \alpha_{seg} r_B]}{(16\pi r_B^4)} \tag{3.119}$$

$$r_e = F_{p1} r_{e1} + F_{p2} r_{e2} \tag{3.120}$$

$$r_B = R_{H0} \tag{3.121}$$

3.5.1.4 Pseudo-Homopolymerization Approximation

Based on the work of [59] and [57], a pseudo-homopolymer probability is defined and used for the calculation of the pseudo-kinetic rate constants, as shown next.

Table 3.2 System conditions used in batch simulation (SAN copolymerization).

Variable	Value	Units
T	60	C
$n_{I,0}$	5.929×10^{-2}	mol
$n_{M_1,0}$	8.00	mol
$n_{M_2,0}$	3.25	mol
F_I	0	$L s^{-1}$
F_{M_2}	0	$L s^{-1}$

Pseudo-homopolymer probability:

$$p_1 = \frac{(k_{p21} + k_{fm21})M_1}{(k_{p21} + k_{fm21}), M_1 + (k_{p12} + k_{fm12})M_2} \quad (3.122)$$

$$p_2 = 1 - p_1 \quad (3.123)$$

Kinetic constants:

$$k_p = p_1(k_{p11} f_{m1} + k_{p12} f_{m2}) + p_2(k_{p21} f_{m1} + k_{p22} f_{m2}) \quad (3.124)$$

$$k_{fm} = p_1(k_{fm11} f_{m1} + k_{fm12} f_{m2}) + p_2(k_{fm21} f_{m1} + k_{fm22} f_{m2}) \quad (3.125)$$

$$k_{fs} = p_1 k_{fs1} + p_2 k_{fs2} \quad (3.126)$$

$$k_{fx} = p_1 k_{fx1} + p_2 k_{fx2} \quad (3.127)$$

$$k_{fp} = p_1(k_{fp11} + k_{fp12}) + p_2(k_{fp21} + k_{fp22}) \quad (3.128)$$

$$k_{td} = p_1(k_{td11} p_1 + k_{td12} p_2) + p_2(k_{td21} p_1 + k_{td22} p_2) \quad (3.129)$$

$$k_{tc} = p_1(k_{tc11} p_1 + k_{tc12} p_2) + p_2(k_{tc21} p_1 + k_{tc22} p_2) \quad (3.130)$$

3.5.2
Illustrative Results

The model equations are implemented on the gPROMS platform in order to obtain a few illustrative plots, resulting from two different situations.

A first simulation is run with the purpose of replicating a batch polymerization, with both monomers and initiator (azo-bis(isoamylnitrile) – AIBN) added in the beginning of the reaction, following the conditions presented in Table 3.2. A second one is run in a semibatch fashion, through adding part of the most reactive monomer (acrylonitrile) along the reaction (see conditions in Table 3.3). The monomer feed rate profile is purposely chosen in such a way that polymer composition remains nearly unaltered throughout the duration of the reaction, which

Table 3.3 System conditions used in batch simulation (SAN copolymerization; semibatch variable: F_{M_2}).

Variable	Value	Units
T	60	C
$n_{I,0}$	5.929×10^{-2}	mol
$n_{M_1,0}$	8.00	mol
$n_{M_2,0}$	2.03	mol
F_I	0	L s^{-1}
F_{M_2}	Fig. 3.8	L s^{-1}

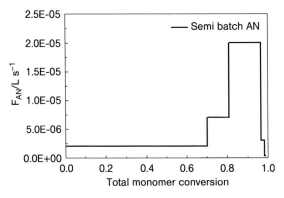

Fig. 3.8 Monomer feed rate profile used in semibatch simulation (SAN copolymerization; semibatch variable: F_{M_2}).

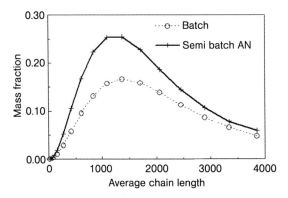

Fig. 3.9 MWD obtained from batch and semibatch simulations (SAN copolymerization; semibatch variable: F_{M_2}).

is usually desirable in an industrial process. The two explored scenarios are materialized in the plots presented in Figs. 3.8–3.11. In these figures it is possible to observe the different evolutions of polymer final MWD, polymer composition and total monomer conversion, according to the conditions set for the simulations.

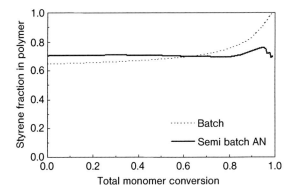

Fig. 3.10 Polymer composition obtained from batch and semibatch simulations (SAN copolymerization; semibatch variable: F_{M_2}).

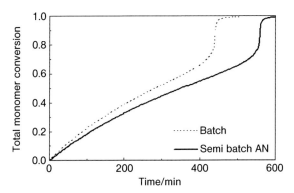

Fig. 3.11 Total monomer conversion obtained from batch and semibatch simulations (SAN copolymerization; semibatch variable: F_{M_2}).

3.6
Conclusions

This chapter has presented some theoretical principals behind polymerization process modeling, applied to free-radical polymer reactions. Comprehensive kinetic schemes encompassing a large number of reactions that might occur during these processes were shown. The specific modeling of polymer molecular properties was highlighted and exemplified for molecular weight distribution (MWD) with different approaches regarding its representation being presented. Several key features related to both homopolymerization and multicomponent polymerization were described and later illustrated through a specific example where the model was used to simulate the styrene–acrylonitrile copolymerization. Other than the kinetic scheme, all mass balances were presented, as well as a concretization of the pseudo-homopolymerization approximation concept formerly described. Two distinct sets of conditions for which the model was tested were described, one involving the addition of most reactive monomer in a semibatch fashion, as an attempt to obtain a near-constant final polymer composition, with the respective results shown in plots. Modeling of polymerization processes is, therefore, a powerful tool allowing researchers and companies to perform a broad variety of simulations, allowing for a multiplicity of activities based on the mathematical translation of real phenomena. These models are also on the origin of more complex, yet crucial, operations such as optimization of process conditions or dynamic process control.

Notation

A_p – proportionality rate constant, cm^3 mol^{-1}
D_I – initiator diffusion coefficient, cm^2 s^{-1}
D_M – monomer diffusion coefficient, cm^2 s^{-1}
D_P – polymer diffusion coefficient, cm^2 s^{-1}
D_n – "dead" polymer chains of chain length n concentration, mol L^{-1}
F – feed volume flow rate, L s^{-1}
F_p – number-average copolymer composition
F_{seg} – probability of two radicals reacting when active centers in close proximity
f_I – initiator efficiency factor
f_m – monomer fraction
G_{12} – binary interaction parameter for viscosity calculation in a mixture
I – initiator concentration, mol L^{-1}
j_c – entanglement spacing
k_t^d – diffusion-controlled termination rate constant, L mol^{-1} s^{-1}
k_t^{res} – residual termination rate constant, L mol^{-1} s^{-1}
k_d – initiator decomposition rate constant, s^{-1}
k_{fm} – chain transfer to monomer rate constant, L mol^{-1} s^{-1}
k_{fp} – chain transfer to polymer rate constant, L mol^{-1} s^{-1}
k_{fs} – chain transfer to solvent rate constant, L mol^{-1} s^{-1}
k_{fx} – chain transfer to modifier rate constant, L mol^{-1} s^{-1}
k_I – initiation rate constant, L mol^{-1} s^{-1}
k_p – propagation rate constant, L mol^{-1} s^{-1}
k_{tc} – combination termination rate constant, L mol^{-1} s^{-1}
k_{td} – disproportionation termination rate constant, L mol^{-1} s^{-1}
M – monomer concentration, mol L^{-1}
M_n – number-average molecular weight, kg mol^{-1}
M_w – weight-average molecular weight, kg mol^{-1}
MW – molecular weight, kg mol^{-1}
N_A – Avogadros's number
P – total "live" polymer chains concentration, mol L^{-1}
p_i – probability of propagation
P_n – "live" polymer chains of chain length n concentration, mol L^{-1}
r – reactivity ratio

$R^\bullet$ – primary radicals concentration, mol L^{-1}
r_B – distance of the chain end from the sphere center,
R_H – hydrodynamic radius, cm
r_{I1} – initiator reaction sphere radius, cm
r_{I2} – initiator diffusion sphere radius, cm
r_m – effective reaction radius, cm
r_t – effective reaction radius, cm
R_T – average number of both monomer sequences occurring in a copolymer, per 100 monomer units
S – solvent concentration, mol L^{-1}
t – time, s
T_{gp} – glass transition temperature, K
V – system volume, L
V_{FH} – average free volume, cm^3 g^{-1}
V_{Pj} – copolymer jumping unit molar volume, cm^3 g^{-1}
X – modifier concentration, mol L^{-1}
X_{c0} – critical degree of polymerization for entanglements

Greek letters

α – probability of propagation
δ – average root-mean-square end-to-end distance per square root of number of monomer units in a chain, Å
ϵ_1 – proportionality constant
η – viscosity, cP
$[\eta]$ – polymer intrinsic viscosity, cm^3 g^{-1}
γ – overlap factor
λ_0 – zeroth moment of the "live" polymer distribution
μ – moments of the "dead" polymer distribution
ω – weight fraction
ϕ – volume fraction
ρ – density, kg dm^{-3}
ξ – ratio of the critical molar volume of the jumping unit to the critical molar volume of the polymer jumping unit

Subscripts

0 – initial conditions
1 – monomer unit of type 1 or polymer chain ending in type 1 monomer
2 – monomer unit of type 2 or polymer chain ending in type 2 monomer
I – initiator
M_1 – monomer of type 1
M_2 – monomer of type 2
mix – mixture

Superscripts

1 – "live" or "dead" polymer chain ending in monomer unit of type 1
2 – "live" or "dead" polymer chain ending in monomer unit of type 2
in – fed to reaction
* – critical

References

1 YOON, W. J., KIM, Y. S., KIM, I. S., CHOI, K. Y., Recent advances in polymer reaction engineering: modeling and control of polymer properties, *Korean Journal of Chemical Engineering* 21 (**2004**), pp. 147–167.

2 KIPARISSIDES, C., Recent advances in polymer reaction engineering: polymerization reactor modeling: a review of recent developments and future directions, *Chemical Engineering Science* 51 (**1996**), pp. 1637–1659.

3 REMPP, P., MERRILL, E. W., *Kinetics of Free Radical Polymerization in Polymer Synthesis*, Hüthig & Wepf Verlag Basel, New York, **1991**.

4 TROMMSDORFF, V. E., KOHLE, H., LAGALLY, P., *Makromolekulare Chemie* 1 (**1947**), p. 169.

5 SCHULZ, G. V., HARBORTH, G., Fitting of polymer distributions of molecular weight by the method of moments, *Makromolekulare Chemie* 1 (**1947**), p. 106.

6 WOLFF, E. H. P., BOS, A. N. R., Modeling of polymer molecular weight distributions in free-radical polymerization reactions. Application to the case of polystyrene, *Industrial and Engineering Chemistry Research* 36 (**1997**), pp. 1163–1170.

7 RUSSELL, G. T., The kinetics of free-radical polymerization: fundamental aspects, *Australian Journal of Chemistry* 55 (**2002**), pp. 399–414.

8 BUBACK, M., EGOROV, M., GILBERT, R. G., KAMINSKY, V., OLAJ, O. F., RUSSELL, G. T., VANA, P., ZIFFERER, G., Critically evaluated termination rate coefficients for free-radical polymerization. 1 – The current situation, *Macromolecular Chemistry and Physics* 203 (**2002**), pp. 2570–2582.

9 DE KOCK, J. B. L., VAN HERK, A. M., GERMAN, A. L., Bimolecular free-radical termination at low conversion, *Journal of Macromolecular Science – Polymer Reviews* C41 (**2001**), pp. 199–252.

10 BALLARD, M. J., NAPPER, D. H., GILBERT, R. G., Kinetics of emulsion polymerization of methyl-methacrylate, *Journal of Polymer Science Part A – Polymer Chemistry* 22 (**1984**), pp. 3225–3253.

11 ZHU, S., TIAN, Y., HAMIELEC, A. E., EATON, D. R., Radical concentrations in free-radical copolymerization of MMA/EGDMA, *Polymer* 31 (**1990**), pp. 154–159.

12 SACKKOULOUMBRIS, R., MEYERHOFF, G., Radical polymerization of methyl-methacrylate over the full range of conversion – stationary and non-stationary experiments for the determination of the rates of propagation and termination, *Makromolekulare Chemie – Macromolecular Chemistry and Physics* 190 (**1989**), pp. 1133–1152.

13 CHIU, W. Y., CARRATT, G. M., SOONG, D. S., A computer-model for the gel effect in free-radical polymerization, *Macromolecules* 16 (**1983**), pp. 348–357.

14 LOUIE, B. M., CARRATT, G. M., SOONG, D. S., Modeling the free-radical solution and bulk-polymerization of methyl-methacrylate, *Journal of Applied Polymer Science* 16 (**1985**), pp. 348–357.

15 BAILLAGOU, P. E., SOONG, D. S., Major factors contributing to the nonlinear kinetics of free-radical polymerization, *Chemical Engineering Science* 40 (**1985**), pp. 75–86.

16 ACHILIAS, D. S., KIPARISSIDES, C., Development of a general mathematical framework for modeling diffusion-controlled free-radical polymerization reactions, *Macromolecules* 25 (**1992**), pp. 3739–3750.

17 ACHILIAS, D., KIPARISSIDES, C., Modeling of diffusion-controlled free-radical poly-

merization reactions, *Journal of Applied Polymer Science* 35 **(1988)**, pp. 1303–1323.

18 LITVINENKO, G. I., KAMINSKY, V. A., Role of diffusion-controlled reactions in free-radical polymerization, *Progress in Reaction Kinetics* 19 **(1994)**, pp. 139–193.

19 ZHU, S., TIAN, Y., HAMIELEC, A. E., EATON, D. R., Radical trapping and termination in free-radical polymerization of MMA, *Macromolecules* 23 **(1990)**, pp. 1144–1150.

20 GARCIARUBIO, L. H., LORD, M. G., MACGREGOR, J. F., HAMIELEC, A. E., Bulk copolymerization of styrene and acrylonitrile – experimental kinetics and mathematical-modeling, *Polymer* 26 **(1985)**, pp. 2001–2013.

21 BHATTACHARYA, D., HAMIELEC, A. E., Bulk thermal copolymerization of styrene p-methylstyrene – modeling diffusion-controlled termination and propagation using free-volume theory, *Polymer* 27 **(1986)**, pp. 611–618.

22 JONES, K. M., BHATTACHARYA, D., BRASH, J. L., HAMIELEC, A. E., An investigation of the kinetics of copolymerization of methyl-methacrylate p-methyl styrene to high conversion – modeling diffusion-controlled termination and propagation by free-volume theory, *Polymer* 27 **(1986)**, pp. 602–610.

23 YARASKAVITCH, I. M., BRASH, J. L., HAMIELEC, A. E., An investigation of the kinetics of copolymerization of methyl-methacrylate p-methyl styrene to high conversion – modeling diffusion-controlled termination and propagation by free-volume theory, *Polymer* 27 **(1987)**, pp. 602–610.

24 GAO, J., PENLIDIS, A., A comprehensive simulator/database package for reviewing free-radical homopolymerizations, *Journal of Macromolecular Science – Reviews in Macromolecular Chemistry and Physics* C36 **(1996)**, pp. 199–404.

25 GAO, J., PENLIDIS, A., A comprehensive simulator database package for reviewing free-radical copolymerizations, *Journal of Macromolecular Science – Reviews in Macromolecular Chemistry and Physics* C38, **(1998)**, pp. 651–780.

26 DUBE, M. A., HAKIM, M., MCMANUS, N. T., PENLIDIS, A., Bulk and solution copolymerization of butyl acrylate/methyl methacrylate at elevated temperatures, *Macromolecular Chemistry and Physics* 203 **(2002)**, pp. 2446–2453.

27 SCORAH, M. J., DHIB, R., PENLIDIS, A., Modelling of free radical polymerization of styrene and methyl methacrylate by a tetrafunctional initiator, *Chemical Engineering Science* 61 **(2006)**, pp. 4827–4859.

28 DHIB, R., GAO, J., PENLIDIS, A., Simulation of free radical bulk/solution homopolymerization using mono- and bi-functional initiators, *Polymer Reaction Engineering* 8 **(2000)**, pp. 299–464.

29 DUBE, M. A., RILLING, K., PENLIDIS, A., A kinetic investigation of butyl acrylate polymerization, *Journal of Applied Polymer Science* 43 **(1991)**, pp. 2137–2145.

30 KRALLIS, A., KOTOULAS, C., PAPADOPOULOS, S., KIPARISSIDES, C., BOUSQUET, J., BONARDI, C., A comprehensive kinetic model for the free-radical polymerization of vinyl chloride in the presence of monofunctional and bifunctional initiators, *Industrial and Engineering Chemistry Research* 43 **(2004)**, pp. 6382–6399.

31 VIVALDO-LIMA, E., HAMIELEC, A. E., WOOD, P. E., Autoacceleration effect in free-radical polymerization – a comparison of the CCS and MH models, *Polymer Reaction Engineering* 2 **(1994)**, pp. 17–85.

32 VIVALDO-LIMA, E., HAMIELEC, A. E., WOOD, P. E., Modeling of the free-radical copolymerization kinetics with crosslinking of methyl methacrylate/ethylene glycol dimethacrylate up to high conversions and considering thermal effects, *Revista de la Sociedad Quimica de Mexico* 47 **(2003)**, pp. 22–23.

33 DUBE, M. A., SOARES, J. B. P., PENLIDIS, A., HAMIELEC, A. E., Mathematical modeling of multicomponent chain-growth polymerizations in batch, semibatch, and continuous reactors: a review, *Industrial and Engineering Chemistry Research* 36 **(1997)**, pp. 966–1015.

34 SMITH, G. B., RUSSELL, G. T., HEUTS, J. P. A., Termination in dilute-solution free-radical polymerization: a composite model, *Macromolecular Theory and Simulations* 12 **(2003)**, pp. 299–314.

35 MOAD, G., SOLOMON, D. H., *The Chemistry of Radical Polymerization*, 2nd edn., Elsevier, The Netherlands, **2003**.

36 TEFERA, N., WEICKERT, G., WESTERTERP, K. R., Modeling of free radical polymer-

ization up to high conversion 2. Development of a mathematical model, *Journal of Applied Polymer Science* 63 **(1997)**, pp. 1663–1680.

37 SMITH, G. B., RUSSELL, G. T., YIN, M., HEUTS, J. P. A., The effects of chain length dependent propagation and termination on the kinetics of free-radical polymerization at low chain lengths, *European Polymer Journal* 41 **(2005)**, pp. 225–230.

38 HEUTS, J. P. A., RUSSELL, G. T., The nature of the chain-length dependence of the propagation rate coefficient and its effect on the kinetics of free-radical polymerization. 1. Small-molecule studies, *European Polymer Journal* 42 **(2006)**, pp. 3–20.

39 OLAJ, O. F., VANA, P., ZODER, M., Chain length dependent propagation rate coefficient $k(p)$ in pulsed-laser polymerization: variation with temperature in the bulk polymerization of styrene and methyl methacrylate, *Macromolecules* 35 **(2002)**, pp. 1208–1214.

40 OLAJ, O. F., ZODER, M., VANA, P., KORNHERR, A., SCHNOLL-BITAI, I., ZIFFERER, G., Chain length dependence of chain propagation revisited, *Macromolecules* 35 **(2005)**, pp. 1944–1948.

41 RUSSELL, G. T., GILBERT, R. G., NAPPER, D. H., Chain-length-dependent termination rate-processes in free-radical polymerizations 2. Modeling methodology and application to methyl-methacrylate emulsion polymerizations, *Macromolecules* 26 **(1993)**, pp. 3538–3552.

42 BUBACK, M., Free-radical polymerization up to high conversion – a general kinetic treatment, *Makromolekulare Chemie – Macromolecular Chemistry and Physics* 191 **(1990)**, pp. 1575–1587.

43 BUBACK, M., HUCKESTEIN, B., KUCHTA, F. D., RUSSELL, G. T., SCHMID, E., Initiator efficiencies in 2,2′-azoisobutyronitrile-initiated free-radical polymerizations of styrene, *Macromolecular Chemistry and Physics* 195 **(1994)**, pp. 2117–2140.

44 BUBACK, M., EGOROV, M., KAMINSKY, V., Modeling termination kinetics in free-radical polymerization using a reduced number of parameters, *Macromolecular Theory and Simulations* 8 **(1999)**, pp. 520–528.

45 BUBACK, M., BARNER-KOWOLLIK, C., EGOROV, M., KAMINSKY, V., Modeling termination kinetics of non-stationary free-radical polymerizations, *Macromolecular Theory and Simulations* 10 **(2001)**, pp. 209–218.

46 SANGWAI, J. S., BHAT, S. A., GUPTA, S., SARAF, D. N., GUPTA, S. K., Bulk free radical polymerizations of methyl methacrylate under non-isothermal conditions and with intermediate addition of initiator: experiments and modeling, *Polymer* 46 **(2005)**, pp. 11451–11462.

47 SANGWAI, J. S., SARAF, D. N., GUPTA, S. K., Viscosity of bulk free radical polymerizing systems under near-isothermal and non-isothermal conditions, *Polymer* 47 **(2006)**, pp. 3028–3035.

48 BHAT, S. A., SARAF, D. N., GUPTA, S., GUPTA, S. K., Use of agitator power as a soft sensor for bulk free-radical polymerization of methyl methacrylate in batch reactors, *Industrial and Engineering Chemistry Research* 45 **(2006)**, pp. 4243–4255.

49 CURTEANU, S., A comparative description of diffusion-controlled reaction-models in free radical polymerization, *Revue Roumaine De Chimie* 48 **(2003)**, pp. 245–262.

50 CURTEANU, S., Modeling and simulation of methyl methacrylate free radical polymerization with intermediate addition of initiator and step change of temperature, *Materiale Plastice* 40 **(2003)**, pp. 96–102.

51 CURTEANU, S., Modeling and simulation of free radical polymerization of styrene under semibatch reactor conditions, *Central European Journal of Chemistry* 40 **(2003)**, pp. 96–102.

52 CURTEANU, S., BULACOVSCHI, V., Free radical polymerization of methyl methacrylate: modeling and simulation under semibatch and nonisothermal reactor conditions, *Journal of Applied Polymer Science* 74 **(1999)**, pp. 2561–2570.

53 QIN, JIGUANG, GUO, WENPING, ZHANG, ZHENG, Modeling of the bulk free radical polymerization up to high conversion – three stage polymerization model I. Model examination and apparent reaction rate constants, *Polymer* 43 **(2002)**, pp. 1163–1170.

54 QIN, JIGUANG, GUO, WENPING, ZHANG, ZHENG, Modeling of the bulk free radical polymerization up to high conversion – three stage polymerization model II.

Model examination and apparent reaction rate constants, *Polymer* 43 (**2002**), pp. 4859–4867.
55 MITA, I., HORIE, K., Diffusion-controlled reactions in polymer systems, *Journal of Macromolecular Science – Reviews in Macromolecular Chemistry and Physics* C27 (**1987**), 91–169.
56 ODIAN, G., *Chain Copolymerization*, Chapter 4, 6th edn., John Wiley & Sons, Inc., New York, **1991**.
57 STORTI, G., CARRA, S., MORBIDELLI, M., VITA, G., Kinetics of multimonomer emulsion polymerization – the pseudo-homopolymerization approach, *Journal of Applied Polymer Science* 37 (**1989**), pp. 2443–2467.
58 SMITH, W. V., EWART, R. H., Kinetics of emulsion polymerization, *Journal of Chemical Physics* 16 (**1948**), pp. 592–599.
59 HAMIELEC, A. E., MACGREGOR, J. F., *Polymer Reaction Engineering*, Hanser Publishers, New York, 1983.
60 XIE, T. Y., HAMIELEC, A. E., Modeling free-radical copolymerization kinetics – evaluation of the pseudo-kinetic rate-constant method 1. Molecular-weight calculations for linear copolymers, *Makromolekulare Chemie – Theory and Simulations* 2 (**1993**), pp. 421–454.
61 MAYO, F. R., LEWIS, F. M., Copolymerization I: a basis for comparing the behavior of monomers in copolymerization, the copolymerization of styrene and methyl methacrylate, *Journal of the American Chemical Society* 66 (**1944**), pp. 1594–1601.
62 ALFREY, T., GOLDFINGER, G., Copolymerization of systems of three and more components, *Journal of Chemical Physics* 12 (**1944**), p. 322.
63 HOCKING, M. B., KLIMCHUK, K. A., Refinement of the terpolymer equation and its simple extension to two- and four-component systems, *Journal of Polymer Science Part A – Polymer Chemistry* 34 (**1996**), pp. 2481–2497.
64 CLAY, P. A., GILBERT, R. G., Molecular-weight distributions in free-radical polymerizations I. Model development and implications for data interpretation, *Macromolecules* 28 (**1995**), pp. 552–569.
65 CANU, P., RAY, W. H., Discrete weighted residual methods applied to polymerization reactions, *Computers and Chemical Engineering* 15 (**1991**), pp. 549–564.
66 ROGOSIC, M., MENCER, H. J., GOMZI, Z., Polydispersity index and molecular weight distributions of polymers, *European Polymer Journal* 32 (**1996**), pp. 1337–1344.
67 CROWLEY, T. J., CHOI, K. Y., Calculation of molecular weight distribution from molecular weight moments in free radical polymerization, *Industrial and Engineering Chemistry Research* 36 (**1997**), pp. 1419–1423.
68 YOON, W. J., RYU, J. H., CHEONG, C., CHOI, K. Y., Calculation of molecular weight distribution in a batch thermal polymerization of styrene, *Macromolecular Theory and Simulations* 7 (**1998**), pp. 327–332.
69 RIVERO, P., Calculation method of molecular weight averages in polymerization with chain-length-dependent termination, *Journal of Polymer Research* 11 (**2004**), pp. 309–315.
70 LAURENCE, R. L., GALVAN, R., TIRRELL, M. V., *Polymer Reaction Engineering*, Chapter 3, VCH Publishers, New York, **1994**.
71 VERROS, G. D., LATSOS, T., ACHILIAS, D. S., Development of a unified framework for calculating molecular weight distribution in diffusion controlled free radical bulk homo-polymerization, *Polymer* 46 (**2005**), pp. 539–552.
72 PARKER, T. S., CHUA, L. O., *Practical Numerical Algorithms for Chaotic Systems*, Springer, New York, **1989**.
73 PRESS, W. H., TEUKOLSKY, S. A., VETTERLING, W. T., FLANNERY, B. P., *Numerical Recipies in Fortran 77: The Art of Scientific Computing*, 2nd edn., Cambridge University Press, New York, **1992**.
74 SALIAKAS, V., CHATZIDOUKAS, C., KRALLIS, A., MEIMAROGLOU, D., KIPARISSIDES, C., Dynamic optimization of molecular weight distribution using orthogonal collocation on finite elements and fixed pivot methods: an experimental and theoretical investigation, *Macromolecular Reaction Engineering* 1 (**2007**), pp. 119–136.
75 KUMAR, S., RAMKRISHNA, D., On the solution of population balance equations by discretization 1. A fixed pivot technique, *Chemical Engineering Science* 51 (**1996**), pp. 1311–1332.
76 PONTES, K. V., MACIEL, R., EMBIRUCU, M., An approach for complete molecular

weight distribution calculation: application in ethylene coordination polymerization, *Journal of Applied Polymer Science* 109 (**2008**), pp. 2176–2186.

77 KERAMOPOULOS, A., KIPARISSIDES, C., Development of a comprehensive model for diffusion-controlled free-radical copolymerization reactions, *Macromolecules* 35 (**2002**), pp. 4155–4166.

78 VRENTAS, J. S., DUDA, J. L., LING, H. C., Self-diffusion in polymer–solvent–solvent systems, *Journal of Polymer Science Part B – Polymer Physics* 22 (**1984**), pp. 459–469.

79 ACHILIAS, D. S., A review of modeling of diffusion controlled polymerization reactions, *Macromolecular Theory and Simulations* 16(4) (**2007**), pp. 319–347.

80 SHEN, J. C., YUAN, T., WANG, G. B., YANG, M. L., Modeling and kinetic study on radical polymerization of methyl-methacrylate in bulk. 1. Propagation and termination rate coefficients and initiation efficiency, *Makromolekulare Chemie – Macromolecular Chemistry and Physics* 192(11) (**1991**), pp. 2669–2685.

4
Modeling and Control of Proton Exchange Membrane Fuel Cells

Christos Panos, Kostas Kouramas, Michael C. Georgiadis, and Efstratios N. Pistikopoulos

Keywords

proton exchange membrane (PEM) fuel cells, model-based control (MPC), membrane-electrode assembly (MEA) modeling, gas diffusion layer (GDL), electroosmotic drag, activation loss

4.1
Introduction

Fuel cells are a promising technology for electrical power generation, widely regarded as a potential alternative for stationary and mobile applications. Fuel cells are electrochemical devices that convert the chemical energy of a fuel to electrical energy. The electrical efficiency of the fuel cells is higher than the most conventional devices for power generation, since they avoid intermediate steps of production of mechanical energy. The transport sector is one of the major contributors to global fossil fuel consumption and carbon emissions. Alternative power devices for automotive applications have been actively studied over the last few years with great emphasis on fuel cells. The primary type of fuel cells for automotive industry application is proton exchange membrane (PEM) fuel cells, due to their suitable properties for vehicle applications such as low sensitivity to orientation favorable power to weight ratio and fast and easy start-up.

A typical fuel cell system consists of a PEM fuel cell stack, a compressor, humidifier, a cooling system to maintain the temperature of the stack, and a hydrogen storage tank (Fig. 4.1). Hydrogen is channeled in the anode side of the fuel cell while air in the cathode side. The compressor and the electric drive motor are used to achieve the desired air mass flow and pressure, while the humidifier has been used to achieve proper humidity of the air in order to minimize the danger of dehydration of the membrane. In addition, a recycling system for hydrogen is usually applied to the system in order to minimize the hydrogen consumption.

If fuel cell systems are to become a credible replacement of existing combustion engines, they must reach a similar level of performance and lifetime. Without taking into account a manufacturing and material costs, the main technical issues toward this goal are operability issues related to ground vehicle propulsion, which

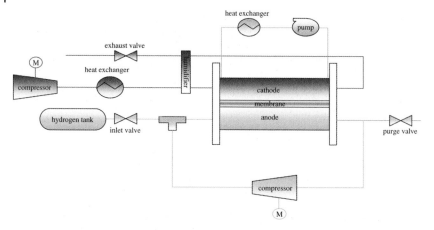

Fig. 4.1 Overall scheme of the fuel cell system.

pose a challenging control problem, especially due to the transient behavior of an integrated fuel cell system such as one shown in Fig. 4.1.

Model-based control (MPC) strategy is a suitable approach to obtain the optimal operation of the fuel cell system, due to its ability to control multiinput–multioutput systems with interactions and disturbances. However, MPC requires an analytical and accurate dynamic mathematical model of the system. The mathematical model should be experimentally validated in several operation conditions since the controller will be designed on this system.

In model predictive control (MPC), an open-loop optimal control problem is solved at regular intervals (sampling instants), given the current process measurements, to obtain a sequence of the current and future control actions up to a certain time horizon (in a receding horizon control fashion), based on the future predictions of the outputs and/or states obtained by using a mathematical representation of the controlled system. Only the first input of the control sequence is applied to the system and the procedure is repeated at the next time instant when the new data are available. Being an online constrained optimization method, MPC not only provides the maximum output of the controlled process but also takes into account the various physical and operational constraints of the system.

The benefits of MPC have long been recognized form the viewpoint of cost and efficiency of operations. Nevertheless, its applications maybe restricted due to increased online computational requirements related to the constrained optimization. In order to overcome this drawback, explicit or multiparametric model predictive control (mp-MPC) was developed [22, 25], which avoids the need for repetitive online optimization. In mp-MPC, the online optimization problem is solved offline with multiparametric programming techniques to obtain the objective function and the control actions as functions of the measured state/outputs (parameters of the process) and the regions in the state/output space where these parameters are valid, that is, as a complete map of the parameters. The control is then applied

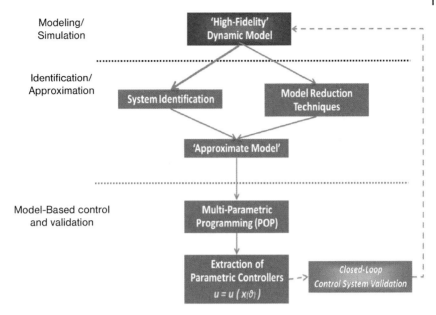

Fig. 4.2 Framework for design explicit/mp-MPC [23].

by means of simple function evaluations instead of typically demanding online optimization computations.

In this chapter, we focus on the mathematical modeling, dynamic optimization, and control issues of PEM fuel cell systems according to the framework showing in Fig. 4.2. The framework that has been used to obtain and validate the control design consists of four key steps [23]:

1. development of a high-fidelity mathematical modeling – used for detailed simulation and (design and operational) optimization studies;
2. development of a reduced order/approximating model, suitable for mp-MPC;
3. design of mp-MPC controllers;
4. validation of the controllers.

Step 1 involves the development of a high-fidelity mathematical model for performing detailed dynamic simulation and design/operational optimization studies. The model is validated using experimental data in several operation conditions in order to guarantee the accuracy of the simulation results. In step 2, a reduced order approximated model is derived by performing system identification or reduced order techniques on the simulation data. Step 3 corresponds to the design of the multiparametric/explicit mp-MPC by applying the available theory and tools of multiparametric programming and control [24, 25]. Finally step 4 involves the offline validation of the derived multiparametric/explicit controllers.

In this chapter, we present a systematic framework for the optimal design and advanced control of the PEM fuel cell system. A detailed mathematical model is first presented and dynamic simulation are performed based on which a reduced order state space (SS) model, suitable for the design of advanced model-based controllers, is derived. Finally, the controller introduced in the actual process and its performance is validated.

4.2
Literature Review

In order to use a fuel cell in an effective way, mathematical models are necessary to be able to analyze the system behavior depending on the system design and operating conditions. Mathematical modeling provides fuel cell designers and users the necessary tools to improve the fuel cell design, reduce the operation cost, and increase the lifetime. The model should predict accurately fuel cell performance under different operating conditions. The cost of savings is tremendous considering the cost of trial and error solutions and fuel cell materials. The models developed in the literature can be classified into three main categories, namely, detailed fuel cell models based on partial differential equations, steady-state fuel cell system based on experimental maps, and dynamic fuel cell system models that neglect spatial variations.

Most of the publications on fuel cell modeling were developed at the cell level and included spatial variations of the fuel cell parameters. Membrane-electrode assembly (MEA) modeling is indeed the base of the entire PEM fuel cell system modeling [21, 27]. Complex electrochemical, thermodynamic, and fluid dynamics principles were used to describe mathematically the entire physical environment of electrochemical reaction, the transport phenomena of gases, water, proton, and electron, as well as the relationships among fuel cell current, voltage, temperature, pressure, and materials (electrode, membrane, and catalyst). The performance of the fuel cell under different steady-state operating conditions can be determined with those models. Some models are based on 2D, steady-state fuel cell modeling [12, 13]. More complex approaches in 3D modeling have also been developed [2], and the main purposes of those detailed models are to design the fuel cell components and to choose the fuel cell operating conditions, but they require large computational calculations. Although they are not suitable for control studies, they establish the fundamental effects of operating parameters such as pressure and temperature on the fuel cell voltage.

Many models predict fuel cell polarization characteristics at different operating conditions. Amphlett *et al.* [3] attempted integrating together mechanistic model and empirical relation to derive advantages from both of them. They developed a simple dynamic model for a PEM fuel cell stack, which predicts fuel cell voltage and stack temperature for a given set of gas feed and operating conditions. Moreover, many publications addressed the water and thermal management of the fuel cell in order to describe the water, thermal, and reactant utilization of fuel cell by means of 2D or 3D models [9]. Fronk *et al.* [8] demonstrated the importance of thermal management while trying to maximize the performance of fuel cell stack used within a vehicle.

The system-level modeling is even more complicated than the MEA modeling and most of the models in the literature are based upon steady-state conditions. Steady-state models are typically used for component sizing, static tradeoff analysis, or cumulative fuel consumption [26]. The vehicle effective inertia is the only dynamic model considering in this type of models. Those models represent each component of the system such as the compressors, heat exchangers, or fuel cell stack voltage as an efficiency map. For instance, the model of Akella *et al.* [1] was used to study the tradeoff between maximum acceleration and auxiliary power sources.

Besides, a majority of the publications are not able to represent transient dynamic phenomena, whereas there are many of them in powertrains applications, such as the large variations underwent by the output power of the fuel cell during acceleration or deceleration. Pathapati *et al.* [21] developed a complete fuel cell system-level dynamic model, incorporating simultaneously three prominent dynamic aspects – the temperature changes of the fuel cell, fluid flow changes through channel, and capacitor effect of charged double layers, which is often neglected.

Despite the large number of developed models in the past 15 years, there is still open research area for the development of models, which are suitable for control and real-time estimation purposes. Current models are either too complex, hence too time consuming regarding computation, or they have not been sufficiently detailed to capture the multiplicity of the fuel cell behavior. This field is in intense development, since such a model is critical for future control development [7, 26]. The concurrence of the current evolution in design of fuel cell systems and of the integrated control techniques in microprocessor systems would allow the development of portable fuel cell applications in which optimized control of fuel cell performance is hence possible.

4.3
Motivation

The basic principle of a proton exchange membrane fuel cell (PEMFC) is very similar to that of a battery. It converts directly the chemical energy into electrical energy and has individual cells combined into stacks. However the difference is that a fuel cell operates as long as fuel is provided and does not need to be recharged, but their efficiency is slightly inferior due to losses of reactants. Fuel cells are characterized by the operating temperature range and the technology used for the membrane and electrolytic separation. For mid-power portable applications and low temperature, PEM fuel cells have become more attractive [6].

The fuel cell was discovered in 1839 by William R. Grove, a British physicist. A PEM fuel cell is composed of two porous carbon electrodes (anode and cathode) separated by a solid polymer electrolyte: the ion-conducting PEM and a thin layer of platinum catalyst are integrated between each electrode and the membrane. The electrodes, catalyst, and membrane together form the MEA. The bipolar flow field plates, which contain the gas channels, are placed on each side of the MEA. The electrodes are connected to an external load circuit.

Both electrodes are composed of a gas diffusion layer (GDL) and a catalyst layer (CATL). In the GDLs, the reactants gases are fed and distributed to the catalyst, and then after reaction, the reaction products (water) are removed, whereas the electrochemical reactions take place in the CATLs [11, 12, 26].

Hydrogen is supplied to the cell through channels in the collection plate on the anode side of the electrolyte membrane and diffuses through the GDL to reach the surface of the electrolyte membrane, where dihydrogen molecules are broken into protons and electrons owing to the action of the catalyst layer. The protons flow from the anode to the cathode through the membrane (proton conductor), while the electrons are transported to the cathode through the external electric circuit, generating electrical work at an external load. The reaction on the anode side can be written as

$$H_2 \rightarrow 2H^+ + 2e^-$$

Similarly, oxygen flows through channels in the collection plate on the cathode side of the electrolyte membrane. Protons and electrons react with oxygen molecules to produce water, and the produced water is removed from the cell through the cathode channels that supply the cell with oxygen. The reaction on the cathode side can be written as

$$\frac{1}{2}O_2 + 2e^- + 2H^+ \rightarrow H_2O$$

Finally the overall reaction of a fuel cell is

$$H_2 + \frac{1}{2}O_2 \rightarrow H_2O$$

PEM fuel cells have the great advantage to generate electricity directly from chemical components. The produced electricity can then be used to power electric engines and other electrical devices. The losses depend on the fuel cell system operating conditions and parameters such as pressure, temperature, water content, and current density. For instance, the resistance in the membrane depends on the water content, which in turn influences the proton conductivity of the membrane.

A fuel cell can be characterized by its polarization curve, which is a plot of cell voltage versus cell current density. Figure 4.3 shows a typical polarization curve for a PEM fuel cell. The fuel cell voltage decreases with the current density drawn from the fuel cell, and thus low-current operations are preferred. However constant operation at low load is not practical in applications, where frequent load changes are required, such as in automobile applications [26].

The losses can be attributed to three mechanisms, each one affecting different part of the polarization curve [26].

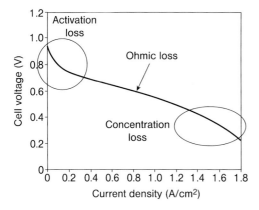

Fig. 4.3 Typical PEM fuel cell polarization curve [26].

- The activation loss or activation overvoltage is due to the energy barrier of reactions. It occurs in driving the chemical reaction that transfers electrons from and to the electrodes. The need to move electrons, and to break and form chemical bonds in the anode and cathode, consumes a part of the available energy. The reaction of oxidation hydrogen is very quick, unlike the oxygen reduction at the cathode. Therefore, the voltage drop related to the activation loss is dominated by the cathode reaction conditions. The activation overvoltage is identifiable at low current densities, where there is no other mechanism of losses (Fig. 4.3).
- The ohmic loss arises from the resistance of the polymer membrane to the transfer of protons and the resistance of the electrodes and collector plates to the transfer of electrons. The ohmic voltage drop is thus proportional to the current density drawn from the system.
- Concentration loss or concentration overvoltage results from the diminution in concentration of the reactants as they are consumed in the reaction. Rapid drop voltage due to transport limitations occurs at high current density (Fig. 4.3).

The fuel cells can be assembled in series or parallel circuits to produce the desired amount of energy, where series yield higher voltage and parallel allows a stronger current to be drawn (*fuel cell stack*).

The fuel cell stack is integrated to several auxiliary components to form a complete fuel cell system. At the minimum, the fuel cell stack requires four flow systems: hydrogen supply to the anode, air supply system to the cathode, deionized water serving as coolant in the stack cooling channel, and deionized water to the humidifier to humidify the hydrogen and the air flows. The reactant flow, the stack temperature and pressure, and the membrane humidity are critical parameters, which require a precise control to the viability, efficiency, and robustness of the fuel cell propulsion systems. The overall system can be partitioned into subsystems, each one interacting and conflicting with the objectives of the other subsystems. This complex system has been the source of various model analysis and model-based control.

4.3.1
Reactant Flow Management

The reactant flow subsystem controls the hydrogen supply and air supply loops. As the vehicle traction motor draws current, hydrogen and oxygen are depleted in the fuel cell stack, so that sufficient reactant flows have to be provided. In case of starvation, the performance decreases and safety issue for the fuel cells can occur. An excess ratio of oxygen with respect to hydrogen generally allows reaching a higher efficiency [5].

Finally the control objective is to provide sufficient reactant flow to ensure fast and safe transient responses and to minimize auxiliary consumption. Indeed the cost, space, weight, and lost power have to be taken into account for the compressor equipment. But quantitatively, the increased pressure raises the exchange current density, which has the effect of lifting the open circuit voltage, described by the Nernst equation. The voltage boost would be

$$\Delta V = \frac{RT}{4F} \ln\left(\frac{P_2}{P_1}\right)$$

That is why it seems interesting to add a compressor for reaching more important power. However, the increase in pressure will entail an increase in temperature as shown in the following expression:

$$T_{out} = T_{atm} + \frac{T_{atm}}{\eta_{cp}} \left[\left(\frac{p_{out}}{p_{atm}}\right)^{\frac{\gamma-1}{\gamma}} - 1\right]$$

where η_{cp} is the efficiency of the compressor and γ the ratio of specific heats of air.

The reactant temperature at the inlet plays an important role in the FC stack temperature management, so in some feeding processes a cooler is added for the cathode inlet in order to have a not too high temperature.

4.3.2
Heat and Temperature Management

The heat and temperature subsystem includes the fuel cell stack cooling system and the reactant temperature system. The reaction is highly exothermic so that as current is drawn by the traction motor, a large quantity of heat is generated within the fuel cell. A passive dissipation by air convection and radiation through the external surface of the stack would not be sufficient and a cooling is required (air or water). Air requires larger volumes; hence water cooling is preferred in the majority of the PEM fuel cell systems.

Moreover, vehicles are routinely subjected to subfreezing temperatures. Nafion's conductivity drops off faster in subfreezing temperatures than it does in above freezing temperatures. This significantly reduces the performance of a fuel cell in these extreme conditions. In fact, cold temperatures have been known to significantly reduce the fuel cell life. That is why transient thermal models are useful for understanding and optimizing stack warm-up to both protect the membranes and provide the power that is required to power a vehicle in extreme conditions.

4.3.3
Water Management

The amount of reactant flow, the current drawn in the stack, and the water injected into the anode and cathode flow streams affect the humidity of the membrane. An insufficient or excessive humidification of the membrane can lead to high polarization losses, 20–40% drop in the voltage. Therefore, the water management to maintain a proper hydration of the polymer membrane and to balance water consumption in the system is essential. The amount of water molecules crossing the membrane is a key factor, as the proton conductivity in Nafion is proportional to the membrane water content. As the current is drawn from the fuel cell, water molecules are produced in the cathode as reaction product and their concentration increases. However, the concentration gradient causes water to diffuse from the cathode to the anode side. This phenomenon is called *back-diffusion* and would be sufficient to moisturize the thin membrane. On the other hand, as current is drawn from the fuel cell, water molecules are also dragged from the anode to the cathode by the hydrogen protons and this phenomenon is called *electro-osmotic drag*. At high current densities, it will lead to the drying out of the anode side even if the cathode is well hydrated.

Additionally for temperature above 60 °C, the air will dry out by evaporation faster than the water is produced. In case of drying out, the resistance across the membrane increases, and eventually the fuel cell can be damaged. Therefore, the water content of the membrane has to be controlled to ensure water to be evaporated at precisely the same rate that it is produced and to avoid the two extreme cases which both lead to significant polarization losses.

4.4
PEM Fuel Cell Mathematical Model

The PEM fuel system model is described analytically in this section. The system under consideration is a PEM fuel cell stack, a compressor, and a cooling system to maintain the temperature of the stack. Hydrogen is channeled in the anode side of the fuel cell while air in the cathode side. The compressor and the electric drive motor are used to achieve the desired air massflow and pressure, while the humidifier has been used to achieve proper humidity of the air in order to minimize the danger of dehydration of the membrane. In addition, a recycling system for hydrogen is applied to minimize the hydrogen consumption.

The proposed model involves three main modules – fluid dynamics model, thermodynamic model, and electrochemical static model. The theoretical equations are combined with experimental/empirical formulations, resulting in a semiempirical dynamical model. Fluid dynamics model is composed of three interconnected modules; anode and cathode flow stream, and the membrane. Thermodynamic model is used to determine the homogeneous temperature of the fuel cell and to design the appropriate cooling system, while electrochemical static model, described in the following section, used to predict stack voltage. The model assumptions are discussed as the equations are presented and a summary of those are presented below:

- ideal gas law was employed for gaseous species;
- laminar flow of the gases;
- isotropic and homogeneous electrolyte, electrode, and bipolar material structures;
- negligible ohmic potential drop in components;
- mass and energy transport is modeled from a macroperspective using volume-averaged conservation equations;
- the product water generated at the cathode is assumed to be in the liquid state;
- uniform stack temperature of the fuel cell due to high thermal conductivity;
- uniform pressure of the stack and anode, cathode;
- constant thermophysical properties;
- current does not change throughout the stack.

The 1D mathematical model, developed in this work, includes mass balances for the anode and cathode side and recirculation, semiempirical equations for the membrane, electrochemical equations, heat balances for the fuel cell, and mass and energy equation for the humidifier, compressor, and the cooling system.

4.4.1
Cathode

The cathode mathematical model describes the air flow behavior in the cathode side of the fuel cell. The equations used to obtain the inlet flow rates as well as the mass balances are the most generally used in the literature. However, the equations of Pukrushpan et al. [26] have been used, particularly to describe the basic model of the mass balances. The model is designed using the mass conservation principle, physicochemical and thermodynamic properties.

The inlet mass flow rates of the three elements, namely oxygen, nitrogen, and vapor, can be calculated using the humidity ratio. Considering a mixture of air and water vapor, the humidity ratio is defined as the ratio of the mass of water vapor to the mass of dry air:

$$w_{v,\text{ca,in}} = \frac{M_{v,H_2O}}{M_{\text{air}}} \frac{\varphi_{\text{ca,in}} p_{\text{sat}}(T_{\text{ca,in}})}{p_{\text{ca,in}} - \varphi_{\text{ca,in}} p_{\text{sat}}(T_{\text{ca,in}})} \qquad (4.1)$$

where ϕ is the relative humidity, p_{sat} the pressure of vapor saturation, M_i (i is O_2, N_2, dry air) the molar mass of the component, and $T_{\text{ca,in}}$ the inlet temperature,

$$\dot{m}_{O_2,ca,in} = x_{O_2,ca,in} \frac{1}{1+w_{v,ca,in}} \dot{m}_{ca,in} \tag{4.2}$$

$$\dot{m}_{N_2,ca,in} = x_{N_2,ca,in} \frac{1}{1+w_{v,ca,in}} \dot{m}_{ca,in} \tag{4.3}$$

$$\dot{m}_{v,ca,in} = 1 - \frac{1}{1+w_{v,ca,in}} \dot{m}_{ca,in} \tag{4.4}$$

with

$$x_{O_2,ca,in} = \frac{m_{O_2,ca,in}}{m_{dry\,air}} = \frac{y_{O_2,ca,in} M_{O_2}}{y_{O_2,ca,in} M_{O_2} + (1 - y_{O_2,ca,in}) M_{N_2}} \tag{4.5}$$

assuming that the dry air mass fraction is equal to the ambient air mass fraction, $x_{N_2,ca,in} = 1 - x_{O_2,ca,in}$ and $x_{O_2,ca,in} = 0.21 M_{O_2} M_{air}^{-1}$ for the air inlet, where $m_{i,ca}$ is the mass flowrate in cathode side (i is O_2, N_2, vapor, dry air), $x_{i,ca,in}$ is molar fraction in the inlet of channel of cathode side (i is O_2, N_2), and $y_{O_2,ca,in}$ is the oxygen molar fraction.

The mass continuity equation is used to balance the mass of the elements inside the cathode

$$\frac{dm_{O_2,ca}}{dt} = \dot{m}_{O_2,ca,in} - \dot{m}_{O_2,ca,out} - \dot{m}_{O_2,reacted} \tag{4.6}$$

$$\frac{dm_{N_2,ca}}{dt} = \dot{m}_{N_2,ca,in} - \dot{m}_{N_2,ca,out} \tag{4.7}$$

$$\frac{dm_{v,ca}}{dt} = \dot{m}_{v,ca,in} - \dot{m}_{v,ca,out} + \dot{m}_{v,memb} + \dot{m}_{evap,ca} \tag{4.8}$$

where $m_{O_2,react}$ is the reacted oxygen, $m_{v,memb}$ is the water mass flowrate across the membrane, m_{evap} is the evaporation mass, and $m_{H_2,react}$ is reacted mass of hydrogen.

It is assumed that no liquid water carried by the inlet air enters the cathode channel according to the suggestion of Del Real et al.; thus $m_{l,in} = 0$. The water produced from the reaction assumed to be in liquid form and has to be evaporated, partially or totally depending on the cathode operating conditions [31]. Consequently, the mass balance of the liquid can be described by the following equation:

$$\frac{dm_{l,ca}}{dt} = \dot{m}_{l,ca,gen} - \dot{m}_{evap,ca} \tag{4.9}$$

where $m_{l,ca,gen}$ is the generated mass of liquid water.

The water quantity obtained through this equation is supposed to be dragged out by the air exhaust.

The dynamic equation for the evaporation inside the cathode channel can be described by the following equation as Del Real et al. [7] have been proposed:

$$\dot{m}_{evap,ca} = \min\left(A_{fc}(p_{sat}(T_{st}) - p_{v,ca})\sqrt{\frac{M_v}{2\pi R T_{st}}}, \dot{m}_{l,ca,gen}\right) \tag{4.10}$$

where A_{fc} is the active area, R is the ideal gas constant, and $p_{v,ca}$ the vapor pressure in the cathode channel.

When the stream pressure is smaller than the saturation pressure, the water is evaporated and the value is positive, while water condenses when the value is negative. As it has been assumed that no liquid water is brought by the air inlet, it is not possible to evaporate more water than produced by the chemical reaction.

Using the ideal gas law, the partial pressure of oxygen, nitrogen, and water vapor inside the cathode channel has been calculated as follows:

$$p_{i,ca} = \frac{R_i T_{st}}{V_k} m_{i,ca}, \quad \text{with } i = O_2, N_2, v \tag{4.11}$$

where $p_{i,ca}$ is the partial pressure and R_i the gas constant of each component (i is O_2, N_2, vapor pressure), and V_k is the volume of the cathode side.

The total cathode pressure can be calculated by adding the partial pressures using the Dalton law,

$$p_{ca} = \sum p_{i,ca} \tag{4.12}$$

Electrochemical principles are used to calculate the rate of oxygen consumption and water production in the fuel cell, and they are functions of the stack current (I_{st}) and the number of cells (N_{fc}),

$$\dot{m}_{O_2,react} = \frac{N_{fc} M_{O_2} I}{4F} \tag{4.13}$$

$$\dot{m}_{l,ca,gen} = \frac{N_{fc} M_v I}{2F} \tag{4.14}$$

where F is the Faraday constant.

The outlet massflow rates for oxygen, nitrogen, and water vapor can be calculated from Eqs. (4.15)–(4.18) as Del Real et al. [7] have been proposed while the assumption of the presence of a valve (back pressure valve) at the air outlet holds. Pukrushpan et al. [26] are using the humidity ratio as used for the inlet flow rates to compute the outlet flowrate, but this has been tested without satisfying results. Equation (4.15) assumes an isentropic expansion and considers the valve as a nozzle with a convergent and a divergent section. The section A_t represents the opening of the valve and will be controlled to regulate the pressure inside the cathode side of the fuel cell:

$$\dot{m}_{ca,out} = A_t \sqrt{2\frac{\gamma}{\gamma-1}\frac{p_{ca}}{v_{air}}\left[\left(\frac{p_{out}}{p_{ca}}\right)^{\frac{2}{\gamma}} - \left(\frac{p_{out}}{p_{ca}}\right)^{\frac{\gamma+1}{\gamma}}\right]}$$

with $P_{out} = P_{atm}$ (4.15)

$$\dot{m}_{O_2,ca,out} = \frac{m_{O_2,ca}}{m_{O_2,ca} + m_{N_2,ca} + m_{v,ca}}\dot{m}_{ca,out} \quad (4.16)$$

$$\dot{m}_{N_2,ca,out} = \frac{m_{N_2,ca}}{m_{N_2,ca} + m_{O_2,ca} + m_{v,an}}\dot{m}_{ca,out} \quad (4.17)$$

$$\dot{m}_{v,ca,out} = \frac{m_{v,ca}}{m_{O_2,ca} + m_{N_2,ca} + m_{v,ca}}\dot{m}_{ca,out} \quad (4.18)$$

where P_{out} is the outlet pressure and v_{air} is the air-specific volume (m^3 kg^{-1}).

4.4.2
Anode

The fuel cell system under consideration assumes storage of hydrogen in metal hydride bed reactor. The optimal operation of the storage tank has significant role in the operation of fuel cell since the pressure and the temperature of the tank have to be regulated in order to achieve the demanded hydrogen mass flowrate. Since the scope of this chapter is not to investigate the modeling and control of the metal hydride tank, we will focus on the fuel cell system. Details of this topic can be found in Kikkinides et al. [15]. The difference between the pressure of the anode and cathode channel should be regulated; hence a valve is needed to adjust the pressure of the fuel and drops it to close to cathode channel. An assumption of pressure drop less than 100 mbar will be considered for the anode channel, which comes in agreement with the supplier specifications. Similar to a cathode model, a dynamic model for the anode channel is described in this section. This model includes dynamic mass balances, partial pressure equations, outlet flow of the anode side, and the recirculation of the unreacted hydrogen.

The equations that model the anode flow are analogous to the ones that model the cathode flow. The mass balance is then expressed as follows:

$$\frac{dm_{H_2,an}}{dt} = \dot{m}_{H_2,an,in} - \dot{m}_{H_2,an,out} - \dot{m}_{H_2,reacted} \quad (4.19)$$

$$\frac{dm_{v,an}}{dt} = \dot{m}_{v,an,in} - \dot{m}_{v,an,out} - \dot{m}_{v,memb} + \dot{m}_{evap,an} \quad (4.20)$$

$$\frac{dm_{l,an}}{dt} = \dot{m}_{l,an,in} - \dot{m}_{l,an,out} - \dot{m}_{evap,an} \quad (4.21)$$

where $\dot{m}_{i,an}$ is the mass flowrate in anode side (i is H$_2$, vapor, liquid), $\dot{m}_{H_2,react}$ is the reacted hydrogen, $\dot{m}_{v,memb}$ is the water mass flowrate across the membrane, and $\dot{m}_{evap,an}$ is the evaporation mass.

The assumption of no liquid water is considered at the inlet of the anode channel, which leads to the conclusion that no liquid water is considered after the recirculation loop using a water separator just before the inlet and that the hydrogen coming from the tank is supposed to be dry ($\dot{m}_{l,an,in} = 0$). Moreover, it is assumed that all the liquid water has been removed by the purge valve through the following equation according to Del Real et al. [7]:

$$\dot{m}_{l,an,out} = \frac{m_{l,an}}{t_{purge}} \tag{4.22}$$

Equations for humidity ratio, inlet flow rates, and the evaporation mass of the anode channel are similar to the cathode:

$$w_{v,an,in} = \frac{M_{v,H_2O}}{M_{H_2}} \frac{\varphi_{an,in} p_{sat}(T_{an,in})}{p_{an,in} - \varphi_{an,in} p_{sat}(T_{an,in})} \tag{4.23}$$

$$\dot{m}_{H_2,an,in} = \frac{1}{1 + w_{v,an,in}} \dot{m}_{an,in} \tag{4.24}$$

$$\dot{m}_{v,an,in} = \frac{1}{1 + w_{v,an,in}} \dot{m}_{an,in} \tag{4.25}$$

$$\dot{m}_{evap,an} = \min\left(A_{fc}(p_{sat}(T_{st}) - p_{v,anch})\sqrt{\frac{M_v}{2\pi R T_{st}}}, 0\right) \tag{4.26}$$

where $p_{v,an}$ is the vapor pressure of the anode channel.

Given that no liquid water enters the anode channel and the water coming from the cathode through the membrane is only vapor, evaporation inside the anode has been limited, allowing only condensation.

The partial pressures inside the channel are also obtained using the ideal gas law

$$p_{i,an} = \frac{R_i T_{st}}{V_k} m_{i,an}, \quad \text{with } i = H_2, v \tag{4.27}$$

and the Dalton law

$$p_{an} = \sum p_{i,an} \tag{4.28}$$

The amount of hydrogen reacted inside the fuel cell is a function of the stuck current and the number of cells,

$$\dot{m}_{H_2,react} = \frac{N_{fc} M_{H_2} I}{2F} \tag{4.29}$$

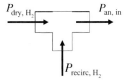

Fig. 4.4 Flows through the T.

4.4.3
Anode Recirculation

There are not many descriptions of the recirculation loop (Fig. 4.4) for the hydrogen in the open literature [4], and the equations have been suggested trying to best fit to the actual behavior in the actual fuel cell. The inlet flow of the anode is obtained by mixing the flow coming from the recirculation and the flow of dry hydrogen coming from the tank inside the T. The pressure at the inlet is calculated using the ideal gas law as follows:

$$P_{an,in} = \frac{(n_{dry,H_2} + n_{recirc,H_2})T_{an,in}}{n_{dry,H_2}\frac{T_{dry,H_2}}{P_{dry,H_2}} + n_{recirc,H_2}\frac{T_{recirc,H_2}}{P_{recirc,H_2}}} \qquad (4.30)$$

where P_i is the pressure, n_i is the moles, and T_i the temperature of the stream i (i is anode inlet, dry hydrogen, and recirculation stream).

The pressure inside the recirculation system has to be higher than the pressure at the inlet in order to avoid counterflow inside the recirculation system. The anode inlet pressure has to be controlled considering cathode channel pressure in order to avoid different pressures of the two inlet streams. Between the tank of dry hydrogen and the inlet anode channel, there is a valve and the mass flowrate coming from the tank, which can be calculated as follows [29]:

$$\dot{m}_{dry,H_2} = A_t \sqrt{2\frac{\gamma}{\gamma-1}\frac{P_{tank}}{v_{H_2}}\left[\left(\frac{P_{dry,H_2}}{P_{tank}}\right)^{\frac{2}{\gamma}} - \left(\frac{P_{dry,H_2}}{P_{tank}}\right)^{\frac{\gamma+1}{\gamma}}\right]} \qquad (4.31)$$

Consequently, the inlet mass flowrate in the anode channel can be calculated by the sum of the massflow rate of the recirculation stream and the hydrogen coming from the metal hydride tank

$$\dot{m}_{an,in} = \dot{m}_{dry,H_2} + \dot{m}_{recirc,H_2} \qquad (4.32)$$

The temperature of this inlet stream is calculated considering a homogeneous combination of both flows. In order to simplify the model, both flows are considered to be pure hydrogen and the specific heat capacities have been assumed constant (i.e., nondependent on the temperature):

$$\int_{T_1}^{T} \dot{m}_1 cp_1 \, dt + \int_{T_2}^{T} \dot{m}_2 cp_2 \, dt = 0 \qquad (4.33)$$

which results to the following equation:

$$T_{an,in} = \frac{\dot{m}_{dry,H_2} c_{p_{dry,H_2}} T_{dry,H_2} + \dot{m}_{recirc,H_2} c_{p_{recirc,H_2}} T_{recirc,H_2}}{\dot{m}_{dry,H_2} c_{p_{dry,H_2}} + \dot{m}_{recirc,H_2} c_{p_{recirc,H_2}}} \qquad (4.34)$$

The temperature of the recirculation flow is assumed to be equal to the temperature of the stack, neglecting a loss of heat in the recirculation pipe.

4.4.4
Fuel Cell Outlet

The anode outlet flowrate $m_{an,out}$ represents the purge of anode side in order to remove accumulated gases and liquid water for the anode side. In the designed PEM, fuel cell system assumed that purge is equal to 0.5 but could vary within the range of zero and one, depending on what specification we want to achieve. With the knowledge of the outlet flowrate of the anode channel, the mass flowrate of the hydrogen and vapor could be calculated similar to the inlet flow:

$$\dot{m}_{H_2,out} = \frac{m_{H_2,an}}{m_{H_2,an} + m_{v,an}} \dot{m}_{an,out} \qquad (4.35)$$

$$\dot{m}_{v,out} = \frac{m_{v,an}}{m_{H_2,an} + m_{v,an}} \dot{m}_{an,out} \qquad (4.36)$$

The outlet flowrate of the stack is determined using an outlet manifold equation

$$\dot{m}_{an,out} = K_{an}(P_{an} - P_{om}) \qquad (4.37)$$

where K_{an} is the outlet manifold coefficient and P_{om} is the anode outlet manifold pressure.

The purge valve added at the outlet of the anode channel is supposed to remove all the liquid from the recirculation stream. The water flowrate through the fuel cell membrane is computed in the following section by the membrane hydration model.

4.4.5
Membrane Hydration Model

The hydration model determines the rate of mass flowrate across the membrane and the water content of the membrane. Semiempirical equations have been proposed to describe the phenomena with different constant coefficients. The following model is the one suggested by Pukrushpan et al. [26] and Müller and Stefanopoulou [18], and proved to represent the real conditions of the membrane. The membrane only allows the transport of vapor water; hence the following equations are only considering gaseous water.

The water transport across the membrane is achieved through two distinct phenomena. Water electro-osmotic drag and back diffusion from the cathode contribute to water transport in the membrane. The former occurs when protons migrate through the membrane from the anode to the cathode and carry water molecules with them. The water dragged from the anode to the cathode runs proportionally to the proton flow and thus this phenomenon increases at higher current density. The latter originates from the diffusion of water through the membrane, driven by the water-concentration gradient in the thickness profile. The phenomenon of water back diffusion across a membrane from the cathode to the anode usually predominates owing to the water produced at the cathode. The combination of those two phenomena results in the following equation, which determines the water flow across the membrane:

$$N_{v,\text{memb}} = nd \frac{I}{A_{\text{fc}} F} - D_w \frac{c_{v,\text{ca}} - c_{v,\text{an}}}{\delta_{\text{memb}}} \qquad (4.38)$$

where I is the stuck current, D_w is the diffusion coefficient of the water in membrane, δ_{memb} is the thickness of the membrane, and $c_{v,\text{ca}}$ and $c_{v,\text{an}}$ are the water concentration in cathode and anode channel, respectively.

The first term represents the net water flow from anode to cathode caused by electro-osmotic drag. The second one represents the net water flow from the cathode to the anode. Thus if $N_{v,\text{memb}}$ is negative, the general transport occurs from the cathode to the anode and opposite if it is positive.

The total stack mass flow rate across the membrane can be computed by the following equation:

$$\dot{m}_{v,\text{memb}} = N_{v,\text{memb}} M_v A_{\text{fc}} N_{\text{fc}} \qquad (4.39)$$

where M_v is the vapor molar mass, N_{fc} is the number of fuel cells, and A_{fc} is the fuel cell active area.

The water concentration at the membrane surfaces on anode and cathode sides depends on the membrane water content and can be obtained using the following equation:

$$c_{v,k} = \frac{\rho_{\text{memb,dry}}}{M_{\text{memb,dry}}} \lambda_k \qquad (4.40)$$

where $\rho_{\text{memb,dry}}$ is the membrane dry density and $M_{\text{memb,dry}}$ is the membrane dry equivalent weight.

The water content in the membrane is defined as the ratio of water molecules to the number of charge sites. These equations are developed based on experimental results:

$$\lambda_k = \begin{cases} 0.043 + 17.81 a_k - 39.85 a_k^2 + 36.0 a_k^3, & 0 < a_k \leq 1 \\ 14 + 1.4(a_i - 1), & 1 < a_k \leq 3 \end{cases} \qquad (4.41)$$

where a_i is the water activity.

The membrane water content can be calculated using the activities of the gas in cathode and anode side, and the average water activity between anode and cathode water activities is calculated below:

$$a_k = \frac{p_{v,k}}{p_{sat,k}} \quad \text{and} \quad a_{memb} = \frac{a_{an} + a_{ca}}{2} \tag{4.42}$$

The electro-osmotic drag coefficient n_d and the water diffusion coefficient D_w are then calculated using the membrane water content [7]

$$D_w = D_{\lambda_{an}} \exp\left(2416\left(\frac{1}{303} - \frac{1}{T_{st}}\right)\right) \tag{4.43}$$

where

$$D_{\lambda_{an}} = \begin{cases} 10^{-6}, & \lambda_{an} < 2 \\ 10^{-6}(1 + 2(\lambda_{an} - 2)), & 2 \leqslant \lambda_{an} \leqslant 3 \\ 10^{-6}(3 - 1.67(\lambda_{an} - 3)), & 3 < \lambda_{an} < 4.5 \\ 1.25 \times 10^{-6}, & \lambda_{an} \geqslant 4.5 \end{cases} \tag{4.44}$$

and

$$n_d = 0.0029\lambda_{an}^2 + 0.05\lambda_{an} - 3.4 \times 10^{-19} \tag{4.45}$$

where T_{st} is the stack temperature.

4.4.6
Electrochemistry

In this section, the steady-state mathematical modeling of the fuel cell voltage is explained. The voltage is a function of stack current, hydrogen and oxygen partial pressure, fuel cell temperature, and membrane humidity ratio. Nernst equation has been used to calculate the stack voltage, which includes the Nernst potential, activation losses, ohmic losses, and concentration losses:

$$V_{st} = N_{fc}(E_{Nernst} - V_{act} - V_{ohm} - V_{conc}) \tag{4.46a}$$

$$E_{Nernst} = 1.23 - 8.5 \times 10^{-4}(T - 298)$$

$$+ \frac{RT}{2F}\left[\ln(p_{H_2}) + 0.5\ln(p_{O_2})\right] \tag{4.46b}$$

Activation loss or activation overvoltage occurs because of the need to move electrons and to break chemical bonds in the anode and cathode. Part of available energy is lost in driving chemical reactions that transfer electrons to and from electrodes. However, the oxidation of H_2 at the anode is very rapid and the reduction of O_2 at the cathode is slow, and then the voltage drop due to activation loss is dominated by the cathode reaction:

$$V_{act} = \xi_1 + \xi_2 T + \xi_3 T \ln(I) + \xi_4 T \ln(C^*_{O_2})$$

$$\text{with } C^*_{O_2} = \frac{p_{O_2,ca}}{5.08 \times 10^6 \exp(\frac{-498}{T_{st}})} \tag{4.47}$$

Ohmic loss arises from resistance of polymer membrane to the transfer of proton and the resistance of the electrode and collector plate to the transfer of electron. Resistance depends on the membrane humidity and cell temperature:

$$V_{ohm} = r_{int} I = (\xi_5 + \xi_6 T + \xi_7 I) I \tag{4.48}$$

Finally *concentration loss* results from the drop in concentration of the reactants as they are consumed in the reaction. This explains the rapid voltage drop at high current density:

$$V_{conc} = -\frac{RT}{2F} \ln\left(1 - \frac{I}{I_{lim}}\right) \tag{4.49}$$

The calculation of the stack voltage includes semiempirical equations; thus many authors have found different values for the coefficients ξ [3, 20, 21]. The chosen coefficients have been taken from [5], defined for fuel cell Ballard Mark V 5 kW.

The gross power produced by the fuel cell is also obtained as follows:

$$P_{st} = V_{st} I_{st} \tag{4.50}$$

To obtain the net power, the power consumed by the auxiliary equipments has to be removed:

$$P_{net} = P_{st} - P_{aux} \tag{4.51}$$

It has to be mentioned here that the compressor is the most power-consuming equipment, using almost 2 kW in average.

4.4.7
Thermodynamic Balance

The temperature significantly affects the performance in the fuel cell by influencing the water removal and reactants activity. Heat is generated by the operation of the fuel cell since the enthalpy that is not converted to electrical energy will instead be converted to thermal energy. Thus the temperature will rise beyond the operating temperature range of the fuel cell and must be kept in the appropriate range using a cooling system.

Two main assumptions are introduced in the heat balances: (i) any fuel energy that is not converted into electrical energy is converted into heat [16] and (ii) the temperatures at the anode and cathode side are equal to the fuel cell stack temperature (uniform temperature). The energy model is using the main terms of the overall energy balance using the following principle:

$$\begin{pmatrix} \text{Energy} \\ \text{Accumulation} \end{pmatrix} = \begin{pmatrix} \text{Inlet} \\ \text{Energy} \end{pmatrix} - \begin{pmatrix} \text{Outlet} \\ \text{Energy} \end{pmatrix} + \begin{pmatrix} \text{Generated} \\ \text{Energy} \end{pmatrix}$$

Mathematical model for fuel cell system with energy balances has been presented by del Real et al. [7], as well as by Müller and Stefanopoulou [18]. The mathematical model for radiation has been suggested by Wishart et al. [31] where the latent heat has been calculated as follows:

$$m_{st} C p_{st} \frac{dT_{st}}{dt} = \dot{Q}_{in} - \dot{Q}_{out} - \dot{Q}_{chem}$$

$$- \dot{Q}_{elec} - \dot{Q}_{cool} - \dot{Q}_{rad} - \dot{Q}_{latent} \tag{4.52}$$

The inlet and outlet heat flowrate by the chemicals elements entering the fuel cell can be calculated by the following equations:

$$\dot{Q}_{in} = \sum_i \dot{m}_{i,in} C p_i T_{in} \tag{4.53}$$

$$\dot{Q}_{out} = \sum_i \dot{m}_{i,out} C p_i T_{st} \tag{4.54}$$

The generated energy by chemical reactions (Q_{chem}), radiation energy (Q_{rad}), the latent energy (Q_{latent}) due to change of water phase, and the heat flowrate in the form of electricity (Q_{elec}) in the PEM fuel cell stack can be calculated as follows:

$$\dot{Q}_{chem} = \dot{m}_{H_2Ogen} \left(\Delta_r H^o(T^0) + C p_{H_2O}(T_{st} - T^0) \right)$$

$$- \dot{m}_{O_2 react} C p_{O_2}(T_{in} - T^0) - \dot{m}_{H_2 react} C p_{H_2}(T_{in} - T^0) \tag{4.55}$$

$$\dot{Q}_{elec} = V_{st} I_{st} \tag{4.56}$$

$$\dot{Q}_{rad} = \varepsilon \sigma A_{rad} \left(T_{st}^4 - T_{amb}^4 \right) \tag{4.57}$$

$$\dot{Q}_{latent} = M_{H_2O} \cdot \dot{m}_{evap} H_{vaporization} \tag{4.58}$$

$$H_{vaporization} = 45070 - 41.9 T_{st} + 3.44 \times 10^{-3} T_{st}^2$$

$$+ 2.54 \times 10^{-6} T_{st}^3 - 8.98 \times 10^{-10} T_{st}^4 \tag{4.59}$$

where ΔH_r^o is the mass-specific enthalpy of formation of liquid water, $C p_{H_2O}$, $C p_{O_2}$, $C p_{H_2}$ are the thermal capacity of water, oxygen, and hydrogen, respectively, ε is the emissivity, σ is the Stefan–Boltzmann constant, A_{rad} is the radiation exchange area, $H_{vaporization}$ is the vaporization enthalpy, T_{amb} is the ambient temperature, T_{in} is the inlet temperature of the reactants, and T^0 is the reference temperature (298 K).

Simulation showed that radiation and latent heat flows are negligible compared to the others. The energy model for the cooling heat flow has been proposed by Yunus [32] considering the fuel cell as a CSTR with cooling coils and then adapting the effectiveness. The same equations are also used to model the heat exchange in a condenser or boiler, vessel with phase changes, and for many characteristics the fuel cell can be considered as one of these vessels with change in phases. Then the coolant energy could be expressed as follows:

$$\dot{Q}_{cool} = \varepsilon \dot{m}_{cool} C p_{cool} (T_{st} - T_{cool,in}) \qquad (4.60)$$

where Q_{cool} is the energy flowrate removed through cooling system, m_{cool} is the coolant mass flowrate, Cp_{cool} is the heat capacity of the coolant, $T_{cool,in}$ is the coolant inlet temperature, and ε is the effectiveness factor ($\varepsilon = 1 - e^{-NTU} = 1 - e^{-\frac{UA}{\dot{m}Cp}}$).

Moreover, we should consider the cooling system as its temperature depends on the exchange heat. Besides, the geometry of the cooling system plays a significant role in the evolution of the temperature and the exchanged heat. The temperature of the cooling liquid is warming along the fuel cell and be calculated by the following equation under the assumption of uniform temperature:

$$\rho_{cool} V_{cool} C p_{cool} \frac{dT_{cool}}{dt}$$
$$= \dot{Q}_{cool} + \dot{m}_{cool} C p_{cool} T_{cool,in} - \dot{m}_{cool} C p_{cool} T_{cool} \qquad (4.61)$$

where ρ_{cool} is the coolant density, V_{cool} is the coolant volume, and T_{cool} is the outlet coolant temperature.

The convection heat flow has not been taken into account as it has been noticed in Del Real et al. [7] that as soon as a cooling system is taken into account, it is possible to substitute the convection heat by the cooling heat flow.

4.4.8
Air Compressor and DC Motor Model

The compressor presented in this chapter is a positive displacement, three-lobe roots type supercharger, and is classified as a constant-volume, variable-pressure machine. This type of compressor is widely described in the literature, and a mathematical model is presented in Tekin et al. [28]. Thus, the mass flow rate and rotation speed relation of the rotary vane compressor are stated as

$$\dot{m}_{comp} = \frac{1}{2\pi} \eta_{vol} \rho_{air} V_{comp/rev} \omega \qquad (4.62)$$

where η_{vol} is the compressor's volumetric efficiency, ρ_{air} is the air density, and V_{comp} is the compressed volume per revolution.

The power consumed by the screw compressor is then calculated for the inputs to the model using the following expression for adiabatic compression

$$\dot{W}_{comp} = \tau_{comp}\omega_{comp} = c_{pair}\dot{m}_{comp,out}\frac{T_{in}}{\eta_{comp}}\left[\left(\frac{P_{im}}{P_{atm}}\right)^{\frac{\gamma-1}{\gamma}} - 1\right] \quad (4.63)$$

where τ_{comp} is the torque required to drive the compressor and ω_{comp} its rotational speed.

It is assumed that there is no mass accumulation inside the compressor such that the mass flow into the compressor is equal to that leaving. The compressor torque can be expressed as

$$\tau_{comp} = \frac{\dot{m}_{comp,out}}{\omega}\frac{T_{atm}c_{pair}}{\eta_{comp}}\left[\left(\frac{P_{im}}{P_{atm}}\right)^{\frac{\gamma-1}{\gamma}} - 1\right] \quad (4.64)$$

Using thermodynamic equations, the temperature of the gas stream leaving the compressor can be calculated from the temperature rise across the compressor:

$$T_{out} = T_{atm} + \frac{T_{atm}}{\eta_{cp}}\left[\left(\frac{p_{out}}{P_{atm}}\right)^{\frac{\gamma-1}{\gamma}} - 1\right] \quad (4.65)$$

Furthermore, the change in humidity inside the compressor has been obtained using the following equation:

$$\varphi_{im} = \frac{\varphi_{atm} P_{im} P_{sat}(T_{atm})}{P_{atm} P_{sat}(T_{im})} \quad (4.66)$$

A lumped rotational parameter model with combined inertia is used to represent the dynamic behavior of the compressor and electric motor speed. According to a Newton's law, the polar second moment of area J multiplied by the derivative of angular speed is equal the sum of all torques about the motor shaft:

$$J\frac{d\omega}{dt} = \sum \tau_i = \tau_{motor} - \tau_{compressor} \quad (4.67)$$

4.4.9
DC Motor

The static model of the Lemco DC motor has been adopted from Grasser [10]. The motor power is calculated based on the voltage applied, which is the control input to the air supply system. The designed model could be used for various different motor technologies, as it is described by Hauer and Moore [14]. The applicable motor equations are as follows:

$$V_{applied}(t) - V_{emf}(t) = L\frac{di}{dt} + Ri(t) \quad (4.68)$$

However, the back-induced electromotive force (V_{emf}) is proportional to the angular speed (ω) seen at the shaft, $V_{emf}(t) = K_b\omega(t)$, the previous equation becomes

$$V_{\text{applied}}(t) = L\frac{di}{dt} + Ri(t) + K_b\omega(t) \tag{4.69}$$

where K_b is the emf constant that depends on the physical properties of the motor, i is the current, R is the nominal resistance (Ω), L is the nominal inductance (H), and J is the inertial load (kg m^2/s^2).

Adopting state-space notation from Grasser [10], the combination of the above equations and using the speed and the current, the equations above are becoming

$$\frac{d}{dt}\begin{bmatrix} i \\ \omega \end{bmatrix} = \begin{bmatrix} -\frac{R}{L} & -\frac{K_b}{L} \\ \frac{K_m}{J} & -\frac{K_f}{J} \end{bmatrix}\begin{bmatrix} i \\ \omega \end{bmatrix} + \begin{bmatrix} \frac{1}{L} \\ 0 \end{bmatrix} V_{\text{applied}}(t) \tag{4.70}$$

where K_m is the armature constant.

Finally, the torque motor (τ_{motor}) is described by the following equation as it is proportional to the current induced by the applied voltage $\tau_{\text{motor}}(t) = K_m i(t)$, K_m being the armature constant.

4.4.10
Cooling System

The cooling system is composed of two main equipments, which are pump and heat exchanger. The cooling water circulates through a loop, so after having been used to cool down the fuel cell, the water is cooled down by a gas-to-liquid heat exchanger (air as a cooling fluid using a fan). The heat exchanger is modeled using the effectiveness or number of transfer units approach. The NTU, which is a characteristic for a given heat exchanger, is computed using NTU = $(U_{\text{rad}} A_{\text{rad}})/c_{\min}$, where U_{rad} is the overall heat transfer coefficient, which can be determined experimentally, and A_{rad} is the heat transfer area. The effectiveness is given by the following equation:

$$\varepsilon = \frac{1 - \exp[-\text{NTU}(1-c)]}{1 - c\exp[-\text{NTU}(1-c)]} \tag{4.71}$$

where $c_{\min}$ is the minimum heat capacity and c is the ratio of the minimum to the maximum heat capacity and can be expressed as follows:

$$c = \frac{(\dot{m}c_p)_{\min}}{(\dot{m}c_p)_{\max}} \tag{4.72}$$

This equation is applicable for counter flow type heat exchanger and should be modified for cross flow type heat exchanger. The heat transfer (Q) between the gases is computed as a function of the maximum heat transfer ($Q_{\max}$) and the effectiveness,

$$\dot{Q} = \varepsilon \dot{Q}_{\max} = \varepsilon \dot{m}_{\min} c_{p,\min}(T_{h,\text{in}} - T_{c,\text{in}}) \tag{4.73}$$

where $T_{h,in}$ and $T_{c,in}$ are the inlet temperature of the hot and cold stream, respectively.

4.5
Reduced Order Model

A model-based control strategy is a suitable approach to obtain a better use of the fuel cell capacities and increase the efficiency of the system, but requires a capable dynamic fuel cell model. Indeed the crucial issue is to be able to balance the two counteracting conditions, namely ensuring acceptable response time for the power demand, while achieving high efficiencies over the entire operating range. Until now, most dynamic two-phase PEM fuel cell models in the literature like Ziegler et al. [30] are too complex for many process control purposes. Recently reduced ones have been derived, which are more suitable for model-based control [7, 11, 12]. To design a controller suitable for our dynamic model, the first task will be to look into a reduced order model suitable for model-based control. The next step will be to design a controller able to regulate the voltage output with respect to a reference value, while the intensity is constant, and the system will be subject to different disturbances in the load.

A reduced order SS model is designed with model identification from the data simulations of the PEM fuel cell process. The input/output data are obtained from simulation with Simulink of the nonlinear system along the given set points, while the SS model parameters are obtained with Matlab Identification Toolbox. The mathematical representation of the SS model without disturbances and with sampling time equal to 1 s is as follows:

$$x(t+1) = Ax(t) + Bu(t), \qquad y(t) = Cx(t) \tag{4.74}$$

where y are the temperature and the voltage of the stack and u is the mass flowrate and the temperature of the coolant, and the voltage of the compressor. The system matrices are given as follows:

$$A = \begin{bmatrix} 0.993 & 0.004 & 0.0009 & 0.001 \\ 0.013 & 0.992 & 0.022 & 0.007 \\ -0.015 & -0.003 & -0.707 & -0.086 \\ -0.001 & -0.005 & 0.0069 & 0.986 \end{bmatrix}$$

$$B = \begin{bmatrix} -2.73\mathrm{e}{-006} & -0.00070165 & 2.090\mathrm{e}{-005} \\ -0.003405 & 0.0050739 & -2.53\mathrm{e}{-005} \\ 0.0016905 & -0.042051 & -0.0001598 \\ -2.5\mathrm{e}{-005} & 0.0022222 & -4.27\mathrm{e}{-006} \end{bmatrix}$$

$$C = \begin{bmatrix} 394.3 & -8.715 & 0.0438 & -0.369 \\ 45.10 & -19.16 & -0.112 & -0.104 \end{bmatrix} \tag{4.75}$$

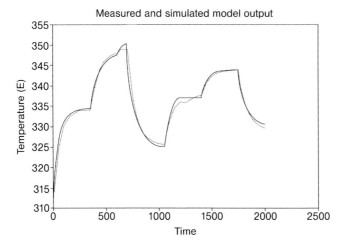

Fig. 4.5 Process and SS model comparison.

A comparison between the reduced order SS model (dotted line) and the high-fidelity dynamic model (black line) is shown in Fig. 4.5. The SS model closely approximates the behavior of the process with a small approximation error (difference between the actual process output and the SS output) of 7%.

The next step in the general framework (Fig. 4.2) involves the design of mp-MPC of the PEM fuel cell system. MPC is based on online optimization of the future control moves, as opposed to other control methods that determine *off-line* a feedback policy. The survey paper of Mayne et al. [17] provides a complete study of stability and optimality of constrained MPC. The main idea of MPC is to use a mathematical model of the system to predict the future effect of the control on the system behavior (output horizon) and then compute the optimal sequence of manipulated variables (input horizon) that minimizes or maximizes the objective function and satisfies the constraints on inputs and outputs.

MPC is implemented as following (Fig. 4.6):

- At time t, the current and past measurements of the output $y(t)$ and/or state variable $x(t)$ ate obtained.
- The optimization problem is solved online over the horizon p to obtain the optimal control policy $u(t+k)$ (optimal sequence of manipulated variables) and the corresponding predicted outputs $y(t+k)$ based on the model equations.
- Only the first component of the control policy $u(t)$ is applied.
- The next measurements are obtained and the procedure is repeated at time $t+1$.

This procedure is repeated at each sampling time to make sure the process will follow the reference profile despite the disturbances (varying operating conditions) and model inaccuracies. MPC is, therefore, a receding horizon control procedure and can be seen as an implicit feedback policy since the control action is obtained for current output and/or state.

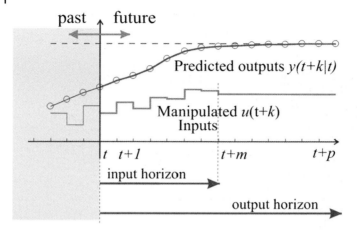

Fig. 4.6 MPC control strategy (Gerogiorgis et al. 2007).

The benefits of MPC have long been recognized for optimal control design. Nevertheless, its application may be restricted due to increased online computational requirements related to the constraint optimization. In order to overcome this drawback, explicit/mp-MPC was developed [24, 25], which avoids the need for repetitive online optimization. In mp-MPC, the optimization problem of the MPC is solved offline by parametric optimization to obtain the optimal solution as an optimal mapping of the current state, output measurements, and reference trajectory instead of demanding online optimization.

The following MPC formulation is considered for the PEM fuel cell control system

$$\min_{u_{t+1},\dots,u_{t+N_u}} J = \sum_{i=1}^{N_y} Q(y_i - y_{\text{ref},i})^2 + \sum_{j=0}^{N_u} (u_j)^2$$

s.t. $\quad x(t+1) = Ax(t) + Bu(t)$

$$y(t) = Cx(t) + d$$

$$\begin{bmatrix} 30 \\ 0.01 \\ 300 \end{bmatrix} \leqslant U_i \leqslant \begin{bmatrix} 150 \\ 0.8 \\ 340 \end{bmatrix} \leqslant Y_i \leqslant \begin{bmatrix} 60 \\ 360 \end{bmatrix}$$

$i = 1, \dots, N_y, \; j = 0, \dots, N_u - 1$

$Q = 100, \; \rho = 0.01, \; R = 1$ (4.76)

where y is the controlled variables, u are the manipulated variables, N_u is the control horizon ($N_u = 2$), and N_y the prediction horizon ($N_y = 10$). The MPC takes into account the operational limitation of the manipulated variables (input constraints) and the controlled variables (output constraints). The problem involves

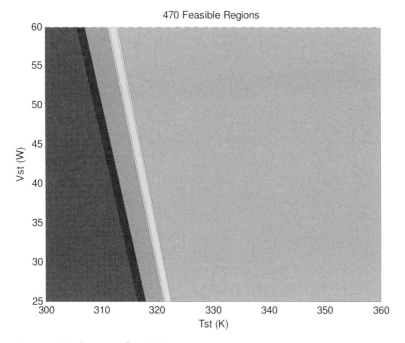

Fig. 4.7 Critical regions of mp-MPC.

six optimization variables ($m_{cool(t+1)}$, $T_{cool(t+1)}$, $V_{comp(t+1)}$, $m_{cool(t+2)}$, $T_{cool(t+2)}$, $V_{comp(t+2)}$) and eight parameters ($\Theta = [x_t, T_{st}, V_{st}, T_{st,sp}, V_{st,sp}]^T$), which represent the states at time zero, the measurements of stack temperature, and voltage and the reference trajectory of the stack temperature and the voltage. The objective function is set to minimize the quadratic norm of the error between the output variables and the reference points.

For the case of constant system matrices, the optimization problem (4.76) is a multiparametric quadratic programming (mp-QP) problem and can be solved with standard multiparametric programming techniques [24]. In our study, Parametric Optimization Software was used [19] to obtain the explicit controller description, which is the optimal map of the control variables as a function of the parameters of the system. This optimal map consists of 470 critical regions (Fig. 4.7) and the corresponding control laws. Each of the critical regions is described by a number of linear inequalities $A_i x \leqslant b_i$ and its corresponding control is piecewise linear $U_f = K_i x + c_i$, where i is the index of solutions. The critical regions are shown in Fig. 4.7, based on a projection of the critical regions on the x_1–x_2 subspaces.

Figures 4.8 and 4.9 depict the simulation results of the mp-MPC implementation for different operating conditions (set points). The controller manages to maintain the variables at the desired reference values for different values of current while satisfying the constraints. The controller showed fast response to temperature and voltage set point changes for different values of current, while managed to keep the output constraints in the feasible range.

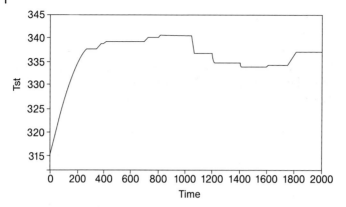

Fig. 4.8 Stack temperature.

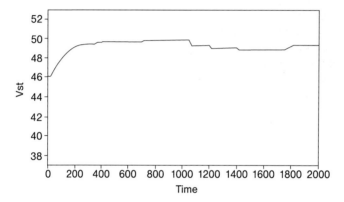

Fig. 4.9 Stack voltage.

4.6
Concluding Remarks

On the environmental point of view, the fuel cell has the key benefit to be a potential zero emission energy source, since the only products of the electrochemical reaction are water and heat. However to compete with internal combustion engines for an application in vehicles, fuel cells have several technical drawbacks to overcome. First, hydrogen is rather an energy transporter than a fuel source, and till now the primary method to produce hydrogen is by steam reforming of natural gas. Second, due to the low power density of hydrogen, a large volume of fuel onboard is required in fuel cell vehicles compared to gasoline ones. Thus hydrogen storage is a problem under active development.

In this chapter, we present a dynamic model for PEM fuel cell stack. To be used on a real system, some parameters would have to be modified to fit properly the PEM fuel cell system under consideration. This dynamic model has the great advantage of lower computation time while providing results consistent with the literature, and well oriented toward control. The first step in the proposed framework is to develop a high-fidelity mathematical model for the PEM fuel cell system. Then a reduced order SS model is designed for optimal control studies. Finally an explicit/mp-MPC controller has been developed to keep the controlled variables close to the set points while taking care of the physical constraints on the manipulated variables, namely the reactant and coolant mass flows. The controller finally selected shows good performance to resist the disturbances in the load.

Notation: Constant Coefficients Used in the Model (value, unit)

Cp_{H_2} – Hydrogen mass specific heat capacity (14 200, J/(kg K))
Cp_{N_2} – Nitrogen mass specific heat capacity (1040, J/(kg K))
Cp_{O_2} – Oxygen mass specific heat capacity (918, J/(kg K))
Cp_{H_2O} – Water mass specific heat capacity (4181, J/(kg K))
F – Faraday's constant (96 500, C/mol)
R – Ideal gas constant (8.314, J/(mol K))
$D_{H_2,ref}$ – Hydrogen diffusion coefficient (9.15e−5, m^2/s)
$D_{O_2,ref}$ – Oxygen diffusion coefficient (2.2e−5, m^2/s)
$D_{H_2O,ref}$ – Water diffusion coefficient (34.5e−6, m^2/s)
$\Delta_r H^o (298\ K)$ – Reaction enthalpy at 298 K (−241 820, J/mol)
M_a – Air molar mass (29e−3, kg/mol)
M_{H_2} – Hydrogen molar mass (2e−3, kg/mol)
M_{N_2} – Nitrogen molar mass (14e−3, kg/mol)
M_{O_2} – Oxygen molar mass (32e−3, kg/mol)
M_{H_2O} – Water molar mass (18e−3, kg/mol)
Cp_{cool} – Coolant mass specific heat capacity (4.181e3, J/(kg K))
ξ_1 – (−0.9514, −)
ξ_2 – (0.00312, −)
ξ_3 – (−0.000187, −)
ξ_4 – (7.4e−5, −)
ξ_5 – (3.30e−3, −)
ξ_6 – (−7.55e−6, −)
ξ_7 – (1.10e−6, −)

References

1 AKELLA, S., SIVASHANKAR, N., GOPASLSWAMY S., Model-based system analysis on a hybrid fuel cell vehicle configuration, in: *Proceedings of 2001 American Control Conference*, vol. 3, **2001**, pp. 1777–1782.

2 AL-BAGHDADI, M. S., AL-JANABI, S., Modelling optimizes PEM fuel cell performance using three-dimensional multiphase computational fluid dynamics model, *Energy Conversion and Management* 48(12) (**2007**), pp. 3102–3119.

3 AMPHLETT, J. C., MANN, R. F., PEPPLEY, B. A., ROBERGE, P. R., RODRIGUES, A., A model predicting transient responses of proton exchange membrane fuel cell, *Journal of Power Sources* 61(1–2) (**1996**), pp. 183–188.

4 BAO, C., OUYANG, M., YI, B., Modeling and control of air stream and hydrogen flow with recirculation in a PEM fuel cell

system-I. Control-oriented modelling, *International Journal of Hydrogen Energy* 31 (**2006**), pp. 1879–1896.

5 Benziger, J. B., Woo, C. H., PEM fuel cell current regulation by fuel feed control, *Chemical Engineering Science* 62(4) (**2007**), pp. 957–968.

6 Caux, S., Lachaize, J., Fadel, M., Shott, P., Nicod, L., Modelling and control of a fuel cell system and storage elements in transport applications, *Journal of Process Control* 15(4) (**2005**), pp. 481–491.

7 Del Real, A. J., Arce, A., Bordons, C., Development and experimental validation of a PEM fuel cell dynamic model, *Journal of Power Sources* 173(1) (**2007**), pp. 310–324.

8 Fronk, M. H., Wetter, D. L., Masten, D. A., Bosco, A., in: *Proceedings of SAE 2000 American Control Conference, PEM Fuel Cell System Solutions for Transportation. Fuel Cell Power for Transportation 2000*, vol. 3, **2000**, pp. 1777–1782.

9 Fuller, T., Newman, J., Water and thermal management in solid-polymer-electrolyte fuel cells, *Journal of Electrochemical Society* 140(5) (**1993**), pp. 1218–1225.

10 Grasser, F., An analytical, control-oriented state space model for a PEM fuel cell system, PhD Thesis, Ecole Polytechnique Federale de Lausanne, **2006**.

11 Grötsch, M., Mangold, M., A two-phase PEMFC model for process control purposes, *Chemical Engineering Science* 63(2) (**2008**), pp. 434–447.

12 Golbert, J., Lewin, D. R., Model-based control of fuel cells: (1) Regulatory control, *Journal of Power Sources* 135 (**2004**), pp. 135–151.

13 Golbert, J., Lewin, D. R., Model-based control of fuel cells: (2) Optimal efficiency, *Journal of Power Sources* 173(1) (**2007**), pp. 298–309.

14 Hauer, K. H., Moore, R. M., Fuel cell vehicle simulation-part 2: Methodology and structure of a new fuel cell simulation tool, *Fuel Cells* 2(3) (**2003**), pp. 95–104.

15 Kikkinides, E. S., Georgiadis, M. C., Stubos, A. K., Dynamic modelling and optimization of hydrogen storage in metal hydride beds, *Energy* 31(13) (**2005**), pp. 2428–2446.

16 Lee, J. H., Lalk, T. R., Modeling fuel cell stack systems, *Journal of Power Sources* 73 (**1998**), pp. 229–241.

17 Mayne, D. Q., Rawlings, J. B., Rao, C. V., Scokaert, P. O. M., Constrained model predictive control: stability and optimality, *Automatica* 36(6) (**2000**), pp. 789–814.

18 Müller, E., Stefanopoulou, A. G., Analysis, modeling, and validation for the thermal dynamics of a polymer electrolyte membrane fuel cell system, *Journal of Fuel Cell Science and Technology* 3(2) (**2006**), pp. 99–110.

19 Parametric Optimixstion Solutions Ltd, *Parametric Optimization (POP) Software*, UK, **2007**.

20 Park, S. K., Choe, S. Y., Dynamic modelling and analysis of a 20-cell PEM fuel cell stack considering temperature and two-phase effects, *Journal of Power Sources* 179 (**2008**), pp. 660–672.

21 Pathapati, P. R., Xue, X., Tang, J., A new dynamic model for predicting transient phenomena in a PEM fuel cell system, *Renewable Energy* 30(1) (**2005**), pp. 1–22.

22 Pistikopoulos, E. N., Dua, V., Bozinis, N. A., Bemporad, A., Morari, M., On-line optimization via off-line parametric optimization tools, *Comput. Chem. Eng.* 26 (**2002**), p. 175.

23 Pistikopoulos, E. N., Perspectives in multiparametric programming and explicit model predictive control, *Journal of American Institute of Chemical Engineering* 55 (**2009**), p. 1918.

24 Pistikopoulos, E. N., Georgiadis, M. C., Dua, V., *Multi-parametric Programming: Theory, Algorithms and Applications*, ISBN-978-3-527-31692-2, Wiley-VCH, Weinheim, **2007a**.

25 Pistikopoulos, E. N., Georgiadis, M. C., Dua, V., *Multiparametric Model-Based Control*, ISBN-978-3-527-31692-2, Wiley-VCH, Weinheim, **2007b**.

26 Pukrushpan, J. T., Stephanopoulou, A. G., Peng, H., *Control of Fuel Cell Power Systems. Principles, Modelling, Analysis and Feedback Design*, Grimble, M. J. et al. (eds.), Springer, London, **2005**.

27 Shan, Y., Choe, S., A high dynamic PEM fuel cell model with temperature effects, *Journal of Power Sources* 145(1) (**2005**), pp. 30–39.

28 TEKIN, M., HISSEL, D., PERA, M. C., KAUFFMANN, J. M., Energy consumption reduction of a PEM fuel cell motor-compressor group thanks to efficient control laws, *Journal of Power Sources* 156 (**2006**), pp. 57–63.

29 THOMAS, P. J., *Simulation of Industrial Processes for Control Engineers*, Elsevier, Amsterdam, **1999**.

30 ZIEGLER, C., YU, H., SCHUMACHER, J., Two-phase dynamic modelling of the PEMFC and simulation of cyclo-voltammograms, *Journal of Electrochemical Society* 152(8) (**2005**), pp. A1555–A1567.

31 WISHART, J., DONG, Z., SECANELL, M., Optimization of a PEM fuel cell system based on empirical data and generalized electrochemical semiempirical model, *Journal of Power Sources* 161 (**2006**), pp. 1041–1055

32 YUNUS A. C., *Heat Transfer: A Practical Approach*, McGraw-Hill, New York, **2003**.

5
Modeling of Pressure Swing Adsorption Processes
Eustathios S. Kikkinides, Dragan Nikolic, and Michael C. Georgiadis

Keywords
pressure swing adsorption (PSA), thermal swing adsorption (TSA), single-bed adsorber, adsorption-layer model, heat transfer, transport properties

5.1
Introduction

Pressure swing adsorption (PSA) is a gas separation process, which has attracted increasing interest over the last three decades because of its low energy requirements as well as low capital investment costs in comparison to the traditional separation processes [1–4]. The commercialization of PSA processes is commonly attributed to Skarstrom and Gurin de Montgareuil in 1957–1958; however many of the basic features of the process have been described in patents as early as 1927 [4]. While initial applications of PSA included gas drying and purification of dilute mixtures, advances in material and process design have broadened significantly the spectrum of applications of this process including small to intermediate scale oxygen and nitrogen production from air (air-separation), hydrogen recovery from different petrochemical processes (steam-methane reforming off gas (SMR) or coke oven gas), and trace impurity removal from contaminated gases.

PSA like all adsorption separation processes requires the use of a fixed bed known as adsorber, which is packed with an adsorbent material that selectively adsorbs one component (or a family of related components) from a gas mixture. This selectivity can be either thermodynamic or kinetic in nature, based on differences in adsorption equilibrium or diffusion rates, respectively. Evidently the adsorbent, and hence the adsorber, will be saturated after a period of time. For this reason, the adsorption step must be accompanied by a regeneration or desorption step, where the preferentially adsorbed species are removed from the adsorbent that can be further used in the next cycle. The adsorption step is terminated well before the more strongly adsorbed species breaks through the bed, while the desorption step is generally terminated before complete regeneration of the adsorber. Therefore, a cyclic steady state (CSS) oscillates about a mean position in the bed, while one needs to operate two or more beds in a proper sequence of steps [4]. The efflu-

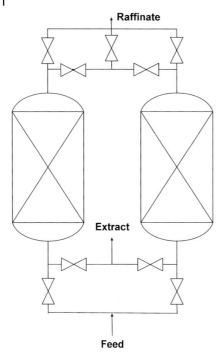

Fig. 5.1 A simple flow chart of a two-bed Skarstrom cycle.

ent during the adsorption step that no-longer contains the preferentially adsorbed species is called the purified or "raffinate" product, while the effluent during the desorption step that contains the more strongly adsorbed species in larger proportions compared to the feed stream is often called the "extract" product. In most PSA processes, only the raffinate product is easily recovered, although there have been several studies where additional steps can be introduced to improve the purity of the extract product [5–7]. A typical schematic of the simple two-bed Skarstrom cycle is shown in Fig. 5.1.

A major advantage of PSA over other types of adsorption processes, such as thermal swing adsorption (TSA), is that in the former, pressure can be changed (swing) much more rapidly than temperature, resulting in much faster cycle operations and thus larger throughput per adsorbent bed volume (when vacuum is used in the desorption step, the process is called vacuum swing adsorption or VSA). However such an operation is restricted to components that are not too strongly adsorbed, otherwise a high vacuum may be required during the regeneration step. In such cases, TSA processes are preferred since even a modest increase in temperature can produce a significant change in the gas–solid adsorption equilibrium characteristics [1–4], as it is illustrated in Fig. 5.2. Moreover, nonselective mass transfer due to the macropore structure of the adsorbent pellet can have a profound effect in the process performance characteristics. It is clear that a proper material design has a major impact on the efficient separation of a gas mixture. Nevertheless, there is also a strong economic incentive for further improvement in the PSA process

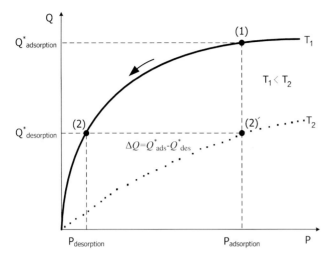

Fig. 5.2 The thermodynamic principle of PSA and TSA processes.

design that is inevitably interlinked with new material design and process applications.

A PSA or VSA process has an important functional difference compared to other separation processes including absorption, distillation, and membrane processes: the process operates under transient conditions, while the above mentioned processes operate at steady state [1–4]. Hence, from an operational point of view, PSA is an intrinsic dynamic process operating in a cyclic manner with each bed undergoing the same sequence of steps. Typical operating steps include pressurization with feed or pure product, high-pressure adsorption, pressure equalization(s), blowdown, and purge. Development of PSA/VSA process models requires an accurate model representation of the adsorption–diffusion process at the microscale pore level, the mass transport characteristics at the particle-pellet level, and a transport model, coupled with the conservation equations of mass, heat, and momentum, at the bed level. The behavior in each bed is described by partial differential and algebraic equations (PDAEs) in space (normally 1D) and time. Moreover, the flow pattern between the beds is needed to describe the way these beds are interconnected with each other.

Simulation of PSA processes is performed by solving the above DAEs for each step repeatedly for many cycles until CSS conditions are achieved. Then process performance is evaluated by calculating several important parameters including product purity, recovery, throughput, and power consumption [4]. Such simulations can replace many expensive and time-consuming laboratory and/or pilot scale experimental studies allowing innovative modifications of existing design configurations at no actual cost [8, 9].

Over the last three decades, several PSA studies have appeared in the literature. An overview of single-bed PSA studies is presented in Table 5.1. Industrial practice indicates that difficult gas separations under high product quality requirements (i.e., purity and/or recovery) rely on complex PSA flowsheets with several intercon-

140 | 5 Modeling of Pressure Swing Adsorption Processes

Table 5.1 Overview of single-bed PSA studies.

Reference	Mass balance[a]	Heat balance	Momentum balance	Isotherm, adsorbent	No. of steps	No. of layers	Application
[10]	LDF	–	–	Linear, CMS	4	1	Air separation
[11]	LDF	Bulk gas	Ergun	Langmuir, AC + 5A	–	2	H_2, CO_2, CH_4, CO
[12]	PD	–	–	Langmuir, alumina	4	1	Air separation
[13]	LEQ, SD, PD	Bulk gas	–	IAS/LRC, AC	5	1	H_2, CO_2, CH_4
[14]	BDPD	Bulk gas	–	Langmuir, 5A	5	1	H_2, CH_4
[15]	PD	–	–	Linear, MS RS-10	4	1	Air separation, theoretical study
[16]	PD	–	–	Linear, MS RS-10	4	1	Air separation, experimental study
[17]	DG	Bulk gas	–	Langmuir, 5A	–	1	Air separation
[18]	LDF	Bulk gas	Ergun	Langmuir, AC + 5A	–	2	H_2, CO_2, CH_4, CO
[19]	DG, LDF–DG, LDF–DGSD, LDF–DGSD	–	–	Langmuir, 4A, 5A, CMS	4	1	Air separation
[20]	LDF	Bulk gas	Darcy	Langmuir, 13X	4	1	CO_2 sequestr.
[21]	LDF	Bulk gas	Ergun	Dual-site Langmuir, 13X	4	1	CO_2 capture
[22]	LDF	Bulk gas	–	Langmuir, AC	4	1	H_2, CH_4
[23]	Bi-LDF	Bulk gas, solid	Ergun	Langmuir, 4A	5	1	Propane, propylene
[24]	PD	–	–	Linear	4	1	Gas drying

a) LEQ, local equilibrium model; LDF, linear driving force model; SD, solid diffusion model; PD, pore diffusion model; DG, dusty gas model.

nected beds and complicated operating procedures. Key literature contributions involving two- and multibed configurations under certain assumptions are summarized in Tables 5.2 and 5.3.

Most of the PSA simulation models that exist in the literature solve the DAEs for each bed/step cycle after cycle using the profiles of all variables at the end of each cycle as initial conditions for the next one, until CSS is achieved, when the profiles of all variables within the bed at the end of the adsorption step remain unchanged (within a predefined tolerance) as the process goes from cycle N to cycle $N + 1$. This is known as the successive substitution approach and it is a stable procedure since it mimics the actual PSA process. However, it can be computationally expensive because it is often slow in convergence requiring hundreds or thousands of cycles to reach CSS. Over the past two decades, a few commercial software platforms have been developed to effectively simulate various PSA/VSA processes [8, 9, 44, 45]. Nevertheless, the design and optimization of PSA processes still remains a challenge due to the complexity of the process applications from an operational and computational point of view, particularly as multiadsorbent, multibed configurations are required to face real-life industrial applications. In such cases, bed interactions and interconnections play an important role in the proper simulation of the relevant PSA/VSA process operation and performance. It has been shown [10] that incorporating a gas valve equation into the PSA model to control flow rate is the best approach to realistically describe bed interactions, and this approach has been adopted on the study of the dynamic behavior of multibed PSA process [40, 43].

It is evident that one can develop various PSA models or different complexity to describe equilibrium and transport properties at bed, particle, and/or pore scale for binary or multicomponent mixtures. Moreover, process complexity increases as more beds are employed in the process design and operation. Thus there has been a strong need for the development of generic multibed PSA models, and there have been several studies in the open literature describing such models [8, 9, 25, 26, 34, 40, 43].

To simulate the behavior of a multibed PSA configuration, two different approaches are commonly employed: the "unibed" and the "multibed" [40]. The "unibed" approach assumes that all beds undergo identical steps so only one bed is needed to simulate the multibed cycle. Information about the effluent streams are stored in data buffers and linear interpolation is used to obtain information between two time points. The "multibed" approach considers a multibed process as a sequence of repetitive stages within the cycle.

In the remainder of this chapter, we present the development of a generic modeling framework for efficient simulation and optimization strategies for the design of PSA processes with detailed adsorption and transport models [43]. The framework is general enough to support an arbitrary number of beds, a customized complexity of the adsorbent bed model, one or more adsorbent layers, automatic generation of operating procedures, all feasible PSA cycle step configurations, and interbed connectivities. The gPROMS modeling and optimization environment have been served as the development platform although the framework can be easily adapted

Table 5.2 Overview of two-bed PSA studies.

Reference	Mass balance	Heat balance	Momentum balance	Isotherm, adsorbent	No. of beds	No. of steps	Bed interactions	No. of layers	Application
[10]	LDF	–	–	Langmuir, CMS/5A	2	4	Valve eq.	1	Air separation
[25]	LDF, PD	–	Darcy	Linear-Langmuir, 5A	1, 2	2, 6	Valve eq.	1	Air separation
[26]	LDF	Bulk gas	Ergun	Dual-site Langmuir, 13X	1, 2	3, 6	Valve eq.	Inert + ads.	Air separation
[27]	LDF, DG	Bulk gas	Darcy	Langmuir	2	4	–	1	Air separation
[28]	LDF, PD	Bulk gas	–	LRC, AC	2	4	–	1	H_2, CH_4
[29]	LDF	–	–	Linear, CMS	2	6	Frozen solid	1	Air separation
[30]	LDF	Bulk gas	Ergun	Langmuir/Freundlich, AC + 5A	2	7	–	2	H_2, CH_4, CO, N_2, CO_2
[31]	LDF	Bulk gas	Ergun	LRC, AC + 5A	2	6	–	1	Air separation
[32]	Exp	Bulk gas	–	X zeolite	2	4	–	1	Air separation

Table 5.3 Overview of multibed PSA studies.

Reference	Mass balance	Heat balance	Momentum balance	Isotherm, adsorbent	No. of beds	No. of steps	Bed interactions	No. of layers	Application
[33]	LEQ	–	–	Langmuir, 5A	4	6	Valve eq.	1	Air separation
[34]	LDF	Bulk gas	–	LRC, AC + 5A	5	8	Linear change	1 or 2	H_2, CH_4, CO, N_2
[35]	LDF	Bulk gas	–	LRC, AC/5A	2, 4	5, 7	–	1	H_2, CH_4, N_2
[36]	LDF	Bulk gas	Ergun	Langmuir, AC + 5A	4	8	Only one bed is simulated; information stored in data buffer	2	H_2, CO_2, CH_4, CO
[37]	LDF	Bulk gas	Darcy	Langmuir, Alumina + AC + zeolite	6	12	Information stored in data buffer	inert + 2 ads.	H_2, CH_4, CO
[38]	–	–	–	AC, zeolite, silica gel	9, 10	9, 11, 6 + 7	–	–	Overview of commercial PSA processes
[39]	LDF	Bulk gas	Darcy	Langmuir, AC	4, 5	7, 9, 12	–	1	H_2, CH_4
[40]	LDF	Bulk gas	Ergun	Dual-site Langmuir, APHP + 5A	5	11	Valve eq.	2	N_2, CH_4, CO_2, CO, H_2,
[41]	LEQ	Bulk gas	–	Ext. Langmuir, 13X	2, 3	4, 6	–	1	CO_2 from flue gases
[42]	LDF	Bulk gas	–	Langmuir, 13X	4, 5	4, 5, 5	Linear change	1	CO_2, N_2, H_2O
[43]	LDF	Bulk gas	–	Langmuir, AC	4–12	4–10	Valve eq.	1	H_2 from SMROG

to other software platforms. To this end, general model equations and boundary conditions are developed, both single- and multibed configurations are studied, and a systematic approach to automatically generate feasible operating procedures in complex PSA flowsheets is proposed.

5.2
Model Formulation

Depending on the general assumptions describing the adsorbent (porous solid)–adsorbate (gas mixture) system, one can employ a broad variety of mathematical models and equations to describe the PSA process [1–4]. The basic assumptions that hold in most PSA models is one dimensional transport in the fixed-bed adsorber along the axial direction in the form of pug flow and axial dispersion, and thermal equilibrium between gas and adsorbent. As a result, one needs to consider the following model equations that describe each step in the process:

- A mass balance for each component in the interstitial fluid (gas phase).
- An energy balance that includes gas, solid, and adsorbed phase.
- A (simplified) momentum balance that predicts the pressure drop along the bed.
- A transport model describing nonselective mass transfer in the macropores of the adsorbent, and surface or activated diffusion of the adsorbates in the micropores of the adsorbent.
- An equilibrium isotherm describing the thermodynamic relationship between the gas and adsorbed phase.

Furthermore, gas valve equations are needed to describe bed interactions, while several auxiliary equations are supplied to determine important parameters that evaluate the process performance, such as product purity, recovery, throughput, and power consumption. The different mathematical models, operating procedures, and auxiliary applications that can be covered by the general modeling framework are presented below.

5.2.1
Adsorbent Bed Models

The mathematical modeling of a fixed-bed adsorber must take into account the simultaneous mass, heat and momentum balances at both, bed and adsorbent particle level, adsorption isotherm, transport, and thermophysical properties of the fluids and boundary conditions for each operating step. The architecture of the developed adsorbent bed model sets a foundation for the problem solution. Internal model organization defines the general equations for the mass, heat, and momentum balances, adsorption isotherm, and transport and thermophysical properties. Phenomena that occur within the particles could be described by using many different transport mechanisms. However, only the mass and heat transfer rates through a particle surface in the bulk flow mass and heat balance equations have to be cal-

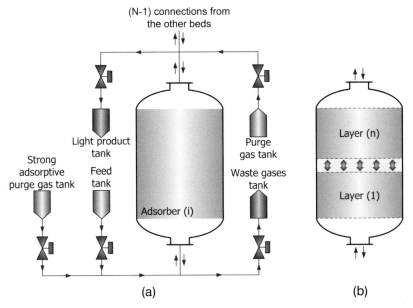

Fig. 5.3 (a) Single-bed adsorber and corresponding connections. (b) Adsorbent column with multiple adsorbent layers.

culated. Hence, the same adsorbent bed model could be used for implementing any transport mechanism as long as it follows the defined interface that calculates transfers through the particle surface. This way, different models describing mass and heat transfer within the particle can be plugged-in into the adsorbent column model. Systems where the mass transfer in the particles is fast enough can be described by the local equilibrium model, thus avoiding the generation of thousands of unnecessary equations. On the other hand, more complex systems may demand mathematical models taking into account detailed heat and mass transfer mechanisms at the adsorbent particle level. They may even employ equations for the phenomena occurring at the molecular level without any changes in the macrolevel model. This is also the case for other variables such as pressure drop, transport, and thermophysical properties of gases that can be calculated by using any available equation. The result of such architecture is the fully customizable adsorbent bed model that can be adapted for the underlying application to the desired level of complexity.

5.2.2
Single-Bed Adsorber

As it is previously shown, a single-bed adsorber is the main building block of a multibed PSA process. Each single-bed adsorbent column is connected to several storage tanks via gas valves (Fig. 5.3(a)). It contains one or more layers of adsor-

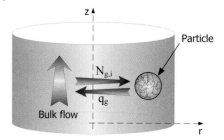

Fig. 5.4 Adsorbent layer.

bents, as can be seen in Fig. 5.3(b). Its main role is to describe the boundary conditions for the various operating steps as well as the connections between adsorbent layers.

5.2.3
Adsorption Layer Model

5.2.3.1 General Balance Equations

An adsorption-layer model must take into the account the simultaneous mass, heat, and momentum balances in the interstitial fluid (bulk gas flow), the transport and thermophysical properties of the gas mixture, and a set of boundary conditions for the inlet and outlet from the layer and for the interface between the bulk gas and particle surface (Fig. 5.4).

The following general assumptions are usually adopted in the derivation of the model equations:

- The flow pattern in the bed is described by axially dispersed plug flow (no variations in radial direction across the adsorber).
- The adsorbent is represented by uniform microporous spheres.

The general balance equation is given in the following form:

$$\frac{D\psi}{Dt} = \psi_g + \nabla D \nabla \psi \tag{5.1}$$

where ψ is concentration of mass, heat, or momentum, D is diffusivity, $D\psi/Dt$ is substantial derivative, ψ_g is generation term, and $\nabla D \nabla \psi$ is diffusion term.

For the *bulk gas flow*, convection and diffusion in x and y directions can be neglected leading to the following general transport equation:

$$\underbrace{\frac{\partial(u\psi)}{\partial z}}_{(1)} + \underbrace{\frac{\partial \psi}{\partial t}}_{(2)} = \underbrace{\psi_g}_{(3)} + \underbrace{\frac{\partial}{\partial z}\left(D\frac{\partial \psi}{\partial z}\right)}_{(4)} \tag{5.2}$$

where the terms (1)–(4) are as follows: (1) convection term, (2) accumulation term, (3) generation term, (4) diffusion term.

5.2.3.2 Mass Balance

Applying general equation (5.2) on the mass transfer in the bulk gas flow ($\psi = C_i$, $D = D_{z,i}$), after few simple mathematical operations, we get

$$\varepsilon_{\text{bed}} \frac{\partial (uC_i)}{\partial z} + \varepsilon_{\text{bed}} \frac{\partial C_i}{\partial t} + (1 - \varepsilon_{\text{bed}}) N_{g,i} = \varepsilon_{\text{bed}} \frac{\partial}{\partial z} \left(D_{z,i} \frac{\partial C_i}{\partial z} \right)$$

$$\forall z \in (0, L), \ i = 1, \ldots, N_{\text{comp}} \tag{5.3}$$

where the term $N_{g,i}$ is the generation term given per unit volume of adsorbent, which quantifies the mass transfer occurring between bulk flow and particles. The actual expression for the generation term depends on the nature of the resistances to the mass transfer (gas film around particles, macro- or micropores, etc.). This analysis will be given in the next section.

5.2.3.3 Heat Balance

Applying general equation (5.2) on the heat transfer in the bulk gas flow ($\psi = \rho c_p T$, $D = a = \lambda / \rho c_p$), after few simple mathematical operations we get

$$\varepsilon_{\text{bed}} \frac{\partial (\rho c_p T u)}{\partial z} + \varepsilon_{\text{bed}} \frac{\partial (\rho c_p T)}{\partial t} = (1 - \varepsilon_{\text{bed}}) q_g + q_{\text{wall}} + \varepsilon_{\text{bed}} \frac{\partial}{\partial z} \left(\lambda \frac{\partial T}{\partial z} \right)$$

$$\forall z \in (0, L) \tag{5.4}$$

where q_g and q_{wall} are generation terms, again given per unit volume of adsorbent. Variable q_g quantifies the heat transfer between bulk flow and particles, while q_{wall} takes into account heat losses through the column walls. Again, the actual expression for q_g depends on the nature of the resistances to the heat transfer and will be given in the next section.

In general, three thermal operating modes exist in PSA: *isothermal*, *nonisothermal*, and *adiabatic*. In the case of gas purification, system can be assumed approximately isothermal. Therefore, the heat balance becomes

$$\frac{\partial T}{\partial z} = 0, \quad \forall z \in (0, L) \tag{5.5}$$

In the case of adiabatic column $q_{\text{wall}} = 0$ and then Eq. (5.4) becomes

$$\frac{\partial (\rho c_p T u)}{\partial z} + \frac{\partial (\rho c_p T)}{\partial t} = \frac{1 - \varepsilon_{\text{bed}}}{\varepsilon_{\text{bed}}} q_g + \frac{\partial}{\partial z} \left(\lambda \frac{\partial T}{\partial z} \right), \quad \forall z \in (0, L) \tag{5.6}$$

If the heat transfer through the column wall cannot be neglected (*nonisothermal* mode), the following equation can be used:

$$q_{\text{wall}} = \frac{2 k_{h,\text{wall}}}{R_{\text{bed}}} (T - T_{\text{wall}}), \quad \forall z \in (0, L) \tag{5.7}$$

where $k_{h,\text{wall}}$ is a heat transfer coefficient between the bulk flow and the wall and T_{wall} is the temperature of the wall.

5.2.3.4 Momentum Balance

The pressure drop is an important variable in modeling of fixed beds having a high impact on the separation quality and operating costs. The hydrodynamics of flow through porous media is the most commonly described by using one of the following correlations for pressure drop:

- Blake–Kozeny (linear-laminar flow)

$$-\frac{\partial P}{\partial z} = 180 \frac{(1-\varepsilon_{\text{bed}})^2}{\varepsilon_{\text{bed}}^3} \frac{\mu u}{(2R_p)^2}, \quad \forall z \in (0, L) \tag{5.8}$$

- Ergun (nonlinear, turbulent flow)

$$-\frac{\partial P}{\partial z} = 150 \frac{(1-\varepsilon_{\text{bed}})^2}{\varepsilon_{\text{bed}}^3} \frac{^2\mu u|u|}{(2R_p)^2} + 1.75 \frac{1-\varepsilon_{\text{bed}}}{\varepsilon_{\text{bed}}^3} \frac{\rho u}{2R_p}, \quad \forall z \in (0, L) \tag{5.9}$$

5.2.3.5 Equation of State

An equation of state is necessary to link concentration with temperature and pressure in the gas mixture. In most cases, the ideal gas law model is employed:

$$P = \sum_{i=1}^{N_{\text{comp}}} C_i RT, \quad \forall z \in [0, L] \tag{5.10}$$

However, in more complex mixtures it may be necessary to employ one of numerous gas equations of state given in general form as

$$P = f(C_1, \ldots, C_{N_{\text{comp}}}, T), \quad \forall z \in [0, L] \tag{5.11}$$

5.2.3.6 Thermophysical Properties

Physical properties of the gas mixture can be assumed constant or calculated using some of the available correlations or thermophysical packages. In either case, physical properties are functions of temperature, pressure, and composition, presented in the following general form:

$$\rho, \lambda, c_p, \mu = f(T, P, C_1, \ldots, C_{N_{\text{comp}}}), \quad \forall z \in [0, L] \tag{5.12}$$

5.2.3.7 Axial Dispersion

Axial concentration and temperature gradients always exist in packed beds. Hence, a diffusive mass and heat transfer will always occur and tend to degrade the performance of the process. An accurate prediction of mass and heat axial dispersion coefficients is, therefore, very important for detailed modeling of the flow through the packed bed. Several correlations for prediction of mass and heat axial dispersion coefficients exist in the literature. Wakao [46, 47] developed one of the most widely used coefficients:

$$D_{z,i} = \frac{D_{m,i}}{\varepsilon_{bed}}(20 + 0.5 Sc\, Re), \quad \forall z \in [0, L], \; i = 1, \ldots, N_{comp} \tag{5.13}$$

$$\lambda_z = \lambda_{bg}(7 + 0.5 Pr\, Re), \quad \forall z \in [0, L] \tag{5.14}$$

The molecular diffusivity for a binary mixture can be calculated by using Chapman–Enskog equation [1, 2]:

$$D_{m,i} = 1.8583 \times 10^{-3} \sqrt{\frac{T^3 \frac{1}{MW_i}}{P\sigma_{12}^2 \Omega_{12}}}, \quad \forall r \in [0, R_p], \; i = 1, 2, \ldots \tag{5.15}$$

where σ_{12}, ε_{12} are constants in Lennard–Jones potential-energy function given by

$$\sigma_{12} = \frac{\sigma_1 + \sigma_2}{2}, \quad \varepsilon_{12} = \sqrt{\varepsilon_1 \varepsilon_2} \tag{5.16}$$

and Ω_{12} is a collision integral which is a function of $k_B T/\varepsilon_{12}$.

5.2.3.8 Transport Properties

Transport properties are assigned constant values or predicted using appropriate correlations. In general, any appropriate correlation can be used. Some of the most commonly used were developed by Wakao [46, 47]:

$$Sh = \frac{k_{f,i} R_p}{D_{m,i}} = \left(2.0 + 1.1 Sc^{0.33} Re^{0.6}\right)$$

$$\forall z \in [0, L], \; i = 1, \ldots, N_{comp} \tag{5.17}$$

$$Nu = \frac{k_h R_p}{\lambda} = \left(2.0 + 1.1 Pr^{0.33} Re^{0.6}\right), \quad \forall z \in [0, L] \tag{5.18}$$

5.2.3.9 Boundary Conditions

At both layer ends, Danckwert's boundary conditions are applied:

- For stream inlet into the layer

$$u\left(C_i - C_i^{in}\right) = D_{z,i}\frac{\partial C_i}{\partial z}, \quad z = 0 \text{ or } z = L, \; i = 1, \ldots, N_{comp} \tag{5.19}$$

$$\rho c_p u\left(T - T^{in}\right) = \lambda_z \frac{\partial T}{\partial z}, \quad z = 0 \text{ or } z = L \tag{5.20}$$

$$u = u^{in}, \quad z = 0 \text{ or } z = L \tag{5.21}$$

- For stream outlet from the layer or closed bed end

$$\frac{\partial C_i}{\partial z} = 0, \quad z = 0 \text{ or } z = L, \ i = 1, \ldots, N_{comp} \quad (5.22)$$

$$\frac{\partial T}{\partial z} = 0, \quad z = 0 \text{ or } z = L \quad (5.23)$$

$$\frac{\partial u}{\partial z} = 0, \quad z = 0 \text{ or } z = L \quad (5.24)$$

where u^{in}, C_i^{in}, and T^{in} are interstitial velocity, concentration, and temperature of the inlet stream into the layer, respectively.

5.2.4
Adsorbent Particle Model

Similar to the bulk gas flow, the adsorption particle model takes into account the simultaneous mass and heat balance, equations for transport and thermophysical properties of the gas mixture, and a set of boundary conditions at the particle surface and particle center (Fig. 5.5). The resulting model equations form a set of differential and algebraic equations that depend on time and radial position in the particle.

The main assumptions made in deriving the model equations are:

- all particles are represented by uniform spheres;
- only changes in the radial direction occur – transport in θ and ϕ directions can be neglected.

Furthermore, it is assumed that the particle temperature is equal to that of the interstitial fluid, an assumption that is commonly used in gas adsorption studies [1–4].

5.2.4.1 General Mass Balance Equations
Starting from the general balance equation (5.2) after a few simple mathematical operations, we get the general expression for mass transfer in the spherical particles:

$$\varepsilon_p \frac{\partial C_i^p}{\partial t} + (1-\varepsilon_p) N_{gi}^p = \varepsilon_p N_i^p, \quad \forall r \in (0, R_p), \ i = 1, \ldots, N_{comp} \quad (5.25)$$

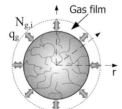

Fig. 5.5 Adsorbent particle.

where C_i^p is concentration of component i in the gas phase, N_{gi}^p is the generation term of the component i, and N_i^p is molar flux of component i. In general, the molar flux includes the sum of different contributions including gas and surface diffusion (SD), convective, and viscous flow. Gas and surface fluxes follow Fick's first law of diffusion, while the other terms can be usually neglected. The equations for the gas and surface molar fluxes within the particle are given by (assuming constant diffusivities):

$$\left(N_i^p\right)_{\text{gaseous}} = D_{e,i}\left(\frac{\partial^2 C_i^p}{\partial r^2} + \frac{2}{r}\frac{\partial C_i^p}{\partial r}\right)$$

$$\left(N_i^p\right)_{\text{surface}} = D_{s,i}\left(\frac{\partial^2 Q_i^p}{\partial r^2} + \frac{2}{r}\frac{\partial Q_i^p}{\partial r}\right) \quad (5.26)$$

where $D_{e,i}$ is the effective gas diffusivity (usually a combined effect of Knudsen and molecular diffusion), $D_{s,i}$ is surface diffusivity, and Q_i^p is the concentration of the gas adsorbed in the particle. The exact expressions used to calculate the different diffusivities depend on the type of the adsorbent–adsorbate system (for the case of SD), the pressure, and the pore size (for the case of gas diffusion).

The mass transfer in the adsorbent particles can be described by several diffusion mechanisms. The most common in the literature are (i) local equilibrium (LEQ), (ii) linear driving force (LDF), (iii) SD, and (iv) pore diffusion (PD) [1–4]. The first two mechanisms (LEQ and LDF) simplify the solution of the model equations significantly by removing the need to solve the mass balance at the particle scale. This is achieved by introducing certain assumptions which result in simplified expressions replacing Eqs. (5.25) and (5.26). The last two mechanisms (SD and PD) are more rigorous approaches that take into account mass balance at both scales (interstitial fluid and adsorbent particle). In the next sections, a derivation for mass balance expressions will be given for all four mechanisms.

5.2.4.2 Local Equilibrium

The main assumptions are that the mass transfer through the particles is instantaneous and that the gas phase is in equilibrium with the solid phase at any radial position. Thus, the gas concentration within the particles is equal to the concentration in the bulk flow:

$$C_i^p = C_i, \quad \forall r \in [0, R_p], \; \forall z \in [0, L], \; i = 1, \ldots, N_{\text{comp}} \quad (5.27)$$

and the gas phase is in equilibrium with the solid phase:

$$Q_i = Q_i^*, \quad \forall r \in [0, R_p], \; \forall z \in [0, L], \; i = 1, \ldots, N_{\text{comp}} \quad (5.28)$$

The generation term in Eq. (5.3) is a sum of amount of gas adsorbed and accumulation in the gas phase within particles:

$$N_{g,i} = \rho^p \frac{\partial Q_i^*}{\partial t} + \varepsilon_p \frac{\partial C_i}{\partial t}, \quad \forall z \in (0, L), \ i = 1, \ldots, N_{comp} \tag{5.29}$$

5.2.4.3 Linear Driving Force (LDF)

LDF approximation is one of the most widely used in adsorption models. In this approach, the gas film resistance is neglected and the overall uptake rate in a particle is expressed as a function of the bulk gas flow concentration:

$$\frac{\partial \bar{Q}_i}{\partial t} = \frac{15 D_{e,i}}{R_p^2}(Q_i^* - \bar{Q}_i), \quad \forall z \in [0, L], \ i = 1, \ldots, N_{comp} \tag{5.30}$$

where $\bar{Q}_i$ is the volume-averaged adsorbed amount per unit volume of sorbent, Q_i^* is adsorbed amount in equilibrium with the gas phase, and $D_{e,i}$ is the effective diffusivity usually measured experimentally. The same equation can be obtained by assuming a parabolic concentration profile within the particle [2]. The generation term in Eq. (5.3) is a sum of amount of gas adsorbed and accumulation in the gas phase within particles:

$$N_{g,i} = \rho^p \frac{\partial \bar{Q}_i}{\partial t} + \varepsilon_p \frac{\partial C_i}{\partial t}, \quad \forall z \in (0, L), \ i = 1, \ldots, N_{comp} \tag{5.31}$$

5.2.4.4 Surface Diffusion

Surface diffusion is the dominant mass transfer mechanism in many adsorptive gas separation processes (as it is a case with diffusion within zeolite microcrystals). Gas phase within particles is neglected except at the particle surface:

$$C_i^p = 0, \quad \forall r \in [0, R_p), \ \forall z \in [0, L], \ i = 1, \ldots, N_{comp} \tag{5.32}$$

Adsorption takes place at the particle surface where the gas phase is in equilibrium with the solid phase:

$$Q_i = Q_i^*, \quad r = R_p, \quad \forall z \in [0, L], \ i = 1, \ldots, N_{comp} \tag{5.33}$$

Accordingly, the adsorbed phase is then transported to the particle center by SD:

$$\frac{\partial Q_i}{\partial t} = \frac{\partial}{\partial r}\left(D_{s,i} \frac{\partial Q_i}{\partial r}\right)$$

$$\forall r \in (0, R_p), \ \forall z \in [0, L], \ i = 1, \ldots, N_{comp} \tag{5.34}$$

Mass transfer between the bulk flow and particles is carried out through the gas film around particles. In that case, the generation term in Eq. (5.3) is given by

$$N_{g,i} = \frac{3 k_f}{R_p}(C_i - C_i^p|_{r=R_p}), \quad \forall z \in [0, L], \ i = 1, \ldots, N_{comp} \tag{5.35}$$

At the particle surface, we have a continuity equation:

$$k_{f,i}\left(C_i - C_i^p\big|_{r=R_p}\right) = D_{s,i} \frac{\partial Q_i}{\partial r}\bigg|_{r=R_p}$$

$$r = R_p, \ \forall z \in [0, L], \ i = 1, \ldots, N_{\text{comp}} \tag{5.36}$$

At the particle center, it is assumed the symmetry condition is employed:

$$\frac{\partial Q_i}{\partial r}\bigg|_{r=0} = 0, \quad r = 0, \ \forall z \in [0, L], \ i = 1, \ldots, N_{\text{comp}} \tag{5.37}$$

5.2.4.5 Pore Diffusion

This model is applicable to the adsorbents with a monodisperse pore structure: activated carbon, alumina, silica gel. Within the pores, gas is transported by combined molecular and Knudsen diffusion. Thus, the mass balance can be given by the following equation:

$$\varepsilon_p \frac{\partial C_i^p}{\partial t} + (1 - \varepsilon_p)\rho^p \frac{\partial Q_i}{\partial t} + \frac{1}{r^2}\frac{\partial}{\partial r}\left(r^2 N_i^p\right)$$

$$\forall r \in (0, R_p), \ \forall z \in [0, L], \ i = 1, \ldots, N_{\text{comp}} \tag{5.38}$$

The gas phase within pores is assumed in equilibrium with the adsorbed phase:

$$Q_i = Q_i^*, \quad \forall r \in [0, R_p], \ \forall z \in [0, L], \ i = 1, \ldots, N_{\text{comp}} \tag{5.39}$$

It is generally agreed that the most convenient way to model multicomponent gas diffusion of N-species in porous media is the dusty gas model (DGM) (see for example the review by Krishna, [48]). The DGM model is derived from the Maxwell–Stefan (MS) equations for a mixture of $N + 1$ species, where the extra species is the solid that consists of giant molecules ("dust") that are immobile and uniformly distributed in space. The DGM equations are written as

$$\frac{N_i^p}{D_{K,i}^e} + \sum_{\substack{j=1 \\ j \neq i}}^{n} \frac{y_j N_i^p - y_i N_j^p}{D_{m,ij}^e} = -\nabla C_i^p, \quad i = 1, \ldots, N_{\text{comp}} \tag{5.40}$$

where $D_{K,i}^e$, $D_{m,ij}^e$ are the effective diffusivities in the molecular and Knudsen regimes for the porous medium, respectively, and are given by the following simple expressions:

$$D_{K,i}^e = \frac{\varepsilon_p D_{K,i}}{\tau_K}, \quad i = 1, \ldots, N_{\text{comp}} \tag{5.41a}$$

$$D_{m,ij}^e = \frac{\varepsilon_p D_{m,ij}}{\tau_m}, \quad i = 1, \ldots, N_{\text{comp}} \tag{5.41b}$$

In the above expressions, τ is the tortuosity factor that expresses the deviation of the effective diffusivities in the porous medium from their respective values at some reference conditions. More specifically for the case of molecular diffusion, we use as reference value the bulk molecular diffusivity calculated by the Chapman Enskog equation (5.15).

The reference value for the Knudsen diffusivity is that of gas in an infinitely long capillary and is given by the following analytical expression (see for example [49]):

$$D_{k,i} = 9.7 \times 10^3 R_{\text{pore}} \sqrt{\frac{TP}{MW_i}}, \quad \forall r \in [0, R_p], \; i = 1, \ldots, N_{\text{comp}} \quad (5.42)$$

Mass transfer between the bulk flow and particles is carried out through the gas film around particles. In that case, the generation term in Eq. (5.3) is given by

$$N_i = \frac{3k_f}{R_p}(C_i - C_i^p|_{r=R_p}), \quad \forall z \in [0, L], \; i = 1, \ldots, N_{\text{comp}} \quad (5.43)$$

At the particle surface, we have a continuity equation

$$k_{f,i}(C_i - C_i^p|_{r=R_p}) = D_{e,i}\frac{\partial C_i^p}{\partial r}\bigg|_{r=R_p}$$

$$r = R_p, \; \forall z \in [0, L], \; i = 1, \ldots N_{\text{comp}} \quad (5.44)$$

At the particle center, it is assumed that there is no mass transfer:

$$\frac{\partial Q_i}{\partial r}\bigg|_{r=0} = 0, \quad r = 0, \; \forall z \in [0, L], \; i = 1, \ldots, N_{\text{comp}} \quad (5.45)$$

5.2.4.6 Gas–Solid Phase Equilibrium Isotherms

The quantitative description of gas–solid interactions at equilibrium conditions in the form of models or correlations is extremely important for the accurate design and simulation of PSA processes. At constant temperature, the model that describes the concentration of a species in the adsorbed phase as a function of the gas composition and pressure is called the adsorption isotherm. Adsorption isotherms are given in the form of algebraic or integral equations that are used to determine the amount of gas adsorbed within the adsorbent particles as a function of pressure, temperature, and composition of the gas. These equations contain several semiempirical parameters that are determined by fitting experimental isotherm data at different temperatures. In general, there are three different categories of single-gas isotherms: single- and dual-site Langmuir isotherms, isotherms based on the Gibbs approach, and isotherms based on the potential-theory approach [1, 2]. For the case of multicomponent mixtures, the more widely used models are linear isotherms, extended Langmuir equations, loading ratio correlations (LRC), and adsorbed solution theory. Each of the above models is briefly described below.

5.2 Model Formulation

Linear Isotherms At sufficiently low concentrations, the amount of adsorbed gas is a linear function of the concentration. This linear relationship is commonly referred to as Henry's law and given by the following equation:

$$Q_i^* = H_i C_i^p, \quad \forall r \in [0, R_p], \; \forall z \in [0, L], \; i = 1, \ldots, N_{\text{comp}} \tag{5.46}$$

where H_i is a function of pressure and temperature:

$$H_i = f(T^p, P) \tag{5.47}$$

An important term in both mass and heat balance equations that has to be calculated is $\partial Q_i^*/\partial t$. Q_i^* is, in general case, a function of pressure, temperature, and component concentrations:

$$Q_i^p = f(T, P, C_i^p), \quad i = 1, \ldots, N_{\text{comp}}$$

Thus, the total derivative per time is given by

$$\frac{\partial Q_i^*}{\partial t} = \frac{\partial Q_i^*}{\partial T} \frac{\partial T}{\partial t} + \frac{\partial Q_i^*}{\partial P} \frac{\partial P}{\partial t} + \sum_k \frac{\partial Q_i^*}{\partial C_k} \frac{\partial C_k}{\partial t}, \quad i, k = 1, \ldots, N_{\text{comp}} \tag{5.48}$$

For constant pressure and temperature, the total derivative simplifies to

$$\frac{\partial Q_i^*}{\partial t} = \sum_k \frac{\partial Q_i^*}{\partial C_k^p} \frac{\partial C_k^p}{\partial t}, \quad i, k = 1, \ldots, N_{\text{comp}} \tag{5.49}$$

In case of linear adsorption isotherm, the time derivative is simple and given by the following equation:

$$\frac{\partial Q_i^*}{\partial t} = H_i \frac{\partial C_i^p}{\partial t}, \quad \forall r \in [0, R_p], \; \forall z \in [0, L], \; i = 1, \ldots, N_{\text{comp}} \tag{5.50}$$

Extended Langmuir Isotherms The single-component Langmuir isotherm can easily be extended for a multicomponent mixture:

$$Q_i^* = Q_{m,i} \frac{b_i P_i}{1 + \sum_j b_j P_j}, \quad i, j = 1, \ldots, N_{\text{comp}} \tag{5.51}$$

$$Q_{m,i} = a_{i,1} + \frac{a_{i,2}}{T^p}, \quad i = 1, \ldots, N_{\text{comp}} \tag{5.52}$$

$$b_i = b_{i,1} \exp\left(\frac{b_{i,2}}{T^p}\right) \tag{5.53}$$

In this case, calculation of the total derivative $\partial Q_i^*/\partial t$ is somewhat more complex. Starting again from Eq. (5.48) and after several mathematical operations, we get the following equation:

$$\frac{\partial Q_i^*}{\partial t} = \frac{Q_{m,i} b_i}{(1 + \sum_j b_j C_j^p)^2} \sum_k Der_k \frac{\partial C_k^p}{\partial t}$$

$$\forall r \in [0, R_p], \ \forall z \in [0, L], \ i, j, k = 1, \ldots, N_{comp} \quad (5.54)$$

where the term Der_k is given by

for $k = i$: $\quad Der_k = 1 + \sum_j b_j C_j^p - b_i C_i^p$

otherwise: $\quad Der_k = -b_k C_i^p \quad (5.55)$

Loading Ratio Correlation Equations The LRC isotherms are based on the extension of the single-component Langmuir–Freundlich isotherm for the case of multicomponent mixtures [1, 2]:

$$Q_i^* = Q_{m,i} \frac{b_i P_i^{\eta_i}}{1 + \sum_j b_j P_j^{\eta_j}}, \quad i, j = 1, \ldots, N_{comp} \quad (5.56)$$

Adsorbed Solution Theory The adsorbed solution theory is a special case in the treatment by Myers and Prausnitz [50], of the mixed adsorbate as a solution in equilibrium with the gas phase. The fundamental thermodynamic equations for liquids are applied to the adsorbed phase. A detailed derivation can be found in Myers and Prausnitz [50], and practical considerations in O'Brien and Myers [51].

In general, any suitable single-component adsorption isotherm can be used together with the adsorbed solution theory. Nevertheless, Langmuir isotherm is the most commonly used one since it can accurately predict the experimental data for single-component gases. Hence, the resulting set of equations needed to be solved given here for the Langmuir isotherm is given below:

$$\pi_i^* = \int_0^{P_i^0} \frac{Q_{pure,i}^*}{P} dP \quad \xrightarrow{\text{Langmuir isotherm}} \quad \pi_i^* = Q_{m,i}(1 + b_i P_i^0)$$

$$\forall z \in [0, L], \ i = 1, \ldots, N_{comp} \quad (5.57)$$

$$\pi_i^* = \pi_{i+1}^*, \quad \forall r \in [0, R_p], \ \forall z \in [0, L], \ i = 1, \ldots, N_{comp} - 1 \quad (5.58)$$

$$Q_{pure,i}^* = Q_{m,i} \frac{b_i P_i}{1 + b_i P_i}$$

$$\forall r \in [0, R_p], \ \forall z \in [0, L], \ i = 1, \ldots, N_{comp} \quad (5.59)$$

$$Q_{m,i} = a_{i,1} + \frac{a_{i,2}}{T}, \quad \forall r \in [0, R_p], \ \forall z \in [0, L], \ i = 1, \ldots, N_{comp} \quad (5.60)$$

$$b_i = b_{i,1} \exp\left(\frac{b_{i,2}}{T^p}\right), \quad \forall r \in [0, R_p], \ \forall z \in [0, L], \ i = 1, \ldots, N_{comp} \quad (5.61)$$

$$X_i^* P_i^0 \gamma_i = X_i P, \quad \forall r \in [0, R_p], \forall z \in [0, L], i = 1, \ldots, N_{\text{comp}} \quad (5.62)$$

$$\sum_i X_i^* = 1, \quad \forall r \in [0, R_p], \forall z \in [0, L], i = 1, \ldots, N_{\text{comp}} \quad (5.63)$$

$$\frac{1}{Q_{\text{total}}} = \sum_i \frac{X_i^*}{Q_{\text{pure},i}^*} + \frac{RTP}{A} \sum_i X_i^* \left(\frac{\partial \ln \gamma_i}{\partial \pi_i^*}\right)_{X_i^*}$$

$$\forall r \in [0, R_p], \forall z \in [0, L], i = 1, \ldots, N_{\text{comp}} \quad (5.64)$$

$$Q_i^* = X_i^* Q_{\text{total}}, \quad \forall r \in [0, R_p], \forall z \in [0, L], i = 1, \ldots, N_{\text{comp}} \quad (5.65)$$

In above equations, π_i^* is spreading pressure, P_i^0 is the equilibrium "vapor pressure" for pure component i at the same spreading pressure and the same temperature as the adsorbed mixture, γ_i is the activity coefficient of component i, X_i^* is the mole fraction in adsorbed phase, and X_i is the mole fraction in gas phase. Many different equations are available for calculating the activity coefficients in Eq. (5.62). For instance, Wilson's equation, coupled with the appropriate equation of state (UNIQUAC), has been successfully used [2].

For the special case of ideal solution where γ_i is unity, we get the ideal adsorbed solution (IAS) theory, Eq. (5.62) reduces to Raoult's law:

$$X_i^* P_i^0 = X_i P, \quad \forall r \in [0, R_p], \forall z \in [0, L], i = 1, \ldots, N_{\text{comp}} \quad (5.66)$$

and Eq. (5.64) becomes

$$\frac{1}{Q_{\text{total}}} = \sum_i \frac{X_i^*}{Q_{\text{pure},i}^*}, \quad \forall r \in [0, R_p], \forall z \in [0, L], i = 1, \ldots, N_{\text{comp}} \quad (5.67)$$

5.2.5
Gas Valve Model

The gas valve model describes a one-way valve. The purpose of the one-way valve is to force the flow only to desired directions and to avoid any unwanted flows. Gas valve equations are shown below:

If $P_{\text{out}} > P_{\text{crit}} P_{\text{in}}$:

$$F = C_v S P P_{\text{in}} \sqrt{\left|\frac{1 - (\frac{P_{\text{out}}}{P_{\text{in}}})^2}{\sum x_i M W_i T}\right|}, \quad i = 1, \ldots, N_{\text{comp}} \quad (5.68)$$

Otherwise:

$$F = C_v S P P_{\text{in}} \sqrt{\left|\frac{1 - (P_{\text{crit}})^2}{\sum x_i M W_i T}\right|}, \quad i = 1, \ldots, N_{\text{comp}} \quad (5.69)$$

Critical pressure:

$$P_{crit} = \left(\frac{2}{1+\kappa}\right)^{\frac{\kappa}{1-\kappa}} \tag{5.70}$$

Power:

$$\text{Power} = \left(\frac{\kappa}{\kappa-1}\right) R T_{feed} \left(\left(\frac{P_{feed}}{P_{low}}\right)^{\frac{\kappa-1}{\kappa}} - 1\right) \frac{N_{feed}}{\tau_{cycle}} \tag{5.71}$$

5.2.6
The Multibed PSA Model

The single-bed PSA model provides the basis for the automatic generation of the flowsheet via a network superstructure of single adsorbent beds. The main building block of the multibed PSA model is illustrated in Fig. 5.3(a), in Section 5.2.2.

The central part of the building block is the adsorbent bed model. Feed, purge gas, and strong adsorptive component streams are connected to the corresponding bed ends via gas valves. This is also the case for light and waste product sinks. The bed is properly connected to all other beds in the system at both ends via gas valves. Such a configuration makes the flowsheet sufficiently general to support all feasible bed interconnections. This way, all possible operating steps in every known PSA process can be supported. The main building block can be replicated accordingly through an input parameter representing the number of beds in the flowsheet. A typical four-bed PSA flowsheet is shown in Fig. 5.6.

Procedures for controlling the operation of multibed PSA processes are highly complex due to the large number of interactions. Hence, an auxiliary program for automatic generation of operating procedures has been developed in [43]. This program generates operating procedures for the whole network of beds according to the given number of beds and sequence of operating steps in one bed. Operating procedures govern the network by opening or closing the appropriate valves at the desired level and changing the state of each bed. This auxiliary development allows the automatic generation of a gPROMS source code for a given number of beds and sequence of operating steps in one bed. It also generates gPROMS tasks ensuring feasible connectivities between the units according to the given sequence. A simplified algorithm illustrating the generation of operating procedures is presented in Fig. 5.7. The program is not limited only to PSA configurations where beds undergo the same sequence of steps, but it can also handle more complex configurations [43].

5.2.7
The State Transition Network Approach

Controlling the execution of a PSA process by using a set of operating procedures has two major drawbacks: (a) the development of operating procedures for

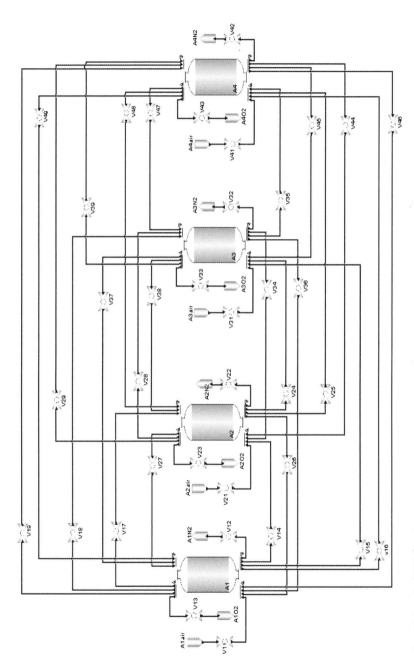

Fig. 5.6 Four-bed PSA flowsheet.

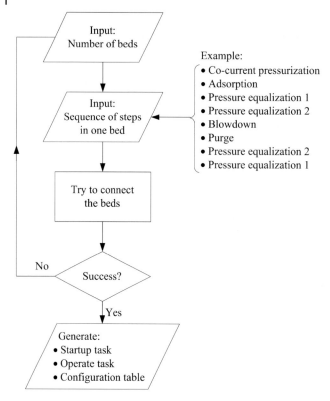

Fig. 5.7 An algorithm for the generation of feasible operating procedures for PSA processes.

multibed PSA flowsheets is a time-consuming task and special software developments need to be developed as outlined above, and (b) it cannot be directly used in optimization studies. For this reason, Nikolic et al. [52] developed a new, more robust method, which relies on a state transition network (STN) representation of the process. States are represented by operating steps (i.e., an adsorbent column can be in one of the operating states such as pressurization, adsorption, purge, etc.), inputs are the time elapsed in the process, the time within the current cycle and several input parameters known a priory at the time of execution. Each state (the operating step) includes a set of boundary conditions and gas valves states (open/closed). A deterministic finite state machine (FSM) is implemented where the next possible state is uniquely determined for a given (current) state and input values. State transitions are decisions when the state change should occur (based on the input values). The start state is commonly co- or counter-current pressurization or pressure equalization (repressurization) step. This way, it is possible to control the execution of the process by specifying a few parameters such as the number of beds, the sequence of steps in one bed, the number of pressure equalization steps, and the start time and duration of each step. An STN graph with all possible state transitions is presented in Fig. 5.8. The overall idea is to identify which state transitions are feasible and conditions under which a particular state

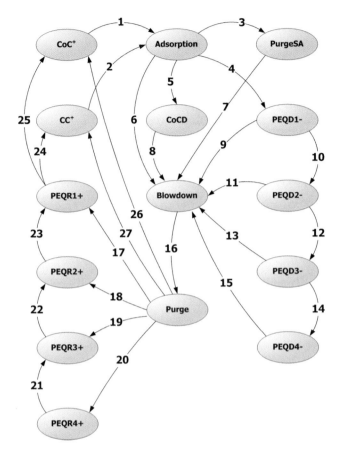

Fig. 5.8 State transition network with all possible state transitions [52].

change occurs under a given current state. For instance, from the pressurization step (either by feed or by light product) it is possible to switch to *adsorption* step only. On the other hand, from the adsorption step it is possible to go to many other states such as co-current depressurization, blowdown, pressure equalization, or purge by strong adsorptive, depending on the step sequence that needs to be simulated. In general, the cycle time and the time elapsed in the process are used to identify the position within the cycle. More specifically, by defining certain process parameters (such as the number of pressure equalization steps), and for a given sequence of steps in one bed and duration of certain steps in the cycle, it is possible to distinguish between allowed and forbidden steps and make the decision about the transition.

Further details on the STN approach and how it is implemented to describe complex multibed PSA process are given in the work of Nikolic *et al.* [52].

5.2.8
Numerical Solution

The modeling equations comprise a system of nonlinear PDAEs. In general, there have been two different approaches proposed to solve these PDAEs at the CSS: direct and iterative methods.

Direct methods employ the concept of complete spatial and temporal discretization using standard discretization schemes [25]. This leads to a large system of nonlinear algebraic equations, which can be solved by large-scale Newton-based equation solvers. This approach allows the straightforward addition of the CSS conditions to discretized equations, therefore leading to the direct solution of the problem at the CSS without needing to go through each cycle during the transient process operation [25]. Unfortunately the direct method has several drawbacks: (a) an effective initialization for a very large number of variables is needed and the Newton-based solver may fail in the face of steep concentration, temperature, and/or velocity wavefronts inside the bed [8]. For this reason, direct methods are rarely used compared to iterative methods, which are employed in the majority of PSA applications and for this reason will be analyzed in more detail below.

Iterative methods solve the model equations using iterative procedures such as successive substitution, where the model equations are solved in sequence, cycle by cycle until CSS is achieved. In this case, a two-step approach, known as the method of lines, is used to discretize the model equations. More particularly, the spatial domains are discretized first using several options from finite difference [13, 14, 20], orthogonal collocation [12, 15, 43], finite volume [26, 53], and Galerkin finite element [7, 54] schemes. Then the resulting system of nonlinear ordinary differential algebraic equations (DAEs) is solved using stiff integrator algorithms based on sophisticated variable-step variable-order numerical schemes, which can efficiently handle model discontinuities and complex operating procedures [55–57]. The advantage of this method is that space and time discretizations are now decoupled to achieve higher-order accuracy, while issues of initialization and convergence are handled more easily and effectively.

It must be noted that in general, the solution of adsorption-based model equations may cause several complications particularly when diffusive or dispersive terms have minimal effect or are absent. This is due to the fact that the resulting partial differential equations become purely hyperbolic causing nonphysical oscillations and numerical instability in the standard discretization schemes. In such cases, the easiest remedy is to employ standard first-order upwind finite difference schemes [58]. However, this will reduce the accuracy of the discretization since the accuracy of first-order differencing schemes scales linearly with the length-step, while the inevitable introduction of numerical dispersion (smearing effect) caused by the use of these schemes will distort the physical picture of the problem, unless one uses an unreasonably large number of discretization points. On the other hand, there exist flux-corrected transport schemes [59] and flux limiter methods [26, 53], where the idea is to use antidiffusion terms and schemes to overcome the effect of excessive numerical dispersion, while at the same time some dispersion will be used to dampen unphysical oscillations. There are also other discretization methods particularly suited for such cases, like the adaptive multiresolution approach [60], which are capable of locally refining the grid in the regions where the solution exhibits sharp features thus allowing nearly constant discretization error throughout the computational domain. This approach has been successfully applied in the simulation and optimization of cyclic adsorption processes [27].

The successive substitution method, also known as Picard iteration, is a stable method with a convergence path that mimics the transient performance of an actual PSA process. However, in many processes, hundreds or thousands of cycles may be needed to reach CSS conditions due to large thermal relaxation times or the presence of recycle streams in the process. There are other iteration methods, besides successive substitution [61–64] that can be used effectively to accelerate convergence to CSS; however one must use them with care as several undesirable issues may take place (see [8] and related references cited therein).

5.3
Case-Study Applications

The developed modeling framework has been applied in a PSA process concerning the separation of hydrogen from SMR off gas using activated carbon as an adsorbent. In this case, the nonisothermal LDF model has been used for simulation purposes. The geometrical data of a column, adsorbent, and adsorption isotherm parameters for activated carbon have been adopted from the work of Park *et al.* [11] and are summarized in the recent study of Nikolic *et al.* [43].

The effective diffusivity, axial dispersion coefficients, and heat transfer coefficient are assumed constant in accordance with [11]. Flowsheets of 1, 4, 8, and 12 beds have been simulated. Two different cycle configurations have been used. The sequence of the steps differs only in the first step. In the first configuration (configuration A), pressurization has been carried out by using the feed stream, whereas in the second (configuration B) by using the light product stream, which is pure H_2. The two configurations are shown schematically in Fig. 5.9.

These configurations differ only in the number of pressure equalization steps introduced. Thus, the one-bed configuration contains no pressure equalization steps, the four-bed involves one, the eight-bed configuration two, and the 12-bed flowsheet involves three pressure equalization steps. The following sequence of steps has been used: pressurization, adsorption, pressure equalization (depressurization to other beds), blowdown, purge, and pressure equalization (repressurization from other beds). The sequence of steps for the one-bed configuration is Ads, Blow, Purge, CoCP, and an example of operating steps employed in the four-bed configuration is presented in Table 5.4, where CoCP stands for co-current pressurization, Ads for adsorption, EQD1, EQD2, and EQD3 are the pressure equalization steps (depressurization to the other bed), blow represents counter-current blowdown, *purge* is the counter–counter purge step, and EQR3, EQR2, and EQR1 are the pressure equalization steps (repressurization from the other bed).

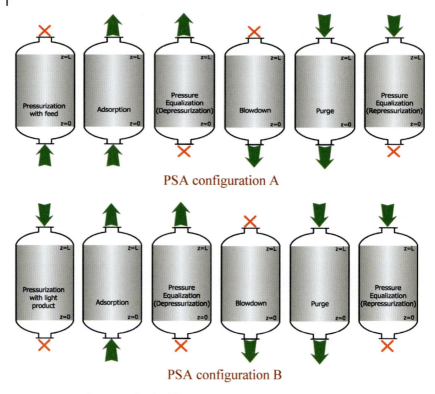

Fig. 5.9 PSA configurations for the different case studies.

Configurations using pressurization by the light product (H_2) are the same apart from the first step. All configurations have been generated using the auxiliary program described in Section 5.2.4. The bed is initially assumed clean. For simulation purposes, the axial domain is discretized using orthogonal collocation on finite elements of third order with 20 elements, although other discretization schemes have been tested.

Three different sets of simulations have been carried out.

Table 5.4 A four-bed six-step PSA configuration.

Bed 1	CoCP	Ads	Ads	EQD1	Blow	Purge	Purge	EQR1
Bed 2	Purge	EQR1	CoCP	Ads	Ads	EQD1	Blow	Purge
Bed 3	Blow	Purge	Purge	EQR1	CoCP	Ads	Ads	EQD1
Bed 4	Ads	EQD1	Blow	Purge	Purge	EQR1	CoCP	Ads

Table 5.5 Simulation results of Run I.

No. of beds	H$_2$ purity (%)	H$_2$ recovery (%)	Power (W)	Productivity (mol/kg)
1	96.87	34.40	5290.2	9.18×10^{-2}
4	99.50	58.06	2794.9	4.85×10^{-2}
8	99.58	68.87	1198.7	2.08×10^{-2}
12	99.36	72.50	1393.6	2.42×10^{-2}

5.3.1
Simulation Run I

In this case, configuration A has been employed (pressurization with feed). The effect of number of beds and cycle time (due to introduction of pressure equalization steps) on the separation quality has been investigated. The following operating conditions have been selected: constant duration of adsorption and purge steps, and constant feed and purge gas flowrates. The input parameters of the simulation are taken from [43], and the simulation results are presented in Table 5.5.

The results clearly illustrate that, as the number of beds increases, significant improvements in product recovery ($\sim$ 38%) and purity ($\sim$ 3%) are achieved. The improved purity can be attributed to the fact that flowsheets with lower number of beds process more feed per cycle (feed flowrate is constant but more feed is needed to repressurize beds from the lower pressure). The noticeable increase in product recovery is the direct result of the pressure equalization steps. On the other hand, power requirements and adsorbent productivity decrease due to the lower amount of feed processed per unit time (the power and productivity of the eight-bed configuration are lower than the power and productivity of the 12-bed configuration due to higher cycle time). The purity of the 12-bed configuration is lower than the purity of the four- and eight-bed ones. This interesting trend can be attributed to the fact that during the third pressure equalization, a small breakthrough takes place thus contaminating the pressurized bed.

5.3.2
Simulation Run II

In simulation Run II, configuration A has been employed (pressurization with feed). The effect of number of beds for constant power requirements and constant adsorbent productivity on the separation quality has been analyzed. The following operating conditions have been selected: constant adsorbent productivity, constant cycle time, and constant amount of feed processed per cycle. The input parameters are taken from [43], and the simulation results are summarized in Table 5.6.

The results show that increasing the number of beds, a slight increase in the product purity ($\sim$ 1%) and a significant increase in product recovery ($\sim$ 52%) are achieved. In this run, the amount of feed processed per cycle is constant and improve in the purity cannot be attributed to the PSA design characteristics and oper-

Table 5.6 Simulation results for Run II.

No. of beds	H$_2$ purity (%)	H$_2$ recovery (%)	Power (W)	Productivity (mol/kg)
1	98.11	20.21	1442.0	2.50×10^{-2}
4	99.29	38.40	1406.9	2.44×10^{-2}
8	99.79	65.25	1409.2	2.45×10^{-2}
12	99.36	72.50	1393.6	2.42×10^{-2}

ating procedure. Due to the same reasons described in Run I, the purity of the 12-bed configuration is lower than the purity of the eight-bed one. The power requirements and adsorbent productivity remain constant due to the constant amount of feed processed per cycle.

5.3.3
Simulation Run III

In this case, configuration B (pressurization with light product) is used. The effect of number of beds for constant power requirements and constant adsorbent productivity on separation quality has been investigated. The following operating conditions have been selected: constant adsorbent productivity, constant duration of adsorption and purge steps, constant feed and purge gas flowrates, constant cycle time, and constant amount of feed processed per cycle. The input parameters are taken from [43], and the simulation results are presented in Table 5.7.

The results illustrate that the number of beds does not affect the product purity, productivity, and power requirements (since the amount of feed per cycle is constant). However, a huge improvement in the product recovery ($\sim 64\%$) is achieved due to three pressure equalization steps. Similar to the results in Runs I and II, the purity of the 12-bed configuration is lower than the purity of the eight-bed one.

The above analysis reveals typical tradeoffs between capital and operating costs and separation quality. Thus, by increasing the number of beds a higher product purity/recovery is achieved while higher capital costs are required (due to a larger number of beds). On the other hand, energy demands are lower due to energy conservation imposed by the existence of pressure equalization steps.

Table 5.7 Simulation results for Run III.

No. of beds	H$_2$ purity (%)	H$_2$ recovery (%)	Power (W)	Productivity (mol/kg)
1	99.87	1.41	1887.8	2.03×10^{-2}
4	99.97	28.47	1180.8	2.05×10^{-2}
8	99.97	50.58	1193.8	2.07×10^{-2}
12	99.83	65.16	1190.9	2.07×10^{-2}

5.4
Conclusions

This chapter discusses recent progress in the area of modeling and simulation of PSA systems. Several research challenges are identified and a generic modeling framework for the separation of gas mixtures using multibed PSA flowsheets is presented. The core of the framework represents a detailed adsorbent bed model relying on a coupled set of mixed algebraic and partial differential equations for mass, heat, and momentum balance at both bulk gas and particle level, equilibrium isotherm equations, and boundary conditions according to the operating steps. The adsorbent bed model provides the basis for building a PSA flowsheet with all feasible interbed connectivities. Operating procedures are automatically generated, thus facilitating the development of complex PSA flowsheet for an arbitrary number of beds. The modeling framework provides a comprehensive qualitative and quantitative insight into the key phenomena taking place in the process. Finally, a case study concerning the separation of hydrogen from SMR off gas is used to illustrate the application and efficiency of the developed framework.

Acknowledgements

Financial support from the PRISM EC-funded Research Training Network (Contract number MRTN-CT-2004-512233) is gratefully acknowledged.

Notation

a_1 – Langmuir isotherm parameter, mol/kg
a_2 – Langmuir isotherm parameter, K
b_1 – Langmuir isotherm parameter, 1/Pa
b_2 – Langmuir isotherm parameter, K
b – Langmuir isotherm parameter, m^3/mol
C – molar concentrations of gas phase in bulk gas, mol/m^3
C^{in} – molar concentrations of gas phase at the inlet of bed, mol/m^3
C^p – molar concentrations of gas phase in particles, mol/m^3
c_p – heat capacity of bulk gas, J/(kg K)
$c_{p,pg}$ – heat capacity of gas within pellets, J/(kg K)
c_p^p – heat capacity of the particles, J/(kg K)
C_v – valve constant
D_e – effective diffusivity coefficient, m^2/s
D_k – Knudsen diffusion coefficient, m^2/s
D_m – molecular diffusion coefficient, m^2/s
d_p – particle diameter, m
D_s – surface diffusion coefficient, m^2/s
D_z – axial dispersion coefficient, m^2/s

F – molar flowrate, mol/s
ΔH_{ads} – heat of adsorption, J/mol
H – Henry's parameter, m^3/kg
k_f – mass transfer coefficient, m^2/s
k_h – heat transfer coefficient, J/(m^2 K s)
$k_{h,wall}$ – heat transfer coefficient for the column wall, J/(m^2 K s)
L – bed length, m
Mw – molecular weight, kg/mol
N_{comp} – number of components, –
N_{feed} – Number of moles in feed stream, mol
N_i – Molar flux through the particle surface, mol/(m^2 s)
Nu – Nusselt dimensionless number, –
Pr – Prandtl dimensionless number, –
P^0 – pressure that gives the same spreading pressure in the multicomponent equilibrium, Pa
P_{in} – pressure at the inlet of the gas valve, Pa
P_{out} – pressure at the outlet of the gas valve, Pa
Q – adsorbed amount, mol/kg
Q^* – adsorbed amount in equilibrium state with gas phase (in the mixture), mol/kg
q_g – heat generation term, heat flux through particle surface, J/(m^2 s)
Q_m – Langmuir isotherm parameter, mol/kg
Q^*_{pure} – adsorbed amount in equilibrium state with gas phase (pure component), mol/kg
Q_{total} – total adsorbed amount, mol/kg
r – radial discretization domain, m
R_{bed} – bed radius, m
Re – Reynolds dimensionless number, –
R_p – pellet radius, m
R_{pore} – radius of the pore, m
Sc – Schmidt dimensionless number, –
Sh – Sherwood dimensionless number, –
sp – stem position of a gas valve
sp_{AdsIn} – stem position of a gas valve during adsorption (inlet valve), –
sp_{AdsOut} – stem position of a gas valve during adsorption (outlet valve), –
sp_{Blow} – stem position of a gas valve during blowdown, –
sp_{PEQ1} – stem position of a gas valve during pressure equalization, –
sp_{PEQ2} – stem position of a gas valve during pressure equalization, –
sp_{PEQ3} – stem position of a gas valve during pressure equalization, –
$sp_{PressCoC}$ – stem position of a gas valve during co-current pressurization, –
$sp_{PressCC}$ – stem position of a gas valve during counter-current pressurization, –
$sp_{PurgeIn}$ – stem position of a gas valve during purge (inlet valve), –
$sp_{PurgeOut}$ – stem position of a gas valve during purge (outlet valve), –
T – temperature of bulk gas, K
T^{in} – temperature of the fluid at the inlet of the bed, K

T^p – temperature of particles, K
T_{wall} – temperature of the wall, K
u – interstitial velocity, m/s
x – molar fraction in gas phase, –
x^* – molar fraction in adsorbed phase, –
Z – compressibility factor, –
z – axial discretization domain, m

Greek letters

ε_{bed} – porosity of the bed, –
ε_p – porosity of the pellet, –
μ – viscosity of bulk gas, Pa s
λ – thermal conductivity of bulk gas, J/(m K)
λ_z – heat axial dispersion coefficient, J/(m K)
λ^p – thermal conductivity of the particles, J/(m K)
π^* – reduced spreading pressure, –
ρ – density of bulk gas, kg/m^3
ρ_{pg} – density of gas within pellets, kg/m^3
ρ^p – density of the particles, kg/m^3
κ – heat capacity ratio, –
τ – time, s
τ_p – tortuosity of the particle, –
τ_{cycle} – total cycle time, s

Subscripts

i – component

Superscripts

p – particle

References

1 RUTHVEN, D. M., *Principles of Adsorption and Adsorption Processes*, Wiley, New York, **1984**.
2 YANG, R. T., *Gas Separation by Adsorption Processes*, Butterworths, Boston, **1987**.
3 SUZUKI, M., *Adsorption Engineering*, Kodansha Elsevier, Tokyo, **1990**.
4 RUTHVEN, D. M., FAROOQ, S., KNAEBEL, K. S., *Pressure Swing Adsorption*, VCH Publishers, New York, **1994**.
5 CEN, P. L., CHEN, W. N., YANG, R. T., Ternary gas mixture separation by pressure swing adsorption – a combined hydrogen methane separation and acid removal process, *Industrial and Engineering*

Chemistry, Process Design and Development 24 **(1985)**, pp. 1201–1208.

6 Suh, S. S., Wankat P. C., Pressure swing adsorption for binary gas separation with Langmuir isotherms, *Chemical Engineering Science* 44 **(1989)**, pp. 2407–2410.

7 Kikkinides, E. S., Yang, R. T., Cho, S. H., Concentration and recovery of CO_2 from flue-gas by pressure swing adsorption, *Industrial and Engineering Chemistry Research* 32 **(1993)**, pp. 2714–2720.

8 Biegler, L. T., Jiang, L., Fox, V. G., Recent advances in simulation and optimal design of pressure swing adsorption systems, *Separation and Purification Reviews* 33 **(2005)**, pp. 1–39.

9 Kumar, R., Fox, V. G., Hartzog, D. G., Larson, R. E., Chen, Y. C., Houghton, P. A., Naheiri, T., A versatile process simulator for adsorptive separations, *Chemical Engineering Science* 49 **(1994)**, pp. 3115–3125.

10 Chou, C. T., Huang, W. C., Incorporation of a valve equation into the simulation of a pressure swing adsorption process, *Chemical Engineering Science* 49 **(1994)**, p. 75.

11 Park, J. H., Kim, J. N., Cho, S. H., Kim, J. D., Yang, R. T., Adsorber dynamics and optimal design of layered beds for multicomponent gas adsorption, *Chemical Engineering Science* 53 **(1998)**, p. 3951.

12 Raghavan, N., Hassan M., Ruthven, D. M., Numerical simulation of a PSA system using a pore diffusion model, *Chemical Engineering Science* 41 **(1986)**, p. 2787.

13 Doong, S. J., Yang, R. T., Bulk separation of multicomponent gas mixtures by pressure swing adsorption: pore/surface diffusion and equilibrium models, *AIChE Journal* 32 **(1986)**, p. 397.

14 Doong, S. J., Yang, R. T., Bidisperse pore diffusion model for zeolite pressure swing adsorption, *AIChE Journal* 33 **(1987)**, p. 1045.

15 Shin, H. S., Knaebel, K. S., Pressure swing adsorption: a theoretical study of diffusion-induced separation, *AIChE Journal* 33 **(1987)**, p. 654.

16 Shin, H. S., Knaebel, K. S., Pressure swing adsorption: an experimental study of diffusion-induced separation, *AIChE Journal* 34 **(1988)**, p. 1409.

17 Serbezov, A., Sotirchos, S. V., Mathematical modelling of multicomponent nonisothermal adsorption in sorbent particles under pressure swing conditions, *Adsorption.* 4 **(1998)**, p. 93.

18 Yang, J., Lee, C. H., Adsorption dynamics of a layered bed PSA for H_2 recovery from coke oven gas, *AIChE Journal* 44 **(1998)**, p. 1325.

19 Mendes, A. M. M., Costa, C. A. V., Rodrigues, A. E., PSA simulation using particle complex models, *Separation and Purification Technology* 24 **(2001)**, p. 1.

20 Ko, D., Siriwardane, R., Biegler, L. T., Optimization of a pressure-swing adsorption process using zeolite 13X for CO_2 sequestration, *Industrial and Engineering Chemistry Research* 42 **(2003)**, p. 339.

21 Ko, D., Siriwardane, R., Biegler, L. T., Optimization of pressure swing adsorption and fractionated vacuum pressure swing adsorption processes for CO_2 capture, *Industrial and Engineering Chemistry Research* 44 **(2005)**, p. 8084.

22 Knaebel, S. P., Ko, D., Biegler, L. T., Simulation and optimization of a pressure swing adsorption system: recovering hydrogen from methane, *Adsorption.* 11 **(2005)**, p. 615.

23 Grande, C. A., Rodrigues A. E., Propane/propylene separation by pressure swing adsorption using zeolite 4A, *Industrial and Engineering Chemistry Research* 44 **(2005)**, p. 8815.

24 Ahn, H., Brandani, S., A new numerical method for accurate simulation of fast cyclic adsorption processes, *Adsorption.* 11 **(2005)**, p. 113.

25 Nilchan, S., Pantelides, C. C., On the optimisation of periodic adsorption processes, *Adsorption.* 4 **(1998)**, p. 113.

26 Jiang, L., Biegler, L. T., Fox, V. G., Simulation and optimization of pressure-swing adsorption systems for air separation, *AIChE Journal* 49 **(2003)**, p. 1140.

27 Cruz, P., Alves, M. B., Magalhaes, F. D., Mendes, A., Cyclic adsorption separation processes: analysis strategy and optimization procedure, *Chemical Engineering Science* 58 **(2003)**, p. 3143.

28 Yang, R. T., Doong, S. J., Parametric study of the pressure swing adsorption process for gas separation: a criterion for pore diffusion limitation, *Chemical Engineering and Communication,* 41 **(1986)**, p. 163.

29 Raghavan, N. S., Hassan, M. M., Ruthven, D. M., Pressure swing air separation on a carbon molecular sieve. II. Investigation of a modified cycle with pressure equalization and no purge, *Chemical Engineering Science* 42 (**1987**), p. 2037.

30 Lee, C. H., Yang, J., Ahn, H., Effects of carbon-to-zeolite ratio on layered bed H_2 PSA for coke oven gas, *AIChE Journal* 45 (**1999**), p. 535.

31 Kim, M. B., Jee, J. G., Bae, Y. S., Lee, C. H., Parametric study of pressure swing adsorption process to purify oxygen using carbon molecular sieve, *Industrial and Engineering Chemistry Research* 44 (**2005**), p. 7208.

32 Reynolds, S. P., Ebner, A. D., Ritter, J. A., Enriching PSA cycle for the production of nitrogen from air, *Industrial and Engineering Chemistry Research* 45 (**2006**), p. 3256.

33 Chou, C. T., Huang, W. C., Simulation of a four-bed pressure swing adsorption process for oxygen enrichment, *Industrial and Engineering Chemistry Research* 33 (**1994**), p. 1250.

34 Warmuzinski, K., Tanczyk, M., Multicomponent pressure swing adsorption. Part I. Modelling of large-scale PSA installations, *Chemical Engineering Process* 36 (**1997**), p. 89.

35 Warmuzinski, K., Tanczyk, M., Multicomponent pressure swing adsorption. Part II. Experimental verification of the model, *Chemical Engineering Process* 37 (**1998**), p. 301.

36 Park, J. H., Kim, J. N., Cho, S. H., Performance analysis of four-bed H_2 PSA process using layered beds, *AIChE Journal* 46 (**2000**), p. 790.

37 Barg, C., Fereira, J. M. P., Trierweiler, J. O., Secchi, A. R., Simulation and optimization of an industrial PSA unit. *Brazilian Journal of Chemical Engineering* 17 (**2000**).

38 Sircar, S., Golden, T. C., Purification of hydrogen by pressure swing adsorption, *Separation Science and Technology* 35 (**2000**), p. 667.

39 Waldron, W. E., Sircar, S., Parametric study of a pressure swing adsorption process, *Adsorption* 6 (**2000**), p. 179.

40 Jiang, L., Biegler, L. T., Fox, V. G., Simulation and optimal design of multiple-bed pressure swing adsorption systems. *AIChE Journal* 50 (**2004**), p. 2904.

41 Chou, C. T., Chen, C. Y., Carbon dioxide recovery by vacuum swing adsorption, *Separation and Purification Technology* 39 (**2000**), p. 51.

42 Reynolds, S. P., Ebner, A. D., Ritter, J. A., Stripping PSA cycles for CO_2 recovery from flue gas at high temperature using a hydrotalcite-like adsorbent, *Industrial and Engineering Chemistry Research* 45 (**2006**), p. 4278.

43 Nikolic, D., Giovanoglou, A., Georgiadis, M. C., Kikkinides, E. S., Generic modelling framework for gas separations using multi-bed pressure swing adsorption processes, *Industrial and Engineering Chemistry Research* 47 (**2008**), pp. 3156–3169.

44 Adsim, User's Manual, Aspen Technology, Cambridge, MA, **2000**.

45 gPROMS User Manual, Ltd., London, UK, **2009**.

46 Wakao, N., Funazkri, T., Effect of fluid dispersion coefficients on particle-to-fluid mass transfer coefficients in packed beds: correlation of Sherwood numbers, *Chemical Engineering Science* 33 (**1978**), p. 1375.

47 Wakao, N., Kaguei, S., Funazkri, T., Effect of fluid dispersion coefficients on particle-to-fluid heat transfer coefficients in packed beds: correlation of Nusselt numbers, *Chemical Engineering Science* 34 (**1979**), p. 325.

48 Krishna, R., The Maxwell–Stefan approach to mass transfer, *Chemical Engineering Science* 52(6) (**1997**), pp. 861–911.

49 Kärger, J., Ruthven, D. M., *Diffusion in Zeolites*, John Wiley & Sons, New York, **1992**.

50 Myers, A. L., Prausnitz, J. M., Thermodynamics of mixed-gas adsorption, *AIChE Journal* 11 (**1965**), p. 121.

51 O'Brien, J. A., Myers, A. L., A comprehensive technique for equilibrium calculations in adsorbed mixtures: the generalized FastIAS method, *Industrial and Engineering Chemistry Research* 27 (**1988**), p. 2085.

52 Nikolic, D., Kikkinides, E. S., Georgiadis, M. C., Optimization of multi-bed pressure swing adsorption processes, *Industrial and Engineering Chemistry Research* 48 (**2009**), pp. 5388–5398.

53 Webley, P. A., He, J. M., Fast solution-adaptive finite volume method for PSA/VSA cycle simulation; 1 single step simulation, *Computers and Chemical Engineering* 23 (**2000**), p. 1701.

54 Teague, K. G., Edgar, T. F., Predictive dynamic model of a small pressure swing adsorption air separation unit, *Industrial and Engineering Chemistry Research* 38 (**1999**), p. 3761.

55 Hindmarsh, A. C., User's manual for LSODE, National Energy Software Center Note, p. 83, **1983**.

56 Petzold, L., Automatic selection of methods for solving stiff and nonstiff systems of ordinary differential equations, *SIAM Journal of Scientific and Statistical Computing* 4 (**1983**), p. 136.

57 Li, S. T., Petzold, L., Software and algorithms for sensitivity analysis of large-scale differential algebraic systems, *Journal of Computational and Applied Mathematics*. 125 (**2000**), p. 131.

58 Kostroski, K. P., Wankat, P. C., High recovery cycles for gas separations by pressure swing adsorption, *Industrial and Engineering Chemistry Research* 45 (**2006**), p. 8117.

59 Boris J. P., Book, D. L., Flux-corrected transport. 1. SHASTA, a fluid transport algorithm that works (reprinted from the *Journal of Computational Physics* 11 (**1973**), pp. 38–69), *Journal of Computational Physics* 135 (**1997**), p. 172.

60 Cruz, P., Alves, M. B., Magalhaes, F. D., Mendes, A., Solution of hyperbolic PDEs using a stable adaptive multiresolution method, *Chemical Engineering Science* 58 (**2003**), p. 1777.

61 Smith, Westerberg, A. W., Acceleration of cyclic steady state convergence for pressure swing adsorption models, *Industrial and Engineering Chemistry Research* 31 (**1992**), p. 1569.

62 Todd, R. S., He, J., Webley, P. A., Beh, C., Wilson, S., Lloyd, M. A., Fast finite-volume method for PSA/VSA cycle simulation – experimental validation, *Industrial and Engineering Chemistry Research* 40(14) (**2001**), pp. 3217–3224.

63 Ding, Y., LeVan, M. D., Periodic states of adsorption cycles III. Convergence acceleration for direct determination, *Chemical Engineering Science* 56(17), (**2001**), pp. 5217–5230.

64 Jiang, L., Biegler, L. T., Fox, V. G., Design and optimization of pressure swing adsorption systems with parallel implementation, *Computers and Chemical Engineering* 29 (**2005**), p. 393.

6
A Framework for the Modeling of Reactive Separations
Eugeny Y. Kenig

Keywords

reactive separations (RS), structured packings, fluid-dynamic approach (FDA), hydrodynamic analogy (HA), rate-based approach (RBA), virtual experiments

6.1
Introduction

Modern chemical process engineering is dominated by the idea of process intensification leading to substantially smaller, cleaner, safer, and more energy efficient process technology [1]. One possible way to realize more intensive operations is to use a symbiosis of reaction and separation in one single unit. Such operations called *reactive separations* (RS) have recently attracted great attention of both industry and academia. However, their complexity is higher than that of traditional nonintegrated processes and, correspondingly, their modeling is also more complex. Moreover, such a modeling requires knowledge on process kinetics, with respect to both transport phenomena and chemistry, and, thus, the equilibrium concept, popular during the last century, cannot be properly used for the design and optimization of RS. The growing computational power as well as the need to understand and implement physicochemical backgrounds of the processes shift the modeling focus to the kinetics-based approach.

On the other hand, application of rigorous theoretical concepts is hindered, due to the complexity of the systems and phenomena under consideration. Multiphase flows, multicomponent systems, complex thermodynamics, complex column unit design and geometry, intricate column links, and a large scale difference between the characteristic dimensions of the phenomena involved make rigorous modeling of RS processes extremely difficult. Instead of trying to develop a unified modeling basis and presumably fighting a losing battle, a contemporary modeling framework can be built up based on the *complementary approach*. Such an approach suggests an efficient combination of different modeling methods. We present here an overview of the complementary approach and illustrate it with several case studies elaborated in our group.

6.2
Reactive Separations

Manufacturing of chemical products from selected feed stocks is based on a variety of chemical reactions. The reaction extend is often limited by the chemical equilibrium between the reactants and products, thus reducing the conversion and selectivity toward the main product. The process must then include the separation of the equilibrium mixture and recycling of the reactants.

Conventionally, each unit separation operation is performed in individual items of equipment, which, when arranged together in sequence, make up the complete process plant. As reaction and separation stages are carried out in discrete equipment units, their equipment and energy costs are added up. However, in recent decades, a combination of separation and reaction inside a single unit has become more and more popular. The potential for capital cost savings is obvious; besides, there are often many other process advantages that accrue from such combinations [2]. Therefore, many new processes called RS have been invented based on this integration principle (see, e.g., [3–10]).

Among the most important examples of RS processes are reactive distillation, reactive absorption, reactive stripping, and reactive extraction. For instance, in reactive distillation, reaction and distillation take place within the same zone of a distillation column. Reactants are converted to products with simultaneous separation of the products and recycle of unused reactants. The reactive distillation process can be both efficient in size and cost of capital equipment and in energy used to achieve a complete conversion of reactants.

Since reactor costs are often less than 10% of the capital investment, the combination of a relatively cheap reactor with a distillation column offers great potential for overall savings. Among suitable reactive distillation processes are etherifications, nitrations, esterifications, transesterifications, condensations, and alcylations [3].

As a rule, RS occur in moving systems, and thus the process hydrodynamics plays an important part. Besides, these processes are based on the contact of at least two phases, and therefore, the interfacial transport phenomena have to be considered. Further common features are multicomponent interactions of mixture components, a tricky interplay of mass transport and chemical reactions, complex process chemistry and thermodynamics.

For all these reasons, the design of RS columns is more sophisticated than that of traditional operations. Above all, the influence of column internals increases significantly. These internals have to enhance both separation and reaction and maintain a sound balance between them. This represents a challenging task since efficient separation requires a large contact area, whereas efficient reaction strives for a significant amount of catalyst.

To solve this problem, a novel generation of column internals, corrugated packings of the regular type, also referred as *structured packings* (SP), has been created. These packings provide enhanced mass transfer performance with relatively low pressure drop and, consequently, have gained a wide acceptance. Since the early

6.2 Reactive Separations | 175

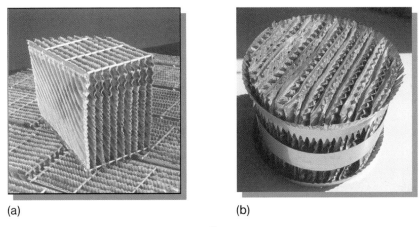

Fig. 6.1 Catalytic structured packings KATAPAK®-S (a) and KATAPAK®-SP-11 (b) by Sulzer Chemtech Ltd.

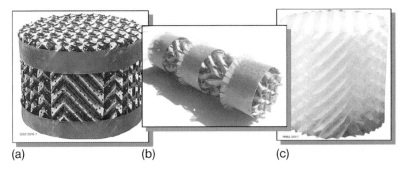

Fig. 6.2 Structured packings metal Mellapak by Sulzer Chemtech Ltd. (a), Montz-Pak A3-500 by Julius Montz GmbH (b), and plastic Mellapak by Sulzer Chemtech Ltd. (c).

1980s, when corrugated sheet metal structured packings appeared in the market, great advances toward the process intensification have been made. Being initially developed for separation of thermally unstable components in vacuum distillation, structured packings have permanently been gaining in popularity and cover a large field of applications in chemical, petrochemical, and refining industries due to their more effective performance characteristics [11].

For heterogeneously catalyzed processes containing solid catalyst phase (e.g., in catalytic distillation and catalytic stripping), structured packings represent complex geometric structures made from gauze wire or metal sheets and containing catalyst pellets (see Fig. 6.1). In this case, both mass transfer area and catalyst volume/surface become important parameters influencing the process performance. For homogeneously catalyzed and autocatalyzed processes (e.g., reactive absorption, reactive distillation, and reactive extraction), the packing function is to provide both sufficient residence time and mass transfer area (Fig. 6.2). In some RS processes, reactive and nonreactive SP are combined within the same column [7, 12].

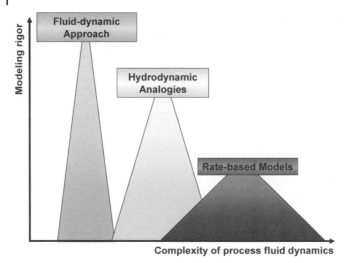

Fig. 6.3 Modeling rigor against fluid dynamics complexity.

The complexity of the coupled phase interactions occurring in RS can only be covered if transport and reaction phenomena are properly understood, and, hence, the development of sound predictive models is required.

6.3
Classification of Modeling Methods

To build a framework for the RS modeling, we first try to classify the modeling methods available for RS, focusing on the kinetics-based techniques. Such a classification can be done based on different principles, depending on the specific aims and criteria. Here, we suggest to classify the models according to their rigor and to relate them to the fluid dynamics complexity of the processes or phenomena under study (Fig. 6.3). This choice can be justified by the crucial importance of the fluid-flow pattern and its decisive influence on the possibility to develop rigorous mathematical process models.

The most rigorous approach to the description of a process or a phenomenon is based on the classical equations of fluid dynamics (fluid-dynamic approach, FDA). These equations are partial differential equations; they provide a local description of the transport phenomena and are supplied by the corresponding initial and boundary conditions. When solved, such models yield local velocity, temperature, and concentration fields that can further be used for the determination of all relevant process characteristics.

Most often, the FDA is realized using modern numerical facilities and tools, thus building a separate and quickly growing application area called computational fluid dynamics (CFD). CFD simulation represents nowadays a powerful approach to many physicochemical problems [13]. However, the simulation of large-scale RS columns still appears too difficult, mostly due to superposition of different scales and largely undetermined position of the phase interface.

Geometrically simple flows seldom occur in industrial separation units in which an intensive phase contact and a highly developed interfacial area are aimed at. This results in a complex unit and internals geometry and, consequently, in an extremely intricate, sometimes virtually chaotic flow pattern (imagine, for instance, bubble columns). Even for the regular geometry provided by corrugated-sheet structured packings, the exact localization of phase interfaces represents a difficult problem, due to intricate interphase interactions. Generally, in such cases, the exact spatial localization of the phase boundaries is not possible, the boundary conditions cannot be applied properly, and, thus, the FDA cannot be realized. Therefore, this approach is associated only with the very left section of the reference frame in Fig. 6.3.

One needs, however, models for the whole spectrum of fluid dynamics complexity that can be encountered in real processes. These models can be developed under additional assumptions with respect to the flow pattern. This results in modeling techniques that are less rigorous, yet they govern more complex phenomena.

One of such approaches proposed recently is called hydrodynamic analogy (HA) [14]. It rests on the analogy between complex flow patterns encountered in industrial separations and geometrically simpler flows, for example, films, cylindrical jets, spherical drops, as well as their combinations. This approach is a compromise between rigor and simplicity and it is related to a section of the reference frame to the right of the FDA (Fig. 6.3).

For the processes in which fluid-dynamic patterns are too complex to be described by either the FDA or by the HA, further assumptions and simplifications of a real fluid-flow picture are required. For instance, large industrial separation units are usually modeled by a proper subdivision of a column unit into smaller elements. These elements (called stages) are linked by mass and energy balance equations. The stages are related to real trays for tray columns and to packing segments for packed columns. They can be described using different theoretical concepts, with a wide range of physicochemical assumptions and accuracy [7].

We will not consider here the equilibrium stage models, as, due to their simplicity, they cannot provide any reasonable link to the apparatus design (regarding, e.g., column internals) and, for kinetically limited processes, are hardly applicable. Instead, the so-called rate-based stage approach or simply rate-based approach (RBA) represents a good choice. Each stage is considered based on a very simple model fluid dynamics, with one-dimensional flow or even stagnant elements. Examples are given by the two-film theory, penetration theory, or surface renewal theory [15]. Model parameters (e.g., film thicknesses in the two-film theory or residence time in the penetration and surface renewal theories) result from experimental investigations and are represented as functions (correlations) of physicochemical properties, operational conditions, and column geometry.

In principle, the RBA can be applied to both complex and simple fluid dynamic patterns, provided that relevant correlations are available. However, it is just the right section of the reference frame in Fig. 6.3 where its use is ultimate. On the contrary, the RBA application for geometrically simpler flows can hardly be considered desirable, since this method largely depends on the quality of experimentally estimated parameters, whereas the FDA and – with minimal exceptions – HA do not. Thus, the allocation of different methods in Fig. 6.3 corresponds to their appropriate application.

The three approaches shown in Fig. 6.3 complement each other in the way that they cover together the whole spectrum of fluid-dynamic complexity of the separation processes. The choice of the approach to be applied depends on the problem under study as well as on particular aims, experience, means, and criteria of the specific user.

6.4
Fluid-Dynamic Approach

In the FDA, velocity, concentration, and temperature fields are calculated explicitly, by using partial differential equations for momentum, mass, and heat transport. In this case, the phase boundaries must be spatially localized, which is possible for simple geometric arrangements. In this respect, first of all, various film-like flows come to mind. They may, for example, occur in tubes, channels, or monolithic structures [16–23]. Similar possibilities exist for other simple geometry flows (e.g., cylindrical jets).

Our group handled various film-like-flow problems, with the focus on the coupled mass and heat transport phenomena in one- and two-phase systems. These problems cover simultaneous heat and mass transfer in nonreacting and reacting mixtures, laminar and turbulent conditions, cocurrent and countercurrent flow regimes shown in Fig. 6.4. The velocity profiles were obtained by solving a simplified form of the Navier–Stokes equations in accordance with the simple flat or cylindrical (nonwavy) film-flow conditions. Mass and heat transfer description was given in a general manner valid for both multicomponent and binary systems. This was done by using the Maxwell–Stefan equations for the description of multicomponent diffusion [24] and a compact matrix-type form of the governing mass-transfer equations [20]. The solution of such highly coupled systems was realized using either analytical matrix-based methods, including a direct matrix generalization of scalar binary problems, or numerical techniques (e.g., finite-difference implicit methods). Besides, certain combinations of numerical and analytical methods were used, for example, the diagonalization of the diffusion matrix, a special treatment of conjugate interfacial boundary conditions, and a semianalytical approach for the reactive systems. Comprehensive reviews of these methods can be found in [19, 20]. Recently, the FDA has been applied to the modeling and simulation of reactive stripping in monoliths with rounded channel cross section [25].

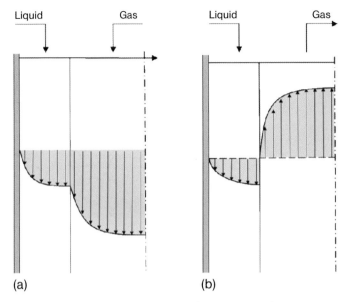

Fig. 6.4 Film-like flows in cocurrent (a) and countercurrent (b) regime.

Another important field of application of the FDA is the investigation of basic phenomena using CFD. As already mentioned, understanding of such phenomena is a critical issue for the improvement of the RS design. In many processes, droplets moving in another (continuous) fluid phase represent the basic flow pattern. Therefore, in last years, theoretical description of the droplet behavior has been in the middle of attention of several research groups [26–28].

To follow the moving droplet interface, two general strategies have been suggested. In the first one, a moving mesh is used to track the interface. As the topology of the interface changes, the mesh is adjusted in accordance with the change. Methods falling under this strategy are called *front tracking methods*. The second approach uses a fixed (Eulerian) mesh, whereas the interface is tracked using different procedures, for example, special markers or functions. Such methods are called *front-capturing methods*.

In the front tracking methods, the interface is resolved directly. It always lies between two neighboring mesh elements. As the interface moves, the neighboring mesh elements deform in order to track the movement of the interface. For the description of the interface movement, an additional explicit equation is necessary. Usually, the latter can be derived from an integral consideration of the transport phenomena or physicochemical conditions around the interface (see, e.g., [29]).

For instance, when the change of the interface position is caused by the mass transfer, the total mass conservation condition can be used as an additional equation describing the mesh deformation. Along these lines, Burghoff and Kenig [27] considered a single droplet in a continuous moving liquid phase and studied in detail the coupling of multicomponent mass transfer and hydrodynamics. This coupling is important for the description of liquid–liquid extraction processes. A rig-

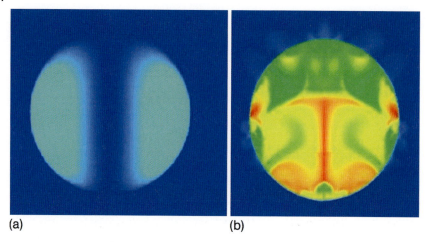

Fig. 6.5 Concentration field of acetone at a single droplet without (a) and with (b) Marangoni convection.

orous model was developed and implemented into the commercial CFD software CFX4.3 (ANSYS, Inc.). The onset of the Marangoni phenomena was accounted for in the numerical simulations, by implementing a term for the calculation of the surface tension variation resulting from mass transfer.

The quaternary system toluene–acetone–methyl isopropyl ketone–water was used for the theoretical analysis. It was found that the onset of interfacial instabilities significantly changed flow conditions at the droplet. The droplet content was mixed intensely that resulted in enhanced mass transfer across the phase interface (cf. Fig. 6.5). The simulation results were compared with the data obtained from experiments, and a good agreement between simulation and experimental results proved the model predictivity [27]. In multicomponent systems, interfacial instabilities were found to appear at lower initial mass fractions of the transferring components and to be stronger than those in the corresponding ternary systems. These results agree with the experimental observations of von Reden [30].

In the front-capturing methods, the interface is described using special markers or functions in a fixed mesh. One of the most popular surface-capturing methods is the *level set* (LS) method. In this method, a function is used to locate the interface that takes positive and negative values on different sides of the interface and zero at the interface. The interface is therefore called zero LS. An overview on the LS methods is presented by Osher and Fedkiw [31]. This method is conceptually simple and easy to implement. Its main drawback is the loss of mass (or volume), especially for significantly deformed interfaces [32].

The *volume of fluid* (VOF) method introduced by Hirt and Nichols [33] represents a typical volume-capturing method. The basic idea of the VOF method is the definition of a volume function that takes values 0 for the first phase, 1 for the second phase, and between 0 and 1 for the cells containing the interface. A detailed review on the VOF methods is published by Rider and Kothe [34]. This approach is capable of handling problems with significant interface topology change and does not suffer from mass (or volume) losses. The extension of the method to 3D simulations is straightforward and no special algorithm is needed for the case of merging or break up of the interface. However, the interface is smeared out and merges "numerically." Thus, up to now, it is not possible to use the VOF method for a rigorous analysis of the droplet or bubble coalescence probability [35]. Thus, the main drawback of this method is the inherent numerical smearing.

Atmakidis and Kenig [36] gave an analysis of the Kelvin–Helmholtz instability encountered in liquid–liquid systems. In this case, hydrodynamics and mass transfer of two immiscible cocurrently moving liquid layers were examined. This is a well known and useful topology for studying interfacial phenomena and flow in two immiscible fluids [37].

A three-component system toluene–acetone–water was chosen, whereas acetone was transferred from a toluene layer to an aqueous layer. To resolve rigorously the movement of the interface, two different front-capturing methods were used – the VOF method and the LS method. In the VOF method, the interface was described using a scalar quantity f that represented the fraction of the volume of a mesh cell occupied by one phase. Thus, $0 < f < 1$ for the cells containing the interface and $f = 0$ or 1 away from the interface [33]. The simulations were performed using the commercial CFD package CFX-10 (ANSYS, Inc.).

In the LS method, a different function is used to track the interface, which is defined as a distance function. In order to take into account density and viscosity discontinuities at the interface and, at the same time, to achieve numerical robustness, the LS is artificially smeared out over a small distance using the Heaviside function [38]. Using this formulation, the smoothed LS function takes the value of 0.5 exactly at the interface and the values 0 or 1 away from the interface. The simulations with the LS method were carried out with the commercial software COMSOL Multiphysics 3.3a by COMSOL AB [39].

To examine the effect of different perturbations on the wave development, two case studies were considered. In the first one, a perturbation was implemented into the initial condition. In the second case, two perturbations were used simultaneously, one as initial and another as inlet perturbation [36]. In Fig. 6.6, the wave evolution is illustrated at three different time points and for different perturbation terms. The organic phase is highlighted using red color (the lower layer) and the aqueous phase using blue color. In both cases, a similar wave development can be observed. In the beginning, the waves develop in the horizontal and afterward in the vertical direction creating typical finger-like structures. When two perturbation terms are used (Fig. 6.6(b)), the initial vertical velocity component of the aqueous phase is higher than in the case with only one perturbation term. This leads to thicker waves with larger amplitudes.

The model was extended to consider mass transfer effects. The transport equation was solved for the transferred component, whereas the interface was assumed to be in equilibrium. The mass transfer study revealed regions where it was promoted due to high convection. This is visible close to the wave crest (Fig. 6.7). In the remaining area, mass transfer is relatively slow and can hardly be visualized after 0.25 s.

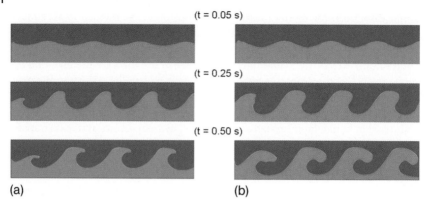

Fig. 6.6 Wave evolution obtained using one perturbation term (a) and two perturbation terms (b).

Similar numerical studies were accomplished with the LS method. Figure 6.8 gives a comparison between the calculated wave amplitudes at different times obtained by the VOF and LS methods. Both methods give very close results and can predict not only the maximum wave amplitude but also the exact time at which the vertical wave development stops and the horizontal development starts.

The FDA can be especially useful for the processes at microscale (see, e.g., [40–42]). For instance, Chasanis et al. [41] numerically investigated carbon dioxide absorption into a sodium hydroxide solution in a microstructured falling-film contac-

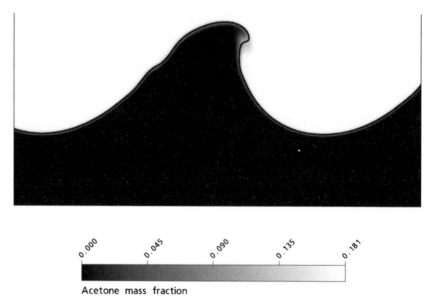

Fig. 6.7 Concentration field of acetone in the two-layer system: changes are visible only in the vicinity of the crest.

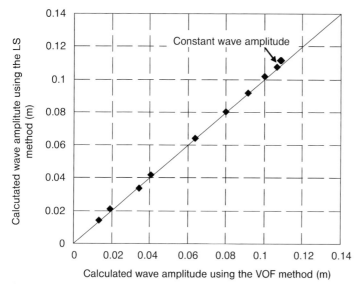

Fig. 6.8 Comparison between the wave amplitudes calculated with the VOF method and the LS method at different times.

tor. A rigorous 2D-model was developed and validated against experimental data from literature. The modeling approach comprised two different steps, namely, an unsteady model in combination with the LS method to capture the interface and a steady-state model to describe the whole absorption process. A good agreement between simulated and experimental data of Zanfir et al. [43] was found. The validated model was used to perform sensitivity studies and to investigate the impact of different process parameters on absorption performance.

Generally, the investigations with FDA significantly extend our understanding of the processes under study, because they are able to deliver local information (velocity, temperature, and concentration fields). Such information can hardly be obtained experimentally.

6.5
Hydrodynamic Analogy Approach

The HA (hydrodynamic analogy) approach is an alternative way to describe the hydrodynamics and transport phenomena in processes in which the exact location of the phase boundaries is not possible, yet the fluid pattern possesses some regularity or structure that can be mirrored by an analogy with more simple flow elements. The basic idea of the approach illustrated by Fig. 6.9 is a reasonable *replacement* of the actual complex hydrodynamics in a column by a combination of *geometrically simpler flow patterns*. Such a geometrical simplification has to be done *in agreement with experimental observations* of fluid flow, which plays a crucial part for the successful application of this approach. Once the observed complex flow is reproduced

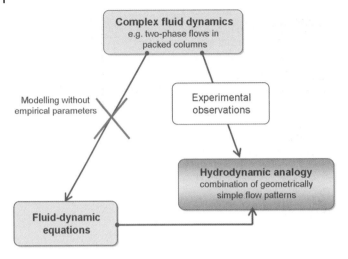

Fig. 6.9 Illustration of the HA approach.

by a sequence of the simplified flow patterns, the partial differential equations of momentum, energy, and mass transfer can be applied to govern the transport phenomena in an entire separation column, similar as described in Section 6.4.

One of the first examples showing how a process can be treated with the use of the HA was given by Kenig and Kholpanov [44] for the *liquid-film pertraction*. This process belongs to the class of liquid membrane separations and is known for its high efficiency [45]. In liquid pertraction, mass transfer occurs between fluid phases; their motion is defined by specified flow conditions. Separation is achieved due to the difference in diffusion properties of the components. The simplest system of this sort consists of three flowing phases, the middle one playing the role of a moving fluid membrane [46].

More complex processes may contain more phases separated by the relevant interfaces. The flat geometry of liquid membranes considerably simplifies the pertraction process modeling and allows an HA between the real fluid dynamics and falling liquid films to be built [20]. In [44], cocurrent flow conditions were considered using an arrangement comprising three liquid films. The governing model consists of the matrix-form convection–diffusion equations written for each of these films, with corresponding boundary conditions including conjugate interfacial relations for both interfaces simultaneously. The solution of this model yields the local concentration profiles of the components and thus retains all the advantages discussed above with respect to the modeling methods by partial differential transport equations [14].

Another example of the HA application was presented in [47] for the process called *zero-gravity distillation*. This name is given because the process can be run in any arbitrary position of the apparatus, from the vertical position, traditional for distillation units, to the horizontal one, traditional for heat pipes [48] because the

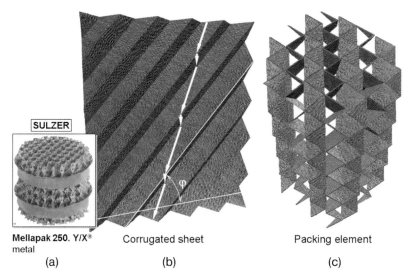

Fig. 6.10 A piece of corrugated metal sheet packing (a), a corrugation sheet with observed liquid flow path (b), and a packing element (c).

driving force of the liquid transport is taken over by the action of capillary pressure gradient at the interface instead of gravity acting in usual distillation.

In zero-gravity distillation, the liquid phase flows slowly through the porous layer. Using the HA between porous and film flow, the actual liquid-phase movement along the apparatus is substituted in [47] by a simpler film-like movement. Once again, this provides an opportunity to apply the matrix-form convection–diffusion equations describing the mass transport in the column, as in the previous example. In [47], a three-component two-phase system ethanol–isopropanol–water was simulated in a horizontally positioned column segment.

A real breakthrough in the HA application has been achieved for distillation processes in columns equipped with corrugated-sheet structured packings. The corrugated sheets are installed countercourse in such a way that they form channels, each being formed by the two wall sides and one open side shared between two neighboring channels (see Fig. 6.10). In line with the HA concept, the basis for the physical model is provided by the observations of fluid flow together with the geometric characteristics of structured packing. According to experimental studies [49–54], these observations can be summarized as follows:

- Gas flow takes place in channels built up by the countercourse assemblage of the corrugated sheets.
- There is a strong interaction between the gas flows in adjacent channels through the open channel side (see Fig. 6.10) responsible for small-scale mixing (i.e., via turbulence) of the gas phase.
- Liquid generally tends to flow in the form of films over the packing surface, whereas the wave formation is largely suppressed due to the corrugations.

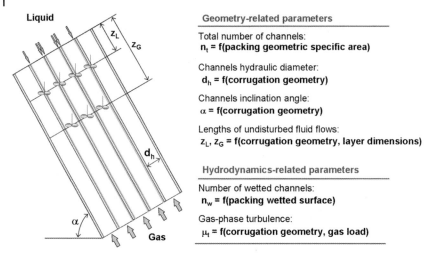

Fig. 6.11 Physical model of structured packing.

- Liquid flows at the minimal angle built by the packing surface and the vertical axis (the so-called gravity-flow angle).
- Abrupt flow redirection at the corrugation ridges together with the influence of intersection points with the adjacent corrugated sheets cause mixing and lateral spreading of the liquid phase.
- Side effects (abrupt flow redirection at the column wall and at transitions between the packing layers) result into large scale mixing of the gas phase.

The physical model represented in Fig. 6.11 comprises all these effects. The pronounced channel flow of the gas phase makes it possible to consider the packing as a bundle of parallel inclined channels with identical cross sections. For simplicity, the circular channel shape is adopted. The number of the channels as well as their diameter are determined from the packing geometric specific area and corrugation geometry, respectively. The gas-flow behavior depends on the operating conditions and varies from laminar to turbulent flow. The liquid flows countercurrently to the gas flow in form of laminar nonwavy films over the inner surface of the channels. Additionally, a uniform distribution of both phases in radial direction is assumed, i.e., no maldistribution is taken into account in the model. The ratio of wetted to total number of channels is the same as the ratio of effective (interfacial) specific area to geometric specific area of the packing. The periodical ideal mixing approximation is necessary to account for real mixing caused by the abrupt change in the flow direction. The length of the laminar flow interval for the liquid phase corresponds to the distance between the two neighboring corrugation ridges, whereas for the gas phase, it is set to be equal to an average channel length [55].

The system of equations describing this arrangement comprises partial differential equations written for the gas and the liquid phases, coupled through the boundary conditions at the phase interface. Numerical solution of this system (finite difference implicit scheme) yields the local temperature and composition fields. The

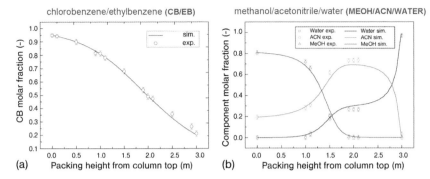

Fig. 6.12 Comparison of simulated (lines) and experimentally measured (circles) liquid composition profiles in the column supplied with the Montz-Pak A3-500 structured packing: a binary system ethylbenzene–chlorobenzene–(EB/CB) (a), a ternary system methanol–acetonitril–water (MEOH–ACN–WATER) (b). Experimental data from [56].

computed average compositions over the packing height were compared with measured data for distillation at total reflux for different structured packings under a variety of operating conditions, and an excellent agreement was established (see, e.g., Fig. 6.12).

This approach has then been further developed and tested for RS operations. In [57], the HA model was extended to govern a reactive stripping process with heterogeneously catalyzed liquid-phase reactions (esterification of hexanoic acid and 1-octanol). The stripping column was packed with either corrugated-sheet structured packing Sulzer™ DX or with one of three different types of film-flow monoliths (with square channels, internally finned channels, and round channels). These internals shown in Fig. 6.13 acted as supports for thin catalyst layers.

To properly account for a porous catalyst sublayer on the catalytic internals, the liquid film flow was subdivided into a part flowing through the layer and a part flowing along it. The liquid-phase reactions were modeled using a quasihomogeneous approach.

To apply the model for monolithic internals with noncylindrical channels, a model extension was suggested that based on an equivalent transformation of the monolith geometries to cylindrical channels. The simulation results were compared with experimental data for all four column internals and a good agreement was found for all studied cases (see Fig. 6.14).

Furthermore, the HA model was successfully applied to different reactive absorption systems, for example, sulfur dioxide and carbon dioxide in aqueous solutions of sodium hydroxide [58]. In this way, reactions occurring solely at the gas–liquid interface and liquid-phase reactions with both infinite and finite reaction rates could be tested.

This proves that a wide class of reactive column internals can be described by the extended HA model. Moreover, the method has a good potential for applications in which mass transfer coefficients are difficult to determine experimentally.

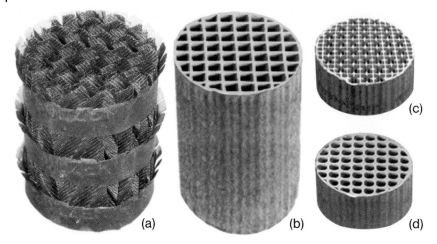

Fig. 6.13 Different catalyst supports; Sulzer™ DX corrugated sheet packing (a), film flow monoliths with square channels (SQ, b), with internally finned channels (IFM, c), and with rounded channels (MRC, d).

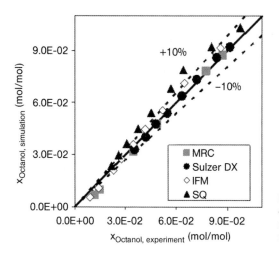

Fig. 6.14 Parity plot of octanol outlet concentration for different monoliths and Sulzer™ DX packing.

6.6
Rate-Based Approach

The RBA represents a well-known way to design large column units [59]. This approach implies a direct consideration of actual rates of multicomponent mass and heat transfer and chemical reactions within a column stage. Mass transfer at the interface between the contacting phases can be described using different theoretical concepts [20, 24]. Most often, the two-film model or the penetration/surface renewal model is used, whereas the model parameters are estimated via empirical correlations. In this respect, the two-film model is advantageous, since there is a broad spectrum of correlations available in the literature, for all types of internals and systems.

In the two-film model, it is assumed that the resistance to mass transfer is concentrated entirely in thin films adjacent to the phase interface and that mass transfer occurs within these films by steady-state molecular diffusion alone. Outside the films, in the fluid bulk phases, the level of mixing is assumed to be so high that there is no composition gradient at all. This means that in the film region, one-dimensional diffusion transport normal to the interface takes place.

In the general case, multicomponent diffusion in the films can be described by the Maxwell–Stefan equations, and thus, the two-phase mass transfer is modeled as a combination of the two-film model presentation and the Maxwell–Stefan diffusion description. In this stage model, the equilibrium is assumed at the interface only.

The film thicknesses represent model parameters that can be estimated using empirical mass transfer coefficient correlations. They govern the mass transport dependence on physical properties, process hydrodynamics, and geometry of the column and are available from the literature. A further important parameter of the film model is the specific contact (interfacial) area, which is also estimated based on experimental data.

The main equations of the RBA model comprise [7, 60]:

- mass and heat balances for the bulk phases (dynamic or steady-state) including source terms in case of reaction;
- mass and heat transfer laws (Maxwell–Stefan, Nernst–Planck, or Fick for diffusion and Fourier for heat conduction);
- thermodynamic equilibrium relations for the interfacial concentrations and temperature;
- continuity equations for the mass and energy fluxes at the phase interface;
- equations for reaction kinetics and equilibrium for reactive systems;
- correlations for process hydrodynamics (hold-up and pressure drop) and for mass and heat transfer coefficients and specific contact area.

The film model can be extended to describe reactive systems (Fig. 6.15). The reaction mechanism can be considered either in the bulk phases (slow reaction), or in the film region (fast reaction), or even in both regions [7].

The case when the reaction mechanism should be taken into account in the film region is especially complex, since the component balance equations including simultaneous mass transfer and reaction in the film become nonlinear (ordinary) differential equations of the second order that have to be solved in conjunction with all other model equations. This can be done either analytically, provided that some further assumptions with respect to the linearization of the diffusion and reaction terms are made [20, 61], or numerically [7, 60, 62].

There is a number of successful applications of the RBA to various practical tasks, both nonreactive and reactive and both steady-state and dynamic [7, 19, 24, 59, 62–65]. Our group has recently managed to model some highly integrated processes with the RBA. Regarding the large computational effort and convergence

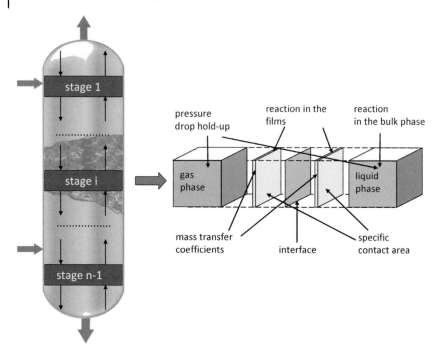

Fig. 6.15 Column stage concept and extended film model for a reactive separation unit stage.

problems typical for these tasks, this is worth mentioning. One of such highly integrated processes is a closed reactive absorption/desorption loop presented in [66]. In this work, an industrial process was considered in which aqueous methyl diethanolamine solution was used for a selective removal of H_2S in presence of CO_2 (reactive absorption) with a subsequent solvent regeneration (reactive desorption).

In [66], both absorber and desorber performances were validated (see Fig. 6.16). It should be born in mind that the rate-based modeling of a reactive absorption unit alone represents a difficult task due to complex reaction schemes and reaction/diffusion lumping in the film region [64]. Therefore, we believe that the modeling of a fully integrated absorption/desorption loop by the RBA in [66] is a remarkable achievement.

Another interesting example is given in [67]. In this work, a novel rate-based description of both nonreactive and reactive dividing wall columns is presented. The advantages of thermally coupled distillation columns with respect to energy saving have long been recognized [68], and their further integration within a single shell by implementing a dividing wall into a single distillation unit [69] represents a modern trend.

Still higher integration can be achieved if reactive distillation occurs in a dividing wall column (Fig. 6.17). This process is known as reactive dividing wall column [67] and, due to its highly integrated configuration, it enables further synergistic effects, for example, overcoming chemical and thermodynamic equilibrium limitations,

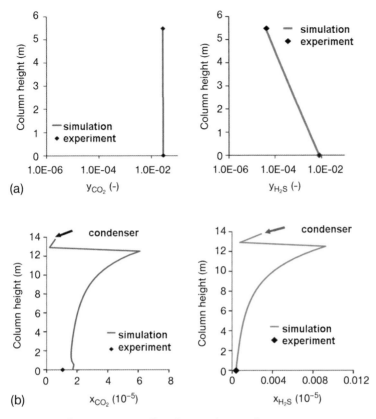

Fig. 6.16 Axial concentration profiles of CO_2 and H_2S and experimental data in the gas phase of the absorber (a) and in the liquid phase of the desorber (b).

ability to separate close boiling components, and reduced number of equipment units.

In [67], the RBA was applied to a nonreactive, ternary alcohol mixture methanol–isopropanol–butanol and successfully validated. Furthermore, in a later work, Mueller and Kenig [70] identified the transesterification of carbonates as an interesting system for the reactive dividing wall column, as this reaction is equilibrium limited and characterized by high conversion but low selectivity. In a further study, it was found that the selectivity and separation of products and nonconverted reactants could be significantly increased by means of the reactive dividing wall column [71].

Another reactive system, hydrolysis of methyl acetate to methanol and acetic acid was studied within the European project INSERT (Integrating Separation and Reaction Technologies, Project No. NMP2-CT-2003-505862), under the sixth Framework Programme of the European Union. Figure 6.18 demonstrates the first successful validation of the rate-based model with the data obtained at BASF SE [72].

6 A Framework for the Modeling of Reactive Separations

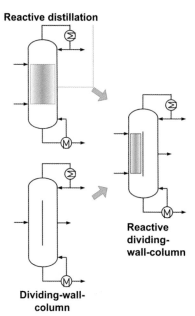

Fig. 6.17 Reactive dividing wall column as an integration of reactive distillation and dividing wall column.

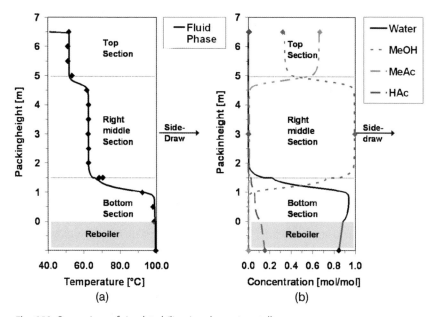

Fig. 6.18 Comparison of simulated (lines) and experimentally measured (points) temperature profiles (a) and concentration profiles (b) for the hydrolysis of methyl acetate (MeOH: methanol, MeAc: methyl acetate, and HAc: acetic acid).

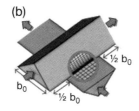

Fig. 6.19 Schematic representation of a periodic element of the corrugated sheet packing (a) and computational domain (b).

6.7
Parameter Estimation and Virtual Experiments

As demonstrated in the previous section, the RBA is capable of modeling and simulating various RS units with reasonable accuracy and predictivity. However, this is only possible if the model parameters are properly determined. These parameters describing the mass transfer and hydrodynamic behavior comprise mass transfer coefficients, specific contact area, liquid hold-up, residence time distribution characteristics, and pressure drop. Usually, the parameters mentioned have to be determined by extensive and expensive experimental estimation procedures and correlated with process variables and specific internals properties. In the literature, one can find numerous correlations presented in different forms; however, their application to particular problems often results in significant discrepancies, so that the choice of a suitable correlation is quite tricky. Thus, the strong dependence of the RBA performance on the model parameters represents its week point [11].

In the nature of things, experiments are performed in really existing equipment units filled with specific column internals. Let us now imagine that one is able to gain a relevant correlation by purely theoretical way, just by simulating the phenomena on and in particular internals for some particular systems. In this case, it would be possible to study the column internals even prior to their manufacturing. Such studies can be considered as *"virtual experiments"* replacing corresponding real experiments for the parameter estimation. Virtual experiments can open the way toward virtual prototyping and manufacturing of column internals and enable computer-aided optimization of both internals and overall processes.

The development of CFD-based virtual experiments for structured packings in RS was one of the main goals of a European project INTINT (Intelligent Column Internals for Reactive Separations, Project No. GRD1 CT1999 10596), within the fifth Framework Programme of the European Union. In this project, universities collaborated with large chemical and petrochemical companies, manufacturers of column internals, and developers of the CFD code (see [73]). A new methodology for the packing optimization was suggested that combined certain CFD procedures with rate-based model simulations accomplished with the help of the software tools developed in INTINT [74, 75]. The INTINT results revealed both advantages and limitations of the suggested approach and were in general encouraging.

For example, *dry pressure drop* of the corrugated sheet packings Sulzer-BX could be estimated with a reasonable accuracy using CFD simulations of a representative small periodical packing element shown in Fig. 6.19. The influence of the appa-

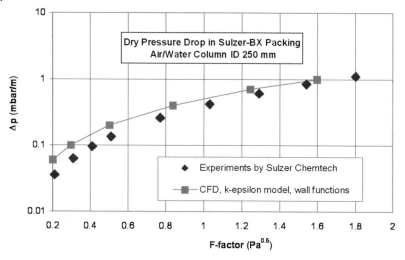

Fig. 6.20 Comparison of simulated (line) and experimentally measured (points) dry pressure drop.

ratus wall was not considered and the flow was treated as established, with the periodic boundary conditions satisfied at the open boundaries.

Numerical experiments performed with the commercial tool CFX 4 by ANSYS® revealed strong pressure-drop sensitivity to the corrugation angle value. Besides, the complicated flow structure necessitated a high-resolution degree of the vortex scales responsible for the turbulence generation. Correlations between the pressure drop and the gas load determined here with a grid of 96,000 control volumes inside a crossover were compared with the corresponding experimental data available from Sulzer Chemtech (see http://www.sulzerchemtech.com/eprise/SulzerChemtech/Sites/designtools/designtools.html). This comparison is presented in Fig. 6.20, and a good agreement can be recognized.

In a further study, the phenomena typical for catalytic packings for reactive distillation were investigated. In particular, liquid flow along an arrangement of catalyst pellets accompanied by mass transfer and chemical reaction was examined. Such arrangements are also typical for fixed bed reactors. In [75], one periodic element (a crossover) of the laboratory-scale version of Katapak®-S was selected for the detailed CFD simulation with the commercial tool CFX-5 by ANSYS®. This solver used the finite volume discretization method in combination with hybrid unstructured grids. Around 1100 spherical particles of 1 mm diameter were included in the computational domain.

A geometry generation procedure for the randomly packed spheres was developed. An adaptive grid technique available in CFX-5 was applied in order to automatically resolve the surface of each grain, thus avoiding unnecessary fine grid far from the surfaces. Several grid adaptation steps were performed until the resulting superficial velocity reached its asymptotic value. Simulations were carried out using pure water as the liquid component. The calculated superficial flow velocity at a load point of 2.2 mm/s agreed well with the experimental results of Moritz and Hasse [76]. The residence time distribution was estimated by analyzing the local velocity field.

In the following work [77], fluid dynamics and *mass transfer phenomena* in fixed beds under liquid flow were investigated using CFX-5. For the description of mass transport phenomena at the catalyst particle surface, the particles were resolved directly. To simulate the mass transport in the chosen system with sufficient accuracy, it was essential to apply a high-density grid, especially near the particle surface. Regarding a high number of grid cells necessary to resolve each particle, one had to restrict the computational domain to avoid prohibitively expensive calculations. Generally, the number of variables increases also with each additional component, and hence, the requirements regarding computer capacity grow.

As a first step, the hydrodynamics in the regular geometry (body-centered cubic and face-centered cubic arrangements) was analyzed using periodic boundary conditions, except for the main flow direction. In this example, CFD is used to determine the liquid–solid mass transfer coefficient correlations that can be used for the design of reactors and (reactive) separation units. Decisive advantages of CFD are that it makes it possible to minimize or even avoid using real experiments, to investigate any arbitrary (and even not truly existing) geometry and to decouple phenomena. In real experiments, for example, an isolated study of external mass transport in the case of porous particles is not possible.

Real experiments for the determination of external mass transfer coefficients were used as an example for virtual experiments with CFD. In particular, experimental studies of Williamson *et al.* [78] and Wilson and Geankopolis [79] on the flow of two liquids (water and a propylene glycol–water mixture) through a packed bed of spherical particles made from solid benzoic acid were applied. As a result, the liquid–solid mass transfer coefficients, so far estimated solely experimentally, were determined in [77] in a purely theoretical way and showed qualitatively good agreement between the theoretical and experimental results.

In a subsequent study, two different methods were applied for the estimation of *residence time distribution* [80]. *The tracer method* imitates the experimental procedure using a nondiffusive tracer, whereas *the postprocessing method* directly calculates the residence time distribution from the local velocity field. Both methods gave similar results. However, the postprocessing method is preferable since it requires less computational power and time compared to the tracer method.

Further, the influence of confining solid walls on pressure drop in packed beds was studied numerically for moderate particle/tube diameter ratios [81, 82]. A regular, face-centered cubic, and an irregular particle configurations were investigated, the latter was created using a ballistic deposition method in combination with the Monte Carlo method. In Fig. 6.21, velocity patterns for both regular and irregular configurations are shown using stream-line representation. To simulate the fully developed flow by neglecting inlet effects, periodic boundary conditions were imposed along the main flow direction. For the regular geometry, only a small element was necessary since this element was repeated periodically. Channeling can be observed in both configurations, but its character is different. In the regular

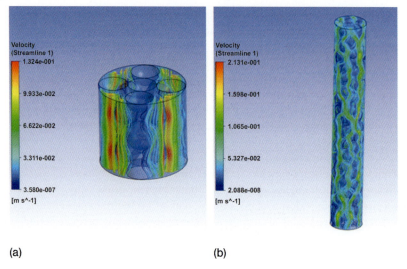

Fig. 6.21 Stream-line representation of the regular (a) and irregular (b) configuration for a tube diameter/particle diameter ratio of 3 and a particle Reynolds number of 25.

configuration, channeling mainly occurs near the wall where the void fraction is substantially high. In the irregular geometry, channeling is more chaotic.

To validate the simulation results, four experimental pressure-drop correlations were used, namely, the Ergun [83], the Carman [84], the Zhavoronkov et al. [85], and the Reichelt [86] correlation. Simulation results for the irregular configuration agree well with the correlations from [85, 86] that take the wall effect into account (Fig. 6.22). On the contrary, the predicted pressure drop for the regular configuration is underestimated, that can be attributed to the structured channeling effect [82]. These results are in line with the reality.

It is obvious that virtual experiments can provide reliable correlations in a purely theoretical way. Such correlations can partly replace experimental correlations necessary for the realization of the RBA. Further advantages of this innovative methodology are discussed in the following section.

6.8
Benefits of the Complementary Modeling

We have now seen that different problems of (reactive) fluid separation processes can be solved using different approaches classified in Fig. 6.3. The choice of the approach to be applied depends on several criteria mentioned in Section 6.3. These different approaches complement each other and cover together the whole spectrum of fluid-dynamic complexity of the separation processes. The overall approach can thus be called *complementary modeling*.

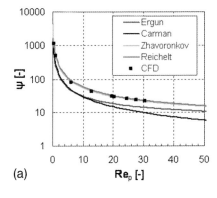
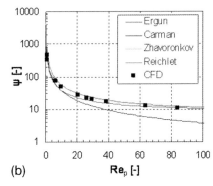

Fig. 6.22 Dimensionless pressure drop ψ as a function of the particle Reynolds number, Re_p: comparison of simulated data (points) and different experimental correlations (lines) for a particle-to-tube diameter of 3 (a) and 5 (b).

Still further complementarity is provided by the links between the different approaches shown in Fig. 6.23. The information necessary for the realization of the RBA (hydrodynamic and mass transfer correlations) can be obtained via the application of modern CFD facilities resulting in theoretically determined parameter correlations, as explained and illustrated in Section 6.7. Most often, such a CFD application is based on the use of periodical representative elements of the real internals, allowing a very fine grid and sufficient resolution of the flow patterns. The subsequent postprocessing provides the required functional dependencies of hydrodynamic and mass transfer parameters.

A similar link can be established between the FDA and the HA (cf. Fig. 6.23). For instance, an extension of the HA model for structured packed columns was found necessary for high gas loads, since the arising intensive gas-phase turbulence could not be properly governed by regular perfect mixing only. The model additionally considered impact of turbulence by implementing *gas-phase turbulent viscosity* [11]. This parameter cannot be estimated within the HA modeling, yet it can be determined by the FDA, based on real packing geometry.

Figure 6.24 illustrates this "collaborative" modeling way. Figure 6.24(a) depicts the velocity field (stream lines) in a single periodic element of a corrugated sheet packing obtained by CFD simulations (CFX™ 11.0), that is, by virtual experiments. Figure 6.24(b) shows the mean integral turbulent viscosities derived from the CFD

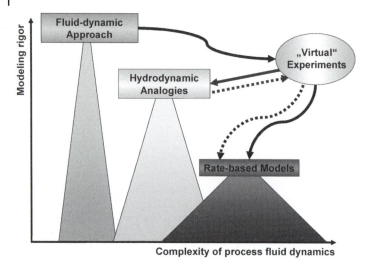

Fig. 6.23 Complementary modeling concept.

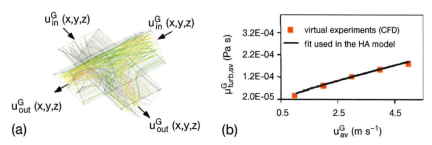

Fig. 6.24 Stream lines in a single periodic element of a corrugated sheet packing (a) and mean integral turbulent viscosities at different gas loads (b); u_{av}^G is the average gas-phase velocity.

results. The corresponding correlation for this parameter (solid line) is then used in the HA simulations [87].

In this manner, not only the number of necessary real hydrodynamic and mass transfer experiments can be reduced but also column internals can be optimized directly "on screen," without the need to manufacture them for validating experiments. Furthermore, virtual experiments can also be performed using the HA (e.g., determination of mass and heat transfer characteristics) and delivered to the RBA in the form of correlations. This link shown in Fig. 6.23 by a dashed line is a topic of an ongoing research.

A methodology for the development of optimal internals based on combining CFD and process simulation in a special optimization algorithm is suggested in [88]. A part of this procedure is sensitivity studies performed with the process simulator (RBA). The internals can then be modified in accordance with the results of these studies, and the modified internals can be further investigated theoretically by means of CFD to derive relevant correlations to be used in the process simulator. After several iterations, internals perfectly fitting the process criteria can be created "on screen" and afterward manufactured. As already mentioned in Section 6.7, in the long-term perspective, this method can be regarded as a way toward virtual prototyping of new, process-specific internals for (reactive) separations, provided that chemical companies and manufactures of internals are interested to collaborate.

6.9
Concluding Remarks

Successful optimization and intensification of chemical engineering processes largely depend on the predictivity and reliability of the process modeling. Because of a large spectrum of process conditions and criteria, physical and mathematical complexities in (reactive) separation processes stretch over a very wide range; thus, no ultimate modeling approach can be found. Therefore, we suggest an idea of complementary modeling based on a reasonable and efficient combination of different approaches. In this chapter, the complementary modeling is highlighted and illustrated with several examples.

We believe that the equilibrium concept, popular during the last century, will loose its significance for design and optimization tasks. It may still be used for some applications and for initialization purposes. However, the computational power as well as the need to understand and implement physicochemical principles of the processes will shift the modeling focus to the kinetics-based modeling approach.

We proposed to classify the kinetics-based models based on the complexity of the process fluid dynamics. For geometrically simple flows, the FDA should be applied as it gives full information about the process in a purely theoretical manner. For very complex flow patterns, the RBA represents a good choice provided that the model parameters are determined properly. The HA approach serves as an intermediate between the FDA and the RBA, and is suitable for processes in which a certain structure or order exists. This extends the application field of rigorous equations of fluid dynamics. In last years, this approach has developed rapidly, and thus, a possibility of its application should be seriously taken into account. Moreover, the HA modeling can be most suitable for some particular tasks, for example, investigation of maldistribution.

Several case studies are presented to highlight the use of all approaches. The latter are complementary in the sense that, together, they are able to govern both simple and very complex process fluid dynamic conditions. Still more important complementarity arises when different methods "collaborate," for instance, by estimating process parameters with a more rigorous method (virtual experiments) and delivering them to the less rigorous one. Along these lines, the individual modeling approaches are cemented into a truly complementary modeling concept.

Acknowledgements

The author would like to thank his former colleagues from the Laboratory of Fluid Separations at the Technical University of Dortmund and all project partners who have been involved in the research activities. He is also grateful to the German Research Foundation (DFG, Grant No. KE 837/3-1 + 2 and KE 837/7-1) and the European Commission (CEC Project No. BE95-1335, GROWTH Projects No. G1RD-CT-1999-00048 and G1RD-CT-2001-00649, NMP Projects NMP2-CT-2003-505862, and Marie-Curie Project No. MRTN-CT-2004-512233).

References

1 Reay, D., Ramshaw, C., Harvey, A., *Process Intensification: Engineering for Efficiency, Sustainability and Flexibility*, Elsevier Science & Technology Books, Amsterdam a.o., **2008**.

2 Noble, T., *Chemical Engineering Progress* 91(11) (**2001**), p. 10.

3 Doherty, M. F., Buzad, G., *Transactions of the IChemE* 70 (**1992**), p. 448.

4 Zarzycki, R., Chacuk, A., *Absorption: Fundamentals and Applications*, Pergamon Press, Oxford, **1993**.

5 Agar, D. W., *Chemical Engineering Science* 54 (**1999**), p. 1299.

6 Bart, H. J., *Reactive Extraction*, Springer, Berlin, **2001**.

7 Noeres, C., Kenig, E. Y., Górak, A., *Chemical Engineering and Processing* 42 (**2003**), p. 157.

8 Stankiewicz, A., Moulijn, J. (eds.), *Re-Engineering the Chemical Processing Plant*, Marcel Dekker, New York, **2003**.

9 Sundmacher, K., Kienle, A., Seidel-Morgenstern, A. (eds.), *Integrated Chemical Processes: Synthesis, Operation, Analysis, and Control*, Wiley-VCH, Weinheim, **2005**.

10 Schmidt-Traub, H., Górak, A. (eds.), *Integrated Reaction and Separation Operations. Modelling and Experimental Validation*, Springer, Berlin, **2006**.

11 Shilkin, A., Kenig, E. Y., Olujic, Z., *AIChE Journal* 52 (**2006**), p. 3055.

12 Sundmacher, K., Kienle, A. (eds.), *Reactive Distillation – Status and Future Directions*, Wiley-VCH, Weinheim, **2002**.

13 Davidson, L., The enterprise-wide application of computational fluid dynamics in the chemicals industry, in: *Proceedings of the 6th World Congress of Chemical Engineering*, Melbourne, **2001**.

14 Kenig, E. Y., *Computers and Chemical Engineering* 21 (**1997**), p. S355.

15 Bird, R. B., Stewart, W. E., Lightfoot, E. N., *Transport Phenomena*, 2nd edn., John Wiley and Sons, Inc., New York, **2002**.

16 Levich, V. G., *Physicochemical Hydrodynamics*, 2nd edn., Prentice-Hall, Englewood Cliffs, NJ, **1962**.

17 Grossman, G., Heat and mass transport in film absorption, in: *Handbook of Heat and Mass Transfer, vol. 2*, Cheremisinoff, N. P. (ed.), Gulf Publishing Company Book Division, Houston, **1986**, p. 211.

18 Yih, S.-M., Modeling heat and mass transport in falling liquid films, in: *Handbook of Heat and Mass Transfer, vol. 2*, Cheremisinoff, N. P. (ed.), Gulf Publishing Company Book Division, Houston, **1986**, p. 111.

19 Kenig, E. Y., *Theoretical Foundations of Chemical Engineering* 28 (**1994**), p. 199, 305.

20 Kenig, E. Y., *Modeling of Multicomponent Mass Transfer in Separation of Fluid Mixtures*, VDI-Verlag, Düsseldorf, **2000**.

21 Boyadjiev, C., Babak, V. N., *Non-linear Mass Transfer and Hydrodynamic Stability*, Elsevier, Amsterdam, **2000**.

22 Heibel, A. K., Heiszwolf, J. J., Kapteijn, F., Moulijn, J. A., *Catalysis Today* 69 (**2001**), p. 153.

23 Killion, J. D., Garimella, S., *International Journal of Refrigeration* 24 (**2001**), p. 755.

24 Taylor, R., Krishna, R., *Multicomponent Mass Transfer*, Wiley, New York, **1993**.

25 Mueller, I., Schildhauer, T. J., Madrane, A., Kapteijn, F., Moulijn, J. A.,

KENIG, E. Y., *Industrial and Engineering Chemistry Research* 46 (**2007**), p. 4149.
26 WAHEED, M. A., HENSCHKE, M., PFENNIG, A., *International Journal of Heat and Mass Transfer* 45 (**2002**), p. 4507.
27 BURGHOFF, S., KENIG, E. Y., *AIChE Journal* 52 (**2006**), p. 4071.
28 WEGENER, M., GRÜNING, J., STÜBER, J., PASCHEDAG, A. R., KRAUME, M., *Chemical Engineering Science* 62 (**2007**), p. 2967.
29 BURGHOFF, S., KENIG, E. Y., Modeling of mass transfer and interfacial phenomena on single droplets using computational fluid dynamics, in: *Proceedings of the ISEC'05 17th International Solvent Extraction Conference, Tucson*, **2005**.
30 VON REDEN, C., Multicomponent mass transfer in liquid–liquid extraction, PhD thesis, University of Dortmund, **1998**.
31 OSHER, S., FEDKIW, R., *Journal of Computational Physics* 169 (**2001**), p. 463.
32 VAN SINT ANNALAND, M., DIJKHUIZEN, W., DEEN, N. G., KUIPERS, J. A. M., *AIChE Journal* 52 (**2006**), p. 99.
33 HIRT, C. W., NICHOLS, B. D., *Journal of Computational Physics* 39 (**1981**), p. 201.
34 RIDER, W. J., KOTHE, D. B., *Journal of Computational Physics* 141 (**1998**), p. 112.
35 BOTHE, D., VOF-simulation of fluid particle dynamics, in: *Proceedings of the 11th Workshop on Two-Phase Flow Predictions*, Sommerfeld, M. (ed.), **2005**.
36 ATMAKIDIS, T., KENIG, E. Y., *The Journal of Computational Multiphase Flows* 2 (**2010**), p. 33.
37 JOSEPH, D. D., RENARDY, Y., *Fundamentals of Two-Fluid Dynamics*, Springer, Berlin, **1993**.
38 OLSSON, E., KREISS, G., *Journal of Computational Physics* 210 (**2005**), p. 225.
39 COMSOL AB, Product information COMSOL Multiphysics, 3.5., Stockholm, **2008**.
40 CHASANIS, P., KENIG, E. Y., HESSEL, V., SCHMITT, S., *Computer-Aided Chemical Engineering* 25 (**2008**), p. 751.
41 CHASANIS, P., LAUTENSCHLEGER, A., KENIG, E. Y., *Chemical Engineering Transactions* 18 (**2009**), p. 593.
42 CHASANIS, P., KERN, J., ZECIROVIC, R., GRUENEWALD, M., KENIG, E. Y., Experimental and numerical investigation of a high performance micro-separation device, in: *Proceeding of the CAMURE-7 7th International Symposium on Catalysis in Multiphase Reactors & ISMR-6 6th International Symposium on Multifunctional Reactors, Montreal, Canada, August*, **2009**.
43 ZANFIR M., GAVRIILIDIS, A., WILLE, C., HESSEL, V., *Industrial and Engineering Chemistry Research* 44 (**2005**), p. 1742.
44 KENIG, E. Y., KHOLPANOV, L. P., *Theoretical Foundations of Chemical Engineering* 27 (**1993**), p. 305.
45 BOYADZHIEV, L., *Separation Science and Technology* 25 (**1990**), p. 187.
46 LAZAROVA, Z., BOYADZHIEV, L., *Talanta* 39 (**1992**), p. 931.
47 TSCHERNJAEW, J., KENIG, E. Y., GÓRAK, A., *Chemie Ingenieur Technik* 68 (**1996**), p. 272.
48 RAMIREZ-GONZÁLEZ, E. A., MARTÍNEZ, C., ALVAREZ, J., *Industrial and Engineering Chemistry Research* 31 (**1992**), p. 901.
49 ZOGG, M., *Chemie Ingenieur Technik* 44 (**1972**), p. 930.
50 ZHAO, L., CERRO, R. L., *International Journal of Multiphase Flow* 18 (**1992**), p. 495.
51 OLUJIC, Z., *Chemical and Biochemical Engineering Quarterly* 11 (**1997**), p. 31.
52 SHETTY, S., CERRO, R. L., *Industrial and Engineering Chemistry Research* 36 (**1997**), p. 771.
53 BEHRENS, M., SARABER, P. P., JANSEN, H., OLUJIC, Z., *Chemical and Biochemical Engineering Quarterly* 15 (**2001**), p. 49.
54 VALLURI, P., MATAR, O. K., HEWITT, G. F., MENDES, M. A., *Chemical Engineering Science* 60 (**2005**), p. 1965.
55 SHILKIN, A., KENIG, E. Y., *Chemical Engineering Journal* 110 (**2005**), p. 87.
56 PELKONEN, S., Multicomponent mass transfer in packed distillation columns, PhD thesis, University of Dortmund, **1997**.
57 BRINKMANN, U., SCHILDHAUER, T., KENIG, E. Y., *Chemical Engineering Science* 65 (**2010**), p. 298.
58 BRINKMANN, U., JANZEN, A., KENIG, E. Y., Modelling of separation processes in reactive gas–liquid systems, in: *Proc. GLS-9 9th Int. Conference on Gas–Liquid and Gas–Liquid–Solid Reactor Engineering, Montreal*, **2009**.
59 SEADER, J. D., *Chemical Engineering Progress* 85(10) (**1989**), p. 41.
60 KENIG, E. Y., PYHÄLACHTI, A., JAKOBSSON, K., GÓRAK, A., AITTAMAA, J., SUNDMACHER, K., *AIChE Journal* 50 (**2004**), p. 322.

61 Kenig, E. Y., Butzmann, F., Kucka, L., Górak, A., *Chemical Engineering Science* 55 (**2000**), p. 1483.
62 Kenig, E. Y., Górak, A., Bart, H.-J., Reactive separations in fluid systems, in: *Re-Engineering the Chemical Processing Plant*, Stankiewicz, A., Moulijn, J. A. (eds.), Marcel Dekker, New York, **2003**, p. 309,
63 Taylor, R., Krishna, R., *Chemical Engineering Science* 55 (**2000**), p. 5183.
64 Kenig, E. Y., Górak, A., Reactive absorption, in: *Integrated Chemical Processes*, Sundmacher, K., Kienle, A., Seidel-Morgenstern, A. (eds.), Wiley-VCH, Weinheim, **2005**, p. 265.
65 Kenig, E. Y., Górak, A., Modeling of reactive distillation, in: *Modeling of Process Intensification*, Keil, F. (ed.), Wiley-VCH, Weinheim, **2007**, p. 323.
66 Huepen, B., Kenig, E. Y., *Industrial and Engineering Chemistry Research* 49 (**2010**), p. 772.
67 Mueller, I., Kloeker, M., Kenig, E. Y., Rate-based modelling of dividing wall columns – a new application to reactive systems, in: *Proceedings of the PRES'2004 7th Conference on Process Integration, Modelling and Optimisation for Energy Saving and Pollution Reduction*, **2004**.
68 Petlyuk, F. B., Platonov, V. M., *International Chemical Engineering* 5 (**1965**), p. 555.
69 Kaibel, G., *Chemical Engineering and Technology* 10 (**1987**), p. 92.
70 Mueller, I., Kenig, E. Y., Modeling of reactive dividing wall columns, in: *Proceedings of the AIChE Annual Meeting*, **2006**.
71 Mueller, I., Kenig, E. Y., *Industrial and Engineering Chemistry Research* 46 (**2007**), p. 3709.
72 Grossmann, C., Kenig, E. Y., *CIT Plus* 5 (**2007**), p. 38.
73 Special Issue, Intelligent column internals for reactive separations, *Chemical Engineering and Processing* 44 (**2005**).
74 Kloeker, M., Kenig, E. Y., Górak, A., *Catalysis Today* 79–80 (**2003**), p. 479.
75 Egorov, Y., Menter, F., Kloeker, M., Kenig, E. Y., *Chemical Engineering and Processing* 44 (**2005**), p. 631.

76 Moritz, P., Hasse, H., *Chemical Engineering Science* 54 (**1999**), p. 1367.
77 Kloeker, M., Kenig, E. Y., Piechota, R., Burghoff, S., Egorov, Y., *Chemical Engineering and Technology* 28 (**2005**), p. 31.
78 Williamson, J. E., Bazaire, K. E., Geankopolis, C. J., *Industrial and Engineering Chemistry, Fundamentals* 2 (**1963**), p. 126.
79 Wilson, E. J., Geankopolis, C. J., *Industrial and Engineering Chemistry, Fundamentals* 5 (**1966**), p. 9.
80 Atmakidis, T., Kenig E. Y., Viva, A., Brunazzi, E., CFD-based modelling of the residence time distribution in structured fixed beds, in: *Proceedings of the PRES'2007 10th Conference on Process Integration, Modelling and Optimisation for Energy Saving and Pollution Reduction*, **2007**.
81 Atmakidis, T., Kenig, E. Y., Kikkinides, S., CFD-based analysis of the wall effect on the pressure drop in packed beds, in: *Proceedings of the ECCE 6th European Congress on Chem. Eng.* **2007**.
82 Atmakidis, T., Kenig, E. Y., *Chemical Engineering Journal* 155 (**2009**), p. 404.
83 Ergun, S., *Chemical Engineering Progress* 48(2) (**1952**), p. 89.
84 Carman, P. C., *Transactions of the Institution of Chemical Engineering* 15 (**1937**), p. 150.
85 Zhavoronkov, N. M., Aerov, M. E., Umnik, N. N., *Journal of Physical Chemistry (Moscow)* 23 (**1949**), p. 342.
86 Reichelt, W., *Chemie Ingenieur Technik* 44 (**1972**), p. 1068.
87 Brinkmann, U., Mitschka, R. P., Kenig, E. Y., Thiele, R., Haas, M., Modelling of structured packed units by the hydrodynamic analogy approach: absorption processes with moderate and high gas loads, in: *Proceedings of the PRES'2008 11th Conference on Process Integration, Modelling and Optimisation for Energy Saving and Pollution Reduction*, **2008**.
88 Kloeker, M., Kenig, E. Y., Hoffmann, A., Kreis, P., Górak, A., *Chemical Engineering Process* 44 (**2005**), p. 617.

7
Efficient Reduced Order Dynamic Modeling of Complex Reactive and Multiphase Separation Processes Using Orthogonal Collocation on Finite Elements

Panos Seferlis, Theodoros Damartzis, and Natassa Dalaouti

Keywords

orthogonal collocation on finite elements (OCFE), rate-based models, reactive distillation, reactive absorption, multiphase distillation, adaptive element partition

7.1
Introduction

Distillation and absorption columns are the most common process units in chemical plants. Separation processes especially those that involve chemical reactions and multiphase systems give rise to complex dynamic models that require significant computational effort for solution. The main objectives for an absorption or distillation dynamic model are the calculation of accurate predictions of the process behavior during dynamic transition and the reliable solution, so that online applications can be facilitated. The calculation of accurate predictions is achieved through the mathematical representation of the major physical and chemical phenomena occurring within the separation column. Multicomponent diffusion, chemical reactions, phase equilibrium, and so forth must be accounted for in a solid and representative way. Nonequilibrium (NEQ) (rate-based) models provide the most comprehensive description of phenomena in industrial columns [1]. Obviously, the level of detail in the description of such phenomena can be determined only based on the ability to perform sensible experiments that would provide sufficient information for an adequate model validation. The derivation of accurate, precise, and uncorrelated estimates for the model parameters is essential in order to ensure good predictive power for the developed dynamic process model [2]. In staged separation processes, these phenomena-based process models are employed in each column stage (i.e., tray). Similarly, packed separation columns can be approximated by an adequate number of discrete equally spaced points within the column resembling a staged analog [3]. The number of physical stages in a staged column or equivalent stages in a packed column is usually quite large resulting in large scale process models that require substantial computational effort for solution. Such computational effort hinders the broad utilization of the process model in online control and optimization applications (e.g., model predictive controllers, optimization of transition column trajectories, and so forth). Therefore, an effective reduction in the size of the process model is sought that would not jeopardize the model accuracy by preserving the phenomenological elements in the model in a condensed model structure.

Several model order reduction methods for nonlinear dynamic separation processes have been developed. Wave propagation theory has been applied to binary and multicomponent distillation [4–6]. The theory identifies patterns in the composition profiles that are approximated by wave equations and give rise to low-order models. The method seems incapable of handling complex phenomena such as chemical reactions and diffusion-controlled distillation. Compartmental models [7] divide the column into compartments and select a representative stage within each compartment, namely, the sensitive stage. The sensitive stage is assumed to describe the dynamics of the associated compartment. Steady-state balances around envelopes initiating from the condenser and the reboiler for the stripping and rectifying sections, respectively, are required to close the dynamic balances. Aggregated models [8] also divide the column into compartments with the total holdup of the stages included in a compartment assigned to the holdup of a single representative "aggregation" stage of the compartment. The balance equations for the other stages in the compartment are assumed to be in quasi-steady-state. Several researchers applied the method to complex distillation columns for high purity air separations [9, 10]. The numerical performance of distillation-aggregated models is thoroughly analyzed in [11]. The compromise between solution speed and model accuracy through the manipulation of the number of aggregated stages used is carefully judged. Singular perturbation analysis of the tray-by-tray column is performed in [12] where a low-order nonlinear model is derived for the slow dynamics that govern the overall dynamic behavior. A balanced empirical grammian with Galerkin projection is used for the model reduction of nonlinear systems with application in distillation in [13]. Most of these methods apply the proposed model reduction techniques in columns that utilize an equilibrium process model with Murphree efficiencies accounting for tray nonideality. A comprehensive overview of fundamental and empirical approaches to the dynamic modeling of distillation columns is available in [14].

Orthogonal collocation on finite elements (OCFE) modeling offers a consistent solution within a unified modeling framework for a wide class of complex separation processes [15]. OCFE techniques in staged columns consider the otherwise discrete column domain as continuous, where composition and temperature profiles are continuous functions of position. Gas and liquid phase material and energy balances are satisfied exactly only at preselected points within the column (i.e., collocation points). The technique has been initially introduced by Stewart and coworkers [16] but adopted by many other researchers for steady state and dynamic simulations of complex separation processes [17–19]. OCFE modeling technique can be applied in conventional and reactive, staged, or packed absorption and distillation columns [15]. The key advantages are the high quality of approximation with a substantial degree of model order reduction that further enhances computational efficiency.

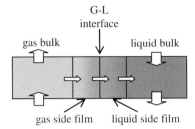

Fig. 7.1 Thin-film model representation.

Section 7.2 offers a detailed description of the modeling equations for staged and packed, nonreactive, reactive, and multiphase absorption and distillation processes. Section 7.3 introduces advanced error monitoring procedures during process dynamic transitions for performance enhancement of the NEQ/OCFE process models. Section 7.4 presents three illustrative case studies of industrial relevance and interest for reactive absorption, reactive distillation, and multiphase reactive distillation units. The case studies not only confirm the merits and strengths but also identify the issues that require special attention in the NEQ/OCFE model formulation. Finally, the epilog summarizes the conclusions and provides the context for future research in the area of reduced dynamic modeling for separation units.

7.2
NEQ/OCFE Model Formulation

The proposed model formulation combines the rigorous NEQ (rate-based) modeling equations [1, 20, 21] with the model reduction properties of the OCFE technique. A rate-based model variation assumes that all the resistance to mass and heat transfer is concentrated in two thin film regions adjacent to a phase interface as shown in Fig. 7.1. Bulk liquid and gas phases are assumed to have uniform properties in all their control volume, whereas chemical reactions may occur in both the bulk and the thin film regions. Thermodynamic equilibrium therefore holds only at the phase interface. The interface structure depends on the number of phases present at any given time instance. In multiphase systems, where a second liquid phase is present, the maximum number of possible interfaces is three (i.e., gas–liquid I, gas–liquid II, and liquid I–liquid II) [22]. Such situation implies that each phase is in contact with all other phases as shown in Fig. 7.2(b). A common situation is that the second liquid phase appears as a dispersed phase within a continuous liquid phase, disallowing the contact between the dispersed liquid and the gas phases. In such a case, only two interfaces are possible (i.e., gas–liquid I and liquid I–dispersed liquid II) as shown in Fig. 7.2(a).

The following assumptions are adopted for the development of the NEQ process model:

7 Efficient Reduced Order Dynamic Modeling

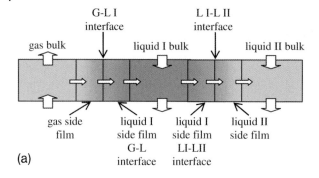

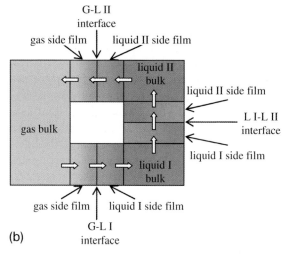

Fig. 7.2 Thin-film model representations for three-phase separation systems. (a) Dispersed liquid phase and (b) two continuous liquid phases.

- One-dimensional mass and heat transport across phase interfaces is assumed.
- Thermodynamic equilibrium holds only at phase interfaces.
- No axial dispersion along the distillation column is considered.
- No entrainment of the liquid phases in the vapor phase occurs.
- Complete mixing of bulk phases is achieved.
- Heat transfer through the thin film layers is negligible.
- Pressure drop inside the column is negligible.
- For three-phase mixtures, no contact between gas and dispersed liquid phase is possible.

Notations at the end of the chapter provide a comprehensive list of the process variables involved in the derivation of the models.

7.2.1
Conventional and Reactive Absorption and Distillation

Following the OCFE formulation as adopted by [18] and later extended to accommodate NEQ reactive distillation [23] and absorption processes [15], the separation column is divided into sections, with each section defined as the part of the column designated by two materials or energy streams entering or leaving the column. Each column section is further divided into smaller subdomains, namely, the finite elements. The number of collocation points within each finite element determines the order of polynomial approximation for the particular column segment. The main feature of the OCFE formulation is that the material and energy balances as resulted from the NEQ rate-based equations are satisfied exactly only at the collocation points. The collocation points are chosen as the roots of the discrete Hahn family of orthogonal polynomials [16]. Such a selection ensures that the collocation points coincide with the location of the actual stages in the limiting case that the order of the polynomial (i.e., number of collocation points) equals the number of actual stages for any given column section. Hence, the OCFE model approaches the complete tray-by-tray model structure as the number of collocation points increases and the two models become identical if the total number of collocation points becomes equal to the number of trays.

Lagrange interpolation polynomials are used within each finite element to approximate the liquid and vapor component flow rates, as well as the liquid and vapor stream enthalpies, as follows:

$$\tilde{L}_i(s) = \sum_{j=0}^{n} W_j^L(s_j)\tilde{L}_i(s_j)$$

$$0 \leq s \leq NT, \text{ with } s_0 = 0, \; i = 1, \ldots, NC \tag{7.1}$$

$$\tilde{G}_i(s) = \sum_{j=1}^{n+1} W_j^G(s_j)\tilde{G}_i(s_j)$$

$$1 \leq s \leq NT+1, \text{ with } s_{n+1} = NT+1, \; i = 1, \ldots, NC \tag{7.2}$$

$$\tilde{L}_t(s)\tilde{H}^L(s) = \sum_{j=0}^{n} W_j^L(s_j)\tilde{L}_t(s_j)\tilde{H}^L(s_j)$$

$$0 \leq s \leq NT, \text{ with } s_0 = 0 \tag{7.3}$$

$$\tilde{G}_t(s)\tilde{H}^G(s) = \sum_{j=1}^{n+1} W_j^G(s_j)\tilde{G}_t(s_j)\tilde{H}^G(s_j)$$

$$1 \leq s \leq NT+1, \text{ with } s_{n+1} = NT+1 \tag{7.4}$$

where NC is the number of the components, NT is the number of actual stages included within a given finite element, n is the number of collocation points selected in a given element, $\tilde{L}_i(s)$ and $\tilde{G}_i(s)$ are the component molar flow rates, $\tilde{L}_t(s)$ and $\tilde{G}_t(s)$ are the total liquid and gas stream flow rates, respectively, and $\tilde{H}^L(s)$ and $\tilde{H}^G(s)$ are the liquid and gas stream molar enthalpies, respectively. The tilde ($\sim$) on the symbols denotes approximation variables. Boundary points of the finite element, $s_0 = 0$ and $s_{n+1} = NT + 1$, are interpolation points for the liquid and vapor phase approximation schemes, respectively. Functions W^L and W^G represent Lagrange interpolation polynomials of the order $n+1$ given by the expressions

$$W_j^L(s) = \prod_{\substack{k=0 \\ k \neq j}}^{n} \frac{s - s_k}{s_j - s_k}, \quad j = 0, \ldots, n \tag{7.5}$$

$$W_j^G(s) = \prod_{\substack{k=1 \\ k \neq j}}^{n+1} \frac{s - s_k}{s_j - s_k}, \quad j = 1, \ldots, n+1 \tag{7.6}$$

Lagrange polynomials $W_j^L(s)$ and $W_j^G(s)$ are equal to zero at collocation points s_k when $k \neq j$ and to unity when $k = j$. Within each finite element, a Lagrange interpolation polynomial of different order can be used. The shape and characteristics of the approximated variable profiles (e.g., linear or irregularly shaped profiles, steep fronts) basically determine the required order of polynomial approximation. Stages that are connected to mass or energy streams entering (e.g., feed streams, streams from a pump-around) or leaving (e.g., product draw streams) the column are treated as discrete equilibrium stages, so that the effects of discontinuities in the mass and energy flows within the column sections do not influence the continuity and smoothness of the interpolating polynomial schemes within the elements (Eqs. (7.1)–(7.4)). Such a requirement can, however, be relaxed when the side material and/or energy streams are distributed uniformly along the length of a finite element or column section (e.g., similar mass and/or heat flows to successive stages within an element) [15]. Hence, the effect of the side streams on the column profiles uniformly spreads throughout the entire element domain and does not diminish the quality of the polynomial approximation.

The conditions at the element boundaries obey zero-order continuity, which is a plausible assumption for staged units with discrete profiles. However, certain smoothness conditions for the concentration and enthalpy profiles may be employed in the case of packed columns (e.g., first derivative continuity at element boundary). Assuming zero-order continuity, the boundary conditions connecting two consecutive elements yield the following relations:

$$\sum_{j=0}^{n_k} W_{j,k}^L(NT_k)\tilde{L}_i(s_{j,k}) = \tilde{L}_i(s_{0,k+1})$$

$$i = 1, \ldots, NC, \quad k = 1, \ldots, NE - 1 \tag{7.7}$$

$$\tilde{G}_i(s_{n_k+1,k}) = \sum_{j=1}^{n_{k+1}+1} W^G_{j,k+1}(1)\tilde{G}_i(s_{j,k+1})$$

$$i = 1, \ldots, NC, \ k = 1, \ldots, NE-1 \tag{7.8}$$

$$\sum_{j=0}^{n_k} W^L_{j,k}(NT_k)\tilde{L}_t(s_{j,k})\tilde{H}^L(s_{j,k}) = \tilde{L}_t(s_{0,k+1})\tilde{H}^L(s_{0,k+1})$$

$$k = 1, \ldots, NE-1 \tag{7.9}$$

$$\tilde{G}_t(s_{n_k+1,k})\tilde{H}^G(s_{n_k+1,k}) = \sum_{j=1}^{n_{k+1}+1} W^G_{j,k+1}(1)\tilde{G}_t(s_{j,k+1})\tilde{H}^G(s_{j,k+1})$$

$$k = 1, \ldots, NE-1 \tag{7.10}$$

In Eqs. (7.7)–(7.10), the first index for symbol s refers to the corresponding collocation point and the second one to the corresponding finite element, whereas n_k is the number of collocation points in the kth element. According to Eq. (7.7), the extrapolated value for the liquid flow rate at the endpoint of the kth element (i.e., $s = NT_k$) is set equal to the flow rate at the interpolation and boundary point in the following $(k+1)$th element (i.e., $s_{0,k+1} = 0$). Similarly, the vapor flow rate at the endpoint of the kth element (i.e., $s_{n+1,k} = NT_k + 1$), which is also an interpolation point for the vapor approximation scheme, is set equal to the extrapolated value for the vapor flow rate at the beginning of the adjacent $(k+1)$th element (i.e., $s = 1$).

Assuming that uniform hydrodynamic conditions prevail within each finite element, the dynamic mass balances employed at the collocation point s_j are described by the following equations:

$$\frac{dm_i^L(s_j)}{dt} = \tilde{L}_i(s_j - 1) - \tilde{L}_i(s_j)$$

$$+ \left(\varphi^L(s_j)R_i^{Lb}(s_j) + N_i^{Lb}(s_j)a^{GL}\right)A^{col}\Delta h$$

$$i = 1, \ldots, NC, \ j = 1, \ldots, n \tag{7.11}$$

$$\frac{dm_i^G(s_j)}{dt} = \tilde{G}_i(s_j + 1) - \tilde{G}_i(s_j)$$

$$+ \left(\varphi^G(s_j)R_i^{Gb}(s_j) - N_i^{Gb}(s_j)a^{GL}\right)A^{col}\Delta h$$

$$i = 1, \ldots, NC, \ j = 1, \ldots, n \tag{7.12}$$

The terms on the left-hand side of Eqs. (7.11) and (7.12) account for the component molar accumulation (i.e., material hold-up) in the liquid and gas bulk phase, respectively. These are related to the available liquid and vapor volumes corresponding to an equivalent stage by the equations:

$$m_i^L(s_j) = \varphi^L(s_j)d^L(s_j)\frac{\tilde{L}_i(s_j)}{\tilde{L}_t(s_j)}A^{\text{col}}\Delta h$$

$$i = 1, \ldots, NC, \ j = 1, \ldots, n \quad (7.13)$$

$$m_i^G(s_j) = \varphi^G(s_j)d^G(s_j)\frac{\tilde{G}_i(s_j)}{\tilde{G}_t(s_j)}A^{\text{col}}\Delta h$$

$$i = 1, \ldots, NC, \ j = 1, \ldots, n \quad (7.14)$$

In the above equations, $d^L(s_j)$ and $d^G(s_j)$ denote the liquid and gas phase density and $\varphi^L(s_j)$ and $\varphi^G(s_j)$ the liquid and gas phase volumetric fraction, respectively, calculated at the conditions prevailing at the collocation points $[\tilde{L}_i(s_j), \tilde{G}_i(s_j), \tilde{T}(s_j), \text{and } \tilde{P}(s_j)]$, while Δh stands for the height of the equivalent stage. Total flow rates and mole fractions in the bulk phases are also defined at the collocation points, as follows:

$$\sum_{i=1}^{NC} \tilde{L}_i(s_j) = \tilde{L}_t(s_j), \qquad x_i^{Lb}(s_j) = \frac{\tilde{L}_i(s_j)}{\tilde{L}_t(s_j)}$$

$$i = 1, \ldots, NC, \ j = 1, \ldots, n \quad (7.15)$$

$$\sum_{i=1}^{NC} \tilde{G}_i(s_j) = \tilde{G}_t(s_j), \qquad y_i^{Lb}(s_j) = \frac{\tilde{G}_i(s_j)}{\tilde{G}_t(s_j)}$$

$$i = 1, \ldots, NC, \ j = 1, \ldots, n \quad (7.16)$$

The dynamic mass balances in the gas and liquid films, considering the effect of chemical reactions on the mass transfer through the film regions, as well as the boundary conditions connecting the gas and liquid bulk phases with the respective films are also satisfied exactly only at the collocation points. The mass balances in the films can be written in the following form:

$$\frac{\partial c_i^{Gf}(s_j)}{\partial t} + \frac{\partial N_i^{Gf}(s_j)}{\partial \eta^{Gf}} - R_i^{Gf}(s_j) = 0$$

$$i = 1, \ldots, NC, \ j = 1, \ldots, n, \ 0 < \eta^{Gf} \leq \delta^{Gf}(s_j) \quad (7.17)$$

$$\frac{\partial c_i^{Lf}(s_j)}{\partial t} + \frac{\partial N_i^{Lf}(s_j)}{\partial \eta^{Lf}} - R_i^{Lf}(s_j) = 0$$

$$i = 1, \ldots, NC, \ j = 1, \ldots, n, \ 0 < \eta^{Lf} \leq \delta^{Lf}(s_j) \quad (7.18)$$

The boundary conditions for Eqs. (7.17) and (7.18) are

$$N_i^{Gb}(s_j) = N_i^{Gf}(s_j)|_{\eta^{Gf}=0}, \qquad y_i^{Gb}(s_j) = y_i^{Gf}(s_j)|_{\eta^{Gf}=0}$$

$$i = 1, \ldots, NC, \ j = 1, \ldots, n \tag{7.19}$$

$$N_i^{Lf}(s_j)|_{\eta^{Lf}=\delta^{Lf}(s_j)} = N_i^{Lb}(s_j), \qquad x_i^{Lf}(s_j)|_{\eta^{Lf}=\delta^{Lf}(s_j)} = x_i^{Lb}(s_j)$$

$$i = 1, \ldots, NC, \ j = 1, \ldots, n \tag{7.20}$$

The diffusion molar flux term $N_i(s_j)$ in Eqs. (7.11), (7.12), (7.17), and (7.18) is estimated by the Maxwell–Stefan equations for multicomponent mixtures, at the conditions prevailing at each collocation point. For ideal gas phase and one-dimensional mass transfer normal to the interface, the Maxwell–Stefan equations for the gas and the liquid thin film sides take the following form:

Gas phase:

$$\frac{\partial y_i^{Gf}(s_j)}{\partial \eta^{Gf}} = -\sum_{\substack{k=1 \\ k \neq i}}^{NC} \frac{[y_k^{Gf}(s_j) N_i^{Gf}(s_j) - y_i^{Gf}(s_j) N_k^{Gf}(s_j)]}{[\tilde{P}(s_j)/R_g \tilde{T}(s_j)] \mathcal{D}_{ik}^G}$$

$$i = 1, \ldots, NC, \ j = 1, \ldots, n, \ 0 < \eta^{Gf} \le \delta^{Gf}(s_j) \tag{7.21}$$

Liquid phase:

$$\sum_{k=1}^{NC-1} \Gamma_{i,k}(s_j) \frac{\partial x_k^{Lf}(s_j)}{\partial \eta^{Lf}} = -\sum_{\substack{k=1 \\ k \neq i}}^{NC} \frac{[x_k^{Lf}(s_j) N_i^{Lf}(s_j) - x_i^{Lf}(s_j) N_k^{Lf}(s_j)]}{c_t(s_j) \mathcal{D}_{ik}^L}$$

$$i = 1, \ldots, NC - 1, \ j = 1, \ldots, n, \ 0 < \eta^{Lf} \le \delta^{Lf}(s_j) \tag{7.22}$$

where the gamma thermodynamic factor is defined as follows:

$$\Gamma_{ik}(s_j) = \delta_{ik} + x_i(s_j) \frac{\partial \ln \gamma_i(s_j)}{\partial x_k(s_j)} \bigg|_{\tilde{T}(s_j), \tilde{P}(s_j), x_k(s_j), k \neq i=1,\ldots,NC-1} \tag{7.23}$$

Terms $R_i(s_j)$ in Eqs. (7.11), (7.12), (7.17), and (7.18) denote the total component reaction rate of the ith component in the gas and liquid bulk and film regions and are estimated at the conditions prevailing at each collocation point by the following equations:

$$R_i(s_j) = \sum_{r}^{NR} \nu_{i,r} r_r(s_j), \quad i = 1, \ldots, NC, \ j = 1, \ldots, n \tag{7.24}$$

where $r_r(s_j)$ denotes the rates of the reactions taking place. OCFE formulation can be tailored to allow reactive and nonreactive sections in the column.

At the gas–liquid interface of each collocation point, thermodynamic equilibrium is assumed, described by the following equation:

$$y_i^{\text{int}}(s_j) = K_i(s_j) x_i^{\text{int}}(s_j), \quad i = 1, \ldots, NC, \ j = 1, \ldots, n \qquad (7.25)$$

The boundary equations at the interface are

$$N_i^{Gf}(s_j)|_{\eta^{Gf} = \delta^{Gf}(s_j)} = N_i^{\text{int}}(s_j) = N_i^{Lf}(s_j)|_{\eta^{Lf} = 0}$$

$$i = 1, \ldots, NC, \ j = 1, \ldots, n \qquad (7.26)$$

$$y_i^{Gf}(s_j)|_{\eta^{Gf} = \delta^{Gf}(s_j)} = y_i^{\text{int}}(s_j), \qquad x_i^{\text{int}}(s_j) = x_i^{Lf}(s_j)|_{\eta^{Lf} = 0}$$

$$i = 1, \ldots, NC, \ j = 1, \ldots, n \qquad (7.27)$$

Neglecting the heat transfer effects along the film regions, the overall dynamic energy balance at each collocation point becomes

$$\frac{dU(s_j)}{dt} = \tilde{L}_t(s_j - 1)\tilde{H}^L(s_j - 1) + \tilde{G}_t(s_j + 1)\tilde{H}^G(s_j + 1) - \tilde{L}_t(s_j)\tilde{H}^L(s_j)$$

$$- \tilde{G}_t(s_j)\tilde{H}^G(s_j) + Q(s_j), \quad j = 1, \ldots, n \qquad (7.28)$$

where

$$U(s_j) = \sum_{i=1}^{NC} \{m_i^L(s_j) u_i^L(s_j) + m_i^G(s_j) u_i^G(s_j)\}, \quad j = 1, \ldots, n$$

Term $Q(s_j)$ is the net heat transferred from the surroundings. Contrary to the basic principle of the OCFE formulation that isolates stages with a mass or energy streams leaving or entering the column and treats them as discrete stages, the systematic and uniform heat exchange that may take place in a given column section can be also formulated as in Eq. (7.28). Deviations from the basic rule are only verified when the heat exchange within a given element is distributed uniformly along the element. For instance, heat losses or heat exchange for a group of neighboring stages can be approximated using balance Eq. (7.28) within finite elements that exclusively contain the stages where the heat exchange takes place. However, pump-around streams that carry significant material and energy amounts from one point of the column to another should be definitely treated as discrete stages.

Empirical correlations are employed for the calculation of the pressure drop in the column, the liquid holdup, the specific interfacial area, and the gas and liquid film thickness. These correlations consider column internals and hydraulics and depend on the type of the column plate (e.g., sieve, bubble cap) or column packing. The calculations are performed for the conditions prevailing at the collocation

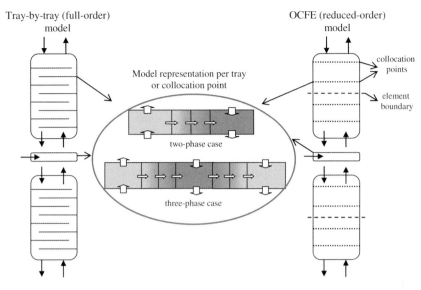

Fig. 7.3 Full- and reduced-order model representation.

points. The column is also subject to other operating limitation such as flooding constraints.

In conclusion, the NEQ/OCFE model formulation preserves the detailed description of the physical and chemical phenomena occurring within an absorption or distillation column as the modeling set of equations for collocation points and discrete stages are equivalent, but uses fewer points for the calculation of the material and energy balances than the full-order model (e.g., tray-by-tray model formulation) as shown in Fig. 7.3. Therefore, a more compact representation is available without diminishing the predictive power of the process model.

7.2.2
Multiphase Reactive Distillation

The appearance of a second liquid phase within the distillation column is imposed by the phase thermodynamic stability. The second liquid phase can be assumed either as a dispersed liquid phase within a continuous liquid phase or a second continuous liquid phase. Therefore, the thin film model representation depends on the assumption about the type of the second liquid phase (Fig. 7.2). In this chapter, the continuous-dispersed model for the two liquid phases is considered as the most likely to occur situation. However, the model is readily expandable to accommodate the more complex three-phase scenario. A unique feature of multiphase distillation is the fact that the boundary between the three-phase and the two-phase region in the column introduces a discontinuity in the concentration and the temperature profiles. Appearance or disappearance of a second liquid phase results in a change in the model structure as the balance equations for the second liquid phase and the corresponding interfaces should be incorporated or removed from the model, respectively. A second drawback is that the phase boundary is not known and its location may change significantly during a column dynamic transition. NEQ process models have been developed in [22, 24]. The derivation for the NEQ/OCFE process balances for each phase proceeds as follows [25]:

Liquid I phase:

$$\frac{dm_i^{L_I}(s_j)}{dt} = \tilde{L}_i^I(s_j - 1) - \tilde{L}_i^I(s_j)$$

$$+ \left[\varphi^{L_I}(s_j)R_i^{L_Ib}(s_j) + N_i^{GLb}(s_j)a^{GL} - N_i^{LL_Ib}(s_j)a^{LL}\right]A^{col}\Delta h$$

$$i = 1, \ldots, NC, \quad j = 1, \ldots, n \tag{7.29}$$

Liquid II (dispersed liquid) phase:

$$\frac{dm_i^{L_{II}}(s_j)}{dt} = \tilde{L}_i^{II}(s_j - 1) - \tilde{L}_i^{II}(s_j)$$

$$+ \left[\varphi^{L_{II}}(s_j)R_i^{L_{II}b}(s_j) + N_i^{L_{II}b}(s_j)a^{LL}\right]A^{col}\Delta h$$

$$i = 1, \ldots, NC, \quad j = 1, \ldots, n \tag{7.30}$$

Gas phase: The gas phase balance is derived from Eq. (7.12).

In the dynamic mass balances (7.29) and (7.30), m_i denotes the component molar accumulation, which may be computed from the following relations:

$$m_i^{L_I}(s_j) = \varphi^{L_I}(s_j)d^{L_I}(s_j)\frac{\tilde{L}_i^I(s_j)}{\tilde{L}_t^I(s_j)}A^{col}\Delta h$$

$$i = 1, \ldots, NC, \quad j = 1, \ldots, n \tag{7.31}$$

$$m_i^{L_{II}}(s_j) = \varphi^{L_{II}}(s_j)d^{L_{II}}(s_j)\frac{\tilde{L}_i^{II}(s_j)}{\tilde{L}_t^{II}(s_j)}A^{col}\Delta h$$

$$i = 1, \ldots, NC, \quad j = 1, \ldots, n \tag{7.32}$$

where φ stands for the phase volumetric holdup fraction and d for the phase molar density at each collocation point. The gas phase holdup is calculated from relation (7.14). The total molar flow rates and molar fractions for each phase are computed from the component flow rates as follows:

7.2 NEQ/OCFE Model Formulation

$$\sum_{i=1}^{NC} \tilde{L}_i^I(s_j) = \tilde{L}_t^I(s_j), \qquad x_i^{I,Lb}(s_j) = \frac{\tilde{L}_i^I(s_j)}{\tilde{L}_t^{II}(s_j)}$$

$$i = 1, \ldots, NC, \; j = 1, \ldots, n \tag{7.33}$$

$$\sum_{i=1}^{NC} \tilde{L}_i^{II}(s_j) = \tilde{L}_t^{II}(s_j), \qquad x_i^{II,Lb}(s_j) = \frac{\tilde{L}_i^{II}(s_j)}{\tilde{L}_t^{II}(s_j)}$$

$$i = 1, \ldots, NC, \; j = 1, \ldots, n \tag{7.34}$$

Under the assumption of negligible heat transfer through the films, the dynamic energy balance attains the form

$$\frac{dU(s_j)}{dt} = \tilde{L}_t^I(s_j-1)\tilde{H}^{L_I}(s_j-1) + \tilde{L}_t^{II}(s_j-1)\tilde{H}^{L_{II}}(s_j-1)$$

$$+ \tilde{G}_t(s_j+1)\tilde{H}^G(s_j+1) - \tilde{L}_t^I(s_j)\tilde{H}^{L_I}(s_j) - \tilde{L}_t^{II}(s_j)\tilde{H}^{L_{II}}(s_j)$$

$$- \tilde{G}_t(s_j)\tilde{H}^G(s_j) + Q(s_j), \quad j = 1, \ldots, n \tag{7.35}$$

In Eq. (7.35), $Q(s_j)$ denotes the net heat rate exchanged between the column and its surroundings, $\tilde{H}$ the stream molar enthalpy, and $U(s_j)$ the total molar energy accumulation, which is calculated from the relation

$$U(s_j) = \sum_{i=1}^{NC} \{m_i^{L_I}(s_j) u_i^{L_I}(s_j) + m_i^{L_{II}}(s_j) u_i^{L_{II}}(s_j) + m_i^G(s_j) u_i^G(s_j)\}$$

$$j = 1, \ldots, n \tag{7.36}$$

The dynamic mass balances in the gas and liquid films at the two possible interfaces (G–L^I and L^I–L^{II}) along with the boundary conditions are as follows:

G–L interface – gas side film:

$$\frac{\partial c_i^{Gf}(s_j)}{\partial t} + \frac{\partial N_i^{Gf}(s_j)}{\partial \eta^{Gf}} = 0$$

$$i = 1, \ldots, NC, \; j = 1, \ldots, n, \; 0 < \eta^{Gf} \le \delta^{Gf}(s_j) \tag{7.37}$$

Boundary conditions:

$$N_i^{Gb}(s_j) = N_i^{Gf}(s_j)|_{\eta^{Gf}=0}, \qquad y_i^{Gb}(s_j) = y_i^{Gf}(s_j)|_{\eta^{Gf}=0}$$

$$i = 1, \ldots, NC, \; j = 1, \ldots, n \tag{7.38}$$

G–L interface – liquid side film:

$$\frac{\partial c_i^{GLf}(s_j)}{\partial t} + \frac{\partial N_i^{GLf}(s_j)}{\partial \eta^{GLf}} - R_i^{GLf}(s_j) = 0$$

$$i = 1, \ldots, NC, \ j = 1, \ldots, n, \ 0 < \eta^{GLf} \leq \delta^{GLf}(s_j) \tag{7.39}$$

Boundary conditions:

$$N_i^{L_1 b}(s_j) = N_i^{GLf}(s_j)|_{\eta^{GLf} = \delta^{GLf}(s_j)}, \qquad x_i^{L_1 b}(s_j) = x_i^{GLf}(s_j)|_{\eta^{GLf} = \delta^{GLf}(s_j)}$$

$$i = 1, \ldots, NC, \ j = 1, \ldots, n \tag{7.40}$$

L–L interface – liquid I side film:

$$\frac{\partial c_i^{LL_1 f}(s_j)}{\partial t} + \frac{\partial N_i^{LL_1 f}(s_j)}{\partial \eta^{LL_1 f}} - R_i^{LL_1 f}(s_j) = 0$$

$$i = 1, \ldots, NC, \ j = 1, \ldots, n, \ 0 < \eta^{LL_1 f} \leq \delta^{LL_1 f}(s_j) \tag{7.41}$$

Boundary conditions:

$$N_i^{L_1 b}(s_j) = N_i^{LL_{II} f}(s_j)|_{\eta^{LL_1 f} = 0}, \qquad x_i^{L_1 b}(s_j) = x_i^{LL_1 f}(s_j)|_{\eta^{LL_1 f} = 0}$$

$$i = 1, \ldots, NC, \ j = 1, \ldots, n \tag{7.42}$$

L–L interface – liquid II side film:

$$\frac{\partial c_i^{LL_{II} f}(s_j)}{\partial t} + \frac{\partial N_i^{LL_{II} f}(s_j)}{\partial \eta^{LL_{II} f}} - R_i^{LL_{II} f}(s_j) = 0$$

$$i = 1, \ldots, NC, \ j = 1, \ldots, n, \ 0 < \eta^{LL_{II} f} \leq \delta^{LL_{II} f}(s_j) \tag{7.43}$$

Boundary conditions:

$$N_i^{L_{II} b}(s_j) = N_i^{LL_{II} f}(s_j)|_{\eta^{LL_{II} f} = \delta^{LL_{II}}(s_j)}$$

$$x_i^{L_{II} b}(s_j) = x_i^{LL_{II} f}(s_j)|_{\eta^{LL_{II} f} = \delta^{LL_{II} f}(s_j)}, \ i = 1, \ldots, NC, \ j = 1, \ldots, n \tag{7.44}$$

Thermodynamic equilibrium is valid at the phase interfaces (G–L^I and L^I–L^{II}) and is described by the following equations:

$$y_i^{GL}(s_j) = K_i^{GL}(s_j) x_i^{GL}(s_j), \quad i = 1, \ldots, NC, \ j = 1, \ldots, n \tag{7.45}$$

$$x_i^{I, LL}(s_j) = K_i^{LL}(s_j) x_i^{II, LL}(s_j), \quad i = 1, \ldots, NC, \ j = 1, \ldots, n \tag{7.46}$$

7.2 NEQ/OCFE Model Formulation

In Eq. (7.46), $x_i^{I,LL}$ and $x_i^{II,LL}$ represent the molar fraction of the first and second liquid phase at the LL interface, respectively. The boundary conditions that arise from the equilibrium equations at the interfaces are:

$$N_i^{Gf}(s_j)|_{\eta^{Gf}=\delta^{Gf}(s_j)} = N_i^{GL}(s_j) = N_i^{GLf}(s_j)|_{\eta^{GLf}=0}$$

$$i = 1, \ldots, NC, \ j = 1, \ldots, n \qquad (7.47)$$

$$y_i^{Gf}(s_j)|_{\eta^{Gf}=\delta^{Gf}(s_j)} = y_i^{GL}(s_j)$$

$$x_i^{GL}(s_j) = x_i^{GLf}(s_j)|_{\eta^{GLf}=0}, \quad i = 1, \ldots, NC, \ j = 1, \ldots, n \qquad (7.48)$$

$$N_i^{LL_I f}(s_j)|_{\eta^{LL_I f}=\delta^{LL_I f}(s_j)} = N_i^{LL}(s_j) = N_i^{LL_{II} f}(s_j)|_{\eta^{LL_{II} f}=0}$$

$$i = 1, \ldots, NC, \ j = 1, \ldots, n \qquad (7.49)$$

$$x_i^{LL_I f}(s_j)|_{\eta^{LL_I f}=\delta^{LL_I f}(s_j)} = x_i^{I,LL}(s_j)$$

$$x_i^{II,LL}(s_j) = x_i^{LL_{II} f}(s_j)|_{\eta^{LL_{II} f}=0}, \quad i = 1, \ldots, NC, \ j = 1, \ldots, n \qquad (7.50)$$

The reaction terms R_i used in the above equations represent the total rate of reaction per component i and are computed from Eq. (7.24). The component molar flux terms N_i at the collocation points are calculated through the multicomponent Maxwell–Stefan relations (Eqs. (7.21) and (7.22) for the gas and liquid phase, respectively).

Equations (7.21) and (7.22) and (7.29)–(7.50) form the NEQ/OCFE model for a three-phase element, whereas Eqs. (7.1)–(7.29) form the NEQ/OCFE model for a two-phase element. One major feature of three-phase distillation units is that the liquid phase split regions are not known before a solution of the modeling equations is obtained. Besides, during dynamic transition, the phase boundary may shift with time. Unless the correct set of modeling equations is used, thus identifying the phase distribution correctly, the simulated results will not represent accurately the column behavior and occasionally fail to reach a feasible solution. The latter is possible if a three-phase element is used to represent a two-phase region. Therefore, a tracking mechanism for the phase boundary is necessary for improved accuracy.

According to the methodology of [26] as later implemented in [25], an element breakpoint is adaptively located at the phase transition boundary. The identification of the phase discontinuity point is performed via a liquid–liquid (LL) flash calculation at the element breakpoints. The LL flash calculation determines whether the formation of the second liquid phase is possible and subsequently specifies whether the element breakpoint becomes a phase transition boundary. The element breakpoint must be allowed to move freely during simulation or optimization studies in order to trace variations in phase allocation. The LL flash at the element breakpoints involves the following additional calculations:

LL flash equations:

$$\Phi(s_0)x_i^{I,Lb}(s_0) + (1 - \Phi(s_0))x_i^{II,Lb}(s_0) = x_i(s_0), \quad i = 1,\ldots, NC \quad (7.51)$$

$$\sum_{i=1}^{NC}(x_i^{I,Lb}(s_0) - x_i^{II,Lb}(s_0)) = 0, \quad i = 1,\ldots, NC \quad (7.52)$$

$$\gamma_i^I x_i^{I,Lb}(s_0) = \gamma_i^{II} x_i^{II,Lb}(s_0), \quad i = 1,\ldots, NC \quad (7.53)$$

where γ_i is the activity coefficient of each component and Φ is the phase split fraction ($\Phi = 1$ for single liquid phase and $0 < \Phi < 1$ for two liquid phases). The LL flash calculation is performed at the top of the finite element (s_0 corresponds to the interpolation point for the liquid phase at the top of each element). Element breakpoints are therefore placed at suitable positions so that the condition $\Phi(s_0) = 1.0$ holds. Similar LL flash calculations are also utilized in a robust algorithm for the tracking of phase distribution changes in [27].

7.3
Adaptive NEQ/OCFE for Enhanced Performance

The quality of the approximation depends on the density of collocation points utilized to describe a given column segment. Column segments with relatively flat composition and temperature profiles would require a lower density of collocation points than a column segment with steep profiles for an accurate model representation. Ideally, two collocation points would suffice to approximate a flat profile. On the contrary, a steep, irregular profile would require a higher density of collocation points around the column segment where such behavior becomes evident. As a general guideline, finite elements with low-order polynomial approximation are preferred from a scheme with fewer elements with higher order polynomials. The density of collocation points within the column can vary based on the appropriate placement of the finite element boundaries. Steep fronts in the column profiles may wander during dynamic transition or new fronts may appear induced by disturbances that upset the column. Therefore, the NEQ/OCFE model must be reinforced with a tool to estimate approximation quality and to further adapt its structure to accurately trace the time variation of profiles.

Several approximation error estimation procedures have been introduced and utilized for OCFE models in steady state and dynamic simulation or optimization [28–32]. These procedures include the evaluation of residuals in the material and energy balance equations at interpolation points obtained from the roots of polynomials [28], at the element midpoint [29], and spline approximation to the derivatives of the interpolation polynomials [30–32].

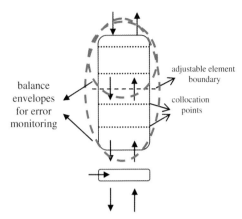

Fig. 7.4 Approximation error monitoring balance envelopes in an NEQ/OCFE model.

The method proposed in [33] is used for the adaptive placement of the element breakpoints for a given total number of collocation points. The method calculates the material and energy balance error around envelopes drawn around certain segments of the column as shown in Fig. 7.4. Each envelope encircles a different number of finite elements. The residuals of the material and energy balances around these envelopes are indicative of the quality of the approximation in the particular column segment. Even though component material balance residuals can be evaluated in conventional distillation or absorption processes, reactive separation processes require the evaluation of the overall material balance residual for each envelope. In dynamic models, the residuals vary with time and are reliable measures of the approximation error during the dynamic transition. The manipulation of the residuals along the column vertical coordinate can be achieved by adaptively changing the finite element partition of the column domain for a given number of finite elements and collocation points. The adaptive placement of element breakpoints alters the density of collocation points in the column according to the quality of the approximation scheme, as deduced from the magnitude of the material and energy balance residuals. Obviously, the composition and temperature profiles along the column during dynamic transition are not known *a priori* but only after the integration of the governing equations. Therefore, the placement of the element breakpoints cannot be decided upon initial column profiles as these may change time as a result of the effects with exogenous disturbances have on the dynamic system.

The main aim is not to obtain a finite element partition that would minimize the magnitude of the balance residuals but rather distribute equally the approximation error among all finite elements in the column [33]. The adaptive placement problem of the finite element breakpoints during a dynamic transition is formed as a dynamic optimization problem. More specifically, the time domain for the simulation is divided into time intervals. Within each time interval, a specific element partition is selected so that the residual equidistribution condition is satisfied for the entire column at each time instance. Hence, the mathematical formulation takes the form:

$$\min_{\substack{NT_{1,j}, NT_{2,j}, \ldots, NT_{NE,j} \\ j=1,\ldots,NP}} f = \text{const}$$

s.t.   Column Dynamic Model

$$Rs_{\text{tot},i} - Rs_{\text{tot},i+1} \leq \varepsilon, \quad i = 1, \ldots, NE - 1 \tag{7.54}$$

where $Rs_{\text{tot},i}$ is the total balance residual for the envelope including the ith element, NT_i is the number of stages included within the boundaries of the ith finite element, indirectly implying the length of the finite element, NP is the number of time intervals that the simulation time domain is divided, and ε is the tolerance that the error equidistribution relations must satisfy. The total residual is calculated as the second norm of the material and energy balance residuals

$$Rs_{\text{tot},i} = \left(Rs_{\text{mat},i}^2 + Rs_{\text{eng},i}^2\right)^{1/2}, \quad i = 1, \ldots, NE \tag{7.55}$$

where $Rs_{\text{mat},i}$ and $Rs_{\text{eng},i}$ are the material and energy balance residuals around the ith envelope. In conventional distillation and absorption, the material balances equal the number of components in the separation process, whereas in reactive columns, only the overall material balance is used (even though individual material balances can be used for inert components). The problem expressed in Eq. (7.54) is similar to an optimal control problem, where the controls (i.e., decision variables) are the element breakpoints that subsequently define the lengths of the finite elements in terms of trays included within the boundaries of each finite element. The size of each finite element (equivalent to the number of stages) can vary within real-valued bounds acting as a continuous decision variable (degree of freedom) in the placement optimization problem. The only constraint that applies on the size of each finite element refers to the approximated total number of real column stages, which must be equal to or greater than the number of collocation points used in the given finite element (i.e., $NT_i \geq n_i$ for the ith element). The performance index related to the approximation error distribution is incorporated in the form of inequality constraints. The partition of the simulation time domain in time intervals is selected arbitrarily and based on the complexity of the resulted optimization problem. Problem (7.54) even though does not guarantee the achievement of the minimum approximation error, it offers a reasonable compromise through the equal distribution of the error throughout the column.

7.4
Dynamic Simulation Results

7.4.1
Reactive Absorption of NO_x

7.4.1.1   Process Description
The reactive absorption of nitrogen oxides (NO_x) from a gas stream by a weak HNO_3 aqueous solution [34–36] is an effective process that removes NO_x from gas

Table 7.1 Chemical reaction mechanism in the NO_x absorption column.

Gas phase	Liquid phase
RR1: $2NO + O_2 \rightarrow 2NO_2$	RR6: $N_2O_4 + H_2O \rightarrow HNO_2 + HNO_3$
RR2: $2NO_2 \leftrightarrow N_2O_4$	RR7: $3HNO_2 \rightarrow HNO_3 + H_2O + 2NO$
RR3: $3NO_2 + H_2O \leftrightarrow 2HNO_3 + NO$	RR8: $N_2O_3 + H_2O \rightarrow 2HNO_2$
RR4: $NO + NO_2 \leftrightarrow N_2O_3$	RR9: $2NO_2 + H_2O \rightarrow HNO_2 + HNO_3$
RR5: $NO + NO_2 + H_2O \leftrightarrow 2HNO_2$	

streams released to the atmosphere and produces nitric acid. Chemical reactions play an important role in this separation system as chemical conversion (e.g., oxidation) of the otherwise insoluble in water components (e.g., NO) to more soluble components (e.g., NO_2) enhances the overall absorption rate. Moreover, nitric acid is produced through a complex reaction mechanism that involves five gas-phase and four liquid-phase reactions provided in Table 7.1 [35]. The oxidation of NO to NO_2 (RR1) is kinetically the slowest and thus the limiting step in the mechanism [34]. The remaining gas-phase equilibrium reactions (RR2)–(RR5) are reversible kinetic reactions. Liquid phase reactions (RR6) and (RR7) are kinetically controlled, whereas reactions (RR8) and (RR9) are considered as reversible kinetic reactions. The temperature-dependent expressions of the equilibrium constants for all the reactions in the kinetic mechanism are shown in Table 7.2 with data taken from [34–36].

The gas-phase diffusion coefficients in the Maxwell–Stefan equations are estimated using the Chapman–Enskog–Wilke–Lee model, whereas the liquid-phase diffusion coefficients are calculated by the method described in [37]. Liquid-phase activity coefficients are calculated by the NRTL activity model, whereas thermodynamic properties such as stream enthalpy and density are calculated by the Soave–Redlich–Kwong equation of state.

A gas stream with high concentration of NO_x that needs to be cleaned enters the bottom of the staged countercurrent-reactive absorption column. The gas stream is washed out by a liquid water (solvent) stream that enters from the top of the

Table 7.2 Kinetic data for the NO_x absorption column.

	Equilibrium and rate constants
RR1	$\log_{10} k_1$: $652.1/T - 0.7356$ $(atm\ s)^{-1}$
RR2	$\log_{10} k_2$: $2993/T - 9.226$ atm^{-1}
RR3	$\log_{10} k_3$: $2003.8/T - 8.757$ atm^{-1}
RR4	$\log_{10} k_4$: $2072/T - 7.234$ atm^{-1}
RR5	$\log_{10} k_5$: $2051.17/T - 6.7328$ atm^{-1}
RR6	$\log_{10} k_6$: $-4139/T + 16.3415$ s^{-1}
RR7	$\log_{10} k_7$: $-6200/T + 20.1979$ $(m^3/kmol)^2\ atm\ s^{-1}$
RR8	K_8: 3.3×10^2 $(kmol/m^3)^{-1}$
RR9	K_9: 3.8×10^9 $(kmol/m^3)^{-1}$

Table 7.3 Inlet stream data for the NO_x absorption column[a].

Gas inlet stream (bottom)		Liquid inlet stream (top)	
NO	21.83 mol/s	H_2O	4.55 mol/s
NO_2	58.08 mol/s	T	293 K
N_2O_4	20.11 mol/s	Side feed stream	
O_2	82.74 mol/s	H_2O	34.62 mol/s
N_2	1016.44 mol/s	HNO_3	6.02 mol/s
T	332 K	NO_2	0.41 mol/s
P	5.6 bar	T	306 K

a) Recycle stream: 4.55 mol/s.

column, whereas a weak solution of nitric acid enters the side of the column. The bottom's liquid stream mainly consisted of an aqueous nitric acid solution is partially recycled in the column for control of the nitric acid product concentration. The nitric acid concentration in the liquid product that leaves the column as a side draw stream is subject to certain quality specifications. The concentration of NO_x gases in the gas stream at the top of the column is subject to composition constraints imposed by environmental regulations for gases released to atmosphere.

According to the absorption column design described in [15], the column is consisted of 44 trays with an internal diameter of 3.6 m. The distance between the trays is equal to 0.9 m, except for the five trays closer to the bottom, where the oxidation reaction (RR1) mostly takes place so that higher tray spacing is used in order to increase the gas phase holdup and subsequently the extent of (RR1). Cooling is provided in the column stages for the removal of the heat of the exothermic oxidation reaction and the control of the column temperature. Temperature control is essential because low column temperatures favor the extent of the oxidation reaction (RR1) and the solubility of NO_2 in water. Column pressure drop, liquid phase holdup, film thickness, as well as stage interfacial area for sieve plates are estimated by empirical correlations found in [38, 39].

The absorption column is partitioned into three sections. The first section lies above the location of the side weak nitric acid solution feed stream, one between the side feed stream and the recycle side stream, and one below the recycle side stream. Stages connected to material streams entering or leaving the column are treated as discrete stages, where the rate-based balance equations are applied. The five oxidation stages closer to the bottom of the column are treated as discrete stages because each stage has a different holdup. The OCFE model formulation in such a case would have miscalculated the holdup for the derivation of the material and energy balances at a collocation point. Each column section is further partitioned into a number of finite elements of a given number of collocation points. Material and energy rate-based balances (Eqs. (7.1)–(7.19)) are then applied at the specified collocation points along the column domain. Gas and liquid inlet stream data for the case study are shown in Table 7.3.

7.4.1.2 Dynamic Simulation Results

The dynamic simulations of the reactive absorption column model consisted of a set of nonlinear differential/algebraic equations are performed in gPROMS® [40] – an integrated process modeling environment. The partial differential equations describing mass transfer in the gas and liquid films (Eqs. (7.16)–(7.19)) are discretized using OCFE. More specifically, fifth- and third-order polynomials are selected using two finite elements in the gas and liquid phase films, respectively.

The NEQ/FULL process model that relates to the tray-by-tray model formulation is assumed to accurately represent the behavior of the actual unit operation. Two different NEQ/OCFE schemes are formed and tested for their accuracy in representing the full-order (tray-by-tray) model with their specific characteristics (i.e., number of elements and collocation points per element) given in Table 7.4. The steady-state behavior of the NEQ/OCFE is in good agreement to the behavior of the NEQ/FULL model as shown in [15].

The dynamic response of the NEQ/OCFE model is investigated through two scenarios. In the first scenario, a 10% increase in the gas inlet stream flow rate is imposed and in the second, a 20% decrease of the cooling water flow rate at all stages with cooling is performed. The dynamic response of two key process variables for the column is shown in Fig. 7.5. In both scenarios, the reduced-order model identifies the qualitative characteristics of the full-order response and in particular the inverse response in the second scenario and the variables overshoot. Comparison of the full-order to the reduced-order dynamic response reveals that the error in the NEQ/OCFE model is within reasonable limits acceptable for control applications. More specifically, the relative error in the maximum overshoot, the steady state, and the rise time calculated for both OCFE formulations does not exceed 5.5% in the estimation for the rise time. However, the reduction in the model size and the computational effort achieved with the NEQ/OCFE model is quite impressive. A model reduction ratio of 0.29 results in a CPU time ratio of 0.14 and an adequately accurate representation of the dynamic response. The slightly faster dynamics obtained by the NEQ/OCFE model may be attributed to the underestimation of the liquid phase holdup in the large top section of the column.

The dynamic behavior of the NEQ/FULL model and the NEQ/OCFE model formulations under closed loop conditions is also explored. A PI controller manipulates the cooling water flow rates in the different column sections so that NO_x outlet concentration remains within a desired range [19]. Figure 7.6 depicts the behavior of the NEQ/FULL and NEQ/OCFE models with the simulated PI controller for a 10% increase in the NO_x content of the inlet gas stream. The overestimation of the new steady state in the outlet gas stream by the NEQ/OCFE model is less than 3% and the error for the composition overshoot is less than 2%.

Real-time control applications require the use of reliable, accurate, and easy to solve dynamic process models. NEQ/OCFE models are good candidates for such applications mainly due to the significant reduction in the number of states involved in the dynamic model, while keeping the key nonlinear features of the process. In addition, the NEQ/OCFE formulation fully preserves the input–output structure of the physical dynamic system, a particularly desirable situation. A situ-

Table 7.4 Model technical specifications and performance characteristics[a].

Model type	Number of stages or finite elements (FE) and collocation points (CP)				Model equations	Model reduction ratio	CPU time ratio	Relative deviation (%) for peak/steady state/rise time
	Top section 29		Middle section 8					
	FE	CP/FE	FE	CP/FE				
FULL	2	4	1	2	42 680	1.0	1.0	– Figs. 7.5(a) and (b)
OCFE A	2	2	1	2	16 720	0.39	0.33	0.06/0.08/2.51 1.24/1.23/3.42
OCFE B	2	2	1	2	12 580	0.29	0.14	0.22/0.23/3.91 1.25/1.02/5.45

a) Note: gPROMS® 3.0.3.

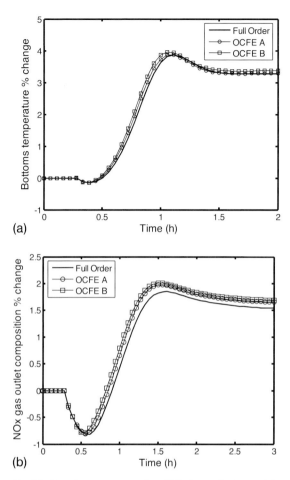

Fig. 7.5 Open loop responses for a 10% change in the gas inlet stream flow rate (a) and a 20% change in the cooling water flow rates (b).

ation that requires special treatment arises when a measurement at specific locations in the column is used in the column control system (e.g., temperature measurements for inferential control schemes). In such a case, the element partition must be modified so that a collocation point is placed at the exact location of the sensor within the column.

7.4.2
Ethyl Acetate Production via Reactive Distillation

7.4.2.1 Process Description
The production of ethyl acetate can be achieved via reactive distillation [41–45, 23], in which ethanol and acetic acid form ethyl acetate through an endothermic liquid

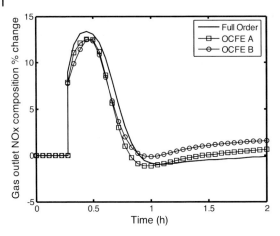

Fig. 7.6 Gas outlet NO$_x$ composition under closed loop conditions.

phase esterification reversible reaction in the presence of sulfuric acid that acts as the catalyst:

$$C_2H_5OH + CH_3COOH \leftrightarrow CH_3COOC_2H_5 + H_2O$$

The case study involves the reactive distillation column design obtained in [23]. Pure acetic acid feeds the column near the top of the column and as the heaviest of the components moves toward the bottom of the column. Ethanol feeds the column near the bottom and as one of the lighter components moves toward the top section. The reaction is assumed to take place only in the liquid phase. The rate of reaction is generally low, and therefore a large residence time in each column stage is required for a reasonable extent of reaction. A second factor that limits the esterification reaction is that ethanol has relatively high volatility and prefers the vapor phase rather than the liquid phase, where the reaction takes place. The expression for the rate of reaction is given as follows:

$$r = k_{01} \exp\left(-\frac{A_1}{T}\right)[ACOOH][EtOH] - k_{02} \exp\left(-\frac{A_2}{T}\right)[EtAc][H_2O] \quad (7.56)$$

The mixture of components exhibits a highly nonideal thermodynamic behavior as demonstrated by the formation of four binary azeotrope mixtures, one ternary azeotrope and one reactive azeotrope when using the Wilson activity model. Unfavorable physical equilibrium constraints the production of an ethyl acetate product stream with high-purity level from a single distillation column. However, the use of a second recovery column operating at a higher pressure allows the separation of the azeotrope mixture to the desired purity level [43]. The recovery distillation column is introduced downstream the reactive column operating at a higher pressure (350 kPa) with the distillate stream mainly containing unreacted reactants recycled to the reactive column.

Table 7.5 Column configuration, technical specifications, and nominal operating point.

Number of stages	
Rectifying/reactive/stripping sections	6/10/6
Stage holdups (m^3)	
Rectifying/reactive/stripping sections	0.15/1.75/1.634
Streams	
Acetic acid feed	Flow rate (kmol/h): 3.21; Temperature (K): 363
Ethanol feed	Flow rate (kmol/h): 3.145; Temperature (K): 363
Recycle feed	Flow rate (kmol/h): 14.5; Temperature (K): 381 Composition EtOH/EtAc/H$_2$O (molar fraction): 0.2856/0.4630/0.2514
Heat duty (MJ/h)	
Condenser/reboiler	−530.08/482.58
Kinetic parameters	
A_1: 7150 K	k_{01}: 1.74 × 10^6 m^3/(kmol h) $A_2 = 7150$ K $k_{02} = 4.428 \times 10^5$ m^3/(kmol h)

The NEQ/OCFE reactive distillation model partitions the column into three sections, the rectifying (the section between the condenser and the acetic acid feed), the reactive (the section between the two side feed streams), and the stripping section (the section from the ethanol feed down to the reboiler). The recycle from the second recovery column and the ethanol feed streams enter the reactive column at the same stage. The stage holdup varies among the three column sections, thus influencing the reaction rate. The column configuration, technical characteristics, and nominal operating point of the column are given in [23] and shown in Table 7.5. The Wilson activity model is used for the liquid phase, whereas an ideal gas phase is assumed. Regressed equations are used for the liquid and vapor molar enthalpy.

7.4.2.2 Dynamic Simulation Results

The dynamic NEQ/OCFE and NEQ/FULL models are developed and solved using gPROMS®. A comparison between the dynamic behavior predicted by NEQ/FULL and the NEQ/OCFE model formulations reveals the excellent prediction properties of the NEQ/OCFE. The dynamic responses for the ethyl acetate bottoms and distillate composition to multiple step changes of the reboiler heat duty are shown in Fig. 7.7. The NEQ/OCFE model retains the ability to accurately represent the dynamic behavior of the process despite the significant decrease in the model size, as deviations from the dynamic response predicted by the full-order model are almost negligible (less than 0.2%). The achieved reduction in the required CPU time for the entire dynamic simulation shown in Table 7.6 clearly demonstrates the definite advantage of the NEQ/OCFE model for real-time control applications.

The outlined procedure for the adaptive placement of the element breakpoints is applied in the NEQ/OCFE model for reboiler heat duty variations. Specifically,

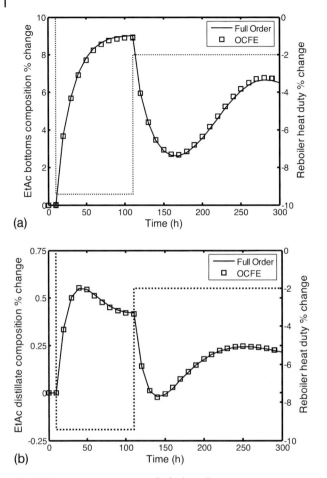

Fig. 7.7 Dynamic responses to reboiler heat duty variation.

several fixed element partitions are tested for the accuracy of their predictions, as estimated by the material balances around envelopes in the column. Attention is focused on the reactive section of the column, where two elements with two collocation points per element are employed representing ten physical stages. Therefore, two envelopes are constructed; the first envelope includes the top element of the reactive section and the second one both elements of the reactive section as shown in Fig. 7.4. The fixed element partitions shown in Fig. 7.8(a) have element lengths of [5, 5] (circle markers) and [4, 6] (square markers), respectively. The adaptive element partition is obtained from the solution of the optimization problem (7.54) with two time intervals starting at time periods 0 and 30 h. The problem is formulated and solved in gPROMS® as a dynamic optimization program. At the solution, the element length partition is obtained as follows: time interval [0–35], element lengths [4.25, 5.25]; time interval [35–50] [4.505, 5.4905]. The reboiler heat duty variations occur at time instances of 10 h (−9.5% from the initial value) and 30 h

Table 7.6 Model technical specifications and performance characteristics[a].

Model type	Number of stages or finite elements (FE), and collocation points (CP)							Model equations	Model reduction ratio	CPU time ratio*	Max relative deviation (%)[†]
	Rectifying section		Reactive section		Stripping section						
	6		10		6						
	FE	CP/FE	FE	CP/FE	FE	CP/FE					
FULL								11 477	1.0	1.0	–
OCFE	1	2	2	2	1	2		4720	0.41	0.29	0.2%

a) Note: gPROMS® 3.0.3.

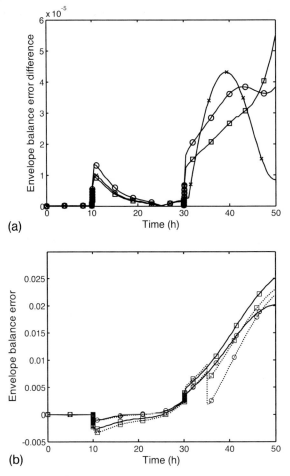

Fig. 7.8 (a) Material balance envelope residual difference time variation for two fixed "o," "□," and an adaptive element partition "x." (b) Material balance envelope residual time variation for fixed (solid) and adaptive (dotted) element partition (first envelope "o" and second envelope "□").

(−2% from the initial value). Both fixed partitions exhibit good error measures after the first step change but peak after the second step change is imposed (Fig. 7.8(a)). The adaptive element partition, however, moves the error peak earlier in the simulation time domain but keeps the time-averaged value better than the achieved value from the fixed element partitions. The time variation of the envelope error difference is shown in Fig. 7.8(a). Figure 7.8(b) shows the individual envelope error time variation for one fixed and the adaptive element partition. It is clear that the adaptive element partition exhibits a lower error value for both envelopes, even though the error difference reaches a peak after the second step change. Overall, the adaptive element placement enhances the approximation performance of the

Table 7.7 Column configuration, specifications, and technical data for the butyl acetate reactive distillation.

Feed stream			
Temperature (K)		291	
Flow rate (kmol/h) BUOH/ACOOH		34.992/34.809	
Column configuration (number stages)			
Rectifying/stripping/total		9/48/60	
Diameter (m)		1.37	
Liquid holdups (m^3)			
Condenser/rectifying/feed/stripping/reboiler		0.676/0.040/0.250/0.197/0.593	
Reflux ratio/reboil ratio		5.52/2.52	
Reaction	k_0 (mol s^{-1})		A (kJ mol^{-1})
Esterification	6.1084×10^4		56.67
Hydrolysis	9.8420×10^4		67.66

NEQ/OCFE model as shown by the achieved approximation error. One of the greatest advantages is that the adaptive placement can monitor the influence of exogenous disturbances automatically and modify the element partition in real time. A possible drawback in the method is associated with the multiplicity of solutions in the dynamic optimization problem as multiple element partitions can satisfy the equidistribution relations in (7.54) that may sometimes cause convergence difficulties.

7.4.3
Butyl Acetate Production via Reactive Multiphase Distillation

7.4.3.1 Process Description

The production of butyl acetate is being carried out in a tray-reactive distillation column via the esterification of n-butanol with acetic acid [46]. The column is divided into two sections separated by the feed stage, where fresh reactants enter the column at stoichiometric ratio. The column characteristics and specifications are given in Table 7.7. For this particular separation system, the conditions prevailing at the top (rectifying) section (i.e., above the feed stage) give rise to a second liquid phase, hence forming a three-phase mixture with two immiscible liquid phases (i.e., an organic and an aqueous phase). The vapor stream from the top of the column enters a total condenser and the condensate is forwarded to a decanter, where the two liquid phases are separated. The organic, rich in butyl acetate, phase returns as a reflux stream into the column, while the aqueous phase is withdrawn as a distillate stream. There is no evidence of a second liquid phase appearance in the section column below the feed stage for all the simulated dynamic cases.

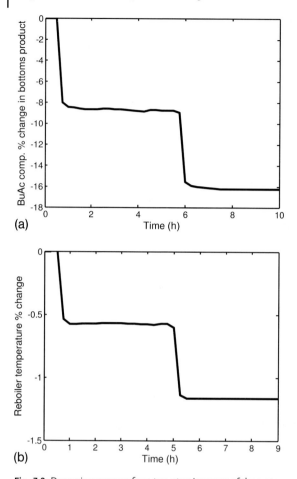

Fig. 7.9 Dynamic response for a two-step increase of the water content in the feed stream.

7.4.3.2 Dynamic Simulation Results

The NEQ/OCFE model employs one element in the rectifying section and four elements in the stripping section with three collocation points per element. The element in the rectifying section uses three-phase balance equations, whereas the stripping section elements use two-phase balance equations. Figure 7.9 shows the response to an increase in the water content of the feed stream performed in two successive steps. In the feed stream initially composed of n-butanol and acetic acid, a significant amount of water is added. The total change in the overall feed stream flow rate is 46%. According to the graphs shown in Fig. 7.9, the butyl acetate composition in the bottoms product undergoes a big drop (around 16%) due to the water addition and the subsequent reduction of the overall reaction rate. The reboiler temperature exhibits a change of around 1%.

Figure 7.10 presents the dynamic profiles for the column products for a 23.5% increase in the water feed content followed by a +10% change in the reboil ratio

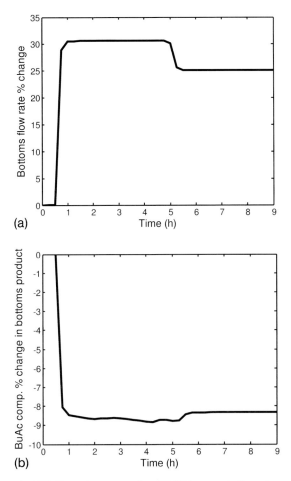

Fig. 7.10 Dynamic response for a 23.5% increase in the water content in the feed stream and a subsequent 10% increase in the reboil ratio.

(the ratio of the vapor stream re-entering the column from the bottom to the product stream leaving the column). As shown in Fig. 7.10, the reboil ratio increase slightly increases the ester product purity in the bottoms product. Therefore, water content in the feed stream should be totally avoided.

The esterification reactions in reactive distillation columns require the presence of catalysts most commonly found as packings of resins located in the section of the column where the reactions take place. The last dynamic simulation case involves a catalyst deactivation scanerio that results in a decrease in the rate of the esterification/hydrolysis reactions. Figure 7.11 shows the dynamic response of the bottoms product composition to a 10% decrease of the reaction rate. In all simulated cases, the distribution of the phases in the column remains unchanged.

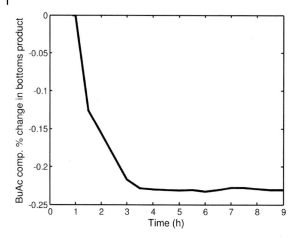

Fig. 7.11 Dynamic response for a 10% decrease in the rate of reaction.

7.5
Epilog

Complex reactive absorption and distillation separation units in reduced order dynamic models based on OCFE have been presented. The dynamic NEQ/OCFE models maintain a detailed description of the major physical and chemical phenomena occurring in a column stage but evaluate the material and energy balances at much fewer points within the column domain than the actual column stages. Therefore, an accurate and reliable model is achieved that requires significantly less computational effort for solution. As a result, NEQ/OCFE models become quite attractive for real-time control applications. The ability of the NEQ/OCFE models to control the approximation error along the column within reasonable levels despite the influence of exogenous disturbances responsible for the formation of steep fronts in the composition and temperature profiles is enhanced through a novel optimization-based finite element partition algorithm. The algorithm places the element breakpoints, subsequently altering the density of collocation points along the column domain, so that the approximation error is equally distributed along the vertical coordinate of the column. Case studies that involve reactive absorption, reactive distillation, and multiphase reactive distillation confirm the strengths of the NEQ/OCFE techniques.

Challenging issues remain the reliable online estimation of model parameters utilizing dynamic data, especially those associated with the thin films (e.g., film thickness) or the kinetic mechanism and the model-based design of experimental runs for the collection of column data rich in information that would facilitate offline model validation. Regarding multiphase separation, the simultaneous adaptive element partition with the monitoring of phase boundaries will enormously enhance the accuracy of the model predictions under dynamic transition.

Notation

a – specific interfacial area (m²/m³)
A^{col} – column stage cross section (m²)
A – reaction activation energy (J/kmol)
C – molar concentration (kmol/m³)
d – molar density (kg/m³)
$Đ$ – Maxwell–Stefan diffusion coefficient (m²/s)
G – gas molar flow rate (kmol/h)
H – stream molar enthalpy (J/kmol)
k_0 – pre-exponential kinetic parameter
K – K-value in phase equilibrium
L – liquid molar flow rate (kmol/h)
m – component molar holdup (kmol)
n – number of collocation points in a finite element
N – molar flux (kmol/(m² s))
NC – number of components
NE – number of finite elements in a column section
NP – number of time intervals in simulation domain
NR – number of chemical reactions
NT – number of physical stages
P – pressure (Pa)
Q – rate of heat loss/heat duty (J/s)
r – rate of reaction (kmol/(m³ s))
R – component rate of reaction (kmol/(m³ s))
R_g – ideal gas constant (8.314 J/(mol K))
s – column position coordinate in OCFE formulation
t – time (h)
T – temperature (K)
u – component internal energy (J)
U – stream internal energy (J)
W – Langrange interpolating polynomial
x – liquid phase mole fraction
y – gas phase mole fraction

Greek letters

γ – activity coefficient
Γ – thermodynamic factor
δ – film thickness (m)
Δh – stage height (m)
η – film coordinate
N – stoichiometric coefficient
φ – phase volumetric holdup fraction (m³ phase/m³ stage)
Φ – phase split fraction

Subscripts

T – Total

Superscripts

G – gas phase
Gb – gas bulk region
Gf – gas film in gas–liquid or gas–liquid I interface
GL – gas–liquid interface
GLf – liquid film in gas–liquid I interface
int – interface
L – liquid phase
Lb – liquid bulk phase
Lf – liquid film in gas–liquid interface
L_I – first liquid I phase (continuous)
$L_I b$ – first liquid I phase (continuous) bulk region
$LL_I f$ – liquid I film in liquid–liquid interface
L_{II} – second liquid II phase (dispersed)
$L_{II} b$ – second liquid II phase (dispersed) bulk region
$LL_{II} f$ – liquid II film in liquid–liquid interface
LL – liquid–liquid interface

References

1 TAYLOR, R., KRISHNA, R., *Multicomponent Mass Transfer*, Wiley, New York, **1993**.
2 KENIG, E. Y., SEFERLIS, P., *Chemical Engineering Progress* 1 (**2009**), p. 65.
3 KENIG, E. Y., SCHNEIDER, R., GORAK, A., *Chemical Engineering Science* 56 (**2001**), p. 343.
4 MARQUARDT, W., *International Chemical Engineering* 30 (**1990**), p. 585.
5 MARQUARDT, W., AMRHEIN M., *Computers and Chemical Engineering* 18 (**1994**), S349.
6 KIENLE, A., *Chemical Engineering Science* 55 (**2000**), p. 1817.
7 BENALLOU, A., SEBORG, D. E., MELLICHAMP, D. A., *AIChE Journal* 32 (**1986**), p. 1067.
8 LÉVINE, J., ROUCHON, P., *Automatica* 27 (**1991**), p. 463.
9 KHOWINIJ, S., HENSON, M. A., BELANGER, P., MEGAN, L., *Separation Purification and Technology* 46 (**2005**), p. 95.
10 BIAN, S., KHOWINIJ, S., HENSON, M. A., BELANGER, P., MEGAN, L., *Separation Purification and Technology* 29 (**2005**), p. 2096.
11 LINHART, A., SKOGESTAD, S., *Computers and Chemical Engineering* 33 (**2009**), 296.
12 KUMAR, A., DAOUTIDIS, P., *Journal of Process Control* 12 (**2002**), p. 475.
13 HAHN, J., EDGAR, T. F., *Computers and Chemical Engineering* 26 (**2002**), p. 1379.
14 ZALIZAWATI, A., NORASHID, A., ZAINAL, A., *Chemical Product Process Model* 2 (12) (**2007**).
15 DALAOUTI, N., SEFERLIS P., *Computers and Chemical Engineering* 30 (**2006**), p. 1264.
16 STEWART, W. E., LEVIEN, K. L., MORARI, M., *Chemical Engineering Science* 40 (**1985**), p. 409.
17 CHO, Y. S., JOSEPH, B., *AIChE Journal* 29 (**1983**), p. 261.
18 SEFERLIS, P., HRYMAK, A. N., *AIChE Journal* 40 (**1994**), p. 813.

19 Dalaouti, N., Seferlis P., *Journal of Cleaner Production* 13 **(2005)**, p. 1461.

20 Kenig, E. Y., Wiesner, U., Gorak, A., *Industrial and Engineering Chemistry Research* 36 **(1997)**, p. 4325.

21 Baur, R., Higler A. P., Taylor, R., Krishna, R., *Chemical Engineering Journal* 73 **(2000)**, p. 33.

22 Higler, A., Chande, R., Taylor, R., Baur, R., Krishna, R., *Computers and Chemical Engineering* 28 **(2004)**, p. 2021.

23 Seferlis, P., Grievink, J., *Industrial and Engineering Chemistry Research* 40 **(2001)**, p. 1673.

24 Lao, M., Taylor, R., Modeling mass transfer in three-phase distillation, *Industrial and Engineering Chemistry Research* 33 **(1994)**, p. 2637.

25 Damartzis, T., Seferlis P., *Industrial and Engineering Chemistry Research* 49 **(2010)**, p. 3275.

26 Swartz, C. L. E., Stewart, W. E., *AIChe Journal* 33 **(1987)**, p. 1977.

27 Brüggemann, S., Oldenburg J., Zhang P., Marquardt, W., *Industrial and Engineering Chemistry Research* 43 **(2004)**, p. 3672.

28 Carey, C. F., Finlayson, B. A., *Chemical Engineering Science* 30 **(1975)**, p. 587.

29 Russel, R. D., Chistiansen J., *SIAM Journal of Numerical Analysis* 15 **(1978)**, p. 59.

30 Cuthrell, J. E., Biegler, L. T., *AIChE Journal* 33 **(1987)**, p. 1257.

31 Renfro, J. G., Computational studies in the optimization of systems described by differential algebraic equations, PhD Thesis, University of Houston, **1986**.

32 De Boor, C., *Dundee Conference on Numerical Solution of Differential Equations, Lecture Notes in Mathematics*, vol. 363, Springer, New York, **1974**, p. 12.

33 Seferlis, P., Hrymak, A. N., *Chemical Engineering Science* 49 **(1994)**, p. 1369.

34 Emig, G., Wohlfahrt, K., *Computers and Chemical Engineering* 3 **(1979)**, p. 143.

35 Joshi, J. B., Mahajani, V. V., Juvekar, V. A., *Chemical Engineering Communication* 33 **(1985)**, p. 1.

36 Suchac, N. J., Jethani, K. R., Joshi, J. B., *AIChE Journal* 37 **(1991)**, p. 323.

37 Siddiqi, M. A., Lucas, K., *Canadian Journal of Chemical Engineering* 64 **(1986)**, p. 839.

38 Zarzycki, R., Chakuc, A., *Absorption: Fundamentals and Applications*, Pergamon Press, New York, **1993**.

39 Rocha, J. A., Bravo, J. L., Fair, J. R., *Industrial and Engineering Chemistry Research* 35 **(1996)**, p. 1660.

40 Process Systems Enterprise Ltd., gPROMS Introductory and Advanced User's Guide, **2005**.

41 Suzuki, I., Komatsu, H., Hirata, H., *Journal of Chemical Engineering Japan* 3 **(1970)**, 152.

42 Venkataraman, S., Chan, W. K., Boston, J. F., *Chemical Engineering Progress* 86 **(1990)**, p. 45.

43 Bock, H., Jimoh, M., Wozny G., *Chemical Engineering Technology* 20 **(1997)**, p. 182.

44 Kenig, E. Y., Bäder, H., Górak A., Bessling, B., Adrian, T., Schoenmakers, H., *Chemical Engineering Science* 56 **(2001)**, p. 6185.

45 Klöker, M., Kenig, E. Y., Górak, A., Markusse, A. P., Kwant, G., Moritz, P., *Chemical Engineering Processing* 43 **(2004)**, p. 791.

46 Steinigeweg S., Gmehling, J., *Industrial and Engineering Chemistry Research* 41 **(2002)**, p. 5483.

8
Modeling of Crystallization Processes
Ali Abbas, Jose Romagnoli, and David Widenski

Keywords
evaporative crystallization, gas antisolvent (GAS) crystallization, supersaturation, Jouyban–Acree model, MOSCED model, thermodynamic nucleation models

8.1
Introduction

Crystallization is an important unit operation used for the production, separation, and purification of particulate solids in the pharmaceutical, fertilizer, fine chemical, and other industries. These processes can vary from traditional crystallization techniques such as evaporation or cooling, to newer techniques such as gas antisolvent or ultrasound-mediated crystallization. Product crystals created from different crystallization methods can have significantly different properties. For example, particles used in medical inhalers are required to be at a specific size range to successfully penetrate the lungs and absorb at the alveoli. Gas antisolvent crystallization is commonly used for this. Cooling and evaporative crystallization are typically used when larger particles are desired. Besides particle size, size distribution is also a key property typically targeted to alleviate downstream processing problems such as filtration blockages. Crystals with uniform size distribution will have similar dissolution rates for drug delivery applications. One large crystal will dissolve much slower in the body than several small crystals of equal mass due to the difference in available surface area. Besides objectives associated with particle properties, crystallization operations must achieve desired product yields.

In order to produce crystals with the intended properties and desired yields, crystallization processes were (and continue to be) operated using rules of thumb developed from years of operating experience. Many experiments are required to determine the effect of different operating conditions on the crystalline product. This can be very laborious and expensive for high-value products resulting in significant delays in bringing new products to market and thus loosing competitiveness. From another aspect, regulatory bodies such as the US Food and Drug Administration require certain fixed operational recipes to be executed and validated. This like rule-of-thumb operations has had the consequence of black-box understanding of crystallization and its operation, resulting ultimately in rigidity and suboptimal process economics. For example, energy usage as in the cooling or evaporation requirements of a crystallizer may not be readily minimized using empirical approaches. Similarly recipe-driven operations can result in economic losses in the form of loss of product batches due to undesired excursions outside recipe limits. It is thus of great significance from the economic point of view to have fundamental knowledge and understanding of crystallization processes. Fundamental crystallization modeling and model-based crystallization approaches have evolved rapidly over the last couple of decades with significant potential shown for replacing the empirical tradition. However, industrial crystallization operations are yet to fully reap the benefits of such model-based technology.

A detailed and validated dynamic crystallization model can be used in several model-based activities ranging from experimental parameter estimation [25, 26, 53, 64, 68] through to simulation analysis [4, 53, 75] and optimization that determines optimal regimes [49, 54, 62, 68]. More advanced use of the crystallization model includes model-based experimental design [16, 67] through to the real-time application in soft-sensing [4] and predictive control [4, 15, 48, 51]. Especially useful is the application of the crystallization model to determine optimal operating policies for different objectives using optimization software in conjunction with the model. These objectives may be to specify a mean size or distribution, or to control the crystallization conditions to produce the desired polymorph. These objectives are typically very difficult to achieve without the use of model-based optimization, and require extraordinary experimental effort to determine optimal operating conditions. Crystallization modeling is thus significantly important for optimal operation of crystallizers, and will continue to be important in the future.

This chapter discusses modeling of crystallization processes. We start with an overview of industrial crystallization, crystallization fundamentals, and mechanisms before we present detailed discussions on crystallization modeling, model solution techniques, and model analysis. We then discuss various model application areas before finishing with two examples, namely, antisolvent crystallization and seeded cooling crystallization.

8.2
Background

Crystallization can be used to make many different solid products. Fine chemicals, agrochemicals, pharmaceuticals, and aerosol particles can all be produced from different crystallization methods. Products created from crystallization are ubiquitous. In the household, common kitchen ingredients such as table salt and sugar as well as active ingredients of drugs in the drug cabinet are products of crystallization. Other examples include rock salt used to deice roadways and sidewalks in cold climates, the silver iodide used to seed clouds to induce rain, crop fertilizers such as ammonium sulfate, calcium chloride used in solarthermal heating, among many other products are produced via crystallization. Products created from crystallization are all around us.

8.2.1
Crystallization Methods

Crystallization can occur through various different modes. One way is reactive crystallization in which the product crystals are formed via a chemical reaction from raw materials. This is used when two liquid solvents react to form a crystalline product. This is experienced in introductory chemistry courses where HCL is neutralized by baking soda ($NaHCO_3$) to produce NaCl crystals. Reactive crystallization is used industrially to produce ammonia. This method can create crystals with undesirable properties. Solution recrystallization is performed to improve crystal properties such as size, size distribution, and shape as well as to increase the product purity.

8.2.1.1 Recrystallization Methods

Evaporation In evaporative crystallization, a saturated solution is heated to cause the solvent to evaporate. The loss of solvent raises the concentration of solute in solution above the solubility limit, which causes supersaturation and the formation of crystals. A key advantage of evaporative crystallization is in its capacity to recover 100% of the solute. The key disadvantage is that it requires a lot of energy to evaporate the solvent. Evaporative crystallization is used for the industrial production of ammonium sulfate and sodium chloride.

Cooling In cooling crystallization, a saturated solution is cooled to a lower temperature. The decrease in temperature lowers the solubility of the solute, which causes supersaturation to happen. The advantage of cooling crystallization is that it has lower energy costs than evaporative crystallization, but the disadvantage is that 100% yield is not attainable. This method is used for compounds that have a solubility that depends on temperature. Cooling crystallization has been performed for production of paracetamol, potassium chloride, and many other compounds.

Salting Out Crystallization Another crystallization method is salting out crystallization (also called drowning out or quench crystallization). This is when another compound is added to the saturated solution to cause crystallization. Liquids and gases are the most commonly used compounds, but solids can be used as well. This method is useful when the crystallizable compounds are heat sensitive, decompose at high temperatures, or have solubilities that do not change significantly with temperature. For these compounds evaporative or cooling crystallization methods cannot be used.

The chosen solvent (gas or liquid) is a solvent in which the target compound (solute) has low solubility, hence the antisolvent description. As the antisolvent is added to the solution, supersaturation is formed which causes the formation of crystals. Organic antisolvents, such as ethanol or methanol, are used for the antisolvent crystallization of inorganic compounds such as NaCl and KCl. Water is used as an antisolvent for the crystallization of organic compounds. Supercritical gases are commonly used as gas antisolvents. One example is supercritical CO_2. Gas antisolvent (GAS) crystallization has several benefits over the use of a liquid antisolvent. GAS has lower separation costs, can produce very fine particles for use as inhalable drugs, and has a higher yield than liquid antisolvents. This method can also be implemented in conjunction with cooling. For example, many organic compounds can be crystallized using a combined cooling-liquid antisolvent crystallization technique. This is particularly applicable for pharmaceuticals.

Ultrasound Crystallization The use of high-intensity ultrasound has proved to be an attractive mechanism for crystallization. This method, also termed sonocrystallization, relies on cavitating bubbles induced by the ultrasound. Very high temperatures and pressures are reported in cavitation zones and these along with the associated supercooling are attributed to the high nucleation rates achieved in sonocrystallization [5]. Consequently, this method is useful for producing crystals with small sizes and narrow distributions [5, 32].

Seeded/Unseeded Crystallization Any crystallization method can be operated either seeded or unseeded. In seeded crystallization, previously formed crystals (seeds) are added to the crystallizer at the beginning of the batch. If sufficient seed is added, nucleation will be suppressed and the dissolved solute will crystallize on seed crystal surfaces and grow the seeds to larger sizes. Experimentally developed seed charts are used to aid in the proper selection of seed size and mass [28, 29, 35]. In unseeded crystallization, the crystals are formed and grown during the batch.

8.2.2
Driving Force

The driving force for crystallization that dictates the rates of nucleation and growth is the difference in chemical potential between the solid and liquid phases, typically expressed in the relative form as

$$S = \exp\left(\frac{\Delta \mu_{cp}}{RT}\right) = \frac{f_l}{f_s} = \frac{\gamma_i C}{\gamma_{i,eq} C_{eq}} \tag{8.1}$$

where S is the relative supersaturation, $\Delta \mu_{cp}$ is the difference in chemical potential, f_l is the fugacity of the liquid phase, f_s is the fugacity of the solid phase, C_{eq} and γ_{eq}, respectively, are the concentration and activity coefficient of the liquid phase at equilibrium, and γ and c are the actual concentration and activity coefficient of the solution, respectively. As an approximation, the ratio of activity coefficients is brought to unity, leading to the practical description of the relative supersaturation defined as the ratio of the solution concentration to the equilibrium concentration:

$$S = \frac{C}{C_{eq}} \qquad (8.2)$$

Another form of relative supersaturation used to model crystallization kinetics is

$$\sigma = \frac{\Delta C}{C_{eq}} = S - 1 \qquad (8.3)$$

Absolute supersaturation is also commonly used and is defined as the difference between the solution concentration and the equilibrium concentration (Eq. (8.4)). This is typically defined in units of g solute/g solvent:

$$\Delta C = C - C_{eq} \qquad (8.4)$$

Supersaturation can be generated by one of three primary methods namely evaporation, cooling, and antisolvent addition. These were discussed briefly in Section 8.2.1. While the supersaturation influences nucleation events and growth phenomena at the molecular scales, it is the rates of evaporation, cooling, and antisolvent addition, which become of importance at the macroscale where manipulation of the process conditions can occur. Advantageously, two or more of these mentioned techniques can be combined in the same operation enabling enhanced results. For instance, adding antisolvent to a cooling crystallization operation provides an extra degree of freedom, where a calculated antisolvent addition can work as a seeding mechanism [19].

8.3
Solubility Predictions

In order to calculate supersaturation in a solution, the equilibrium solubility needs to be determined. This can be done using a solubility model that must be dependent on what manipulated variable(s) the crystallization method is changing. If cooling is used, then the solubility model needs to be temperature dependent. However, if antisolvent crystallization is used then the solubility model needs to be solvent composition dependent.

8.3.1
Empirical Approach

The simplest of solubility models are empirical models. These are simply mathematical equations fitted to experimental data. Empirical models relate solubility to a measured experimental variable. These experimental variables can be tempera-

Table 8.1 Empirical solubility model examples.

System	Solubility model	Dependency	Reference
Paracetamol in ethanol[a]	$C_{eq} = C_1 \exp(C_2 T)$	Temperature	Worlitschek and Mazzotti [74]
Paracetamol in acetone/water[b]	$C_{eq} = C_1 + C_2 w^5 + C_3 w^4 + C_4 w^3 + C_5 w^2 + C_6 w + C_7$	Solvent composition	Zhou et al. [77]
Paracetamol in isopropanol/water[c]	$C_{eq} = (C_1 x_3^2 + C_2 x_3 + C_3) \exp[(C_4 x_3^2 + C_5 x_3 + C_6) T]$	Temperate and solvent composition	

a) Model reliable between 10 and 50 °C.
b) Model reliable between 30 and 100 solute free mass percent water at 16 °C.
c) Model reliable between 60 and 80 solute free mass percent water between 40 and 10 °C.

ture, solvent composition, density, conductivity, absorbance, etc. Several empirical models for various systems are shown in Table 8.1.

8.3.2
Correlative Thermodynamic

The next class of solubility models is correlative models. Correlative thermodynamic solubility models are models that have thermodynamic meaning and are fit to experimental data. The most common of these models are excess Gibbs energy models. These are further simplified to activity coefficient models. The most common activity coefficient models are the van Laar, Wilson, NRTL, and UNIQUAC models. These are displayed in Table 8.2. All these models have binary interaction parameters that are fit to experimental data.

8.3.3
Predictive Thermodynamic

Predictive thermodynamic models are generalized models used to predict solubility behavior of different compounds. These models shown in Table 8.3 are developed from extensive experimental data for many different chemical systems. The parameters for these models are correlated depending on the chemical structure or properties of the compounds. With the database of parameters, solubility data can be predicted for systems not used to create the model. Some predictive models are Jouyban–Acree, MOSCED, NRTL-SAC, and UNIQUAC.

8.3 Solubility Predictions | 245

Table 8.2 Correlative thermodynamic solubility models.

Model		Parameters	
van Laar	$\ln \gamma_i = \dfrac{B_{VL}}{(1+\frac{B_{VL}x_i}{A_{VL}x_j})}, \quad 1 = x_i + x_j$	Binary interaction	A_{VL}, B_{VL}
Wilson	$\ln \gamma_i = -\ln(x_i + \Lambda_{ij}x_j)$ $- x_j (\dfrac{\Lambda_{ji}}{x_j + \Lambda_{ji}x_i} - \dfrac{\Lambda_{ij}}{\Lambda_{ij}x_j + x_i})$	Binary interaction	$\Delta \lambda_{ij}$
	$\Lambda_{ji} = \dfrac{v_j}{v_i}\exp(-\dfrac{\Delta \lambda_{ij}}{RT})$ $1 = x_i + x_j$	Molar volume	v_i
NRTL	$\ln \gamma_i = x_j^2 [\tau_{ji}(\dfrac{G_{ji}}{x_i + x_j G_{ji}})^2 + \dfrac{\tau_{ij}G_{ij}}{(x_j + x_i G_{ij})^2}]$	Binary interaction	Δg_{ij}
	$G_{ij} = \exp(-\alpha \tau_{ij})$	Nonrandomness	α
	$\tau_{ji} = \dfrac{\Delta g_{ij}}{RT}, \quad 1 = x_i + x_j$		
UNIQUAC	$\ln \gamma_i = \ln \gamma_i^C + \ln \gamma_i^R$	Binary interaction	a_{ij}
	$\ln \gamma_i^C = \ln \dfrac{\Phi_i}{x_i} + \dfrac{z}{2}q_i \ln \dfrac{\theta_i}{\Phi_i} + l_i$ $- \dfrac{\Phi_i}{x_i}\sum_j x_j l_j$		
	$\ln \gamma_i^R = q_i'[1 - \ln(\sum_j \theta_j' \tau_{ji})$	Surface of interaction	q_i'
	$- \sum_j \dfrac{\theta_j' \tau_{ij}}{\sum_k \theta_k' \tau_{kj}}]$		
	$l_i = \dfrac{z}{2}(r_i - q_i) - (r_i - 1)$		
	$\theta_i = \dfrac{q_i x_i}{\sum_j q_j x_j}, \quad \Phi_i = \dfrac{r_i x_i}{\sum_j r_j x_j}$	Area structural	r_i
	$\theta_i' = \dfrac{q_i' x_i}{\sum_j q_j' x_j}$		
	$\tau_{ij} = \exp[-\dfrac{a_{ij}}{T}], \quad 1 = x_i + x_j$	Volume structural	q_i

8.3.3.1 Jouyban–Acree Model

The Jouyban–Acree is a model used to estimate the solubility of organic compounds in a binary solvent mixture [30]. The model has three constants that are fitted to a wide range of organic solutes and solvents. This model requires the solubilities of both pure components in a binary solute–solvent system, and predicts the solubility of a solute in a solvent mixture.

8.3.3.2 MOSCED Model

One predictive thermodynamic model is the MOSCED (modified separation of cohesive energy density) model. The MOSCED model is a model used to calculate infinite dilution activity coefficients. The advantage of the MOSCED model is that no experimental data are needed to calculate the infinite dilution activity coefficients. The MOSCED model further calculates temperature-dependent infinite dilution activity coefficients, such that a temperature-dependent activity coefficient model is not required.

The representation of the MOSCED model shown in Table 8.3 is used to find substance j's infinite dilution activity coefficient in substance i. Similarly, the model can be used to find the infinite dilution activity coefficient for substance i in j by

Table 8.3 Predictive solubility models.

MOSCED	$\ln \gamma_{i,j}^{\infty} = \frac{v_i^0}{RT}[(\lambda_j - \lambda_i)^2 + \frac{q_j^2 q_i^2 (\tau_j - \tau_i)^2}{\psi_j} + \frac{(\alpha_j - \alpha_i)(\beta_j - \beta_i)}{\xi_j}] + d_{ji}$ $d_{ji} = \ln(\frac{v_i^0}{v_j^0})^{aa} + 1 - (\frac{v_i^0}{v_j^0})^{aa}, \quad aa = 0.953 - 0.002314[(\tau_i)^2 + \alpha_i \beta_i]$ $\psi_j = POL + 0.002629 \alpha_j \beta_j$ $\xi_j = 0.68(POL - 1) + [3.24 - 2.4 e^{(-0.002687(\alpha_j \beta_j)^{1.5})}](\frac{293}{T})^2$ $POL = q_j^4[1.15 - 1.15 e^{(-0.002337(\tau_j))^3}] + 1, \quad \alpha_j = \alpha_j^0 (\frac{293}{T(K)})^{0.8}$ $\beta_j = \beta_j^0 (\frac{293}{T(K)})^{0.8}, \quad \tau_j = \tau_j^0 (\frac{293}{T(K)})^{0.4}$
NRTL-SAC	$\ln \gamma_i = \ln \gamma_i^C + \ln \gamma_i^R, \quad \ln \gamma_i^C = \ln \frac{\phi_i}{x_i} + 1 - r_i \sum_j \frac{\phi_j}{r_j}$ $\ln \gamma_i^R = \sum_m r_{m,i}(\ln \Gamma_m^{lc} - \ln \Gamma_m^{lc,i})$ $\ln \Gamma_m^{lc} = \frac{\sum_j x_j G_{jm} \tau_{jm}}{\sum_k x_k G_{km}} + \sum_{m'} \frac{x_{m'} G_{mm'}}{\sum_k x_k G_{km'}} (\tau_{mm'} - \frac{\sum_n x_n G_{nm'} \tau_{nm'}}{\sum_k x_k G_{km'}})$ $\ln \Gamma_m^{lc,i} = \frac{\sum_n x_{n,i} G_{nm} \tau_{nm}}{\sum_k x_{k,i} G_{km}} + \sum_{m'} \frac{x_{m',i} G_{mm'}}{\sum_k x_{k,i} G_{km'}} (\tau_{mm'} - \frac{\sum_n x_{n,i} G_{nm'} \tau_{nm'}}{\sum_k x_{k,i} G_{km'}})$ $x_n = \frac{\sum_j x_j r_{n,j}}{\sum_i \sum_m x_i r_{m,i}}, \quad x_{n,i} = \frac{r_{n,i}}{\sum_m r_{m,i}}$ $r_i = \sum_m r_{m,i}, \quad \phi_i = \frac{r_i x_i}{\sum_j r_j x_j}, \quad G_{km} = e^{-\alpha_{km} \tau_{km}}$
UNIFAC	$\ln \gamma_i = \ln \gamma_i^C + \ln \gamma_i^R n$ $\ln \gamma_i^C = \ln \frac{\Phi_{i,U}}{x_i} + \frac{z}{2} q_i \ln \frac{\theta_i}{\Phi_{i,U}} + l_i - \frac{\Phi_{i,U}}{x_i} \sum_j x_j l_j$ $\ln \gamma_i^R = \sum_k v_k^{(i)}(\ln \Gamma_k - \ln \Gamma_k^{(i)})$ $\ln \Gamma_k = \ln \Gamma_k^{(i)} = Q_k[1 - \ln(\sum_m \Theta_m \Psi_{mk}) - \sum_m \frac{\Theta_m \Psi_{km}}{\sum_n \Theta_n \Psi_{nm}}]$ $l_i = \frac{z}{2}(r_{i,U} - q_{i,U}) - (r_{i,U} - 1), \quad \theta_i = \frac{q_{i,U} x_i}{\sum_j q_{j,U} x_j}$ $\Phi_{i,U} = \frac{r_{i,U} x_i}{\sum_j r_{j,u} x_j}, \quad \Theta_m = \frac{Q_m X_m}{\sum_n Q_n X_n}, \quad \Psi_{mn} = e^{(-\frac{a_{mn}}{T})}$ $r_{i,U} = \sum_k v_k^{(i)} R_k, \quad q_{i,U} = \sum_k v_k^{(i)} Q_k$
Jouyban–Acree	$\log x_{2,\text{mix}}(T) = f_1 \log x_{2,1}(T) + f_3 \log x_{2,3}$ $+ f_1 f_3 [\frac{C_0}{T} + \frac{C_1(f_1 - f_3)}{T} + \frac{C_2(f_1 - f_3)^2}{T}]$

switching the subscripts *i* and *j* around. The MOSCED model contains five parameters: λ, α, β, q, τ, and v corresponding to dispersion, hydrogen bond acidity, hydrogen bond basicity, induction, polarity, and molar volume, respectively. Details on these parameters and their correlated values for various compounds are given in Lazzaroni et al. [36].

Once the two infinite dilution activity coefficients are calculated from the MOSCED model, they can be substituted into an excess Gibbs energy model to find the binary interaction parameters for that system. With the MOSCED model, no experimental data is needed to calculate these binary interaction parameters.

8.3.3.3 NRTL-SAC Model

The next predictive thermodynamic model considered is the NRTL segment activity coefficient model (NRTL-SAC) developed by Chen and Song [13]. The NRTL-

SAC model is derived from the polymer NRTL model with similar segment theory. The NRTL-SAC model breaks down each molecule into three different segments: hydrophobicity (X), polarity (Y−, Y+), and hydrophilicity (Z).

8.3.3.4 UNIFAC Model

Another predictive thermodynamic model is the UNIFAC model developed by Fredenslund, Jones, and Prausnitz [23]. The UNIFAC model, though similar to the UNIQUAC model, has one important difference. The UNIQUAC model is a correlative model that has adjustable parameters that are unique for each binary system, and the UNIFAC model is a predictive model that has two different parameters for each functional group. Some examples of functional groups are CH_3, OH, and CHO. Each functional group has an area and volume structural parameter. Also, each functional group pair has two unique binary interaction parameters associated with that pair. Predicting activity coefficients with the UNIFAC model is easy. All that is needed is to decompose the chemicals into their substituent groups, and look up the group parameters in the literature.

8.3.3.5 Solubility and Activity Coefficient Relationship

In order to determine equilibrium solubility from an activity coefficient model, the solid solubility equation needs to be used. To determine the equilibrium solubility, the heat of fusion (ΔH_{fusion}), change in heat capacity from liquid to solid phase (ΔC_p), and the melting temperature of the compound (T_{melt}) are needed,

$$\ln(x_i \gamma_i) = \frac{\Delta H_{fusion}}{R}\left(\frac{1}{T_{melt}} - \frac{1}{T}\right) - \frac{\Delta C_p}{R}\left(\ln\left(\frac{T_{melt}}{T}\right) - \frac{T_{melt}}{T} + 1\right) \quad (8.5)$$

8.3.4 Solubility Examples

The first example contrasting the use of different solubility models will be for the paracetamol–ethanol system in the cooling crystallization mode. The solubility models thus need to be temperature dependent. The empirical model chosen has the following form:

$$C_{eq} = C_1 \exp(C_2 T) \quad (8.6)$$

with constants C_1, C_2 correlated from data in Worlitschek and Mazzotti [74].

The correlative models chosen are the van Laar, Wilson, and NRTL models. The binary interaction parameters in Table 8.4 were calculated using data from Worlitschek and Mazzotti [74].

The predictive solubility models chosen were the MOSCED, NRTL-SAC, and UNIFAC models. The parameters for the MOSCED [36], NRTL-SAC [14], and UNIFAC [57] models are listed in Tables 8.5, 8.6, and 8.7, respectively.

All the models were then plotted against the experimental data. Figure 8.1 shows that the empirical and correlative models are all extremely good fits to the data. However, the predictive models are much poorer fits. The best of these is the UNI-

Table 8.4 Estimated binary interaction parameters for solubility models.

Solubility model	Binary interaction parameters		
	α	$A_{VL}/\lambda_{12}/\Delta g_{12}$	$B_{VL}/\lambda_{21}/\Delta g_{21}$
van Laar	–	−184.2	0.2186
Wilson	–	1858 J/mol K	−1181 J/mol K
NRTL	0.3777	2403 J/mol K	−1351 J/mol K

Table 8.5 MOSCED model parameters.

	MOSCED parameter					
	v^0	λ	τ^0	q	α^0	β^0
Acetone	73.8	13.7	8.3	1.0	0	11.1
Ethanol	58.6	14.4	2.5	1.0	12.6	13.3
Water	36.0	10.6	10.5	1.0	52.8	15.9
Acetaminophen	105.4	18.5	2.7	0.9	16.2	13.2

Table 8.6 NRTL-SAC parameters.

	Acetone	Ethanol	Water	Acetaminophen
Hydrophobicity (X)	0.131	0.256	0	0.498
Polarity (Y−)	0.109	0.081	0	0.487
Polarity (Y+)	0.513	0	0	0.162
Hydrophilicity (Z)	0	0.507	1	1.270

Table 8.7 UNIFAC model parameters.

	Acetone	Ethanol	Water	Acetaminophen
Area structural parameter ($r_{i,U}$)	2.5735	2.5755	0.92	5.7528
Volume structural parameter ($q_{i,U}$)	2.336	2.5880	1.40	4.5840

FAC model, followed by the NRTL-SAC and MOSCED models. Even though the predictive models are not as good as the empirical or correlative ones, they do reasonably approximate the experimental data.

The second example is a ternary acetone–water–acetaminophen system in isothermal liquid antisolvent crystallization. In this example, the solubility models will be dependent on solvent composition. For this example, only empirical and predictive models will be compared. The empirical model chosen is

$$C_{eq} = C_1 + C_2 w^5 + C_3 w^4 + C_4 w^3 + C_5 w^2 + C_6 w + C_7 \tag{8.7}$$

8.3 Solubility Predictions

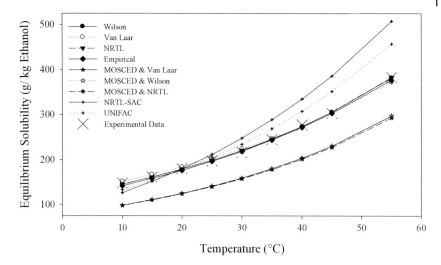

Fig. 8.1 Equilibrium solubility predictions for acetaminophen in ethanol.

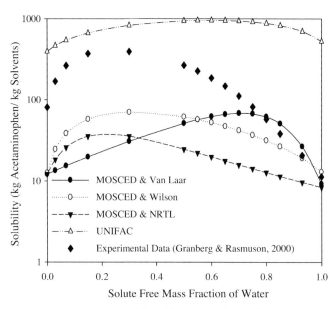

Fig. 8.2 MOSCED and UNIFAC equilibrium solubility predictions.

with constants listed in Zhou *et al.* [77]. The predictive models chosen are Jouyban–Acree, MOSCED, NRTL-SAC, and UNIFAC. Since the MOSCED model predicts infinite dilution activity coefficients, it will be paired with the van Laar, NRTL, and Wilson activity coefficient models.

Figure 8.2 shows that the UNIFAC and all the MOSCED models give poor solubility predictions. The UNIFAC model greatly overpredicts the equilibrium solubility, while the MOSCED model underpredicts the equilibrium solubility. These

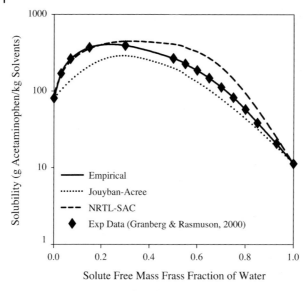

Fig. 8.3 NRTL-SAC, empirical, and Jouyban–Acree equilibrium solubility predictions.

models give poor solubility predictions for this system. Figure 8.3 shows that the NRTL-SAC and Jouyban–Acree give much better solubility predictions than the MOSCED or UNIFAC models. These predictive models were designed to be used for pharmaceutical organics, such as paracetamol, which is why these are the superior predictive models. However, no model matches the accuracy of the empirical model which fits the experimental data very well.

8.3.5
Solution Concentration Measurement Process Analytical Tools

There are several different ways to measure solution concentration to obtain experimental data to fit the solubility models and subsequently obtain a measure of supersaturation. Some of these methods are Raman spectroscopy, ATR-FTIR spectroscopy, ATR-UV-VIS spectroscopy, densitometry, conductivity, gravimetry, and optical refractive index. The spectroscopic methods measure solution concentration by measuring absorbance peaks at various wavelengths. These absorbance peaks are correlated to experimental data to determine concentration. Similarly, densitometry, conductivity, and refractive index measure solution density, conductivity, refractive index, and these values are correlated to experimentally measured solubility data. Gravimetry is commonly used to determine the experimental solubility data that these measurement techniques are correlated to. We will not discuss these concentration and supersaturation measurement techniques in detail in this chapter; however, the reader may refer to reviews by others on this topic [17, 21, 40, 66].

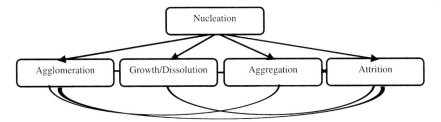

Fig. 8.4 Crystallization kinetic mechanisms.

8.4
Crystallization Mechanisms

During crystallization processes, many different mechanisms occur. The first one to occur is nucleation. This is followed by several different phenomena such as growth, agglomeration, aggregation, and attrition. The various different crystallization mechanisms are shown schematically in Fig. 8.4.

8.4.1
Nucleation

Nucleation is the crystallization phenomenon that causes the formation of nuclei. In order for nucleation to occur, the solution's concentration must exceed the metastable limit. This limit will be different for homogeneous and heterogeneous nucleation. These nuclei are formed via two mechanisms – primary and secondary nucleation (Fig. 8.5). Primary nucleation is when nuclei are formed in the absence of already formed crystals. This can occur through either homogeneous or heterogeneous nucleation. Homogeneous nucleation occurs when the nuclei reaches a critical nucleus size and causes spontaneous nucleation. Heterogeneous nucleation occurs when nuclei are formed on particles, such as dust, in solution. Heterogeneous nucleation occurs at lower supersaturation levels than homogeneous nucleation, so this will be the preferred nucleation mechanism if particles are in the solution.

Secondary nucleation is when new nuclei are created from preexisting crystals. These nuclei are removed from the crystals' surface. This can be caused by fluid shear (surface nucleation) or from crystals contacting each other (contact nucleation).

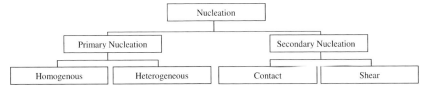

Fig. 8.5 Various nucleation types.

Table 8.8 Empirical nucleation models.

Equation	Reference
$B = k_b S^b$	[9, 70]
$B = k_b (\Delta C)^b$	[9, 10, 22, 46]
$B = k_b \sigma^n M_t^i$	[76, 78]
$B = A e^{-k(\log S)^{-2}}$	[55]
$B = A e^{\frac{k}{\ln^2 S}}$	[9, 18, 45, 69]
$B = k_b G^i M_T^i N^m e^{-\frac{E}{RT}}$	[31, 65]
$B = k_b G^i M_T^j N^m$	[31]
$B = p_3 [n(L,t) L^{P_5} dL]^{P_1} (C - C_s)^{P_3}$	[20]

Table 8.9 Thermodynamic nucleation models.

Model	Reference/type of nucleation
$B = E \frac{k_a \mu_2 D}{d_m^4} \exp[-\pi(\frac{\gamma_{sl} d_m^2}{kT}) \frac{1}{\ln S}]$	[74] (Secondary nucleation)
$B_{\text{hom}} = 1.5 \times D_{AB}(CN_A)^{7/3} \sqrt{\frac{\gamma_{CL}}{kT}} \frac{1}{C_c N_A}$ $\times \exp(-\frac{16}{3}\pi(\frac{\gamma_{CL}}{kT})^3 (\frac{1}{C_c N_A})^2 \frac{1}{(v \ln S)^2})$	[42] (Homogeneous nucleation)
$B_{\text{het}} = (\frac{1}{2\pi} a_{for} d_m H E_{ad} (CN_A)^{7/3} \sqrt{\frac{f \gamma_{CL}}{kT}} V_m)$ $\times (\frac{D_{\text{surf}} \sin \theta}{r_c} H E_{ad} d_m^{3/2} (CN_A)^{1/6}$ $+ 3\pi D_{AB}(1 - \cos \theta))$ $\times \exp(-\frac{4}{3}\pi f \frac{\gamma_{CL}}{kT} (\frac{2M_W \gamma_{CL}}{\rho_c RT v \ln S})^2)$	[42] (Heterogeneous nucleation)
$B_{\text{surf}} = 9E \frac{D_{AB}}{d_m^4 L_{32}} \varphi_T \exp(-\pi \frac{(\gamma_{CL} d_m^2/kT)^2}{v \ln S})$	[42] (Secondary surface nucleation)
$B = A \exp[-\frac{16 \pi \sigma^3 v^2}{3 k^3 T^3 \ln^2(S)}]^{0.5}$	[44] (Homogeneous nucleation)

Nucleation equations can be empirical based or have some thermodynamic meaning. Most empirical nucleation equations use a power law formula. To better model nucleation, the model should be dependent on the manipulated variables of the experiment. Several empirical nucleation models are listed in Table 8.8.

Thermodynamic nucleation models can be generated from classical nucleation theory. Several different thermodynamic models are listed in Table 8.9.

8.4.1.1 Modeling Nucleation

To determine a proper nucleation model, the type of model needs to be chosen. Will the model be empirical, thermodynamic, or a combination of the two? Second, what type of nucleation needs to be modeled, i.e., primary or secondary nucleation? This is important because there are fundamental differences between these two nucleation types. Secondary nucleation models need to take into account crystal holdup, while primary models do not. The second step is to decide what variables need to be parameterized to experimental data. Third, are the standard parameters sufficient or do they need to be expanded?

The first example will be to make an empirical model of secondary nucleation for a cooling crystallization process. First, a basic nucleation model will be formed:

$$B = k_b \mu_2 S^b \tag{8.8}$$

The constants k_b and b can be fit as they are, or they can be made functions of composition and temperature.

The second example will be to make a thermodynamic model of primary homogeneous nucleation. The thermodynamic equation in Mersman [42] will be used,

$$B_{\text{hom}} = 1.5 \times D_{AB}(CN_A)^{7/3} \sqrt{\frac{\gamma_{CL}}{kT} \frac{1}{C_c N_A}}$$

$$\times \exp\left(-\frac{16}{3}\pi \left(\frac{\gamma_{CL}}{kT}\right)^3 \left(\frac{1}{C_c N_A}\right)^2 \frac{1}{(\nu \ln S)^2}\right) \tag{8.9}$$

In this model, the process parameters, process variables, and constants need to be determined. The constants are Avogadro's number (N_A), Boltzmann constant (k), molar crystal density (C_c), and the dissociation index (ν). The process variables are temperature (T), solution concentration (C), and relative supersaturation (S). The process parameters are the diffusion coefficient (D_{AB}) and interfacial tension (γ_{CL}). These parameters may be handled in several different ways. First, the parameter can be parameterized simply as a single parameter. Second, it can be parameterized as a thermodynamic function. Third, it can be parameterized empirically.

Method 1:

$$D_{AB} = a$$

$$\gamma_{CL} = b$$

Method 2:

$$D_{AB} = \frac{kT}{2\pi \eta \sqrt[3]{\frac{1}{C_c N_A}}}$$

$$\gamma_{CL} = kTK(C_c N_A)^{2/3} \ln\left(\frac{C_c}{C^*}\right)$$

Table 8.10 Empirical growth models.

Equation	Reference
$G = k_g \Delta C^g$	[9, 24]
$G = k_g \sigma^g$	[8, 9, 45, 47]
$G = k_g N \exp(\frac{-E}{RT}) \Delta C^g$	[65]

Method 3:

$$D_{AB} = f(\text{temperature, composition})$$

$$\gamma_{CL} = f(\text{temperature, composition})$$

Some of these methods may give substantially better experimental fits than the others, so the appropriate one should be selected for use in the crystallization model.

8.4.2
Growth and Dissolution

Once crystals are formed in solution, they increase in size through various crystal growth mechanisms. Dissolution is the opposite of crystal growth and describes the process of the crystal dissolving in the solution. Crystal growth is analogous to heterogeneous catalytic reactions. Growth can either be mass transfer or surface integration limited. If growth is mass transfer limited, as soon as the solute crosses the boundary layer the solute integrates into the crystal very quickly. If growth is surface integration limited, then the solute crosses the boundary layer quickly, but integrates slowly into the crystal. Crystal growth can be modeled with many different equations, but the distinction between mass transfer limited and surface integration limited growth is the exponent on the supersaturation term. If the exponent is 1, then it is mass transfer limited, and if the exponent is greater than 1, then it is surface integration limited growth. The classic crystal growth equation is shown below:

$$G = k_g \Delta C^g \tag{8.10}$$

where k_g can take on many different forms. k_g can be a constant, can depend on temperature, on composition, or a combination of these. g is usually a constant but can depend on composition as well. Common empirical growth models are listed in Table 8.10.

Growth can be modeled in a similar manner as nucleation, first either an empirical or functional model needs to be chosen. Second, the modifiable process parameters are chosen. Third, the functional form of the parameters is selected.

The first growth modeling example will be for joint cooling-antisolvent crystallization. An empirical model will be chosen because the type of crystal growth is unknown.

The general growth model is

$$G = k_g \Delta C^g \tag{8.11}$$

with parameters k_g and g. These parameters will be expanded to be dependent on temperature and composition, and the exponent dependent on composition:

$$k_g = \left(k_1 w_3^2 + k_2 w_3 + k_3\right) \exp\left(\frac{-(E_a + k_4 w_3)}{RT}\right) \tag{8.12}$$

$$g = g_1 w_3 + g_2 \tag{8.13}$$

which combines to

$$G = \left(k_1 w_3^2 + k_2 w_3 + k_3\right) \exp\left(\frac{-(E_a + k_4 w_3)}{RT}\right) \Delta C^{g_1 w_3 + g_2} \tag{8.14}$$

The second example will be for mass-transfer limited growth. The base model is

$$G = k_g \Delta C \tag{8.15}$$

The parameter k_g will be parameterized as a mass transfer coefficient. This mass transfer coefficient can be of any form found in the literature. k_g could also be parameterized in an empirical form. Dissolution is mass transfer limited and is modeled the same way as mass transfer limited growth.

8.4.3
Agglomeration and Aggregation

Another crystallization phenomenon that can occur is agglomeration. Agglomeration is when crystals attach to each other through strong interparticle forces such as chemical bonds. Agglomeration can be both beneficial and detrimental. Agglomeration can allow for the formation of large crystal complexes at the expense of a well-defined crystal morphology and habit. Aggregation is similar to agglomeration with the difference in that the particles are attached through weak bonds such as van der Waals forces. Aggregated particles can be separated by shear or solvents.

8.4.4
Attrition

Attrition is when crystals collide together or with the crystallizer components (such as agitator and baffles) and break into multiple smaller fragments. These collisions can also cause secondary nucleation. The factors that affect attrition are crystal hardness, impeller size and speed, and suspension density. Equation (8.16) is an attrition model developed by Mersman [42]:

$$B_{\text{attrit}} = 7 \times 10^{-4} \varphi_T \frac{H_V^5}{\mu^3} \left(\frac{\Gamma}{K}\right)^{-3} \frac{\pi^2 \rho_c \bar{\varepsilon} N_v}{2k_v \, Po} \frac{N_{a,\text{eff}}}{N_{a,\text{tot}}} \eta_w^3 \eta_g \tag{8.16}$$

Like agglomeration, attrition is not normally modeled unless the experimental data support a significant presence of attrition.

8.5
Population, Mass, and Energy Balances

8.5.1
Population Balance

Crystallization is a particulate process, so a balance is required to track the crystals in the system. This balance is called a population balance that was presented for crystallizer systems by Hulbert and Katz [27]. This balance tracks the number of crystals in every size class as they are formed through nucleation in the smallest size class, and change in size either through growth, agglomeration, or attrition. The population balance equation is

$$\frac{\partial n(z,t)}{\partial t} + \nabla \cdot \left(n(z,t)\bar{G}\right) - (B - D) = \sum_k n_k(z,t) Q_k \tag{8.17}$$

where n is the crystal density, B is the birth rate of crystals, D is the death rate of crystals, $\bar{G}$ is the 3D growth rate of crystals, and n_k is the number of crystals entering or leaving the crystallizer of stream Q_k. To simplify the population balance, a few assumptions are commonly used. First, crystal growth can be assumed to be independent of crystal size (follows McCabe's law). Second, attrition and agglomeration can be assumed for some systems to be negligible. Third, the crystal can be characterized by one size axis. Fourth, since this is a batch process there is no entry or exit of crystals from the crystallizer. With these simplifying assumptions, the population balance becomes Eq. (8.18) with initial and boundary conditions:

$$\frac{\partial n(L,t)}{\partial t} + G \frac{\partial n(L,t)}{\partial L} - B = 0$$

$$n(L,0) = n_0(L)$$

$$B_0 = n(0,t) G \tag{8.18}$$

8.5.2
Solution Methods

8.5.2.1 Method of Moments

This population balance can be solved through various different methods. The first method is known as the method of moments. The method of moments presented by Hulbert and Katz [27] and later by Randolph and Larson [60] is commonly used. This method reduces the partial differential population balance equation into a set of coupled ordinary differential equations (ODEs). The advantage becomes the ease of solution, as ODE solvers are readily available. This method has a disadvantage that a unique CSD cannot be reconstructed from a finite set of moment equations as different CSDs may be formulated from the same moment equations. A second disadvantage is that population balances carrying size dependent growth functions may not be solved.

The population density is a function of crystal size and we can use the density function, $n(L)$, to express the number of crystals, N_i, in a size class L_i to L_{i+1} per unit volume of slurry as

$$N_1 = \int_{L_i}^{L_{i+1}} n(L)\, dL \tag{8.19}$$

From Eq. (8.19), a series of relationships evolves representing the moments of the density function. We write the jth moment of the density function as

$$N_j = \int_0^\infty L^j n(L)\, dL \tag{8.20}$$

where the zeroth moment, μ_0, of the distribution, at $j = 0$, and expressed as

$$\mu_0 = \int_0^\infty n(L)\, dL \tag{8.21}$$

characterizes the total number of crystals. The first, second, and third moments also provide information about the distribution and represent the total length, the total area, and the total volume of crystals, respectively. That is,

$$\mu_1 = \int_0^\infty Ln(L)\, dL \tag{8.22}$$

$$\mu_2 = \int_0^\infty L^2 n(L)\, dL \tag{8.23}$$

$$\mu_3 = \int_0^\infty L^3 n(L)\, dL \tag{8.24}$$

8.5.2.2 Discretization Method

The population balance can also be solved through various discretization methods. One way is through the method of lines where the size axis is discretized through finite-differences. It has been shown that the backward finite difference method is more stable than central finite difference discretization method [4].

An example showing the discretization method is shown below:

$$\frac{\partial n(L,t)}{\partial t} + G\frac{\partial n(L,t)}{\partial L} - B = 0 \qquad (8.25)$$

$$\frac{dn_1}{dt} = B - G\frac{n_1}{2\delta_1} \qquad (8.26)$$

$$\frac{dn_i}{dt} = G\left(\frac{n_{i-1}}{2\delta_{i-1}} - \frac{n_i}{2\delta_i}\right), \quad i = 2,\ldots,\zeta \qquad (8.27)$$

where ζ is the number of discretization intervals, L_i is the length of the ith interval, b is a geometric constant, L_0 is the nucleate size, L_{max} is the largest size axis length, and δ_i is the length of the ith interval given by

$$\delta_i = L_i - L_{i-1}, \quad i = 1,\ldots,\zeta \qquad (8.28)$$

The individual discretization lengths are chosen using a geometric series

$$L_i = L_0 b^i, \quad i = 0,\ldots,\zeta \qquad (8.29)$$

$$b = \left(\frac{L_{max}}{L_0}\right)^{\frac{1}{\zeta}} \qquad (8.30)$$

Other discretization methods that have been used to solve the population balance are through the use of finite elements and wavelets [41, 58].

Selection of Discretization Intervals When solving the population balance through discretization methods, it is important to select the appropriate number of size intervals. If too few intervals are used, the crystal distribution will be modeled poorly. If too many are used, the computational time will be inefficient.

The following example is given to illustrate the optimization of the number of intervals using the method of lines discretization technique. The simulation software gPROMS was used for this analysis. Previous crystallization simulations from Nowee et al. [52–54] have utilized the software gPROMS to solve the population balance of ammonium sulfate and sodium chloride. In solving the crystallization population balance, they use a geometric distribution to discretize the length axis

$$L_i = L_0 b^i, \quad i = 0,\ldots,\zeta \qquad (8.31)$$

$$b = \left(\frac{L_{max}}{L_0}\right)^{\frac{1}{\zeta}} \tag{8.32}$$

$$\delta_i = L_i - L_{i-1} \tag{8.33}$$

where ζ is the number of discretization intervals, L_i is the length of the ith interval, b is a geometric constant, L_0 is the nucleate size, L_{max} is the largest size axis length, and δ_i is the length of the ith interval.

The disadvantage of this method is in the representation of the CSD. This will skew the CSD to favor larger crystal sizes. If the distribution is not unimodal, the larger sizes will have an overemphasized percentage than the smaller sizes. For distributions that may show nonunimodal behavior, it is preferred to use a discretization that has equally distributed intervals

$$L_i = L_1 + (i-1)\left(\frac{L_{max} - L_1}{\zeta - 1}\right), \quad i = 2, \ldots, \zeta \tag{8.34}$$

$$\delta = \left(\frac{L_{max} - L_1}{\zeta - 1}\right) \tag{8.35}$$

where L_1 is the nucleate size and δ is the interval length. Since one of the benefits of discretization is the visualization of the CSD, it would be good if that is as accurate as possible. This causes the equally distributed method to be the discretization of choice for CSD visualization. This discretization method is then applied to the population balance for a constant volume batch crystallizer with no agglomeration and where crystal growth follows McCabe's law:

$$\frac{\partial n(L,t)}{\partial t} + G\frac{\partial n}{\partial L} - B = 0 \tag{8.36}$$

where G is the growth rate, B is the nucleation rate, L is the length, and $n(L,t)$ is the crystal distribution. The population balance is then discretized in a backward finite difference manner:

$$\frac{dn_1}{dt} = B - G\frac{n_1}{2\delta_1} \tag{8.37}$$

$$\frac{dn_i}{dt} = G\left(\frac{n_{i-1}}{2\delta_{i-1}} - \frac{n_i}{2\delta_i}\right), \quad i = 2, \ldots, \zeta \tag{8.38}$$

This finite difference discretization technique is a simple straightforward technique compared to other discretization techniques in the literature. In order to test the dependency of the simulation results, a previously developed gPROMS crystallization model, for the cooling crystallization of potassium chloride was used. To test the grid dependency of the model, all the variables were fixed except for the grid interval lengths. The size axis was fixed from 0.05 to 1000 μm, and the interval lengths were varied by varying the number of intervals from 10 to 2000. In

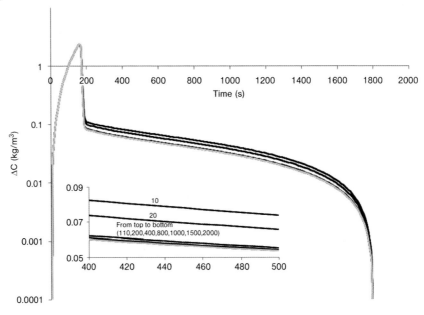

Fig. 8.6 Supersaturation dependency.

investigating the grid dependency, four crystallization variables were reported; the absolute supersaturation, the number weighted mean size, the volume weighted mean size, and the final number percent CSD.

8.5.2.2.1 Grid Dependency Analysis Supersaturation is the most important variable in crystallization processes, and dictates the growth and nucleation mechanisms. It is thus the first variable analyzed. It is imperative that the supersaturation not be a function of interval length. Otherwise the growth and nucleation results will be incorrect. Due to the fact that the supersaturation profile crosses several orders of magnitude, the supersaturation is plotted on a semi-log plot so that the interval dependency can be more easily examined. Figure 8.6 illustrates the dependency of the supersaturation profile on the discretization intervals. At low discretization interval numbers such as 10 or 20, the supersaturation profile is slightly larger than the supersaturation profile for the higher interval number discretizations. However, increasing the number of discretization intervals past 110 does not significantly increase the accuracy of the results as can be seen in the inset of Fig. 8.6.

The next important variable analyzed is the average size of the crystals. One way to represent the average size is through the number weighted mean size. By looking at Fig. 8.7, it can be seen again that the lower intervals of 10 and 20 slightly underestimate the number mean size. It appears that 110 discretization intervals would again be adequate, and by looking at the inset of Fig. 8.7 there is slight improvement in the number mean size by increasing the number of intervals from 110 to 2000. The number mean size increases a micron from 111 to 112 µm. Using

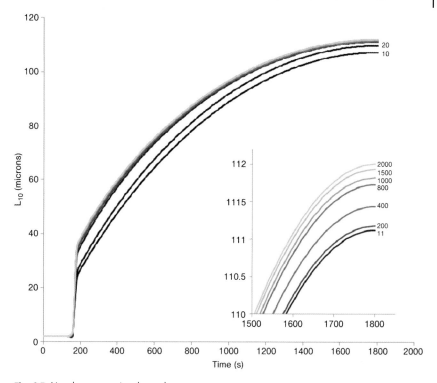

Fig. 8.7 Number mean size dependency.

110 discretization intervals introduces less than 1% error in the number mean size average than if 2000 discretization intervals were used.

The third variable analyzed is the volume mean size. Figure 8.8 shows that there is a greater interval dependency by this variable than the previous two variables. Using intervals less than 400 results in overestimated average volume mean sizes. The zoomed inset in Fig. 8.8 shows that using 400 intervals is not sufficient. 800 intervals are required to reasonably approach the accuracy afforded by 2000 intervals.

The last variable analyzed is the number percent CSD. Since the magnitude of the number percent is a function of the number of intervals, each distribution was normalized for comparative purposes. Figure 8.9 shows why the low interval numbers were so bad at estimating the previous variables. The distributions representing 10 and 20 discretization intervals are extremely poor estimations of the CSD. Once the number of discretization intervals approaches 110, the CSD starts resembling the higher interval number CSDs. Further increasing the number of intervals to 800 results in the interval dependency becoming less sensitive. It can also be seen why the volume mean size is more sensitive to the number of discretization intervals than the number mean size. This is due to the width of the distribution. Even though there is the same amount of crystals both above and below the mean, the total volume of crystals is not the same. The higher volume of

262 | 8 Modeling of Crystallization Processes

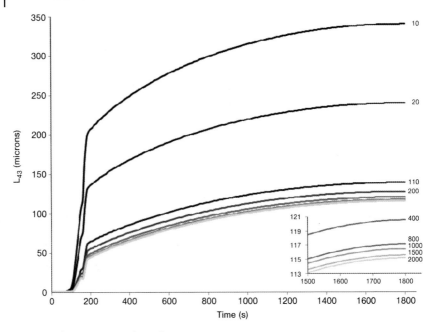

Fig. 8.8 Volume mean size dependency.

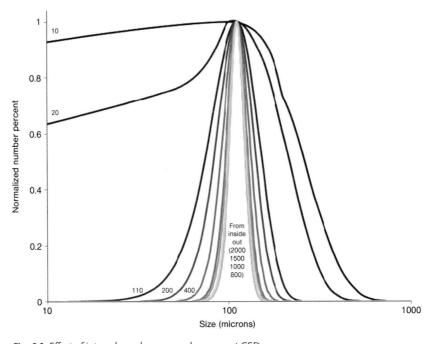

Fig. 8.9 Effect of interval number on number percent CSD.

8.5 Population, Mass, and Energy Balances

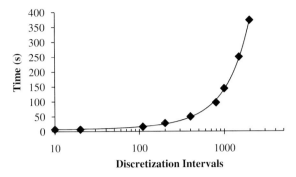

Fig. 8.10 Effect of interval number on simulation time.

crystals larger than the number mean size results in a bias in the volume mean size to larger sizes because they have much larger volumes than the smaller sizes.

Interval Effect on Computational Time To ensure the most accurate results, one would use the highest number of intervals possible. Why not then use more than 2000 intervals such as 5000 or 10 000? This can be explained by looking at Fig. 8.10. As the number of intervals increase, the computational time penalty becomes more severe. To be able to estimate the computational time required for using more than 2000 intervals, the number of discretization intervals versus computational time was fit to a second order polynomial with an R^2 value of 0.999 shown in the equation

$$t = 4.98 \times 10^{-0.5} \zeta^2 + 8.48 \times 10^{-0.2} \zeta + 6.27 \tag{8.39}$$

For example, the estimated computational time to do 5000 intervals would be almost 30 min, while 100 min would be needed for 10 000 intervals.

The time required to use more than 2000 intervals is not worth the computational time required for minimal improvements in the simulation results. In addition, simulation engines occasionally have stability problems with intervals more than 2000.

The number of intervals required to give accurate results depends on several factors. First is the distance between the largest and smallest size. The potassium chloride discretization was done from 0.05 to 1000 μm, which resulted in an interval length dependent only on the number of intervals. If the size limits are changed, then the grid will become coarser or finer than what it was previously. If the maximum size increased from 1000 to 2000 μm, then the number of intervals would be doubled to ensure the same interval length. Similarly if the maximum size becomes 500 μm then the number of intervals could be halved, which would decrease the computational time required. This means that carefully chosen size limits can make the simulation more computationally efficient. Second, it depends on what variables are of importance. If only the supersaturation and number mean size variables are important to report, then a coarser grid can be used than that would have been required if the volume mean size is the variable of interest. Since the volume mean size is often a reported variable, it would follow that at least 800 discretization intervals should be used for appropriate accuracy.

8.5.3
Mass and Energy Balances

To track the total mass of solute in solution/suspension, a mass balance is needed. A mass balance for a batch crystallizer is

$$\frac{d(Cm_s)}{dt} = -3p_c k_v GV \int_0^\infty n(L,t) L^2 \, dL \tag{8.40}$$

where C is the solution concentration (kg solute/kg solvents), m_s is the mass of solvents (kg), ρ_c is the solid crystal density (kg/m^3), k_v is the volumetric shape factor, G is the growth rate (m/s), $n(L,t)$ is the crystal density, and L is the size of crystals. This can be further simplified by noticing that the integral is the second moment of the distribution (μ_2):

$$\frac{d(Cm_s)}{dt} = -3p_c k_v GV \mu_2 \tag{8.41}$$

The last balance required for a crystallizer is an energy balance. The energy balance must include the enthalpy that is created from the production of crystals (heat of crystallization, ΔH_c), the enthalpy of the solution, heat loss from the system (Q_{lost}), energy added from mixing (W_s), heating/cooling jacket ($UA\Delta T_{lm}$), and the transient term. This is given in Eq. (8.42),

$$\frac{d(\rho C_p VT)}{dt} = -3\Delta H_c \rho_c \alpha_v V G \mu_2 - UA\Delta T_{lm} - Q_{\text{lost}} + W_s \tag{8.42}$$

8.6
Crystal Characterization

8.6.1
Crystal Shape

Crystals can be formed in much different morphology. The crystal's shape can be cubic, monoclinic, orthorhombic, needle, or some other shape. In addition, many compounds display polymorphism where they can have more than one crystal structure. Paracetamol, for example, can crystallize into multiple forms depending on the crystallization conditions, and can crystallize into either octahedral or monoclinic forms. Another crystal shape property is the crystal habit. Habit relates to the dimensions of the crystal faces. Two crystals with the same crystal structure may have different shapes. This difference is taken into account by the crystal habit.

If a crystal has a shape that is not a perfect cube, then it cannot be accurately described by just one size axis. Since it is inconvenient to track the particle growth

Table 8.11 Definition of crystal mean sizes.

Mean size	Moment definition	Meaning
D_{10} or L_{10}	μ_1/μ_0	Number mean crystal size
D_{21} or L_{21}	μ_2/μ_1	Length mean crystal size
D_{32} or L_{32}	μ_3/μ_2	Area mean (Sauter) crystal size
D_{43} or L_{43}	μ_4/μ_3	Volume mean crystal size

of multiple size dimensions, a characteristic length (L) is chosen. Once the characteristic length is chosen, then different crystal shape factors have to be determined. The most commonly used shape factors are area and volume shape factors defined, respectively, as

$$k_a = \frac{\text{Area of actual structure}}{L^2} \quad (8.43)$$

$$k_v = \frac{\text{Volume of actual structure}}{L^3} \quad (8.44)$$

These shape factors are equal to 1 for a cubic structure.

8.6.2
Crystal Size

There are several different ways to measure the mean size of a crystal batch. These are all calculated from the moments of the population balance. The commonly used ones are the number mean size and the volume mean size. The various sizes and their definitions are displayed in Table 8.11. In order to determine the variation of crystal sizes in a sample, the coefficient of variation (COV or CV) is used. This along with the mean size can give an idea what the sample size distribution is. Equation (8.45) is the generic form of the moment equation. All the mean sizes reported in Table 8.11 are calculated using Eq. (8.45). For example, if volume mean crystal size is desired, then $i = 4$ and $j = 3$ so $D[i, j] = D[4, 3]$ and is computed for that (i, j) pair,

$$D[i, j] \text{ or } L[i, j] = \frac{\mu_i}{\mu_j} = \frac{\int_0^\infty L^i n(L, t)\, dL}{\int_0^\infty L^j n(L, t)\, dL} \quad (8.45)$$

8.6.3
Crystal Distribution

However, to determine the actual distribution of crystals, several different distribution functions are used. There are two main types of distributions – cumulative and percent. The first distribution is the cumulative number distribution. The cumulative number distribution defines the fraction of particles that are up to a specific size. Similarly, there is a cumulative mass distribution. This defines the fraction of particle mass up to that specific size. The other commonly used distributions are number and volume percent distributions.

8.6.4
Particle Measurement Process Analytical Tools

There are also several different ways to measure particle size and size distributions. These analytical techniques include laser diffraction, focused beam reflectance measurement (FBRM), and image analysis via online microscopy, and offline microscopy (light microscopes, or scanning electron microscopes). Laser diffraction is used offline to determine particle size and size distributions. Several different companies produce equipment that use laser diffraction for particle size analysis. These include Beckman Coulter, CILAS, and Malvern. In laser diffraction, a laser is shone on the particles, and the diffraction pattern is read from the detectors. Using Mie theory, the particle size and distribution are calculated from the diffraction data. The downside to laser diffraction is that the particles are assumed to be spherical. Focused beam reflectance measurements are used by the Mettler–Toledo Lasentec FBRM. This apparatus is used *in situ* in the crystallizer. It can be used to determine particle size (as chord length), particle chord length distribution, and to determine the onset of nucleation. The FBRM does not assume a particle shape when calculating particle size, but since it calculates chord length and not particle size, a correlation must be used to convert chord length to particle size. Image analysis is used to analyze microscopic pictures of particles to determine particle size and distributions. The advantage of image analysis is that the actual crystals being measured are seen. A review of these techniques is given by Abbas *et al.* [1, 11] and Li *et al.* [38].

8.7
Solution Environment and Model Application

In order to create a complete crystallization model, various different components, entities, and procedures must be combined into a single modeling/simulation environment. The selection of such environment is a critical issue and is discussed next.

8.7.1
Simulation Environment

Some popular modeling environments suitable for modeling crystallization processes are Matlab, and PARSIVAL. Matlab allows for the use of various toolboxes to add required functionality such as partial differential equation solvers, parameter estimation, and optimization. Matlab has been used by several workers to model crystallization processes [39, 64, 68, 77]. PARSIVAL is a stand-alone environment for modeling and solving every aspect of the crystallization process [74].

Another model development and solution software environment very suitable and extensively used for crystallization modeling is gPROMS® [2, 7, 33, 34, 56]. The general process modeling system (gPROMS®) is an equation-oriented high-level declarative modeling, simulation, and optimization package. It allows the development of hierarchical models of arbitrary depth involving a range of process models including distributed systems and processes with discontinuities. gPROMS directly supports simulation, parameter estimation, and optimization. Additionally, process model libraries are available for a number of processes, including crystallization, batch reactor operation, tray distillation, and process control devices (PID controllers), among others.

Model development in gPROMS comprises the creation of a series of entities through a friendly user interface (ModelBuilder), each of which is structured in a particular way and serves a particular purpose. The MODEL entity contains the set of equations (algebraic, differential, linear, nonlinear, or combination of them, etc.) defining the system. On the other hand, the PROCESS entity represents a simulation run; it defines the initial conditions and other variables values necessary to complete the degrees of freedom. All model-based activities require both these entities.

gPROMS simulates models of unprecedented complexity and offers numerical methods to solve the activities needed for crystallization processes. These facts, in addition to the capability of accepting high-level declarative language input files, contributed to the selection of gPROMS as the solution engine for our crystallization modeling.

8.7.2
Experimental Design

Experimental design is an essential step because data from experiments are needed to estimate the parameters of the kinetic submodels. This can be done either through model-developed experimental design or by statistical design of experiments. Model-developed experimental design uses the crystallization model to approximate parameter guesses. The designed experiments aim at maximizing the information received from the experimental data to calculate the unknown kinetic parameters. This is equivalent to minimizing the variance in the parameters. Experimental design determines what the initial conditions should be, when to take measurements, the experiment duration, and what the manipulated variables change over the experiment.

The other simpler method is to initially use a few different operating conditions that vary within the known limits of the crystallization operating variables. This is useful when more than one kinetic model is being postulated, or if the appropriate model is unknown. If the model is unknown, then model-based experimental design cannot be used. The operating conditions can be comprised of different evaporation rates, different cooling rates, or different antisolvent feed rates. Once a

Table 8.12 Considered growth models.

Model 1	$G = k_g(\Delta C)^g$
Model 2	$G = k_g(\frac{\Delta C}{C})^g$
Model 3	$G = (k_0 + k_1 z_3 + k_2 z_3^2)(\Delta C)^{g_0+g_1 z_3}$
Model 4	$G = (k_0 + k_1 z_3 + k_2 z_3^2)(\frac{\Delta C}{C})^{g_0+g_1 z_3}$

model is created using this method, it can be used to plan further experiments using model-developed experimental design. A detailed discussion on model-based experimental design for crystallization processes can be found in [16, 67].

8.7.3
Parameter Estimation

In recent years, rigorous parameter estimation methods have been employed in crystallization kinetics identification but of all techniques, posing the estimation as an optimization problem has been regarded as most relevant [43, 61]. Nonlinear optimization techniques such as maximum likelihood are now available paving the way for more efficient estimation and with the advances in power of modern computers, optimization methods are becoming more prevalent in the literature. The mathematical background behind these techniques is intricate and there are several text available detailing the mathematical derivations [12, 37] for the reader to refer to.

The crystallization model described before consists of a coupled set of nonlinear integro-differential equations and can be represented by $Y(t) = [\bar{L}(t), \rho(t)]$ being the differential output variables – mean size and solution density, respectively, $U(t) = [F(t)]$ being the time-varying input variable which is the feed rate of antisolvent, and θ being the vector of parameters to be identified which in this case is the set $[k_b, k_g, b, g]$ when considering k_g and g to hold constant values, or the set $[k_b, k_0, k_1, k_2, b, g_0, g_1]$ when considering k_g and g to be functions of antisolvent mass fractions in solute-free media (see Table 8.12). We can then represent the model as follows:

$$f(Y(t), U(t)) = 0 \qquad (8.46)$$

Equation (8.46) is accompanied by a set of initial conditions. Measured variables from experiments are essential in the parameter estimation exercise and are denoted by the vector $\hat{y}(t) = [\hat{L}(t), \hat{\rho}(t)]$. The maximum likelihood criterion that describes the highest probability of the model predicting the real data is given by the following objective function:

Table 8.13 Calculated model parameters.

Parameter	Model 1	Model 2	Model 3	Model 4
B	3.464	1.494	1.937	0.9655
k_b	2.11E+07	2.12E+07	9.66E+06	9.68E+06
G	2.576	0.2302	–	–
k_g	3.10E−05	7.67E−07	–	–
g_0	–	–	1.051	0.8006
g_{1n}	–	–	4.875	0.7379
k_0	–	–	1.86E−05	9.88E−05
k_1	–	–	0	−0.00241
k_2	–	–	0	0.01681

$$\Phi(k,\theta) = \frac{M}{2}\ln(2\pi)$$

$$+ \frac{1}{2}\min_{k,\theta}\left(\sum_{i=1}^{\alpha}\sum_{j=1}^{\beta_i}\sum_{k=1}^{\gamma_{ij}}\left[\ln(\sigma_{ijk}^2) + \frac{\hat{Y}_{ijk} - Y_{ijk}}{\sigma_{ijk}^2}\right]\right) \quad (8.47)$$

where M is the total number of measurements taken during all experiments, α, β_i and γ_{ij} are, respectively, the number of experiments, the number of variables measured in the ith experiment and the number of measurements of the jth variable in the ith experiment, σ_{ijk}^2 is the variance of the kth measurement of variable j in experiment i. σ_{ijk}^2 can be described by any of a number of models available [59] when the error structure of the data is known.

In previous works, we successfully applied this method to the identification of ammonium sulfate kinetics for crystallizations under cooling mode of operation [3, 52]. The same method was applied with differences being modifications in the model and experimental conditions to describe antisolvent crystallization under constant temperature [54]. Again, in this antisolvent crystallization case, the implementation of the maximum likelihood optimization is carried out using the gEST facility in gPROMS. The kinetic parameter set that provides the highest probability of the model predicting the real data is identified.

The considered growth kinetic models are given in Table 8.12, included in the same nucleation model:

$$B - k_b \Delta C^b M_T \quad (8.48)$$

The estimated parameters for each of the four models are shown in Table 8.13.

8.7.4
Validation

Now that each model's parameters have been calculated, each model can be compared against each other to see which is best at modeling the experimental data.

From the figures shown in Tables 8.14 and 8.15, the figures show that growth model 4 most closely matched the experimental data.

8.8
Optimization

A key advantage of crystallization modeling is that it allows for the calculation of optimal operating profiles. Before the use of dynamic crystallization models, most crystallization operations were operated by experience and rules of thumb. With the recent proliferation of crystallization modeling, model-based optimization is becoming more common which gives more rapid and vastly superior results than previous methods.

In order to perform a crystallization optimization, an objective function needs to be chosen. This objective function can be related to the crystal properties such as size and coefficient of variation [49, 52, 54], or for minimizing nucleation [62, 68]. The objective function can also be related to the operation of the crystallizer such as to minimize total energy costs. Multiobjective optimization can also be performed combining different objective functions [62, 63, 68], for example, to maximize crystal mean size while minimizing energy costs. Frequently, when performing multiobjective optimizations there will not be one unique solution, but there will be a set of solutions forming a Pareto optimal set. With Pareto optimal sets, there is a tradeoff between the multiple objective functions. If one objective function is improved, the other will become less optimal.

The next important part of optimization is to define the optimization constraints. These are usually physical constraints that cannot be exceeded. Constraints can be related to process time, equipment limitations, product quality, and other limitations. The next step is to determine what process variables the optimization will change. Common manipulated process variables are seed mass, cooling rate, and antisolvent feed rate. If the manipulated variable changes during the experiment, then control intervals need to be chosen.

In the next two sections, we present model-based optimization results from two different crystallization applications: the first focuses on the calculation of optimal antisolvent feedrate, while the focus of the second is on the determination of the optimal seeding in cooling crystallization.

8.8.1
Example 1: Antisolvent Feedrate Optimization

The crystallization model presented above was used to perform two different optimizations in an antisolvent crystallization application [54]. One for the maximization of volume mean size, and the other to produce a specific mean size. First, the manipulated variable is defined as the antisolvent feedrate. The next step is to define the control horizon for the manipulated variable. For both objective functions, the control horizon is split into 10 equal divisions. The antisolvent feedrate

Table 8.14 Validated solution concentration and volume mean size profiles.

Antisolvent feed rate	Solution concentration	Volume mean size
50 mL/h		
98.4 mL/h		

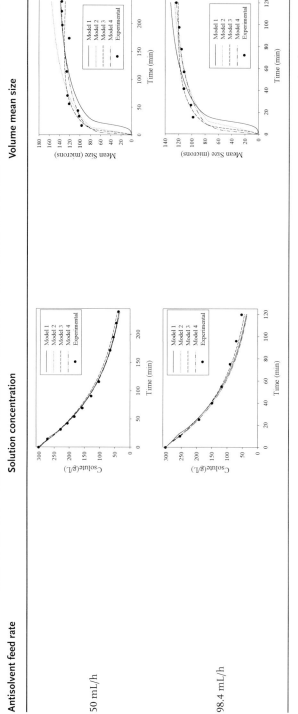

(continued on next page)

272 | *8 Modeling of Crystallization Processes*

Table 8.14 (continued.)

Antisolvent feed rate	Solution concentration	Volume mean size
194 mL/h		

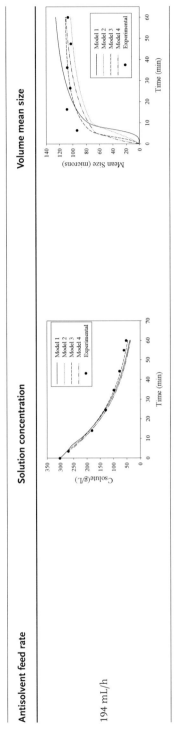

Table 8.15 Validated final volume percent crystal size distributions.

Antisolvent flowrate	Final volume percent CSD
50 mL/h	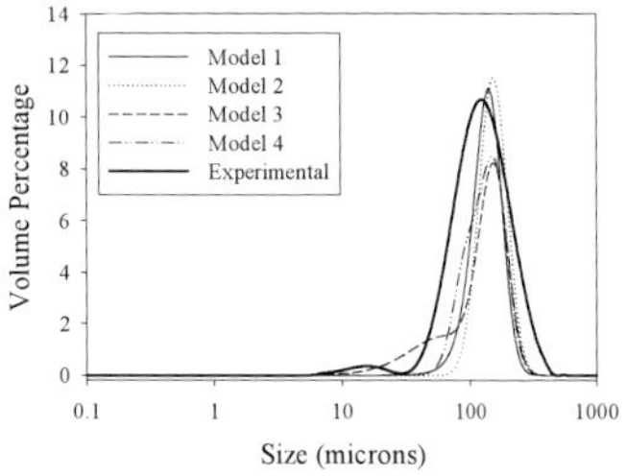
98.4 mL/h	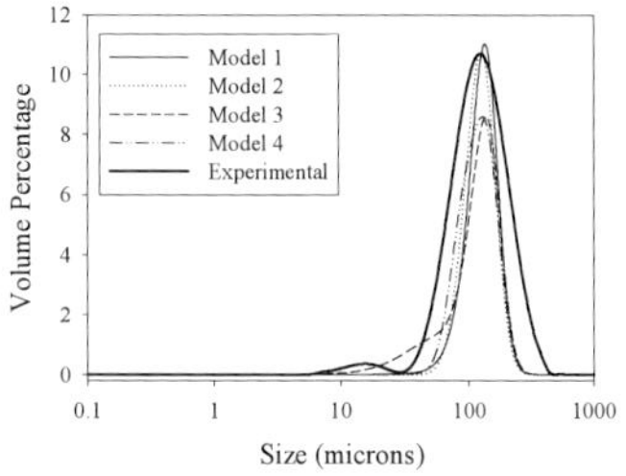
194 mL/h	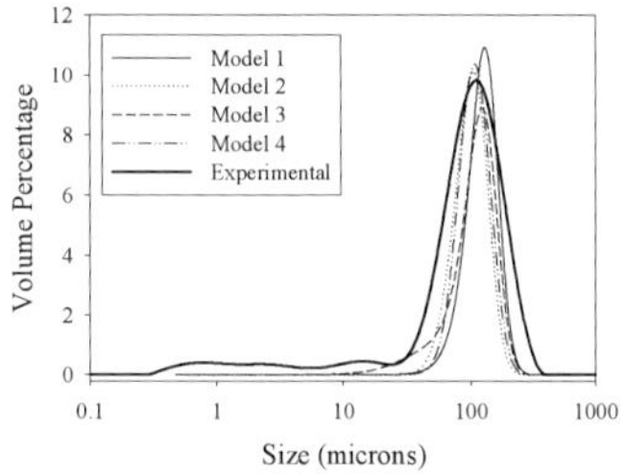

is controlled in a piece-wise constant manner. Next the constraints are set. There are specified constraints for the process time, antisolvent feedrate, crystallizer final volume, and crystal yield. Finally, the initial conditions are set, being the initial solvent composition, temperature, and seed loading. The complete optimization formulation is shown below for the maximization of volume mean size objective function:

$$OBJ = \max D_{43} \begin{cases} 1800 \leq t_f \leq 16\,000 \text{ s} \\ V_f \leq 500 \text{ mL} \\ \text{Yield} \geq 20\% \\ 0.375 \leq \frac{dV}{dt} \leq 125 \frac{\text{mL Ethanol}}{\text{min}} \end{cases} \quad (8.49)$$

$$T = 25\,°\text{C}, \quad w_i = 0\%, \quad n_i(L, 0) = 0, \quad C_i = C_i^* \quad (8.50)$$

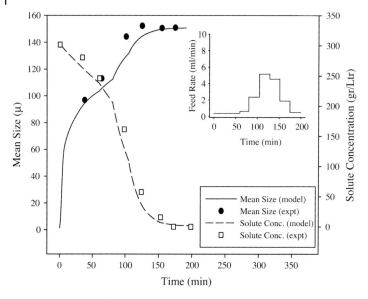

Fig. 8.11 Experimental and simulation trajectories for the volume mean size maximization optimization objective.

The optimization was performed using the previously developed model using the gPROMS entity gOPT. The calculated optimal antisolvent flowrate is shown in the inset of Fig. 8.11.

The calculated optimal profile was then both simulated using the crystallization model in gPROMS and implemented experimentally. As seen in Fig. 8.11, the simulated results match the experimental data very closely.

This was repeated for the second objective function, which varies from the first objective function with a volume mean size constraint between 90 and 110 µm. This limits the volume mean size around the objective of 100 µm. This optimization was once again performed using gPROMS and the calculated optimal antisolvent flow rate is displayed in Fig. 8.12 in the inset. As before, the simulated results closely match the experimental data. Significantly, this second optimization formulation allows one to target and control the operation toward specific crystal sizes rather than maximize or minimize the crystal particle size.

8.8.2
Example 2: Optimal Seeding in Cooling Crystallization

The next optimization example is for a seeded cooling crystallizer as previously presented in Nowee et al. [52]. For a seeded crystallizer, the optimization control variables are the seed distribution, seed mass, and the cooling profile. A similar crystallization model was formulated and validated for ammonium sulfate in water.

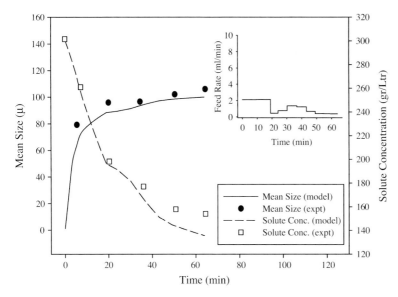

Fig. 8.12 Experimental and simulation trajectories for the 90–110 μm volume mean size optimization objective.

The population balance includes nucleation (B), disappearance (D_a), dissolution (D), and growth (G) terms:

$$\frac{\partial n(L,t)}{\partial t} = (B - Da) + (D - G)\frac{\partial n(L,t)}{\partial L} \tag{8.51}$$

The crystallization kinetics were described using a power law and are shown below:

$$B = k_b \Delta C^b \tag{8.52}$$

$$G = k_g \Delta C^g \tag{8.53}$$

$$D_a = k_a \Delta C^a \tag{8.54}$$

$$D = k_d \Delta C^d \tag{8.55}$$

with kinetic parameters ($k_a, k_b, k_d, k_g, a, b, c, d$) estimated and validated in a similar way shown in Section 8.7.4. With the crystallization model complete, model-based optimization was done. There were two separate objective functions – one to maximize volume mean size and the other to minimize the variance of the final CSD. The control variables were the seed mass, seed distribution, and the temperature profile. The seed mass and distribution were combined into a single vector, N_0. The temperature was discretized into 10 control intervals in a piecewise linear manner. The optimization also had seed, temperature, and concentration constraints. The seed number must be less than 9×10^8, which corresponds to 10% of the final yield. The temperature had a preset operating range, final value range,

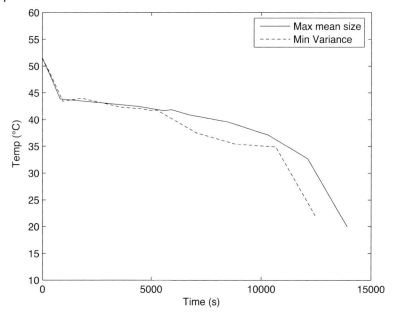

Fig. 8.13 Calculated optimal temperature profiles for two optimization objective functions: maximum mean size and minimum variance.

and cooling and heating rate limits. The final concentration also must be less than a prescribed value. These constraints are listed in the optimization formulation as

$$\underset{N_0, T(t)}{\text{Optimize } \Phi} \begin{cases} 0 \leq N_0 \leq 9 \times 10^8 \text{ No/m}^3 \\ 20\,°\text{C} \leq T(t) \leq 55\,°\text{C} \\ -0.006\,°\text{C/s} \leq \frac{dT(t)}{dt} \leq 0.006\,°\text{C/s} \\ 20\,°\text{C} \leq T_f(t_f) \leq 25\,°\text{C} \\ C_s(t_f) \leq 30 \text{ kg/m}^3 \\ \Psi_{f,\min} \leq \Psi_f(t_f) \leq \Psi_{f,\max} \end{cases} \quad (8.56)$$

The calculated optimal profiles for both objective functions are shown below. The mean crystal size increased to over 600 μm from a seed size of 220 μm for the "maximize mean size" objective function. As seen in Figs. 8.13–8.15, the experimental results match the simulation results for both optimizations.

8.9
Future Outlook

Crystallization modeling continues to advance on several fronts helping improve crystallization batch time, efficiency, and product quality. Improvement in predictive solubility models will allow for the selection of the proper solvent and antisol-

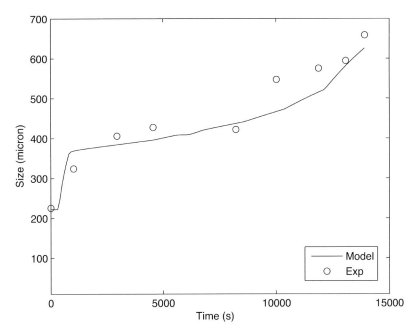

Fig. 8.14 Simulated and experimental volume mean size trajectory for volume mean size optimization objective.

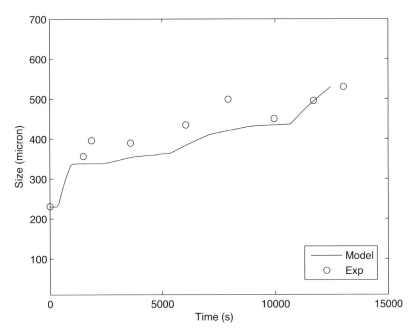

Fig. 8.15 Simulated and experimental volume mean size trajectory for variance optimization objective.

vent for a given process. These predictive models will also lessen the need for solubility data when deriving crystallization models for simulation and optimization [71, 72]. More advanced growth and nucleation equations are always under development. New innovative ways to solve the population balance are being derived. Crystallization models are being combined with state-of-the-art CFD modeling.

The continued development of new process analytical tools allows for *in situ* measurement of crystallization properties that were previously impossible. Novel ways to use online process analytical tools are being found for both parameter estimation and process control [6, 39, 50, 73, 77]. The use of these online process analytical tools in conjugation with advanced models will allow for robust optimal operation of many crystallization processes. This will allow for the rejection of disturbances that is difficult to accomplish with offline recipes. The availability of online image analysis will allow for better understanding and control of particle shape and polymorphism.

It is important that companies keep up to date with current technology and model development. These technologies can be migrated on to the factory floor from the research labs. The technicians can be trained by the researchers on how to interpret the data, and keep the crystallizer operating optimally. However, the researchers must continue to modify and improve the models to keep up with changes in operating conditions, product quality requirements, and technological advances. Thus, for optimal operation, crystallization modelers are required for the implementation, improvement, and maintenance of these crystallization models.

Nomenclature

a_{for} – area of foreign substance (m^2/m^3)
a_{mn} – UNIFAC Interaction Parameter between groups m and n (K^{-1})
aa – MOSCED parameter (dimensionless)
A_{VL} – van Laar binary activity coefficient (dimensionless)
B – discretization parameter (dimensionless)
B – nucleation rate (crystals/s m^3)
B_{attrit} – attrition secondary nucleation (#crystals/m^{3}s)
B_{hom} – homogeneous nucleation (#crystals/m^{3}s)
B_{het} – heterogeneous nucleation (#crystals/m^{3}s)
B_{surf} – surface nucleation (#crystals/m^{3}s)
B_{VL} – van Laar binary activity coefficient (dimensionless)
C – solution concentration (kg/kg solvent)
C_c – crystal molar density (m^3/kmol)
C_{eq} – equilibrium concentration (kg/kg solvent)
C_{eq}^{ref} – reference equilibrium concentration (kg/kg solvent)
d_{ji} – MOSCED parameter (dimensionless)
d_m – molecular diameter (m)
D – diffusion coefficient (m/s)
D_{AB} – diffusion coefficient (m/s)

D_{surf} – surface diffusion (m/s)
E – nucleation effectiveness factor (dimensionless)
E_a – crystal growth energy of activation (J/mol)
F – geometric correction factor (dimensionless)
f_l – fugacity of liquid phase (dimensionless)
f_s – fugacity of solid phase (dimensionless)
G crystal growth exponent (dimensionless)
G – crystal growth (m/s)
G_{ij} – NRTL parameter (dimensionless)
G_{km} – NRTL-SAC parameter (dimensionless)
HE_{AD} – adsorption constant (dimensionless)
H_V – Vicker's hardness (N/m^2)
K – Boltzmann constant (J/K)
k_a – crystal surface shape factor (dimensionless)
k_d – dissolution mass transfer coefficient (m/s)
k_g – crystal growth preexponential factor (m/s (m^3/kmol)$^{1.9}$)
k_v – crystal volumetric shape factor (dimensionless)
K – interfacial tension correlation constant (dimensionless)
l_i – UNIFAC parameter (dimensionless)
L – crystal length (m)
L_{32} – Sauter mean size (m)
L_{43} – volume averaged mean size (m)
L_i – crystal discretization length at the ith interval (mm)
L_0 – crystal nuclei size (mm)
L_{max} – maximum crystal size (mm)
m_s – mass of solvent (kg)
M_w – molecular weight of KCL (kg/kmol)
$n(L,t), n$ – crystal density function (#crystals/m^3)
n_j – discretized number density (#crystals/m^3)
N_A – Avogadro's number (molecules/mol)
$N_{a,\text{eff}}$ – effective attrition fragments (#crystals)
$N_{a,\text{tot}}$ – total attrition fragments (#crystals)
N_V – flow number (dimensionless)
OBJ – optimization objective function (varies)
Po – power number (dimensionless)
POL – MOSCED parameter (dimensionless)
q_i – MOSCED induction parameter (dimensionless)
$q_{i,U}$ – UNIFAC area structural parameter of component i (dimensionless)
Q_k – UNIFAC volume structural parameter of group k (dimensionless)
r_c – nuclei critical radius (m)
r_i – total segment number of component i (dimensionless)
$r_{i,U}$ – UNIFAC volume structural parameter of component i (dimensionless)
$r_{m,i}$ – number of segment species m in component i (dimensionless)
R – gas constant (J/mol K)
R_k – UNIFAC area structural parameter (dimensionless)

S – relative supersaturation (dimensionless)
Sc – Schmidt number (dimensionless)
T – time (s)
T – temperature (K)
T_{melt} – acetaminophen melting temperature (K)
x_i – mole fraction of component i (dimensionless)
$x_{l,i}$ – segment-based mole fraction of segment species l in component i (dimensionless)
x_i – mole fraction of component i (dimensionless)
X_m – UNIFAC mole fraction of group m (dimensionless)
V_m – molecular volume of KCl (m³/particle)
w_i – solute-free antisolvent mass fraction (dimensionless)
Z – UNIFAC coordination number (dimensionless)
A – NRTL nonrandomness parameter (dimensionless)
α_{km} – NRTL-SAC nonrandomness parameter (dimensionless)
α_i – MOSCED hydrogen bond acidity parameter ((J/cm³)$^{0.5}$)
α_i^0 – MOSCED hydrogen bond acidity parameter at 293 K ((J/cm³)$^{0.5}$)
B – MOSCED hydrogen bond basicity parameter ((J/cm³)$^{0.5}$)
β_i^0 – MOSCED hydrogen bond basicity parameter at 293 K ((J/cm³)$^{0.5}$)
γ_i – activity coefficient of component i (dimensionless)
$\gamma_{i,j}^\infty$ – infinite dilution activity coefficient of i in j (dimensionless)
$\gamma_{i,eq}$ – equilibrium activity coefficient of component i in solution (dimensionless)
γ_i^R – residual activity coefficient of component i (dimensionless)
γ_i^C – combinatorial activity coefficient of component i (dimensionless)
γ_{sl} – interfacial tension (J/m²)
Γ_k – UNIFAC residual activity coefficient of group k (dimensionless)
Γ_k^i – UNIFAC residual activity coefficient of group k in component i (dimensionless)
Γ_m^{lc} – NRTL-SAC activity coefficient of segment species m (dimensionless)
$\Gamma_m^{lc,i}$ – NRTL-SAC activity coefficient of segment species m in component i (dimensionless)
Γ/K – fracture resistance (J/m²)
δ_i – ith discretization interval (mm)
Δc – absolute supersaturation (m³/kmol)
ΔC_p – change in heat capacity from liquid to solid phase (J/mol K)
Δg_{ij} – NRTL binary interaction parameter (J/mol K)
ΔH_{fusion} – heat of fusion (kJ/mol)
$\Delta \lambda_{ij}$ – Wilson binary interaction parameter (J/mol K)
$\Delta \mu_{cp}$ – chemical potential difference (J/mol)
$\bar{\varepsilon}$ – mean specific power input (W/kg)
Z – number of discretization intervals (dimensionless)
η_l – kinematic viscosity of the solvent (m²/s)
η_g – geometric target efficiency (dimensionless)
η_w – velocity target efficiency (dimensionless)
θ_i – UNIFAC area fraction of component i (dimensionless)

Θ_m – UNIFAC area fraction of group m (dimensionless)
Λ – MOSCED dispersion parameter $((J/cm^3)^{0.5})$
Λ_{ij} – Wilson parameter (dimensionless)
μ – shear modulus (N/m^2)
μ_i – ith moment ($\mu m^i/m^3$)
V – ion correction factor (dimensionless)
v_i^0 – MOSCED molar volume parameter (cm^3/mol)
v_i – molar volume of component i (m^3/mol)
$v_k^{(i)}$ – number of k UNIFAC groups in component i (dimensionless)
ξ_j – MOSCED parameter (dimensionless)
ρ_c – crystal density (kg/m^3)
T – MOSCED parameter $((J/cm^3)^{0.5})$
τ_j^0 – MOSCED parameter at 293 K $((J/cm^3)^{0.5})$
τ_{ij} – NRTL parameter (dimensionless)
τ_{nm} – NRTL-SAC parameter (dimensionless)
Φ_i – segment mole fraction of component i (dimensionless)
$\Phi_{i,U}$ – UNIFAC volume fraction of component i (dimensionless)
ψ_j – MOSCED parameter (dimensionless)
ψ_{nm} – UNIFAC group interaction parameter between groups n and m (dimensionless)

References

1 ABBAS, A., NOBBS, D., ROMAGNOLI, J. A., Investigation of on-line particle characterization in reaction and cooling crystallization systems. Current state of the art, *Measurement Science and Technology* 13(3) (2002), pp. 349–356.

2 ABBAS, A., ROMAGNOLI, J. A., Modelling environment for the advanced operation of crystallisation processes, process systems engineering 2003, Pts A and B', in: *Computer-Aided Process Engineering*, vol. 15, Chen, B., Westerberg, A. W. (eds.), Elsevier, Kunming, China, pp. 1250–1255, 2003.

3 ABBAS, A., ROMAGNOLI, J. A., DCS implementation of optimal operational policies: a crystallisation case study, *International Journal of Computer Applications in Technology* 25 (2006), pp. 198–208.

4 ABBAS, A., ROMAGNOLI, J. A., Multiscale modeling, simulation and validation of batch cooling crystallization, *Separation and Purification Technology* 53(2) (2007), pp. 153–163.

5 ABBAS, A., SROUR, M., TANG, P., CHIOU, H., CHAN, H. K., ROMAGNOLI, J. A., Sonocrystallisation of sodium chloride particles for inhalation, *Chemical Engineering Science* 62 (2007), pp. 2445–2453.

6 ABU BAKAR, M. R., NAGY, Z. K., SALEEMI, A. N., RIELLY, C. D., The impact of direct nucleation control on crystal size distribution in pharmaceutical crystallization processes, *Crystal Growth and Design* 9(3) (2009), pp. 1378–1384.

7 ALI, M. I., Struvite crystallization in fed-batch pilot scale and description of solution chemistry of struvite, *Chemical Engineering Research and Design* 85(A3) (2007), pp. 344–356.

8 ANGERHOFER, M., Understanding the kinetics of barium sulfate crystallisation, PhD Dissertation, Technical University Munich, **1994**.

9 Aoun, M., Plasari, E., David, R., Villermaux, J., A simultaneous determination of nucleation and growth rates from batch spontaneous precipitation, *Chemical Engineering Science* 54(9) (**1999**), pp. 1161–1180.

10 Baldyga, J., et al., Mixing precipitation model with application to double feed semibatch precipitation, *Chemical Engineering Science* 50 (**1995**), pp. 1281–1300.

11 Barrett, P., Smith, B., Worlitschek, J., Bracken, V., O'Sullivan, B., O'Grady, D., A review of the use of process analytical technology for the understanding and optimization of production batch crystallization processes, *Organic Process Research and Development* 9 (**2005**), pp. 348–355.

12 Beck, J. V., Arnold, K. J., *Parameter Estimation in Engineering and Science*, Wiley, New York, **1977**.

13 Chen, C. C., Song, S., Solubility modeling with a nonrandom two-liquid segment activity coefficient model, *Industrial and Engineering Chemistry Research* 43 (**2004**), pp. 8354–8362.

14 Chen, C. C., Crafts, P. A., Correlation and prediction of drug molecule solubility in mixed solvent systems with the nonrandom two-liquid segment activity coefficient (NRTL-SAC) model, *Industrial and Engineering Chemistry Research* 45 (**2006**), pp. 4816–4824.

15 Christofides, P. D., El-Farrac, N., Li, M., Mhaskar, P., Model-based control of particulate processes, *Chemical Engineering Science* 63(5) (**2008**), pp. 1156–1172.

16 Chung, S. H., Ma, D. L., Braatz, R. D., Optimal model-based experimental design in batch crystallization, *Chemometrics and Intelligent Laboratory Systems* 50 (**2000**), pp. 83–90.

17 Cornel, J., Lindenberg, C., Mazzotti, M., Quantitative application of *in situ* ATR-FTIR and Raman spectroscopy in crystallization processes, *Industrial and Engineering Chemistry Research* 47 (**2008**), pp. 4870–4882.

18 Dirksen, J. A., Ring, T. A., Fundamentals of crystallisation kinetic effects on particle size distribution and morphology, *Chemical Engineering Science* 46 (**1991**), pp. 2389–2427.

19 Doki, N., Kubota, N., Yokota, M., Kimura, S., Sasaki, S., Production of sodium chloride crystals of uni-modal size distribution by batch dilution crystallization, *Journal of Chemical Engineering of Japan* 35 (**2002**), pp. 1099–1104.

20 Eek, R. A., Control and dynamic modeling of industrial suspension crystallizers, PhD Dissertation, Technical University of Delft, The Netherlands, **1995**.

21 Feng, L., Berglund, K. A., ATR-FTIR for determining optimal cooling curves for batch crystallization of succinic acid, *Crystal Growth and Design* 2(5) (**2002**), pp. 449–452.

22 Fitchett, D. E., Tarbell, J. M., Effect of mixing on the precipitation of barium sulphate in an MSMPR reactor, *AIChE Journal* 36 (**1990**), pp. 511–522.

23 Fredenslund, A., Jones, R. L., Prausnitz, J. M., Group-contribution estimation of activity coefficients in nonideal liquid mixtures, *AIChE Journal* 21(6) (**1975**), pp. 1086–1099.

24 Gunn, D. J., Murthy, M. S., Kinetics and mechanisms of precipitations, *Chemical Engineering Science* 27 (**1972**), pp. 1293–1313.

25 Hu, A., Rohani, S., Jutan, A., Modelling and optimization of seeded batch crystallizers, *Computers and Chemical Engineering* 29 (**2004**), pp. 911–918.

26 Hu, Q., Rohani, S., Wang, D. X., Jutan, A., Nonlinear kinetic parameter estimation for batch cooling seeded crystallization, *AIChE Journal* 50(8) (**2004**), pp. 1786–1794.

27 Hulbert, H. M., Katz, S., Some problems in particle technology: a statistical mechanical formulation, *Chemical Engineering Science* 9 (**1964**), pp. 555–574.

28 Jagadesh, D., Kubota, N., Yokota, M., Sato, A., Tavare, N. S., Large and monosized product crystals from natural cooling mode batch crystallizer, *Journal of Chemical Engineering of Japan* 29(5) (**1996**), pp. 865–873.

29 Jagadesh, D., Kubota, N., Yokota, M., Doki, N., Sato, A., Seeding effect on batch crystallization of potassium sulphate under natural cooling mode and a simple method of crystallizer, *Journal of Chemical Engineering of Japan* 32(4) (**1999**), pp. 514–520.

30 Jouyban, A., Chan, H. K., Chew, N. Y. K., Khoubnasabjafari, M., Acree, W. E.,

Jr., Solubility prediction of paracetamol in binary and ternary solvent mixtures using Jouyban–Acree model, *Chemical and Pharmaceutical Bulletin* 54(4) (**2006**), pp. 428–431.

31 KILPIO, T., NORDEN, H. V., Simulation and control of continuous crystallisation, *Acta Polytechnica Scandinavica*, vol ch. 235 (**1996**), pp. 1–49.

32 KIM, S., WEI, C., KIANG, S., Crystallization process development of an active pharmaceutical ingredient and particle engineering via the use of ultrasonics and temperature cycling, *Organic Process Research and Development* 7 (**2003**), pp. 997–1001.

33 KOUGOULOS, E., JONES, A. G., WOOD-KACZMAR, M. W., Process modelling tools for continuous and batch organic crystallization processes including application to scale-up, *Organic Process Research and Development* 10(4) (**2006**), pp. 739–750.

34 KRAMER, H. J. M., BERMINGHAM, S. K., VAN ROSMALEN, G. M., Design of industrial crystallisers for a given product quality, *Journal of Crystal Growth* 199(1) (**1999**), pp. 729–737.

35 KUBOTA, N., DOKI, N., YOKOTA, M., SATO, A., Seeding policy in batch cooling crystallization, *Powder Technology* 121 (**2001**), pp. 31–38.

36 LAZZARONI, M., BUSH, D., ECKERT, C., Revision of MOSCED parameters and extension to solid solubility calculations, *Industrial and Engineering Chemistry Research* 44 (**2005**), pp. 4075–4083.

37 LEONARD, T., HSU, J. S. J., *Bayesian Methods: An Analysis for Statisticians and Interdisciplinary Researchers*, Cambridge University Press, Cambridge, UK, **1999**.

38 LI, M., WILKINSON, D., PATCHIGOLLA, K., Comparison of particle size distributions measured using different techniques, *Particulate Science and Technology* 23 (**2005**), pp. 265–284.

39 LINDENBERG, C., KRATTLI, M., CORNEL, J., MAZZOTTI, M., BROZIO, J., Design and optimization of a combined cooling/antisolvent process, *Crystal Growth and Design* 9(2) (**2009**), pp. 1124–1136.

40 LIOTTA, V., SABESAN, V., Monitoring and feedback control of supersaturation using ATR-FTIR to produce an active pharmaceutical ingredient of a desired crystal size, *Organic Process Research and Development* 8 (**2004**), pp. 488–494.

41 LIU, Y., CAMERON, I. T., A new wavelet-based method for the solution of the population balance equation, *Chemical Engineering Science* 56(18) (**2001**), pp. 5283–5294.

42 MERSMAN, A., *Crystallization Technology Handbook*, 2nd edn., Marcel Dekker, New York, **2001**.

43 MIGNON, D., MANTH, T., OFFERMANN, H., Kinetic modelling of batch precipitation reactions, *Chemical Engineering Science* 51 (**1996**), pp. 2565–2570.

44 MULLIN, J. W., *Crystallisation*, 3rd edn., Butterworth-Heinemann, London, **1993**.

45 NIELSEN, A. E., *Kinetics of Precipitation*, Pergamon, Oxford, **1964**.

46 NIELSEN, A. E., Nucleation and growth of crystals at high supersaturation, *Krist. Technik* 4 (**1969**), pp. 17–38.

47 NIELSEN, A. E., TOFT, J. M., Electrolyte crystal growth kinetics, *Journal of Crystal Growth* 67(2) (**1984**), pp. 278–288.

48 NAGY, Z. K., BRAATZ, R. D., Open-loop and closed-loop robust optimal control of batch processes, *Journal of Process Control* 14(4) (**2004**), pp. 411–422.

49 NAGY, Z. K., FUJIWARA, M., WOO, X. Y., BRAATZ, R. D., Determination of the kinetic parameters for the crystallization of paracetamol from water using metastable zone width experiments, *Industrial and Engineering Chemistry Research* 47 (**2008**), pp. 1245–1252.

50 NAGY, Z. K., FUJIWARA, M., BRAATZ, R. D., Modelling and control of combined cooling and antisolvent crystallization processes, *Journal of Process Control* 18 (**2008**), pp. 856–864.

51 NAGY, Z. K., Model based robust control approach for batch crystallization product design, *Computers and Chemical Engineering* 33(10) (**2009**), pp. 1685–1691.

52 NOWEE, S. M., ABBAS, A., ROMAGNOLI, J. A., YEO, P., Optimization in seeded cooling crystallization: a parameter estimation and dynamic optimization study, *Chemical Engineering and Processing* 46 (**2007**), pp. 1096–1106.

53 NOWEE, S. M., ABBAS, A., ROMAGNOLI, J. A., Antisolvent crystallization: model identification, experimental validation and dynamic simulation, *Chemical Engineering Science* 63 (**2008**), pp. 5457–5467.

54 Nowee, S. M., Abbas, A., Romagnoli, J. A., Model-based optimal strategies for controlling particle size in antisolvent crystallization operations, *Crystal Growth and Design* 8(8) (**2008**), pp. 2698–2706.

55 Pamplin, B. R., *Crystal Growth*, Pergamon Press, Oxford, **1975**.

56 Parisi, M., Terranova, A., Chianese, A., Pilot plant investigation on the kinetics of dextrose cooling crystallization, *Industrial and Engineering Chemistry Research* 46(4) (**2007**), pp. 1277–1285.

57 Poling, B. E., Prausnitz, J. M., O'Connell, J. P., *The Properties of Gases and Liquids*, 5th edn., McGraw-Hill, New York, **2000**.

58 Qamar, S., Elsner, M. P., Angelov, I. A., Warnecke, G., Seidel-Morgenstern, A., A comparative study of high resolution schemes for solving population balances in crystallization, *Computers and Chemical Engineering* 30 (**2006**), pp. 1119–1131.

59 PSE Ltd., *gPROMS Advanced User Guide – Release 2.1*, Process Systems Enterprise Ltd., London, **2001**.

60 Randolph, A. D., Larson, M. A., *Theory of Particulate Processes*, 2nd edn., Academic Press, San Diego, **1998**.

61 Rawlings, J. B., Miller, S. M., Witkowski, W. R., Model identification and control of solution crystallisation processes: a review, *Industrial and Engineering Chemistry Research* 32 (**1993**), pp. 1275–1296.

62 Sarkar, D., Rohani, S., Jutan, A., Multi-objective optimization of seeded batch crystallizations, *Chemical Engineering Science* 61 (**2006**), pp. 5282–5295.

63 Sheikhzadeh, M., Trifkovic, M., Rohani, S., Real-time optimal control of an anti-solvent isothermal semi-batch crystallization process, *Chemical Engineering Science* 63 (**2008**), pp. 829–839.

64 Tadayon, A., Rohani, S., Bennett, M. K., Estimation of nucleation and growth kinetics of ammonium sulphate from transients of a cooling batch seeded crystallizer, *Industrial and Engineering Chemistry Research* 41 (**2002**), pp. 6181–6193.

65 Tavare, N. S., Ammonium sulfate crystallization in a cooling batch crystaliser, *Separation Science and Technology* 27(11) (**1992**), pp. 1469–1487.

66 Thompson, D. R., Kougoulos, E., Jones, A. G., Wood-Kaczmar, M. W., Solute concentration measurement of an important organic compound using ATR-UV spectroscopy, *Journal of Crystal Growth* 276 (**2005**), pp. 230–236.

67 Togkalidou, T., Braatz, R. D., Johnson, B. K., Davidson, O., Andrews, A., Experimental design and inferential modeling in pharmaceutical crystallization, *AIChE Journal* 47(1) (**2001**), pp. 160–168.

68 Trifkovic, M., Sheikhzadeh, M., Rohani, S., Kinetics estimation and single and multi-objective optimization of a seeded anti-solvent, isothermal batch crystallizer, *Industrial and Engineering Chemistry Research* 47 (**2008**), pp. 1586–1595.

69 Van der Leeden, et al., Induction time in seeded and unseeded precipitation, in: *Advances in Industrial Crystallization*, Garside et al. (eds), Butterworth-Heinemann, Oxford, **1991**, pp. 31–46.

70 Van Leeuwen, M. L. J., Bruinsma, O. S. L., Van Rosmalen, G. M., 3-Zone approach for precipitation of barium sulfate, Paper No. O402.02, in: *XI International Conference on Crystal Growth*, The Netherlands, **1995**.

71 Widenski, D., Abbas, A., Romagnoli, J., Effect of the solubility model on antisolvent crystallization predicted volume mean size, *Chemical Engineering Transactions* 17 (**2009**), pp. 639–644.

72 Widenski, D., Abbas, A., Romagnoli, J., Evaluation of the effect of the solubility model on antisolvent crystallization optimization, in: *Proceedings of the International Symposium on Advanced Control of Chemical Processes*, Istanbul, Turkey, **2009**, pp. 212–217.

73 Woo, X. J., Nagy, Z. K., Tan, R. B. H., Braatz, R. D., Adaptive concentration control of cooling and antisolvent crystallization with laser backscattering measurement, *Crystal Growth and Design* 9(1) (**2009**), pp. 182–191.

74 Worlitschek, J., Mazzotti, M., Model-based optimization of particle size distribution in batch-cooling crystallization of paracetamol, *Crystal Growth and Design* 4 (**2004**), pp. 891–903.

75 Xie, W., Rohani, S., Phoenix, A., Dynamic modeling and operation of a seeded batch crystallizer, *Chemical Engineering Communications* 187 (**2001**), pp. 229–249.
76 Yokota, M., Sato, A., Kubota, N., New effective-nuclei concept for simplified analysis of batch crystallization, *AIChE Journal* 45(9) (**1999**), pp. 1883–1891.
77 Zhou, G. X., Fujiwara, M., Woo, X. Y., Rusli, E., Tung, H. H., Starbuck, C., Davidson, O., Ge, Z., Braatz, R. D., Direct design of pharmaceutical antisolvent crystallization through concentration control, *Crystal Growth and Design* 6(4) (**2006**), pp. 892–898.
78 Zumstein, R. C., Rousseau, R. W., Utilisation of industrial data in the development of a model for crystallizer simulation, *AIChE Symposium Series* 83(253) (**1987**), p. 130.

9
Modeling Multistage Flash Desalination Process – Current Status and Future Development
Iqbal M. Mujtaba

Keywords
freshwater demand, desalination, multistage flash (MSF) process, model-based techniques, desalination plant, brine heater fouling factor

9.1
Introduction

Water is essential to all living species on earth and quality water must be available in abundance to all species. Potable water supply is becoming a limited resource nowadays due to the increase in size and improved standard of living of communities. In addition, agriculture and industry require sustainable water supplies throughout the world [57]. At the beginning of 2000, about 1.1 billion people, one-sixth of the world's population had no access to improved water supply. By the year 2015, it is estimated that an additional 1.5 billion people will need access to water supply. The disputes over water will inevitably become more common, as 220 river basins globally are shared by two or more countries [59]. As the price of imported water goes up, many countries (e.g., Singapore) dependent on the potable water supply from other countries (e.g., Malaysia) would like to turn to new technology to make potable water from reclaimed water [61].

Global water shortages will become so catastrophic over the next decades that two in three people on the planet will face regular depletion of water supplies. Global thirst will turn millions into water refugees. Water refugees are likely to become commonplace leading to hydrological poverty. Millions of villagers in India, China, and Mexico may have to move because of lack of water [60].

Increase in population (Fig. 9.1) and standards of living (together with water pollution) are diminishing the quantity of naturally available freshwater while the demand for it is increasing continuously. Freshwater consumption is increasing at the rate of 4–8% per year, 2.5 times the population growth [27].

About 70% of the earth is covered in water. Out of 1.4 billion km^3 of the total amount of water, 97.5% is saltwater and 2.5% is freshwater. Of the available freshwater, 70% is locked up in the form of glacial ice, permafrost, or permanent snow. Groundwater and soil moisture account for about 29%. Freshwater lakes

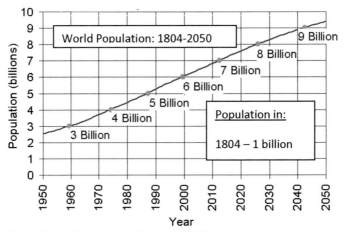

Fig. 9.1 World population 1804–2050. Adopted from El-Dessouky and Ettouney [13].

and marshlands hold about 0.99%. Rivers, the most visible form of freshwater, account for about 0.01% of all forms of freshwater [source: Environment Canada: http://www.ec.gc.ca/WATER].

As more than 94% of the world's water is saline, desalination technology is vital for sustaining human habitation (including agriculture and industry) in many parts of today's world [8]. London's water supply is so fragile and its population is growing so fast that without the desalination plant, regeneration in the city could be undermined [39], http://www.guardian.co.uk/environment/2006/may/24/water.uknews. The commonly used industrial desalination processes can be classified broadly into two groups: (a) heat-consuming or thermal processes and (b) power-consuming or membrane processes; thermal process being the oldest and most dominating for large-scale production of freshwater in today's world (Fig. 9.2). Today, multistage flash (MSF) process is the largest sector (about 50%) in desalination. The process capacity has increased from 500 (1950) to 75 000 m^3/day [32].

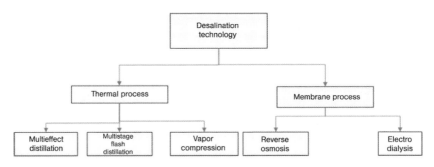

Fig. 9.2 Different types of desalination processes.

While there is no shortage of research work in desalination for the last few decades, the ongoing objective has still been to improve the design, operation, and control of desalination processes (mainly thermal) to ensure quality water at cheaper price with lower environmental impact. While membrane-based processes are becoming cheaper with new development of membranes, thermal processes are still very energy intensive (energy being supplied by steam) and are cause of concern for environment. In addition, with seasonal seawater temperature variation the common industrial practice is to operate the thermal-based plants at high temperature in summer. This leads to the use of increased amount of antiscalant to reduce fouling and corrosion of heat exchangers (and plant equipment) [32]. Although this reduces frequent shutdown of the plant but causes further environmental problem associated with the added chemicals.

Model-based techniques are extensively used in process systems engineering activities for optimal design, operation, control, and even for designing experiments. This is due to the fact that model-based techniques are less expensive compared to any experimental investigation for finding best design, operation, control, etc. The yearly event of *European Symposium on Computer Aided Process Engineering* (since 1991 and the Proceedings published by Elsevier) and 3-yearly event of *International Symposium on Process Systems Engineering* (since 1985) and the *Computers and Chemical Engineering Journal* (since 1979 by Elsevier) cover design, operation, control, process integration of many processes but desalination processes (very limited such as [21, 54, 55, 58, 62]). Interestingly, most reported literatures on desalination are experimentally based and are mostly published in *Desalination Journal* (since 1966 by Elsevier).

There are only few published works dealing with rigorous mathematical modeling, mathematical optimization, and model-based control on MSF desalination processes (since 1957) [1, 2, 5, 10–14, 21, 22, 24–27, 29, 34, 49] (are the major ones in the last 45 years). However, the main focus of many of these works was to develop tailor-made solution algorithms with simplifying assumptions and was not to exploit the full potential of the model-based techniques by utilizing many available commercial process modeling (and optimization) tools such as gPROMS [18].

With this in mind, this chapter highlights the state-of-the-art in mathematical modeling and future challenges for the computer-aided process engineering (CAPE), process systems engineering community, and the practitioners of desalination to address sustainable freshwater issue of tomorrow's world via desalination using model-based techniques.

9.2
Issues in MSF Desalination Process

Figures 9.3 and 9.4 show a typical MSF desalination process with detailed features of stages. An MSF process consists of three main sections – brine heater, recovery section with NR stages (flash chambers), and a rejection section with NJ stages. Seawater enters into the last stage of the rejection stages and passes through a

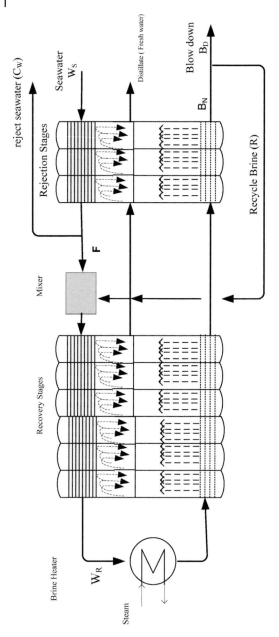

Fig. 9.3 A typical MSF desalination process [20].

series of tubes to remove heat from the stages. Before the rejection section, seawater is partly discharged to the sea to balance the heat. The other part is mixed with the recycled brine from the last stage of the rejection section and fed before last stage of the recovery section. Seawater is flowing through the tubes in differ-

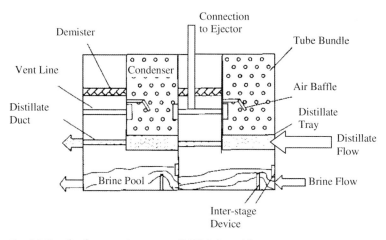

Fig. 9.4 Details of consecutive stages of MSF (adopted from [13]).

ent stages to recover heat from the stages and the brine heater raises the seawater temperature to the maximum attainable temperature (also known as top brine temperature, TBT or T_{B0}). After that, it enters into the first flashing stage and produces flashing vapor. This process continues until the last stage of the rejection section. The concentrated brine from the last stage is partly discharged into the sea and the remaining is recycled as mentioned before. Note, many alternative configurations are possible depending on the way the seawater is fed and brine is recycled [13, 14, 50].

With the basis of given *freshwater production rate, seawater composition*, and *temperature*, the main issues in an MSF process are to determine the following:

- *Design parameters*: Number of stages, width and height of the stages, heat transfer area (number of tubes in the condensers), materials of construction, vent line orifice (air and noncondensable), demister size and materials, interstage brine transfer device, brine heater area, etc.
- *Operation parameters*: Steam flow, TBT, brine recycle rate, and seawater rejection rate.
- *Cost*: Capital, operating (utilities, cleaning), pretreatment, and posttreatment (chemicals).

The purpose of any study (experimental or model based; past, current, and future) is to optimize design and operating parameters so as to either maximize profitability of operation or minimization of cost or minimization of external energy input or maximizing the recovery ratio, maximizing the plant performance (GOR – gained output ratio) ratio, etc. Economically and environmentally, the net energy consumption and added chemicals to reduce heat exchangers scaling are major issues in MSF processes. Development of economically and environmentally efficient desalination processes is the future challenge, which can be achieved by better design, operation, and control, and model-based techniques can play a significant role in this.

9.3
State-of-the-Art in Steady-State Modeling of MSF Desalination Process

Flash distillation existed from the beginning of the 19th century. Office of Saline Water, USA, was established in 1952 to develop economical flash distillation based desalination process [9]. First patent of MSF process for desalination was obtained by Silver in 1957 [46]. Table 9.1 shows the journal and the year of publication of MSF processes for the first time. Although London-based journal *Engineering* had been publishing technical papers on desalination processes since 1958, the dedicated journal *Desalination* was created in 1966.

As can be seen from Table 9.1, the model-based technique in MSF desalination was first applied in 1970. Referring to Figs. 9.3 and 9.4, an MSF process model should include an accurate description (via mathematical equations) of (i) mass/material balance, (ii) energy balance, (iii) thermal efficiency, (iv) physical properties (such as heat capacity, density, boiling point temperature elevation (TE) due to salinity, heat of vaporization), (v) heat transfer coefficients (taking into account fouling, noncondensable gases (NCGs)), (vi) pressure and temperature drop, (vii) geometry of brine heater, demister, condenser, stages, vents, (viii) interstage flow (orifice), (ix) thermodynamic losses including the nonequilibrium allowance and demister losses, and (x) kinetic model for salt deposition and corrosion.

Table 9.2 describes the evolution of steady-state MSF process models over the last half century (included only the major developments published in international journals).

A typical MSF process model includes mass and energy balances, the geometry of the stages, and physical properties which are functions of temperature and salinity. With reference to Figs. 9.3 and 9.5, the model equations used by Helal *et al.* [21], and Rosso *et al.* [42] are given below.

The following assumptions are made in the model:

- the distillate from any stage is salt free;
- heats of mixing are negligible;
- no subcooling of condensate leaving the brine heater;
- there are no heat losses; and
- there is no entrainment of mist by the flashed vapor.

The model equations for stage number j (Fig. 9.5) are given in the following (some symbols are defined in the nomenclature and the rest can be found in [21] and [42]).

(a) *Stage model.*

Mass balance in the flash chamber:

$$B_{j-1} = B_j - V_j \tag{9.1}$$

Salt balance:

$$B_{j-1} \times CB_{j-1} = B_j \times CB_j \tag{9.2}$$

Table 9.1 First appearance of MSF desalination process in different journals.

Journal, Yr	Author	Task	Model based
Engineering, 1958	Silver	Design calculations	No
Desalination, 1966	Clelland & Stewart	Design and optimization	No
Industrial & Engineering Chemistry, 1967	Cadwallader	Scaling issues	No
Chemical Engineering Science, 1970	Mandil & Ghafour	Optimization	Yes
Computers & Chemical Engineering, 1986	Helal *et al.*	Modeling and simulation	Yes

Table 9.2 Steady-state models for MSF since 1970.

References	Type/description of model
[29]	Approximate Lumped Parameter Model, constant physical properties (independent of seawater composition and temperature), constant heat transfer coefficients (HTC), constant stage temperature drop
[11]	Simple but Stage to Stage Model, constant specific heat capacity of feed water in condenser, linear and simplified TE (boiling point TE) correlation for each temperature range (<112 °C, 112–168 °C, >168 °C), very high temperature operation (steam temperature 268 °C), seawater temperature, 38 °C, constant HTC in condensers, no fouling/scaling, model equations linearized, reformulated for easy sequential or iterative solution
[6]	Model similar to Coleman [11], unidirectional information flow by state inversion [65], mathematical information flow is opposite to physical flow thus allowing stage-by-stage calculation independently from cold end to the hold end, brine heater considered to be last (mathematical) stage
[21]	Detailed Stage to Stage Model, nonlinear TE correlation and other physical properties as the function of (temperature, seawater composition), temperature loss due to demister included, HTC via polynomial fit (fouling included), very high temperature operation (steam temperature 174 °C), TBT = 162.7 °C, seawater temperature, 63 °C, model equations linearized, reformulated to be solved by well-known iterative TDM solution method (developed for distillation models in late sixties, Wang and Henke [64]; Naphthali and Sandholm [35])
[4]	Model similar to Helal *et al.* but developed orthogonal collocation based solution techniques
[42]	Model and solution techniques similar to Helal *et al.* but carried out different simulation studies (see Table 9.7)
[58]	Configuration slightly different than Helal *et al.* but solution techniques similar. Model validation with plant data
[15]	Model based on Helal *et al.* [21] but included: heat losses to the surroundings, constant inside/outside tube fouling factors, pressure drop across demister, number of tubes in the condenser and tube material, constant nonequilibrium allowance (measure of stage thermal efficiency), stage-by-stage calculation based on El-Dessouky and Ettouney [13]
[12]	Model similar to Helal *et al.* [21] and El-Dessouky *et al.* [15]. Developed iteration-based solution algorithm to solve nonlinear algebraic equations by reordering them into three subsets
[48, 52, 53, 56]	Model based on Helal *et al.* [21] but included NN (neural network) based correlation for TE calculation. Modeled and solved using equation oriented mathematical software gPROMS [18]

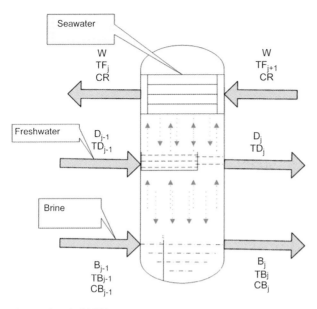

Fig. 9.5 A typical MSF stage.

Mass balance for the distillate tray:

$$D_j = D_{j-1} - V_j \tag{9.3}$$

Enthalpy balance on flash brine:

$$B_j = (h_{Bj-1} - h_{vj})/(h_{Bj} - h_{vj}) \times B_{j-1} \tag{9.4}$$

$$h_{vj} = f(T_{Sj}) \tag{9.5}$$

$$h_{Bj} = f(C_{Bj}, T_{Bj}) \tag{9.6}$$

Overall enthalpy balance:

$$\begin{aligned}
W_R &\times S_{Rj} \times (T_{Fj} - T_{Fj+1}) \\
&= D_{j-1} S_{Dj-1} \times (T_{Dj-1} - T^*) + B_{j-1} S_{Bj-1} \\
&\quad \times (T_{Bj-1} - T^*) - D_j S_{Dj} \times (T_{Dj} - T^*) \\
&\quad - B_j S_{Bj} \times (T_{Bj} - T^*) \quad \text{(recovery stage)}
\end{aligned} \tag{9.7}$$

$$\begin{aligned}
W_s &\times S_{Rj} \times (T_{Fj} - T_{Fj+1}) \\
&= D_{j-1} S_{Dj-1} \times (T_{Dj-1} - T^*) + B_{j-1} S_{Bj-1} \\
&\quad \times (T_{Bj-1} - T^*) - D_j S_{Dj} \times (T_{Dj} - T^*) \\
&\quad - B_j S_{Bj} \times (T_{Bj} - T^*) \quad \text{(rejection stage)}
\end{aligned} \tag{9.8}$$

$$S_{Rj} = f(T_{Fj+1}, T_{Fj}, C_R) \quad \text{(recovery stages)} \tag{9.9}$$

$$S_{Rj} = f(T_{Fj+1}, T_{Fj}, C_R) \quad \text{(rejection stages)} \tag{9.10}$$

$$S_{Dj} = f(T_{Dj}) \tag{9.11}$$

$$S_{Bj} = f(T_{Bj}, C_{Bj}) \tag{9.12}$$

Heat transfer equation:

$$W_R \times S_{Rj} \times (T_{Fj} - T_{Fj+1})$$
$$= \frac{U_j \times A_j \times (T_{Fj} - T_{Fj+1})}{\ln \frac{(T_{Dj} - T_{Fj+1})}{(T_{Dj} - T_{Fj})}} \quad \text{(recovery stage)} \tag{9.13}$$

$$W_s \times S_{Rj} \times (T_{Fj} - T_{Fj+1})$$
$$= \frac{U_j \times A_j \times (T_{Fj} - T_{Fj+1})}{\ln \frac{(T_{Dj} - T_{Fj+1})}{(T_{Dj} - T_{Fj})}} \quad \text{(recovery stage)} \tag{9.14}$$

Overall heat transfer coefficient:

$$U_j = f\left(W_R, T_{Fj}, T_{Fj+1}, T_{Dj}, D_j^i, D_j^o, L_j^i, f_j^i\right) \quad \text{(recovery stage)} \tag{9.15}$$

$$U_j = f\left(W_s, T_{Fj}, T_{Fj+1}, T_{Dj}, D_j^i, D_j^o, L_j^i, f_j^i\right) \quad \text{(recovery stage)} \tag{9.16}$$

Distillate and flashing brine temperature correlation:

$$T_{Bj} = T_{Dj} + TE_j + EX_j + \Delta_j \tag{9.17}$$

Distillate and flashed steam temperatures correlation:

$$TS_j = T_{Dj} + \Delta_j \tag{9.18}$$

$$TE_j = f(T_{Dj}, C_{Bj}) \tag{9.19}$$

$$\Delta_j = f(T_{Dj}) \tag{9.20}$$

$$EX_j = f(H_j, w_j, T_{Bj}) \tag{9.21}$$

(b) *Brine heater model.*

$$B_0 = W_R \tag{9.22}$$

$$C_{B0} = C_R \tag{9.23}$$

Overall enthalpy balance:

$$B_0 \times S_{RH} \times (T_{B0} - T_{F1}) = W_{steam} \times \lambda_S \tag{9.24}$$

$$\lambda_S = f(T_{steam}) \tag{9.25}$$

Heat transfer equation:

$$W_R \times S_{RH} \times (T_{B0} - T_{F1}) = \frac{U_H \times A_H \times (T_{B0} - T_{F1})}{\ln \frac{(T_{steam} - T_{F1})}{(T_{steam} - T_{B0})}} \tag{9.26}$$

$$U_H = f\left(W_R, T_{B0}, T_{F1}, T_{steam}, D_H^i, D_H^o, f_H^i\right) \tag{9.27}$$

$$S_{RH} = f(T_{B0}, T_{F1}) \tag{9.28}$$

(c) *Splitters model.*

Blowdown splitter:

$$B_D = B_{NS} - R \tag{9.29}$$

Reject seawater splitter:

$$C_W = W_S - F \tag{9.30}$$

(d) *Makeup mixers model.*

Mass balance:

$$W_R = R + F \tag{9.31}$$

Salt balance:

$$R \times C_{BNS} + F \times C_S = W_R \times C_R \tag{9.32}$$

Overall enthalpy balance:

$$W_R \times h_W = R \times h_R + F \times h_F \tag{9.33}$$

$$h_W = f(T_{Fm}, C_R) \tag{9.34}$$

$$h_F = f(T_{FNR+1}, C_F) \tag{9.35}$$

$$h_R = f(T_{BNS}, C_{BNS}) \tag{9.36}$$

Note, T^* in Eqs. (9.8) and (9.9) refers to a reference temperature $= 0\,°C$. Simulation results will vary slightly if a different reference temperature is used.

Table 9.3 Physical and chemical properties equations.

Specific enthalpy of saturated water
$h_D = 1.8 \times (-31.92 + 1.0011833 T_{steam}$
$- 3.0833326 \times 10^{-5} \times T^2_{steam}$
$+ 4.666663 \times 10^{-8} \times T^3_{steam}$
$+ 3.1333334 \times 10^{-10} \times T^4_{steam})$.

Specific enthalpy of vapor
$h_{vj} = 596.912 + 0.46694 \times T_s$
$- 0.000460256 \times T^2_s$.
For brine heater,
$h_v = h_{vj}$, $T_s = T_{steam}$ in F.
Latent heat of vapor: $\lambda_s = h_v - h_D$.

Specific heat capacity of seawater/brine
$S_{Bj} = [1 - C_B \times (0.011311 - 1.146 \times 10^{-5} T_B)] \times S_d$.
For $S_{Rj} = S_{Bj}$,
$C_B = C_S$ in wt%, $C_B = C_R$ in wt%, $T_B = T_{Fj+1}$ in F.
For brine heater, $S_{RH} = S_{Bj}$, $C_B = C_R$ in wt%, $T_B = T_{F0}$ in F.

Specific heat capacity of pure water
$S_{Dj} = 1.0011833 - 6.1666652 \times 10^{-5} T_D + 1.3999989 \times 10^7 \times T^2_D + 1.3333336 \times 10^{-9} \times T^3_D$.
For brine, seawater $S_{RH} = S_{Dj}$, $T_D = T_{Bj}$ in F.

Specific enthalpy of seawater/brine
$h_{Bj} = 4.186 * ((4.185 - 5.381 \times 10^{-3} \times C_j + 6.26 \times 10^{-6} \times C^2_j) \times T_{Bj} - (3.055 \times 10^{-5}$
$+ 2.774 \times 10^{-6} \times C_j - 4.318 \times 10^{-8} \times C^2_j) \times T^2_{Bj}/2 + (8.844 \times 10^{-7}$
$+ 6.527 \times 10^{-8} \times C_j - 4.003 \times 10^{-10} * C^2_j)/T^3_{Bj}/3)$
For mixer, $h_W = h_{Bj}$, $T_{Bj} = T_{Fm}$, $C_{Bj} = C_R$, $h_F = h_{Bj}$, $T_{Bj} = T_{FNR}$, $C_{Bj} = C_R$,
$h_R = h_{Bj}$, $T_{Bj} = T_{BN}$, $C_{Bj} = C_{BNS}$.
Density of brine $\rho_j = 16.01846 \times (62.707172 + 49.364088 \times C_{Bj} - 0.43955304 \times 10^{-2} \times T_B$
$- 0.032554667 \times C_{Bj} \times T_B - 0.46076921 \times 10^{-4} \times T^2_B + 0.63240299 \times 10^{-4} \times C_B \times T^2_B)$.

Temperature loss due to demister and nonequilibrium
$\Delta_j = \exp(1.885 - 0.02063 \times T_D)/1.8$,
$EX_j = (195.556 \times H^{1.1}_j (\varpi_j \times 10^{-3})^{0.5})/(\Delta T^{2.5}_{Bj} \times T^{2.5}_{Sj})$,
$\omega_j = W/w$, $\Delta T_{Bj} = (T_{Bj} - T_{Bj-1})$ in F, where $W = W_R$ (recovery stage) in lb/h,
$W = W_S$ (reject stage) in lb/h = (kg/h).

Overall heat transfer coefficient
$U_j = \dfrac{4.8857}{(y+z+4.8857 \times f^i_j)}$, where $y = [(v \times D^i)^{0.2} \times [(160 + 1.92 \times T_F) \times v]$,
$v = f(L_j, W_R/W_S, \rho, D^i, D^o_j,)$, $z = 0.1024768 \times 10^{-2} - 0.7473939 \times 10^{-5} \times T_D$
$+ 0.999077 \times 10^{-7} \times T^2_D - 0.430046 \times 10^{-9} \times T^3_D + 0.6206744 \times 10^{-12} \times T^4_{Dj}$.
For brine heater, $U_H = U_j$, $T_F = T_{B0}$ in F, $T_{steam} = T_D$ in F,
$f^i_j = f^i_H$, $D^i_j = D^i_H$, $L_j = L_H$, $D^i = D^i_H$, $D^o_j = D^o_H$.

The correlations for physical and chemical properties (except the calculation of TE) are shown in Table 9.3. Adequate knowledge of the total heat transfer area, the length of the flash chamber, are needed for modeling, design, and scale up of MSF processes. These parameters are dependent on/interrelated with TBT [51]. Also accurate estimation of TE due to salinity is important in developing a reliable process model. Several correlations for estimating the TE exist in the literature (Fig. 9.6). In addition, Tanvir and Mujtaba [52] developed several neural network

Correlation 1: Bromley et al. [7]; experimental data source: Bromley et al. [7]
$TE = 13832.0 \times x \times BPT^2 \times \begin{pmatrix} 1+0.001373 \times BPT - 0.00272 \times BPT \times \sqrt{\dfrac{x}{100}} + \dfrac{17.86 \times x}{100} - \\ \dfrac{0.0152 \times x}{100 \times BPT} \times \dfrac{(BPT-225.9)}{(BPT-236)} - \dfrac{2583 \times x}{100 \times BPT} \times \left(1-\dfrac{x}{100}\right) \end{pmatrix}.$
where, *BPT* = boiling point of pure water at °K and x is in weight percent
Correlation 2: Helal et al. [21]
$TE = \begin{pmatrix} \left(\dfrac{565.757}{T} - 9.81559 + 1.54739 \times \ln T\right) - (337.178/T - 6.41981 + 0.922753 \ln T) \times C \\ + \left(\dfrac{32.681}{T} - 0.55368 + 0.079022 \times \ln T\right) \times C^2 \end{pmatrix}$ $\times \left(\dfrac{C}{\left(\dfrac{266919.6}{T^2} - \dfrac{379.669}{T} + 0.3341 + 69\right)}\right).$
Correlation 3: Fabuss [17]; data source: Fabuss [17]
$TE = \alpha_0 \times C + \beta_0 \times C^2$, where C is the technical concentration factor of seawater, $\alpha_0 = \alpha_1 + \alpha_2 \times BPT + \alpha_3 \times BPT^2$ and $\beta_0 = \beta_1 + \beta_2 \times BPT + \beta_3 \times BPT^2$ where, $\alpha_1 = 0.2009$, $\alpha_2 = 0.2867\text{E-}2$ and $\alpha_3 = 0.002\text{E-}4$; $\beta_1 = 0.0257$, $\beta_2 = 0.193\text{E-}2$ and $\beta_3 = 0.0001\text{E-}4$ BPT = boiling point of pure water at °C
Correlation 4: El-Dessouky and Ettouney [13]
$TE = Ax + Bx^2 + Cx^3$. where $A = 8.325 \times 10^{-2} + 1.883 \times 10^{-4} \times T + 4.02 \times 10^{-6} \times T^2$; $B = -7.625 \times 10^{-4} + 9.02 \times 10^{-5} \times T + 5.2 \times 10^{-7} \times T^2$ $C = 1.522 \times 10^{-4} - 3 \times 10^{-6} \times T^{-3} \times 10^{-8} \times T^2$; T = BPT in °C and x = salinity in weight percent

Here, TE = temperature elevation

Fig. 9.6 Empirical correlations for TE estimation.

based correlations for estimating TE (output) as shown in Fig. 9.7 based on two inputs to the neural network (X, salinity; BPT, boiling point temperature).

9.3.1
Scale Formation Modeling

Although there is a huge amount of resources on the study of scaling and corrosion [5, 37, 45, 63], a limited number of research can be found on the modeling (or attempts to modeling) of scale formation (calcium carbonate, magnesium hydroxide,

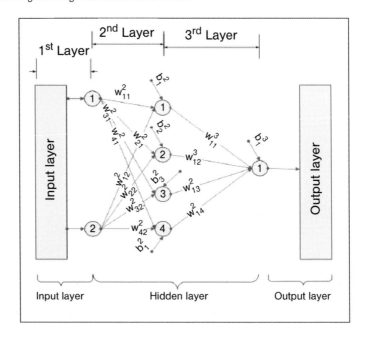

$TE_j = TE_{scaleup} std\_TE + mean\_TE$ $\quad TE_{scaleup} = a_1^3 = w_{11}^3 a_1^2 + w_{12}^3 a_2^2 + w_{13}^3 a_3^2 + w_{14}^3 a_4^2 + b_1^3$

$a_1^2 = \tanh(w_{11}^2 x_{scaleup} + w_{12}^2 BPT_{scaleup} + b_1^2) \quad a_2^2 = \tanh(w_{21}^2 x_{scaleup} + w_{22}^2 BPT_{scaleup} + b_2^2)$

$a_3^2 = \tanh(w_{31}^2 x_{scaleup} + w_{32}^2 BPT_{scaleup} + b_3^2) \quad a_4^2 = \tanh(w_{41}^2 x_{scaleup} + w_{42}^2 BPT_{scaleup} + b_4^2)$

$x_{scaleup} = (x - mean\_x)/std\_x \quad BPT_{scaleup} = (BPT - mean\_BPT)/std\_BPT$

std_x = 2.169   std_BPT = 21.02   std_TE = 0.352

mean_x = 2.169   mean_BPT = 91.549   mean_TE = 0.606

2nd layer:

$w_{11}^2 = 0.917 \quad w_{21}^2 = 0.213 \quad w_{31}^2 = 0.514 \quad w_{41}^2 = 0.580 \quad w_{12}^2 = 1.395 \quad w_{22}^2 = 0.087$

$w_{32}^2 = -0.174 \quad w_{42}^2 = 0.225 \quad b_1^2 = 2.448 \quad b_2^2 = 0.829 \quad b_3^2 = 0.409 \quad b_4^2 = -2.398$

3rd layer: $w_{11}^3 = 0.005 \quad w_{12}^3 = 6.364 \quad w_{13}^3 = 0.466 \quad w_{14}^3 = -1.797 \quad b_1^3 = 2.312$

Note $BPT = T_{Di}, x \text{ (wt\%)} = C_{Bi} \text{ (wt/wt)} \times 100$

Fig. 9.7 NN architecture and NN-based correlation for TE.

calcium sulfate) in MSF process. Table 9.4 lists the major development in that area since 1967.

Most of these models have been developed and studied on their own and have not been a part of the models described in the previous section (except Wagnick and the most recent work of Hawaidi and Mujtaba [20], Said et al. [43] which are described briefly in the following and in the case study section).

Table 9.4 List of modeling work on scale formation.

References	Purpose/findings
[9]	Use more CO_2 to reduce scale
[63]	Correlations for calculating overall heat transfer coefficient taking into account the presence of NCGs
[33]	Kinetic model of carbonates formation. Antiscalant eliminates/reduces the rate of scale formation
[3]	Solubility of CO_2 decreases at high temperature, allowing scaling by carbonates. Modeling the release of CO_2 at high temperature
[20]	Development of linear dynamic profile for brine heater fouling factor (due to scaling) and study the performance of MSF process
[43]	Embed Wagnick's correlations in Rosso et al.'s [42] MSF model to study the impact of NCGs on plant performance

9.3.1.1 Estimation of Dynamic Brine Heater Fouling Profile

Figures 9.8 and 9.9 show the variation of actual fouling factor (m^2 K/kw) with time (h) of the brine heater section [19]. Figure 9.8 represents the case where an antiscale dosing of 0.8 ppm polyphosphate was used with TBT = 90 °C, while Fig. 9.9 represents the case where an antiscale dosing of 3 ppm polyphosphate was used with TBT = 108 °C. Using regression analysis, a simple linear relationship was obtained by Hawaidi and Mujtaba [20], which was used for calculating dynamic overall heat transfer coefficient for the brine heater using Eq. (9.27),

$$f_{bh} = 2 \times 10^{-5} t + 0.0509 \tag{9.37}$$

The constant 0.0509 in Eq. (9.37) represents the initial fouling of the brine heater section (f_{bh}, m^2 hK/kcal) at $t = 0$ (say January, at the beginning of the operation after plant overhauling).

9.3.1.2 Modeling the Effect of NCGs

In MSF, the NCGs entering with the feed water are liberated during the evaporation process and have to be removed by adequate venting. NCGs are a serious

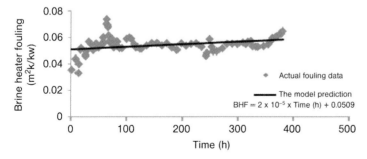

Fig. 9.8 Dynamic brine heater fouling factor (antiscale dosing of 0.8 ppm polyphosphate and TBT = 90 °C).

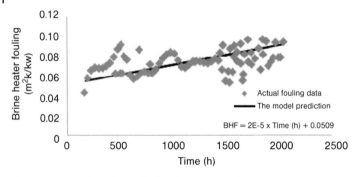

Fig. 9.9 Dynamic brine heater fouling factor (antiscale dosing of 3 ppm polyphosphate and TBT = 108 °C).

problem in MSF, which mainly consist of air (N_2 and O_2) and CO_2. In addition, the presence of NCGs is caused by the leakage of ambient air through flanges, manholes, instrumentation nozzles, into the parts of the evaporating brine. NCG gases even at low concentrations reduce performance, efficiency, and hence increase the cost in most thermal desalination units [26]. Carbon dioxide dissolves in the condensate and lowers its pH value. In the presence of O_2, this may cause corrosion of the condenser tubes and may lead to reduced plant lifetime [36] and to frequent plant shutdown for maintenance. This will in turn lead to loss of freshwater production.

Study considering the effect of NCG gases on the overall heat transfer coefficient, energy consumption, plant design, and production capacity of MSF process is almost nonexistent. The presence of NCGs reduces the overall heat transfer coefficient for the condensing vapor and the temperature at which it condenses at given pressure in the vapor space [63]. Said et al. [43] used correlations reported by Wagnick [63] as shown in Fig. 9.10 which takes into considerations the effect of the presence of NCGs and fouling factors on the overall heat transfer coefficient in the heat recovery section, heat rejection section, and brine heaters. These correlations calculate the inside and outside overall heat transfer coefficients, which depend on the fouling factors, flowrate, temperature, and physical properties such as thermal conductivity, viscosity, density, and specific heat of the condensing vapor and the brine inside the condenser.

9.3.1.3 Modeling of Environmental Impact

Due to increasing environmental legislation, the activities in the area of assessing/quantifying environmental impact are gaining importance in MSF. Earlier focus was on reducing metal corrosion to improve plant life rather than quantifying the effect of corrosion on the environment. For example, Oldfield and Todd [38] evaluated the level of contamination due to corrosion of copper-based alloys and stainless steel but no further link on environmental impact was established. Table 9.5 shows some of the major work since 1999 on quantifying environmental impact and on providing guidelines to reduce water pollutants.

Steam Side Heat Transfer Coefficient

$$h_o = .725 \left(\frac{K_l^3 \rho_l (\rho_l - \rho_v) g \rho \lambda_v}{\sigma_o \mu \Delta T} \right)^{.25} C_1 C_1$$

$C_1 = 1.23795 + .353808 N_1 - .0017035 N_1^2$, $C_2 = 1 - 34.313 X_{nc} + 1226.8 X_{nc}^2 - 14923 X_{nc}^3$

$N_1 = .564 \sqrt{N_t}$, $N_t = 4M_f / \pi \sigma_i^2 \rho_f V_f$

Water Side Heat Transfer Coefficient

$$h_i = (3293.5 + T(84.24 - .1714T) - x(8.471 + .1161x + .2716T)) / \left(\left(\frac{\sigma_i}{.017272} \right)^{.2} \right) (.656V)^{.8} (\sigma_i / \sigma_o)$$

Overall heat transfer coefficients

$$\frac{1}{U_o} = \frac{1}{h_i \left(\frac{d_i}{d_o} \right)} + \frac{1}{h_o} + \frac{1}{h_i} + r_t + r_{fi} + r_{fo}$$

Fig. 9.10 Overall heat transfer coefficient correlations.

Table 9.5 Studies on environmental impact.

References	Purpose
[22]	Five steps *Procedures* (groupings of critical chemicals, corresponding marine ecosystems, etc.) for Environmental Impact Assessment
[44]	Quantitative guidelines on the reduction of chlorine injection
[23]	Pollution by excess chlorination around seawater intake (to avoid marine organisms getting into the process)
[23]	Impact of chemical loading (corrosion + antiscalant) in Red Sea – vulnerable to ecological damage
[47]	Model linking improvement in plant efficiency and environmental impact

Sommariva et al. [47], for the first time, attempted to establish relations between improvement in plant performance ratio (efficiency) and environmental impact (and therefore, the paper was without a single reference) in thermal desalination system. Figure 9.11 shows the system boundary around (dotted line) which the energy balance (Fig. 9.12) was carried out for the purpose of assessing the environmental impact and connecting it with the plant performance ratio.

9.4
State-of-the-Art in Dynamic Modeling of MSF Desalination Process

Compared to the research using steady-state model for simulation and optimization, the research using dynamic model is only handful. Table 9.6 provides a list

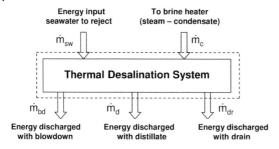

Fig. 9.11 System boundary for energy balance and environmental impact assessment [47].

Energy balance (around the system boundary, Figure 11):

$$\dot{m}_d h_d + \dot{m}_{dr} h_2 + \dot{m}_{bd} h_2 = \dot{m}_{sw} h_1 + \dot{m}_c \Delta H$$

Performance Ratio, GOR: $\eta = \dfrac{\dot{m}_d}{\dot{m}_c}$

Thermal Impact (Energy Dissipated): $E_{diss} = (\dot{m}_{bd} + \dot{m}_{dr}) \Delta H_{sw}$

Thermal Impact vs Performance Ratio: $E_{diss} = \dot{m}_d \left(\dfrac{\Delta H}{\eta} - \Delta h_d \right)$

Fig. 9.12 Energy balance, thermal impact, and plant performance [47].

Table 9.6 Use of dynamic model in MSF.

References	Purpose/software
[24, 25]	Simulation/SPEEDUP
[30]	Control/SPEEDUP
[58]	Simulation/SPEEDUP
[31]	Simulation/LSODA
[48]	Simulation/gPROMS
[41]	Simulation

of some of the major works in this area. Interestingly the dynamic models used in all these works are an extension of the steady-state model of Helal et al. [21]. Stage mass, concentration, and temperature were used as state (dynamic) variables in the models (as shown below). The assumptions are same as those listed for steady-state model. Most recently, Sowgath [48] used NN-based correlation for estimating TE in the dynamic model.

(a) *Stage model.*

The flashing brine model.
Mass balance:

$$\frac{d}{dt} M_j^B = B_{j-1} + B_j - V_j \qquad (9.38)$$

Salt balance:

$$\frac{d}{dt}\left(M_j^B \times X_j^B\right) = B_{j-1} \times CB_{j-1} - B_j \times CB_j \qquad (9.39)$$

Enthalpy balance for the flashing brine:

$$M_j^B \times h_j^B = B_{j-1} \times \left(h_{j-1}^B - h_j^B\right) - V_j \times \left(h_j^V - h_j^B\right) \qquad (9.40)$$

Mass balance for the distillate tray:

$$\frac{d}{dt} M_j^D = D_{j-1} + D_j - V_j \qquad (9.41)$$

$$h_{vj} = f(T_{Sj}) \qquad (9.42)$$

$$h_{Bj} = f(C_{Bj}, T_{Bj}) \qquad (9.43)$$

$$h_{Dj} = f(T_{Dj}) \qquad (9.44)$$

The cooling brine tube.
Mass balance:

$$W_j = W_{j+1} = W_T \qquad (9.45)$$

Salt balance:

$$W_{j-1} \times CF_{j-1} = W_j \times CF_j \qquad (9.46)$$

Enthalpy balance:

$$M_j^F \frac{d}{dt} h_j^F = U_j \times A_j \times \frac{(T_j^F - T_{j+1}^F)}{\ln \frac{(T_j^D - T_{j+1}^F)}{(T_j^D - T_j^F)}} - W_T \left(h_j^F - h_{j+1}^F\right) \qquad (9.47)$$

$$M_j^D \times \frac{dh_j^D}{dt} + \left(h_j^V - h_j^D\right) \times \frac{dM_j^B}{d}$$

$$= D_{j-1} \times h_{j-1}^D + B_{j-1} \times h_{j-1}^B$$

$$- D_j \times h_j^D - B_j \times h_j^B - W_T \times \left(h_j^F - h_{j+1}^F\right) \qquad (9.48)$$

$$S_{Rj} = f(T_{Fj+1}, T_{Fj}, C_R) \quad \text{(recovery/rejection stage)} \tag{9.49}$$

$$S_{Dj} = f(T_{Dj}) \tag{9.50}$$

$$S_{Bj} = f(T_{Bj}, C_{Bj}) \tag{9.51}$$

Overall heat transfer coefficient:

$$U_j = f\left(W_R, T_{Fj}, T_{Fj+1}, T_{Dj}, D_j^i, D_j^o, L_j^i, f_j^i\right) \quad \text{(recovery stage)} \tag{9.52}$$

$$U_j = f\left(W_s, T_{Fj}, T_{Fj+1}, T_{Dj}, D_j^i, D_j^o, L_j^i, f_j^i\right) \quad \text{(rejection stage)} \tag{9.53}$$

Distillate and flashing brine temperature correlation:

$$T_{Bj} = T_{Dj} + TE_j + EX_j + \Delta_j \tag{9.54}$$

Distillate and flashed steam temperatures correlation:

$$T_{Sj} = T_{Dj} + \Delta_j \tag{9.55}$$

Temperature drop due to demister correlation:

$$\Delta_j = \exp(1.885 - 0.02063 \times T_j^D) \tag{9.56}$$

$$EX_j = f(H_j, w_j, T_{Bj}) \tag{9.57}$$

The relationship for the evaluation of the pressure (atm) of the stage P_j

$$\log 10 \frac{P_C}{P_j} = \frac{X}{T_j^V} \left(\frac{a + b \times X + c \times X}{d \times X} \right) \tag{9.58}$$

where $a = 3.2437814$, $b = 5.86826 \times 10^{-3}$, $c = 1.1702379 \times 10^{-8}$, $d = 2.1878462 \times 10^{-3}$, $X = T_c - T_j^S$, $P_c = 218.167$ atm, and $T_c = 647.27$ K are the critical pressure and temperature of water, respectively.

Orifice equations for distillate and brine flow rate.
Empirical relationship for orifice model is taken from [28, 40].
Empirical relationship between brine orifice and brine flow rate:

$$M_j^B = \rho_j^B \times Oh_j^B \times L_j \times V_j^B \tag{9.59}$$

$$V_j^B = \sqrt{\frac{2 \times g \times (L_j - L_{j+1} + (P_{j+1} - P_j))}{1 - \left(\frac{Cc_j^B \times Oh_j^B \times Ow_j^B}{L_j \times Q_B}\right)}} \tag{9.60}$$

$$\Delta P = P_{j+1} - P_j + 0.098 \times \left(L_j - Cc_j^B \times Oh_j^B\right) \tag{9.61}$$

$$Cc_j^B = 0.61 + 0.18 \times r_{Bj} - 0.58 \times r_{Bj}^2 + 0.7 \times r_{Bj}^3 \qquad (9.62)$$

$$r_{Bj} = \frac{g \times \frac{\rho_j^B}{1000} \times Oh_j^B}{100 \times (P_{j-1} - P_j) \times g \times \frac{\rho_j^B}{1000} \times L_j} \qquad (9.63)$$

Empirical relationship between distillate orifice and distillate flow rate:

$$M_j^D = \rho_j^D \times Oh_j^D \times L_j \times V_j^D \qquad (9.64)$$

$$V_j^D = \sqrt{\frac{2 \times g \times (L_j - L_{j+1} + (P_{j+1} - P_j))}{1 - \left(\frac{Cc_j^D \times Oh_j^D \times Ow_j^D}{L_j \times Q_D}\right)}} \qquad (9.65)$$

$$\Delta P = P_{j+1} - P_j + 0.098 \times \left(L_j - Cc_j^D \times Oh_j^D\right) \qquad (9.66)$$

$$Cc_j^D = 0.61 + 0.18 \times r_{Dj} - 0.58 \times r_{Dj}^2 + 0.7 \times r_{Dj}^3 \qquad (9.67)$$

$$r_{Bj} = \frac{g \times \frac{\rho_j^D}{1000} \times Oh_j^D}{100 \times (P_{j-1} - P_j) \times g \times \frac{\rho_j^D}{1000} \times L_j} \qquad (9.68)$$

$$M_j^F = \rho_j^F \times Oh_j \times A_j \qquad (9.69)$$

(b) *Brine heater model.*

Overall mass balance:

$$B_{BH} = W_T \qquad (9.70)$$

Salt mass balance:

$$C_{BH} = C_R \qquad (9.71)$$

Enthalpy balance of the cooling brine:

$$M_{BH} \times \frac{d}{dt} h_{BH} = U_H \times A_H \times \frac{T_{BH} - T_1^F}{\ln \frac{T_{steam} - T_1^F}{T_{steam} - T_{BH}}} - W_T \times \left(h_{BH} - h_1^F\right) \qquad (9.72)$$

Enthalpy balance of the condensing vapor:

$$W_{steam} \times \lambda_{steam} = U_H \times A_H \times \frac{T_{BH} - T_1^F}{\ln \frac{T_{steam} - T_1^F}{T_{steam} - T_{BH}}} \qquad (9.73)$$

The correlations for physical and chemical properties are given in Table 9.3, and in Figs. 9.6 and 9.7.

9.5
Case Study

9.5.1
Steady-State Operation

With the process models described in Section 9.3, a number of simulation studies have been reported in the literature. Some of them are highlighted in Table 9.7.

Each set of specifications [21, 42] satisfying the degrees of freedom requires reordering of the model equations and different (or tailored) solution algorithm. However, that was not the case by Tanvir and Mujtaba [53, 54], Hawaidi and Mujtaba [20], and Said et al. [43] as the equation-oriented package gPROMS [18] (which automatically takes care of any equation orientation required) was used in their work. The capabilities of such packages are enormous and should be fully explored for better design and operation of MSF processes.

For a given TBT (T_{B0}), the observations made by Tanvir and Mujtaba [53, 54, 56] and Sowgath and Mujtaba [49] are shown in Fig. 9.13, which shows about 15% drop in freshwater production in summer compared to that in winter. However,

Table 9.7 Simulation studies in MSF (1986–2009).

References	Parameter specifications (purpose)
[21]	(i) R, C_W, F, T_S (performance calculation)
	(ii) D_N, T_{B0}, F, C_W (fixed product demand)
	(iii) W_S, T_{B0}, F, C_W/R (fixed energy input)
[42]	(i) W_S, T_S, R, C_W (performance calculation)
	(ii) Effect of T_S (iii) Effect of T_seawater
	(iv) Effect of number of stages
[58]	(i) W_S, T_{B0}, F, C_W/R (performance calculation and validation with plant data)
[47]	Plant performance vs. environmental impact
[53]	(i) W_S, T_S, R, C_W (performance calculation)
	(ii) Effect of T_S
	(iii) Effect of T_seawater
[54]	(i) D_N, T_{B0}, W_S, C_W (fixed product demand)
	(ii) Effect of T_S
	(iii) Effect of T_seawater
	(iv) Effect of number of stages
	(v) Effect of heat exchanger area
[20]	(i) D_N, T_{B0}, W_S, C_W (fixed product demand)
	(ii) Effect of T_seawater
	(iii) Effect of dynamic scaling
[43]	(i) T_S, T_{B0}, W_S, C_W
	(ii) Effect of T_seawater
	(iii) Effect of NCGs

Seawater Temperature (°C)	Freshwater (kg/h)	% drop
23 (winter)	1.09 E6	---
35 (summer)	9.31 E5	14.6

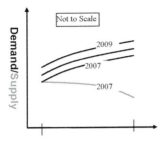

Fig. 9.13 Water supply at fixed design, operation, and demand forecast [53].

the demand of water in summer is usually higher and it increases year by year (Fig. 9.13; also [27]). As the design and configuration of almost all industrial plants is fixed, even to supply freshwater at a fixed rate throughout the year or to increase the production at any time of the year, the common industrial practice is to operate the plant at high temperature [16, 52–54]. However, this results in increased fouling and corrosion of heat exchangers and plant equipment, as illustrated in Fig. 9.14. The end results are [16] as follows:

- Frequent plant shutdowns, interrupting freshwater supply.
- Or the use of increased amount of antiscalant when operating the plant at high temperatures, which drives up the cost.

Said et al. [43] carried out simulation of MSF at different seawater temperature to see the effect of NCGs on the performance of MSF processes. The process configuration is same as used by Rosso et al. [42] and Tanvir and Mujtaba [52] with a total of 13 stages in the recovery and three stages in the rejection sections. The

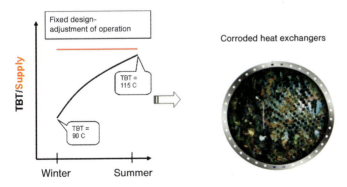

Fig. 9.14 High-temperature operation leading to heat exchanger corrosion.

Table 9.8 Constant parameters and input data [43].

	A_j/A_H	D_j^i/D_H^i	D_j^o/D_H^o	f_j^i/f_H^i	$w_j/L_j/L_H$	H_j
Brine heater	3530	0.022	0.0244	1.86×10^{-4}	12.2	Chapter 3
Recovery stage	3995	0.022	0.0244	1.4×10^{-4}	12.2	0.457
Rejection stage	3530	0.024	0.0254	2.33×10^{-5}	10.7	0.457
W_s 31 416.67 kg/s	T_{steam} 97 °C	$T_{seawater}$ Variable	C_s 5.7 wt%	R 1763.88 kg/s	C_w 1561.11 kg/s	

specifications (satisfying the degrees of freedom) are same as those used by Tanvir and Mujtaba [52], which are shown in Table 9.8.

Table 9.9 shows the effect of different amount of NCGs on the plant performance in terms of freshwater production, steam consumption, and GOR,

$$\text{GOR} = \text{Total distilled } (D_{NS})/\text{steam needed } (W_{steam}).$$

For a given NCG (0.015 wt%) and steam temperature (T_{steam}) (97 °C), simulations are carried out at different seawater temperature ($T_{seawater}$). The results are presented in Table 9.10 with the total amount of freshwater produced (D_{NS}), GOR, TBT, and final bottom brine temperature (BBT). Comparison of the results with

Table 9.9 Effect of NCGs on MSF plant performance.

NCGs (wt%)	D_{NS} (kg/s)	W_{steam} (kg/s)	GOR
0.015	270.517	35.721	7.571
0.02	270.480	35.734	7.569
0.03	270.460	35.737	7.567
0.04	270.400	35.747	7.560
0.05	270.061	35.806	7.542
0.06	266.241	36.445	7.305

Table 9.10 Effect of $T_{seawater}$ and T_{steam} on D_{NS}, GOR, TBT, and BBT.

	$T_{seawater}$ (°C)	D_{NS} (kg/s)	W_{steam} (kg/s)	GOR
[43]	35	270.0	35.7	7.57
[54]	35	258.6	33.1	7.82
[43]	45	230.0	30.4	7.56
[54]	45	218.9	28.3	7.72

those of Tanvir and Mujtaba [52] clearly shows the effect of NCGs. The amount of steam consumption goes very high due to change in overall heat transfer coefficient. Although the freshwater production rate improves, the value of GOR goes down.

9.5.2
Dynamic Operation

Using the dynamic model presented earlier (with NN-based correlation for TE calculation), Sowgath [48] reported the following three steady-state conditions of an MSF process (Table 9.11).

The MSF process is assumed to be at steady-state condition at $T_{seawater} = 23\,°C$ and $T_{steam} = 97\,°C$ (Case 1, Table 9.11), and the model is simulated at that condition for 5 s (by setting all differential variables of the dynamic model to zero). An external disturbance of seawater temperature is considered where it increases from $23\,°C$ to $45\,°C$ (Case 2, Table 9.11 and T_{steam} remaining at $97\,°C$). Note, in reality the plant will not experience such a big step change in a short period of time. However, Sowgath [48] considered this case to test the robustness of the dynamic model in terms of handing large step change. With the change in proposed seawater temperature, the plant will reach to a different steady-state condition.

Figure 9.15 shows the steady-state conditions in terms of freshwater production (the values are very close to those reported in Table 9.11). The dynamic model is now subjected to another step change but in terms of T_{steam} which is changed from to $97\,°C$ to $116.5\,°C$ (Case 3, Table 9.11). This takes the freshwater production level back to the first steady-state level (i.e., $T_{seawater} = 23\,°C$ and $T_{steam} = 97\,°C$) (as desired). For fixed water demand, Sowgath [48] carried out further simulation using the dynamic model.

Hawaidi and Mujtaba [20] calculated the brine heater fouling factor at discrete time interval using the dynamic estimator given in Eq. (9.37) and used the steady-state model to evaluate steady-state performance at discrete time intervals. Seasonal variation of seawater temperature is considered (based on Abdel-Jawed and AL-Tabtabael [1]). For different seawater temperatures, corresponding brine heater fouling factors are calculated using Eq. (9.37) and reported as average for the month concerned (time zero is at the day 1 of January). With a fixed freshwater demand $D_j = 945\,000$ kg/h, TBT = $90\,°C$ with antiscaling (polyphosphates) rate of 0.8 ppm the performance of MSF process is presented in Table 9.12.

Table 9.11 Effect of $T_{seawater}$ and T_{steam} on D_{NS}.

Case	$T_{seawater}$	T_{steam}	D_{NS} (kg/h)
1	23	97	1.09E+06
2	45	97	7.88E+05
3	45	116.5	1.09E+06

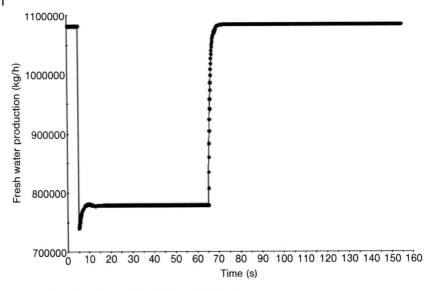

Fig. 9.15 The dynamic model prediction of freshwater production in MSF subject to seawater and steam temperature disturbance.

Table 9.12 Performance of MSF at discrete time intervals.

Month	$T_{seawater}$	f_{bh}	T_{steam}	Antiscalant (kg/h)	W_{steam}	GOR
Jan	15	0.065	93.6	2252.2	116 955	8.08
Mar	20	0.093	94.4	2539.9	120 122	7.86
May	28	0.121	95.6	3159.0	127 307	7.42
Jul	32	0.150	96.7	3542.0	132 241	7.14
Aug	35	0.164	97.4	3891.4	136 718	6.91
Oct	30	0.192	97.4	3340.0	129 986	7.27
Dec	20	0.221	97.1	2538.6	120 535	7.84

9.6 Future Challenges

9.6.1 Process Modeling

Figure 9.16 combines the state-of-the-art in process modeling with what needs to be done in future. What has, more or less, been done is shown in green and challenges for future are shown in orange. Kinetic model for fouling, corrosion model for material selection, model for impact assessment, fluid flow of noncondensables, fluid mixing, etc. are yet to be fully reflected in a reliable process model.

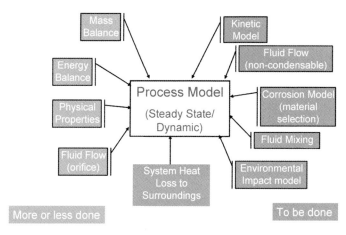

Fig. 9.16 Challenges in process modeling.

9.6.2
Steady-State and Dynamic Simulation

Figure 9.17 shows the current state of how model-based simulations are carried out (a) and the future opportunities (b). Tailor-made algorithms are time consuming as they require representation of the model equations for each new set of specifications for performance calculations [21]. This is not the case with many current commercial software such as gPROMS, SPEEDUP, and ASPEN. Full potential of these packages are yet to be explored.

9.6.3
Tackling Environmental Issues

Currently (Table 9.5), issues related to environmental impact are dealt in a reactive mode as shown in Fig. 9.18(a). The environmental impact from an existing process is assessed and based on the current environmental legislation, the operations are adjusted. The preventive mode (Fig. 9.18(b)) requires that new design and operations are achieved based on a set/desired environmental targets. While trial and error based on experimental studies is time consuming and expensive, studying

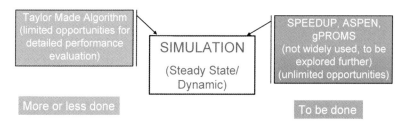

Fig. 9.17 Opportunities in MSF process simulation.

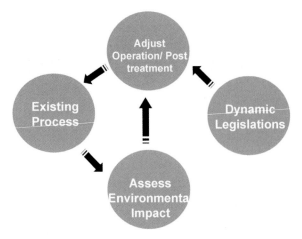

(a) Reactive Mode for Tackling Environmental Problem

(b) Preventive Mode for Tackling Environmental Problem

Fig. 9.18 Tackling environmental problem.

these via mathematical techniques (modeling and optimization) is less time consuming and inexpensive and remains a future challenge.

9.6.4
Process Optimization

Figure 9.19 combines the state-of-the-art in process optimization with future challenges and opportunities. Typical process engineering problems (such as process design or plant operation) have many, and possibly an infinite number of, solutions. The term *optimization* is freely used to describe the complete spectrum of techniques from the basic multiple run approach of trial and error to highly complex numerical strategies. As mentioned earlier, trial and error experimental approach for optimal design and operation of MSF process will be extremely time consuming and expensive. Model-based techniques, in comparison, are less time consuming and inexpensive. This can evaluate thousands of alternative design and operation scenario cost effectively.

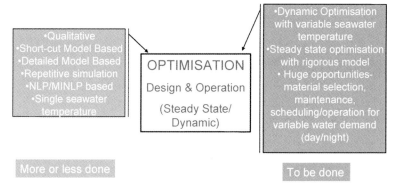

Fig. 9.19 Challenges in process optimization.

9.7 Conclusions

This chapter highlights the state-of-the-art in modeling of MSF desalination process. Changing environmental legislations is continuously taking the operations of existing MSF plants to its limit in terms of energy consumption and profit. With strict but set environmental targets, new designs and operations by trial and error are very expensive and time consuming. While model-based techniques could be an answer to that, the existing models are far away from the reality. For example, model for the mixing in MSF stages, detailed model of the noncondensable flow paths affecting heat transfer, kinetic model for corrosion and scaling, and model for assessing environmental impact are missing.

Exploitation of full economic benefit of replacing time-consuming and expensive experimental studies of MSF processes requires development of an accurate mathematical model. Future challenges in meeting sustainable freshwater demand dictate development of rigorous, sophisticated MSF process model and applying advanced mathematical techniques for optimum (energy efficient and environment friendly) design and operation of MSF processes.

Nomenclature

B_j – brine flow leaving stage j, kg/h
B_D – blow down mass flow rate, kg/h
C_{Bj} – brine concentration, wt/wt
C_W – rejected seawater flow rate, kg/h
C_S – seawater salt concentration, wt/wt
C_R – seawater salinity in the recovery stages, wt/wt
D_j – distillate flow from stage j, kg/h
F – make-up seawater flow rate, kg/h
R – recycle stream flow rate, kg/h

T_{Bj} – temperature of flashing brine leaving stage j, °C
T_{Dj} – temperature of distillate leaving stage j, °C
T_{Fj} – seawater temperature leaving stage j, °C
W_s – seawater mass flow rate, kg/h
W_R – seawater flow in the recovery section, kg/h

References

1 ABDEL-JAWED, M., AL-TABTABAEL, M., Impact of current power generation and water desalination activities on Kuwaiti marine environment, in: *Proceedings of IDA World*, **1999**.
2 ALI, E., ALHUMAIZI, K., AJBAR, A., *Desalination* 121 (**1999**), pp. 65–85.
3 AL-ANEZI, K., HILAL, N., *Desalination* 204 (**2007**), pp. 385–402.
4 AL-MUTAZ, S., SOLIMAN, M. A., *Desalination* 74(1/3) (**1989**), pp. 317–326.
5 AL-SOFI, M. A., *Desalination* 126 (**1999**), pp. 61–76.
6 BEAMER, J. H., WILDE, D. J., *Desalination* 9 (**1971**), p. 259.
7 BROMLEY, L. A., et al., *AIChE Journal* 20 (**1974**), p. 326.
8 BUROS, O. K., *The Desalting ABC's for International*, Desalination Association, Topsfield, MA, **1990**.
9 CADWALLADER, E. A., *Industrial and Engineering Chemistry* 59(10) (**1967**), p. 42.
10 CLELLAND, D. W., STEWART, J. M., *Desalination* 1(1) (**1966**), pp. 61–76.
11 COLEMAN, A. K., *Desalination* 9 (**1971**), pp. 315–331.
12 EL-DESSOUKY, H. T., BINGULAC, S., *Desalination* 107 (**1996**), pp. 171–193.
13 EL-DESSOUKY, H. T., ETTOUNEY, H. M., *Fundamentals of Salt Water Desalination*, Elsevier Science, Amsterdam, **2002**.
14 EL-DESSOUKY, H. T., ETTOUNEY, H. M., AL-ROUMI, Y., *Chemical Engineering Journal* 73 (**1999**), pp. 173–190.
15 EL-DESSOUKY, H., SHABAN, H. I., AL-RAMADAN, H., *Desalination* 103 (**1995**), pp. 271–287.
16 ELMOUDIR, W., ELBOUSIFFI, M., AL-HENGARI, S., *Desalination* 222 (**2008**), pp. 431–440.
17 FABUSS, B. M., in: *Principles of Desalination*, vol. 1, Spiegler, K. S., Laird, A. D. (eds.), Academic Press, New York, **1980**.
18 gPROMS, Introductory User Guide, Process System Enterprise Ltd. (PSE), http://www.psenterprise.com/gproms/, **2005**.
19 HAMED, O. A., AL-SOFI, M. A. K., MUSTAFA, G. M., DALVI, A. G., Performance of different anti-scalants in multi-stage flash distillers, *Desalination* 123 (**1999**), pp. 185–194.
20 HAWAIDI, E. A. M., MUJTABA, I. M., Internal Report, University of Bradford, UK, **2009**.
21 HELAL, A. M., MEDANI, M. S., SOLIMAN, M. A., FLOWER, J. R., *Computers and Chemical Engineering* 10 (**1986**), pp. 327–342.
22 HOEPNER, T., *Desalination* 124 (**1999**), pp. 1–12.
23 HOEPNER, T., LATTEMANN, S., *Desalination* 152 (**2002**), pp. 133–140.
24 HUSSAIN, A., et al., *Desalination* 92 (**1993a**), pp. 21–41.
25 HUSSAIN, A., et al., *Desalination* 92 (**1993b**), pp. 43–55.
26 KHAN, R., Effect of non-condensable in seawater evaporators, *Chemical Engineering Progress* 68 (**1972**), pp. 79–80.
27 LIOR, N., in: *Proceedings of 17th International Congress of Chemical and Process Conference, 27–31 August, Praha, Czech Republic*, **2006**.
28 LIOR, N., LIZEE, V., MIYATAKE, O., Correlations for predicting the flow through MSF plant interstage orifices, in: *IDA Madrid Proceedings*, vol. V, IDA, Spain, Madrid, **1997**, pp. 169–182.
29 MANDIL, M. A., GHAFOUR, E. E. A., *Chemical Engineering Science* 25 (**1970**), pp. 611–621.
30 MANIAR, V. M., DESHPANDE, P. B., *Journal of Process Control* 6 (**1995**), pp. 49–66.
31 MAZZOTTI, M., et al., *Desalination* 127 (**2000**), pp. 207–218.

32 Mjalli, F. S., Abdel-Jabbar, N., Ettouney, H., Qiblawey, H. A. M., *Chemical Product and Process Modeling* 2 (**2007**), p. A29.
33 Mubarak, A., *Desalination* 120 (**1998**), pp. 33–39.
34 Mussati, S. F., Marcovecchio, M. G., Aguirre, P. A., Scenna, N. J., *Desalination* 182 (**2005**), p. 123.
35 Naphthali, L. M., Sandholm, D. P., *AIChE Journal* 17 (**1971**), p. 148.
36 Oldfield, J. W., Vapour side corrosion in MSF plants, *Desalination* 66 (**1987**), pp. 171–187.
37 Oldfield, J. W., Todd, B., *Desalination* 38(1–3) (**1981**), pp. 233–245.
38 Oldfield, J. W., Todd, B., *Desalination* 108 (**1996**), pp. 27–36.
39 Public Inquiry, The Guardian, 24 May, **2006**.
40 Reddy, K. V., Husain, A., Woldai, A., Al-Gopaisi, D. M. K., Dynamic modelling of the MSF desalination process, in: *IDA Abu Dhabi Proceedings*, vol. IV, IDA, Spain, Madrid, **1995**, pp. 227–242.
41 Rimawi, M. A., Ettouney, H. M., Aly, G. A., *Desalination* 74 (**1989**), pp. 327–338.
42 Rosso, M., et al., *Desalination* 108 (**1996**), pp. 365–374.
43 Said, S. A., Mujtaba, I. M., Emtir, M., in: *Proceedings of International Conference on Energy, Water, Desalination and Environment, 7–8 December, Libya*, **2009**.
44 Shams El Din, A. M., Arain, R. A., Hammoud, A. A., *Desalination* 129 (**2000**), pp. 53–62.
45 Shams El Din, A. M., Mahmoud, E. A., *Desalination* 71 (**1989**), p. 313.
46 Silver, R. S., British Patent No. 829,819, **1957**.
47 Sommariva, C., Hogg, H., Callister, K., *Desalination* 167 (**2004**), pp. 439–444.
48 Sowgath, M. T., PhD Thesis, University of Bradford, UK, **2007**.
49 Sowgath, T., Mujtaba, I. M., *The Chemical Engineer, IChemE* 6 (**2008**), pp. 28–29.
50 Spiegler, K. S., *Salt-Water Purification*, Plenum Press, New York, **1977**.
51 Spiegler, K. S., Liard, A. D. K., *Principles of Desalination*, Academic Press, New York, **1980**.
52 Tanvir, M. S., Mujtaba, I. M., *Desalination* 195 (**2006a**), pp. 251–272.
53 Tanvir, M. S., Mujtaba, I. M., in: *Proceedings of International Water Conference, Porto, Portugal, 12–14 June*, **2006b**, pp. 300–308.
54 Tanvir, M. S., Mujtaba, I. M., *European Symposium on Computer Aided Process Engineering*, in: *Computer Aided Chemical Engineering*, vol. 21A, Marquardt, W., Pantelides, C. (eds.), Elsevier, Amsterdam, **2006c**, pp. 315–320.
55 Tanvir, M. S., Mujtaba, I. M., *European Symposium on Computer Aided Process Engineering*, in: *Computer Aided Chemical Engineering*, vol. 24, Plesu, V., Agachi, P. S. (eds.), Elsevier, Amsterdam, **2007**, pp. 763–768.
56 Tanvir, M. S., Mujtaba, I. M., *Desalination* 222 (**2008**), pp. 419–430.
57 Technical Roadmap, Section 3.5, *IChemE* (**2007**) p. 24.
58 Thomas, P. J., et al., *Computers and Chemical Engineering* 22 (**1998**), pp. 1515–1529.
59 The Independent, Issue 23, March **2001**, London, UK.
60 The International Herald Tribune, February 14, **2004**.
61 The New Sunday Times, Issue 14, July **2002**, Kuala Lumpur, Malaysia.
62 Voros, N., Maroulis, Z. B., Marinos-Kouris, D., *Computers and Chemical Engineering* 20 (**1996**), p. S345.
63 Wagnick, K., in: *Proceedings of the IDA World Congress on Desalination and Water Science*, vol. 2, Abu Dhabi, **1995**, pp. 201–218.
64 Wang, J. C., Henke, G. E., *Hydocarbon Process* 45(8) (**1966**), p. 155.
65 Wilde, D. J., Beightler, C. S., *Foundations of Optimization*, Prentice-Hall, Englewood Cliffs, NJ, **1967**.

Part II
Biological, Bio-Processing and Biomedical Systems

10
Dynamic Models of Disease Progression: Toward a Multiscale Model of Systemic Inflammation in Humans

Jeremy D. Scheff, Panagiota T. Foteinou, Steve E. Calvano, Stephen F. Lowry, and Ioannis P. Androulakis

Keywords

infection, inflammation, heart rate variability, human, mathematical modeling, proinflammatory mediators, human endotoxin

10.1
Introduction

The rapid progress of molecular biology in the wake of the identification of DNA's structure [1] serves as a reminder of the power of reducing a system to its smallest possible components and studying them. Yet, although reductionism is powerful, its scope is limited. This is widely recognized in the study of complex systems whose properties are greater than the sum of their constituent parts [2]. Recognizing that complexity, the emerging field of systems biology attempts to harness the power of mathematics, engineering, and computer science to analyze and integrate data with the ultimate goal of creating models of entire biological systems. Thus, the recent growth of interest in systems biology reflects the increasing importance that integrative initiatives are being accorded in the biological sciences. Eloquently, Mesarovic presents two important roles that systems theory could play in biology: (i) to develop general systems models that can be used as "the first step toward arriving at a more detailed representation of the biological system," and (ii) to provide "a basis for communication between different fields since the formal concepts of behavior (adaptation, evolution, robustness, etc.) are defined in a precise manner and in setting of minimal mathematical structure reflecting, therefore, the minimal degree of special features of the real-life system from which the formal concept has been abstracted" [3].

Many problems in postgenomic biology are now converging to the challenges facing engineers building complex "networks of networks," and systems theory is undergoing a revolution as radical as biology's [4]. Networks, defined as an interconnected group of systems, are potentially characterized by a critical property of complexity: emergence [5]. In the context of a biological system, the implication is that the macroscopic response (phenotype) of a system is the result of propagating information, in the form of disturbances, across an intricate web of interacting modules, forming a complexity pyramid [6, 7]. At the lowest level, there are interactions between molecular components of a cell, such as genes, RNA, proteins, and metabolites. These interactions define elementary building blocks that are organized into pathways and regulatory motifs, which in turn are integrated, through appropriate interactions, into interacting modules that eventually give rise to an organism's response. The emergent behavior of a biological system, whether it relates to the control of the expression of a single gene [8] or the manifestation of a disease [9] is the result of the coordinated action of network elements. As such, deciphering the connectivity and the dynamics of emerging network architectures becomes a critical question in the analysis of biological systems.

In this chapter, we will discuss the potential role of integrative initiatives in the quest to better understand and model complex physiological responses. The unifying hypothesis is that the observed response is the outcome of the orchestrated interactions of critical modules in the form of a network. Thus, a systems-based approach is proposed that aims at exploring the emergence of interaction networks and functional modules linking processes that span various scales from the cellular (low) level to the systemic (high) host response level. We focus our analysis on a critical physiological response: inflammation.

10.2
Background

More than 40 million major surgical operations are performed annually in the United States, of which as many as two million are complicated by surgical site infections [10]. Surgical adverse effects contribute significantly to postoperative morbidity by perturbing the immune system toward a severely suppressed state that promotes sepsis [11]. Surgical patients account for approximately 30% of all sepsis patients [12] and the present incidence of acquired surgical site infections is likely to continue to increase among nontrauma surgical patients [13]. Sepsis is a syndrome resulting from massive, acute activation of the systemic inflammatory response (SIRS) in the setting of severe infection and it remains one of the leading causes of death in United States [12, 14]. The manifestation of SIRS criteria is the common clinical phenotype of stressed surgical patients and reflects the presence of consequential systemic inflammation [13]. However, systemic inflammation is not inherently detrimental. Inflammatory processes are required for immune surveillance and regeneration after injury, during which multiple cell types are deployed to locate pathogens, recruit cells, and eventually eliminate the offenders and restore homeostasis [15]. It is the dysregulation of the resolution of inflammation that has undesirable implications.

Despite numerous supportive preclinical studies, most generated hypotheses related to the management and treatment of severe human inflammation have failed clinical testing [16]. Even the improved capacity to acquire quantitative data in a clinical setting has generally failed to improve outcomes in acutely ill patients.

Thus, the intricacies in translating basic research to clinical practice are recognized as a challenge impeding the successful transfer of information from the preclinical to the clinical stage [17, 18]. These failures have been attributed to invoking the single variable assumption in a clinical scenario [19]. Mathematical models integrating the interacting elements of the unified inflammatory response offer the opportunity to establish a causal inference relationship through the manipulation of the corresponding dynamic elements [20]. As a result, there is a growing research effort toward the development of systems-based, quantitative models of the inflammatory response at various degrees of complexity [21]. Such models are critical enablers in advancing the translational potential of clinical research, a subject recently reviewed [22–24].

10.2.1
In-Silico Modeling of Inflammation

The modeling approaches fall broadly in two categories: those based on explicit dynamic [25, 26] and agent-based models which are discrete in time and space [165]. One of the earliest mathematical models of inflammation dates back to the early 1980s when Lauffenburger and co-workers [27] described the local tissue inflammatory response to bacterial invasion. In this model, the leukocytes are continuously distributed while their accumulation and efficiency in localization within the inflammatory lesion coupled with their phagocytic activity determine the resolution of infection. This model expresses the dynamic interaction between the invader and a homogeneous leukocyte population using a two-variable model that consists of bacterial and leukocyte densities. An extension of this model replaced the single-cell target with a density number associated with the target population [28]. In this model, the principal goal is to address the effect of factors such as chemotaxis, cell speed, and persistence on target elimination dynamics. Further attempts [29] explored the interaction of the immune system with a target population (bacteria, viruses). Such analyses explore nonlinear interaction rules between the immune and target cells that determine the outcome of the immune response. Alternative modeling approaches placed emphasis on simulating interactions at the cellular level in response to an infection [30].

Among the simplest, yet very informative, inflammation models incorporating measured quantities is the one proposed by Kumar *et al.* [31] as shown in Eq. (10.1). The model tracks three basic variables indicative of the onset, progress, and resolution of the inflammatory response, which include the pathogen (p) along with early (m) and late (L) proinflammatory mediators:

$$\frac{dp}{dt} = k_p p(1-p) - k_{pm} mp \qquad (10.1\text{a})$$

$$\frac{dm}{dt} = (k_{mp} p + L)m(1-m) - m \qquad (10.1\text{b})$$

$$\frac{dL}{dt} = k_{lm}f(m) - k_l L \qquad (10.1c)$$

$$f(m) = 1 + \tan h\big((m-\theta)/w\big) \qquad (10.1d)$$

In this reduced model, early proinflammatory mediators are activated in response to an invading pathogen. The pathogen is under logistic growth and is also eliminated at a rate proportional to the product (mp), which represents how often the two agents (m) and (p) interact. The early proinflammatory mediator is also under logistic growth and is stimulated by the pathogen and late proinflammatory mediators. The early proinflammatory mediators also activate late proinflammatory mediators by the nonlinear sigmoidal function $(f(m))$. Both classes of proinflammatory mediators have intrinsic death rates, enabling the resolution of inflammation after the mediators have been activated. There is a positive feedback relationship between the early and late proinflammatory mediators, resulting in the possibility of a feedback loop causing unresolved inflammation. In general, this system behaves like a predator-prey model with a delayed response due to the distinction between early and late proinflammatory mediators.

Despite the simplicity of this model, its dynamics are interesting. Various physiological states and responses can be simulated by changing model parameters and initial conditions: healthy, persistent noninfectious inflammation, persistent infectious inflammation, severe immune-deficiency, and a periodic recurring infection. While these results are purely qualitative, they illustrate the insight that can be gained from the mathematical analysis of biological systems. Therefore, further work has been done toward producing more detailed and accurate models of inflammation.

Later it was suggested [25, 26] that the outcome of a healthy inflammatory response is determined by a balanced regulation in the dynamics of pro- and anti-inflammation. In a further refinement of this model [19], the dynamic evolution of effector cells (macrophages, neutrophils) is distinguished from the corresponding activation of effector cytokines and there is emphasis on the importance of modeling crucial signaling pathways (e.g., complement activation). An extension of this research effort focused on the development of more generalized inflammatory models accounting for a diverse array of initiating events [32, 33]. Models that describe the dynamics of the immune system in response to other infectious agents have also been proposed [34], characterizing the rates of various processes contributing to the progression of the disease while focusing specifically on the control of the infection by innate and adaptive immunity [35]. Recently, innovative computational approaches were proposed to integrate community-wide *in silico* models using the framework of agent-based modeling [165, 166]. Such collaborative frameworks synthesize partial information into a unified model that explores the complexity of the inflammatory response [20, 164], where the underlying principle is to establish the rules among the actors (agents) of the biological response [36]. Agents represent entities, such as cells and cytokines, which interact through local rules on a spatial grid of various probabilities [37]. Such models shed useful insight on the interacting elements that compose the host's heterogeneity.

The key characteristic of these models is the *a priori* postulation of certain components that are consistent with prior biological knowledge. This raises a very interesting question: what constitutes a critical component of the inflammatory response? This question becomes particularly relevant since technological developments give us the ability to measure the effect of environmental perturbations at the molecular and cellular level [38–40], thereby generating data at enormous rates. A critical question is how to determine, based on large amounts of experimental data, which components constitute critical state variables that capture the essence of the response, reminiscent of the minimal model introduced early on [41]. Once the state space has been identified, we need to develop the appropriate wiring architectures that convolute a multitude of external signals. Therefore, these approaches raise two critical questions: (a) what constitutes an underlying dynamic response, and (b) what is an appropriate inflammation model. In order to address these pressing issues, we have undertaken an integrated approach that aims at attacking the problem from different angles and at different scales. Thus, the goal of this research is to develop a systems-based model of human endotoxemia, as a prototype model for systemic inflammation in humans that establishes direct communication links among processes that capture essential aspects of the multiscale nature of the response.

10.2.2
Multiscale Models of Human Endotoxemia

Inflammation can be studied in the absence of complex pathophysiology and co-morbidities of human sepsis by using surrogate models. Human endotoxin challenge is one well-accepted surrogate model for studying the acute inflammatory response as it captures many of the clinically observed features of systemic inflammation [9, 42–45]. Endotoxin, a major component of the outer membrane of Gram-negative bacteria, activates the innate immune system, leading to inflammation. This moiety can be a complicating factor in a variety of situations including trauma, burns, invasive surgery, and organ-specific illnesses. The prototypical examples of endotoxin are *lipopolysaccharide*s (LPS). The response following endotoxin administration in human subjects includes core temperature, cardiac, vasomotor, hematologic, metabolic, hormonal, acute phase reactant, and cytokine components that have been well described [43, 46–48]. In particular, innate immune cell activation leads to production and release of proinflammatory cytokines, which are proximal mediators of the SIRS.

Mechanisms aiming at the regulation of the inflammatory response involve not only the local release of anti-inflammatory cytokines, but also hormonal and autonomic influences whose effectiveness dissipates during prolonged stressful conditions [13]. Recent studies indicate that the central nervous system (CNS) is a pivotal regulator of the immune response through its control of inflammation at various levels [49–53]. The primary stress response pathway by which the CNS regulates the immune system is the hypothalamic-pituitary-adrenal axis (HPA), through the production of glucocorticoids. Further, activation of the sympathetic division (SNS) of the autonomic nervous system regulates immune function primarily via the release of adrenergic neurotransmitters [54]. Such an enhanced endocrine hormone profile has been demonstrated during the early phase response to endotoxin injury [43]. Although most research has focused on the sympathetic immunomodulatory output, it recently became clear that the other arm of CNS, the parasympathetic division (PNS), is also involved in the reflex regulation of inflammation [55]. All these functions are centrally integrated through an extensive and coherent circuitry involving all areas of the medulla, hypothalamic regions and the autonomic system [56].

Disruptions may occur at any level of autonomic function and thereby have a deleterious effect on the host. Heart rate variability (HRV), assessed by evaluating the standard deviation of normal to normal interbeat intervals, is a noninvasive means of quantifying the cardiac autonomic input and is a predictor of trauma patient outcome [57, 58]. Decreases in HRV, i.e., increases in regularity, have also been extensively studied and characterized as generalized responses to human endotoxemia [59, 60]. It has been hypothesized that the reduction in HRV represents an increased isolation of the heart from other organs [61]. This hypothesis, originally introduced by Godin and Buchman [62], suggests that healthy organs behave like biological oscillators coupled to one another. The SIRS initiates a communication disruption and uncoupling. Loss of high-level signal variability reflects a detectable systemic-level loss of adaptability and fitness [16]. Thus, the development of surrogate multiscale, nonlinear, dynamic, complex models gains ever-increasing acceptance as a means of deciphering the intricacies of critical illness, emphasizing the loss of normal rich variability (decomplexification) in critically ill patients [63].

Dissecting the role of the CNS in controlling inflammatory processes requires an understanding of the complex relationship between central autonomic activity and the immune response. A vital enabler in that respect is the development of a systems-based approach that integrates human data across multiple scales and subsequently models the emerging host response as the outcome of orchestrated interactions of critical modules. The emergence of methods that enable such analysis is due to the tremendous advances in monitoring changes at the cellular level, driven primarily by developments in measuring gene expression at the genome-wide scale. With the maturation of this technology, what started as an attempt to classify patterns [64] of gene expression has evolved into sophisticated analyses providing semimechanistic pharmacogenomic models [65]. In an attempt to study the underlying complexity of an *in vivo* human response to endotoxin, we discuss the process of developing a clinically relevant, mechanism-based human inflammation model that bridges the initiating signal (LPS) and phenotypic expressions (HRV) through semimechanistic host response models that include transcriptional (cellular) dynamics, signaling cascades, and physiological components. Thus, this modeling effort intends to establish direct communication links from the cellular to the systemic host response level integrating bidirectional influences between central systems and physiological systems that should not be viewed as distinct functional domains. Such a modeling attempt could potentially provide invaluable insights into how disruption within these compartments contributes to morbidity and mortality in severely stressed patients, thus making it a critical enabler for the generation of alternative testable hypotheses that could clarify how cellular events and inflammatory processes mediate the links between patterns of autonomic control (HRV) and clinical outcomes.

10.2.3
Data Collection

A great deal about the initial human response to infection challenges has been learned from the elective administration of endotoxin [43], LPS, a major component of the Gram-negative bacteria outer membrane. This model has led to numerous relevant publications [9, 42, 43, 45] related to elucidating changes in metabolism and in the production of pro- and anti-inflammatory cytokines, as well as a wide range of gene expression and plasma concentration alterations as the result of intravenous administration of LPS.

The data used in this study are integrated across multiple scales including the cellular (gene expression) level, data at the level of circulating hormones, and finally at the systemic level, namely HRV. At the gene expression level, the data were generated as part of the *Inflammation and Host Response to Injury Large Scale Collaborative Project* funded by the USPHS, U54 GM621119 [66]. Human subjects were injected intravenously with endotoxin (CC-RE, lot 2) at a dose of 2 ng/kg bodyweight (endotoxin-treated subjects) or 0.9% sodium chloride (placebo treated subjects). Following lysis of erythrocytes and isolation of total RNA from leukocyte pellets [9], biotin-labeled cRNA was hybridized to the Hu133A and Hu133B arrays containing a total of 44,924 probes for measuring the expression level of genes that can be either activated or repressed in response to endotoxin. A group of 5093 probe sets were characterized by significant variation (corresponding to 0.1% false discovery rate) across the time course of the experiment using the SAM software [67]. The data are publicly available through the Gene Expression Omnibus database (http://www.ncbi.nlm.nih.gov/geo/) under the accession number GSE3284.

In addition to transcriptional profiling analysis, blood samples are also collected and analyzed to determine the plasma concentration of counter-regulatory hormones, such as epinephrine, after endotoxin administration at the following four time points: 0, 2, 4, and 6 h [68]. The purpose of this study was broader; it attempted to assess the effect of acute hypercortisolemia on hormonal and cytokine responses to human endotoxemia. Further, human volunteers were injected with the same amount of LPS while vital signs, including HRV indices, were recorded [69]. In particular, human subjects received an initial recording of heart rate and an electrocardiogram (ECG) to screen for any arrhythmic patterns or irregular heartbeats. During the analysis of HRV, parameters and interbeat intervals were collected using ECG data at a rate of 256 samples/s. Therefore, in a continuous ECG record, each QRS complex (which corresponds to the depolarization of the ventricles) was detected and the normal-to-normal (NN) intervals were tabulated. The overall HRV was assessed by calculating the standard deviation of normal interbeat intervals (SDNN). Therefore, at the systemic level HRV measurements are gathered from [69] while plasma cortisol concentrations are also recorded under the systemic inflammatory manifestations of human endotoxemia. Therefore, at the level of circulating hormones, the plasma concentration of stress hormones including cortisol and epinephrine are employed [68, 69]. The data have been appropriately de-identified, and appropriate IRB approval and informed, written consent were obtained by the glue grant investigators [9].

10.3 Methods

10.3.1 Developing a Multilevel Human Inflammation Model

Considering the leukocytes as a well-defined system, the purpose of human studies like the ones described in the previous section is to qualitatively characterize the cellular dynamics. The purpose of a systems biology approach, on the other hand, is to reverse engineer quantifiable representations of the intracellular dynamics [70] by identifying (i) appropriate constitutive elements; (ii) the topology of the interactions among these elements; and (iii) the quantitative relations among these elements. The analysis of gene expression data has provided a potential solution to the first issue. The following issues require the construction of the topology and dynamics of the underlying network describing the dynamics at a single scale, i.e., that of the leukocytes. Vodovotz and coworkers recently discussed the state of the art in mechanistic simulations of inflammation [21]. Therefore, the development of a cellular semimechanistic model is explored, particularly focusing on three unique aspects. *First*, the essential responses characterizing the cellular (transcriptional) dynamics are identified by the analysis of the leukocyte gene expression data. *Second*, the concept of physicochemical modeling [71] is used to express the intertwined relations and dynamics that connect extracellular signals and intracellular signaling cascades, eventually leading to emergent transcriptional dynamics (identified in the previous step). *Finally*, the pharmacodynamic concept of indirect response (IDR) [72] is applied to establish implicit interactions among signaling molecules and the emerging transcriptional responses.

10.3.1.1 Identification of the Essential Transcriptional Responses

Upon intravenous administration of endotoxin, the circulating leukocyte population undergoes a dynamic and reproducible set of changes, particularly in gene expression patterns, followed by a return to baseline within 24 h [9]. This response to endotoxin in humans includes changes in core temperature, cardiac, vasomotor, hematologic, metabolic, hormonal, acute phase reactant, and cytokine components that have been well described in the literature [43, 46–48]. Innate immune cells are activated and initiate the release of proinflammatory cytokines, which are the proximal mediators of the SIRS. Although the bulk of this proinflammatory mediator release likely originates in cells of the reticuloendothelial system [73], leukocytes in peripheral blood are also activated and are available for sampling with minimal invasiveness.

Model-based approaches have been proposed as a means to study the underlying complexity of the dynamics of inflammation and to establish quantifiable relationships among the various components of the inflammatory response [32, 35, 74]. A number of excellent prior studies [25, 26, 31–33, 75] have approached the simulation of inflammation based on the kinetics of well-known components of the inflammatory response. In other words, these models make *a priori* assumptions about certain components that are known to play a major role in triggering the inflammatory response based on our current biological knowledge [32, 35]. In line with the goal of creating a model with reduced complexity, made up only of the most essential components of the inflammatory response, it is desirable to identify a limited number of time-dependent interactions between key elements that are highly sensitive to specific modes of initiation and modulation of the response.

Clearly, the primary problem that needs to be addressed is the systematic identification of representative biological features that adequately represent the complex dynamics of a host undergoing an inflammatory response. This requires the decomposition of the nonlinear dynamics of inflammation into an elementary set of responses that can serve as a surrogate for the collective behavior of the system. This problem can be addressed by studying gene expression data concerning the host response to an inflammatory insult, consistent with the idea that cellular responses correspond to dynamically converging high-dimensional transcriptional trajectories [76]. Decomposing the dynamics of the entire system into a reduced set of intrinsic responses enables us to understand the complex dynamics of the system by studying the properties of its essential components. Given data generated by the transcriptional profiling of human blood leukocytes, we hypothesized that the genes that are most responsive to LPS have concerted changes in their expression profiles which are governed by specific mechanisms.

Based on previous work, a microclustering technique that symbolically transforms the gene expression time series data [77] based on the SAX method [78] is applied. This procedure is performed by averaging the replicates together at each time point, normalizing those average values so they have the same standard deviation for each gene, and setting breakpoints such that each symbol has an equal probability of being selected for random data. Therefore, genes with similar expression patterns will cluster to the same symbolic representation, which results in the enrichment of certain clusters. Enrichment is defined by calculating a *p*-value based on the cluster size and filtering out those expression motifs that are highly likely to be generated by random noise; this cutoff is set at $p = 1/$(number of motifs). The existence of enriched motifs is interesting because, in response to a stimulus, genes tend to be activated in concerted groups [79]. Therefore, the overrepresented motifs identified by this method likely represent biologically relevant signals.

This set of overrepresented clusters can be further pruned with the goal of finding the minimal set of clusters that maximally deviate from baseline conditions. This requires an exploration of the concept of "transcriptional state" (TS) that was introduced in [77]. The TS of the system is defined as the overall distribution of expression values at a specific time point; then, the deviation of the TS can be quantified at each time point with respect to a baseline distribution ($t = 0$ h) by applying the Kolmogorov–Smirnov test [80]. Given the aforementioned metric, the next step is to identify the reduced set of expression motifs which characterize the maximum deviation of the TS of the system. This defines a combinatorial optimization problem, which is addressed by a stochastic optimization algorithm based on simulated annealing [81].

The basic assumption is that due to an external disturbance, LPS administration in our case, the system is perturbed from homeostasis and eventually, once the LPS is cleared and the inflammatory mediators are eliminated, the host returns to the original state. Due to global nature of the transcriptional measurements and the fact that we do not *a priori* select a limited set of responsive genes, the entirety of the transcriptional response is expected to exhibit a rather Gaussian response with no clear defining responses [82]. However, through the use of the concept of TS it is possible to tease out the essential components of the cellular response to an external disturbance.

Three critical expression motifs are identified, and each is enriched in critical and relevant biological pathways: (i) early upregulation response (proinflammatory component, P) – genes in this major temporal class are important in activating transcription factors that act synergistically with proinflammatory transcription factors; (ii) late upregulation response (anti-inflammatory component, A) – genes in this functional class that counter-react the inflammatory response; and (iii) downregulation response (energetic component, E) – the downregulated essential response is characterized by a set of genes which are mainly involved in the cellular bioenergetic processes. These three intrinsic responses are then used as key components in the development of a mathematical model of inflammation.

10.3.1.2 Modeling Inflammation at the Cellular Level

In our injury model, the inflammatory response is initiated when endotoxin binds to its pathogen recognition receptors, TLR4 [83], which ultimately triggers the activation of proinflammatory transcription factors. Our inability to precisely model the complex signaling events that characterize system's adaptation to environmental changes makes IDR modeling appealing [84, 85]. In this type of situation, it is useful to apply the basic principles of IDR modeling, which are widely used in developing pharmacodynamic and pharmacogenomic models [65, 86]. The underlying assumption is that external signals indirectly affect the synthesis and/or degradation term of the response of interest. Therefore, in order to establish quantifiable relationships among the previously identified essential components of human inflammation, an IDR model was proposed in [22–24, 87]. The proposed model couples receptor-mediated phenomena with transcriptional effects based on ligand–receptor kinetics while each transcriptional motif is assumed to be the manifestation of a process involving a 0th order synthesis rate and a first-order degradation term.

In IDR models, external signals can either stimulate or inhibit the production and degradation rates of the response. The underlying assumption of this modeling effort is that intracellular signaling cascades activate inflammation-specific transcriptional responses [88]. Although a large number of transcription factors are known to be involved in inflammation, one a particular family, NF-κB, is of particular importance. First, the nuclear factor κB family is known to be a major player in the inflammatory response [94] and as such it has been widely studied. Second, this breadth of research on NF-κB has led to the development of numerous independent modeling approaches in order to quantify the expected response of its signaling cascade [93]. This leads to the hypothesis that NF-κB can serve as a proxy for the inflammation-specific transcription factor that initiates the expression of proinflammatory genes, and that this transcription factor is activated by the binding of LPS to its receptor. The binding interaction between endotoxin (LPS) and its receptor (R) is assumed to be a standard ligand–receptor interaction [92].

The proinflammatory response is the first-line transcriptional response after the recognition of LPS [88]. Proinflammation will serve as the signal that further stimulates the downregulation of genes that are associated with the cellular energetic processes [89]. We hypothesize that the proinflammatory response acts as a stimulatory factor for the energetic response, while a dysregulation in the cellular bioenergetics can serve as a positive feedback signal for the proinflammatory response. The anti-inflammatory response is the essential immunoregulatory signal that aims at restoring homeostasis in the host defense system. Thus, it will be stimulated by the activation of the inflammatory components, which are the proinflammatory and energetic responses. Furthermore, it will serve as the inhibitory signal on the production rate of the proinflammation and the energetic response.

The dynamics of LPS, the inflammatory stimulus, are described in (10.2a) by two terms: a logistic function with growth rate $k_{\text{lps},1}$ and a first-order elimination with rate $k_{\text{lps},2}$ [90]. Despite the presence of various mediators that are activated in response to LPS (e.g., LPS binding to the LBP plasma protein during its recognition by the host), we model LPS with a single compartment pharmacodynamic model, assuming a homogeneous circulating blood compartment. The logistic function is a sigmoidal function that is commonly used to model a variety of physical situations in which a quantity's growth is self-limited; specifically, initial growth is approximately exponential, but as the carrying capacity is approached, the growth monotonically goes to zero [90]. Therefore, by adjusting the relative magnitude of the two rate parameters of the clearance of LPS, we can simulate situations where the bacterial concentration is not fully eliminated. In human subjects, the endotoxin is cleared within the first 1–2 h after LPS administration with an approximate average half time $\tau_{1/2} \sim$ 8–15 min [91]. The two parameters $k_{\text{lps},1}$ and $k_{\text{lps},2}$ have been independently estimated so that the LPS profile decays within 2 h in the absence of any complications [91].

The dynamics of the TLR4 receptor (R), (10.2b), depend on the association and dissociation parameters of the ligand–receptor interaction [92]. These parameters, k_1 and k_2, are based on literature values [95]. The translation of R to its surface protein mRNA$_{,R}$ is governed by the rate k_{syn}, which describes the dynamic evolution of synthesis of new receptors; hence, this parameter is estimated so that the surface-free receptor is downregulated based on the premise that under the inflammatory stimulus, the surface-free receptors are occupied.

The dynamics of the gene transcript of the receptor, (10.2c), are characterized by a production rate ($K_{in,mRNA,R}$) and a degradation rate ($K_{out,mRNA,R}$) and are assumed to be indirectly stimulated by the proinflammatory signal. The measured $mRNA_{,R}$ is characterized by an upregulation for the first 4 h post-LPS administration followed by a return to baseline. Therefore, its two parameters are estimated so that we can best fit the available $mRNA_{,R}$ data. The equilibrium (LPSR) complex, (10.2d), is characterized by the binding parameters k_1 and k_2 as well as k_3, which shows the rate of formation of the activated signal (IKK) that eventually leads to the upregulation of the NF-κB transcription factor. Therefore, by this link we introduce the NF-κB signal transduction cascade, which serves to regulate the expression of proinflammatory genes.

Numerous signaling molecules and reactions participate in the NF-κB signaling pathway [93]; however, sensitivity analysis [96] demonstrated that the activity of NF-κB is primarily modulated by only IKK and IκBa. As such, [97] proposed a minimal model of NF-κB that accounts for the dynamic oscillations in NF-κB activity. Instead of simulating the kinase activity as a constant parameter and incorporating saturation and degradation rates as discussed in [97], we model IKK as a transient signal. Thus, the cellular surface complex (LPSR) induces the activation of kinase activity (IKK) with a rate k_3 while being eliminated with a rate k_4 (10.2e). The nonlinear Hill-type component is an essential functional form used to achieve a bistable response [98–101]. Such a bistability is an essential characteristic of the nonlinear dynamics of inflammation, as suggested by various animal studies [98–102]. In chronic inflammatory diseases, several cytokines might be responsible for perpetuating and amplifying inflammation through IKK [103]. Therefore, we simulate this interaction by the presence of a positive feedback loop in kinase (IKK) activity.

Because NF-κB is active as a transcription factor only when it is located in the nucleus, it is important to make the distinction in localization in the model. Assuming that NFκB$_n$ (10.2f) is the fraction of total NF-κB that is in the nucleus, the term (1-NFκB$_n$) denotes the available free cytoplasmic NF-κB; therefore, the terms nuclear concentration (NFκB$_n$) and nuclear activity are used interchangeably. The import rate of cytoplasmic NF-κB into the nucleus depends on the availability of free cytoplasmic NF-κB (1-NFκB$_n$), which is stimulated by IKK. However, the degradation rate of NFκB$_n$ depends on the presence of IκBa, its primary inhibitor, as it retrieves nuclear NF-κB by forming an inactive complex in the cytoplasmic region [104].

The dynamics of the gene transcript of IκBa ($mRNA_{,I\kappa Ba}$) (10.2g) are characterized by a zero-order production rate ($K_{in,I\kappa Ba}$) and a first-order degradation rate ($K_{out,I\kappa Ba}$), which is stimulated by NFκB [103]. The protein inhibitor IκBa (10.2h) is generated by the translation of its gene transcript ($mRNA_{,I\kappa Ba}$) and it degrades at a rate $k_{1,2}$, which is stimulated by IKK activity. IκBa forms a complex with the available cytoplasmic NF-κB; mathematically we express this as the product $(1\text{-NF}\kappa B_n)I\kappa Ba$. Finally, in order to achieve a zero steady state for the protein inhibitor $I\kappa Ba$, we need the additional negative term $-k_{I,1}$.

At the transcriptional response level, the nuclear activity of NF-κB (NFκB_n) serves as the activating signal that indirectly stimulates the essential proinflammatory response (P), which quantitatively is expressed by the linear function $H_{P,\text{NF}\kappa B_n}$ (10.2i). We are also assuming that the energetic response potentiates inflammation through $H_{P,E}$ [89]. The essential anti-inflammatory signaling component (10.2j) inhibits the production rate of the proinflammatory component; is stimulated by the activated proinflammatory response ($H_{A,P}$) as well as by the energetic response ($H_{A,E}$); and decays with the rate $k_{\text{out},A}$. The energetic variable (E) (10.2k) is stimulated by the proinflammatory response (P) and we are also assuming that the crucial anti-inflammatory component (A) counteracts both the proinflammatory and energetic responses of the system.

Thus, the human inflammation model consists of eleven (11) state variables: (i) the inflammatory instigator (LPS), (ii) the endotoxin signaling free protein receptor (R, TLR4), (iii) the mRNA of TLR4 (mRNA$_R$), (iv) the formed complex (LPSR), (v) the active kinase activity (IKK), (vi) the nuclear concentration of NF-κB (NFκB_n), (vii) the mRNA of $I\kappa$Ba (mRNA$_{I\kappa Ba}$), (viii) the translated protein inhibitor of N-FκB ($I\kappa$Ba), and finally the transcript abundance of the essential transcriptional responses of (ix) proinflammation (P), (x) anti-inflammation (A), and (xi) the energetic response (E). Given that anti-inflammatory drugs, such as corticosteroids, play a pivotal role in modulating the progression of inflammation, their contribution is further studied.

The contribution of corticosteroids, to be used as a template for assessing anti-inflammatory intervention strategies, is integrated into the model by incorporating appropriate PD/PK models. This is accomplished by drawing upon the significant prior research that has worked toward elucidating the mechanisms behind corticosteroid activity [105–112]. By taking glucocorticoid signaling into account, their pharmacogenomic effects can be simulated [113, 114]. To demonstrate the capability of our model to incorporate interacting modules, we opt to integrate the regulatory signaling information with the anti-inflammatory mechanism of corticosteroids. The corticosteroid intervention envelope consists of a set of interactions that involve (i) the binding of the corticosteroid drug (F) to its cytosolic receptor (GR); (ii) the subsequent formation of the drug–receptor complex (FR); (iii) the translocation of the cytosolic complex to the nucleus ($FR(N)$), where it regulates the expression of numerous genes; and finally (iv) the autoregulation of the gene transcript of the glucocorticoid receptor (R_m).

The drug is modeled either with a mono-exponential kinetic model that quantifies the plasma concentration of the steroid (F) (10.2l) under conditions of intravenous injection or with a time invariant parameter (R_{in}) that reflects the constant infusion rate. In both cases, the drug is eliminated with a rate k_{el}. The dynamics of the gene transcript of the corticosteroid drug (R_m) (10.2m) are characterized by a zero-order production rate ($k_{\text{syn\_Rm}}$) and a first-order degradation rate (k_{deg}). The active drug–receptor complex ($FR(N)$) exerts an inhibitory effect on the mRNA of the glucocorticoid receptor. The parameter IC_{50_Rm} is the concentration of the nuclear drug–receptor complex $FR(N)$ at which the synthesis rate of the receptor drops to half of its baseline value. The dynamics of the free cytosolic receptor density, GR, are shown in (10.2n) where $k_{\text{syn\_R}}$ is the synthesis rate of receptor, R_f is the fraction of the drug that is recycled, k_{re} represents the recycling of drug from the nucleus to the cytosol, and k_{on} is associated with the drug–receptor binding. In addition, $k_{\text{dgr\_R}}$ is the degradation rate of GR. The formed cytosolic complex (FR) (10.2o) is characterized by the binding interaction k_{on} of the ligand (F) with its receptor (GR) and on its translocation rate k_T to the nucleus. Therefore, the translocation of the drug–receptor complex to the nucleus is represented by the nuclear receptor complex $FR(N)$, which is the active complex that regulates various genes in the nucleus. By incorporating an appropriate PK/PD model, we can study the corticosteroid-dependent regulation of either the inhibitor of NFκB (10.2p) or the anti-inflammatory component of the response (10.2q). We recently demonstrated that this indirect model properly captures the onset and resolution of inflammation; additionally, it is able to predict a number of clinically relevant responses that are beyond the range of data used in parameter estimation [22–24]. This justifies the fundamental assumptions used in the establishment of functional relationships between the individual components that define the architecture of the model. The mathematical representation of this model is succinctly presented in Eq. (10.2) as follows:

$$\frac{d\text{LPS}}{dt} = k_{\text{lps},1} \cdot \text{LPS} \cdot (1 - \text{LPS}) - k_{\text{lps},2} \cdot \text{LPS} \tag{10.2a}$$

$$\frac{dR}{dt} = k_{\text{syn}} \cdot \text{mRNA},R + k_2 \cdot (\text{LPSR}) - k_1 \cdot \text{LPS} \times R - k_{\text{syn}} \cdot R \tag{10.2b}$$

$$\frac{d(\text{LPSR})}{dt} = k_1 \cdot \text{LPS} \times R - k_3 \cdot (\text{LPSR}) - k_2 \cdot (\text{LPSR}) \tag{10.2c}$$

$$\frac{d(\text{mRNA},R)}{dt} = K_{\text{in,mRNA},R} \cdot (1 + H_{P,R})$$

$$\quad - K_{\text{out,mRNA},R} \cdot \text{mRNA},R \tag{10.2d}$$

$$\frac{d\text{IKK}}{dt} = k_3 \cdot (\text{LPSR})/(1 + \text{I}\kappa\text{Ba}) - k_4 \cdot \text{IKK} + P \cdot \left(\frac{\text{IKK}^2}{1 + \text{IKK}^2}\right) \tag{10.2e}$$

$$\frac{d\text{NF}\kappa\text{B}_n}{dt} = \frac{k_{\text{NF}\kappa\text{B},1} \cdot \text{IKK} \cdot (1 - \text{NF}\kappa\text{B}_n)}{(1 + \text{I}\kappa\text{Ba})} - k_{\text{NF}\kappa\text{B},2} \cdot \text{NF}\kappa\text{B}_n \times \text{I}\kappa\text{Ba} \tag{10.2f}$$

$$\frac{d\text{mRNA}_{\text{I}\kappa\text{Ba}}}{dt} = K_{\text{in,I}\kappa\text{Ba}} \cdot (1 + k_{\text{I}\kappa\text{Ba},1} \cdot \text{NF}\kappa\text{B}_n)$$

$$\quad - K_{\text{out,I}\kappa\text{Ba}} \cdot \text{mRNA}_{\text{I}\kappa\text{Ba}} \tag{10.2g}$$

$$\frac{d\mathrm{I}\kappa\mathrm{Ba}}{dt} = k_{\mathrm{I},1} \cdot \mathrm{mRNA}_{\mathrm{I}\kappa\mathrm{Ba}}$$

$$- k_{\mathrm{I},2} \cdot (1 + \mathrm{IKK}) \cdot (1 - \mathrm{NF}\kappa\mathrm{B}_n) \cdot \mathrm{I}\kappa\mathrm{Ba} - k_{\mathrm{I},1} \quad (10.2\mathrm{h})$$

$$\frac{dP}{dt} = K_{\mathrm{in},P} \cdot (1 + H_{P,\mathrm{NF}\kappa\mathrm{B}_n}) \cdot (1 + H_{P,E})/A - K_{\mathrm{out},P} \cdot P \quad (10.2\mathrm{i})$$

$$\frac{dA}{dt} = K_{\mathrm{in},A} \cdot (1 + H_{A,P}) \cdot (1 + H_{A,E}) - K_{\mathrm{out},A} \cdot A \quad (10.2\mathrm{j})$$

$$\frac{dE}{dt} = K_{\mathrm{in},E} \cdot (1 + H_{E,P})/A - K_{\mathrm{out},E} \cdot E \quad (10.2\mathrm{k})$$

$$\frac{dF}{dt} = \begin{cases} R_{\mathrm{in},F} - k_{\mathrm{el}} \cdot F, & \text{infusion} \\ -k_{\mathrm{el}} \cdot F, & \text{injection} \end{cases} \quad (10.2\mathrm{l})$$

$$\frac{dR_m}{dt} = k_{\mathrm{syn\_Rm}} \cdot \left(1 - \frac{FR(N)}{IC_{50\_\mathrm{Rm}} + FR(N)}\right) - k_{\mathrm{deg}} \cdot R_m \quad (10.2\mathrm{m})$$

$$\frac{dR_F}{dt} = k_{\mathrm{syn\_R}} \cdot R_m + r_f \times k_{\mathrm{re}} \cdot FR(N)$$

$$- k_{\mathrm{on}} \cdot F \times R_F - k_{\mathrm{dgr\_R}} \times R_F \quad (10.2\mathrm{n})$$

$$\frac{dFR}{dt} = k_{\mathrm{on}} \cdot F \cdot R_F - k_T \times FR$$

$$\frac{dFR(N)}{dt} = k_T \times FR - k_{\mathrm{re}} \cdot FR(N) \quad (10.2\mathrm{o})$$

$$\begin{cases} \frac{d\mathrm{mRNA}_{\mathrm{I}\kappa\mathrm{Ba}}}{dt} = K_{\mathrm{in},\mathrm{I}\kappa\mathrm{Ba}} \cdot (1 + k_{\mathrm{I}\kappa\mathrm{Ba},1} \cdot \mathrm{NF}\kappa\mathrm{B}_n) \cdot (1 + FR(N)) \\ \quad - K_{\mathrm{out},\mathrm{I}\kappa\mathrm{Ba}} \cdot \mathrm{mRNA}_{\mathrm{I}\kappa\mathrm{Ba}} \\ \frac{dA}{dt} = K_{\mathrm{in},A} \cdot (1 + H_{A,P}) \cdot (1 + H_{A,E}) \cdot (1 + FR(N)) - K_{\mathrm{out},A} \cdot A \end{cases} \quad (10.2\mathrm{p})$$

$$H_{i,j} = k_{i,j} \cdot J \quad (10.2\mathrm{q})$$

10.3.1.3 Modeling Inflammation at the Systemic Level

While the leukocyte transcriptional response to endotoxin is important, it is not the only effect of endotoxin treatment. In addition, a neuroendocrine response characteristic of acute injury and sepsis is induced [115]. A rise in circulating endocrine hormones is manifested 2–4 h after endotoxin administration [43]. Moreover, diminished HRV, which represents autonomic dysfunction, is also induced in human endotoxemia [16]. The physicochemical host response model described in Eq. (10.2) demonstrates how we can link the initiating signal (LPS) with cellular (leukocyte) transcriptional dynamics. However, in order to simplify the model, the immunomodulatory role of autonomic (hormonal) influences, including endogenous cortisol, is not accounted for. Therefore, the model can be extended by considering critical aspects of the neuroendocrine-immune system axis. We opt, therefore, to extend our prior modeling effort by incorporating components of the neuroendocrine-immune crosstalk that connects the cellular response level with neural-based pathways, whose disturbance is assessed by HRV.

10.3.1.4 Modeling Neuroendocrine–Immune System Interactions

A low-dose endotoxin injection leads to stress hormone secretion [43]. The CNS regulates the immune system in response to stress primarily by the HPA axis, which is activated to produce glucocorticoids. The HPA axis, one of the peripheral limbs of the stress system, responds to many circadian and blood-borne signals [116]. These signals include proinflammatory cytokines produced by immune-mediated inflammatory reactions [117]. As such, in inflammation, corticotrophin-releasing hormone (CRH) is secreted form the hypothalamus, which subsequently stimulates the expression of adrenocorticotrophic hormone. This hormone ultimately moves to the adrenal glands where it increases the expression and release of glucocorticoids [118]. Therefore, the acute production of glucocorticoids in the adrenal cortex is the main effector end point of this neuroendocrine system [119].

While glucocorticoids are important in the regulation of immune function and inflammation, other counter-regulatory hormones associated with distress, including catecholamines (epinephrine, EPI), can also modulate a range of immune functions [118]. Such hormones are secreted by the sympathetic nervous system (SNS) and interact with adrenergic receptors on immune cells. Considerable evidence suggests that the proinflammatory response (P) acts as the signal that stimulates central components of the stress system through the afferent vagus nerve [54]. It is reasonable to assume that the proinflammatory response (P) acts as the signal that not only stimulates the secretion of cortisol but also stimulates the secretion of catecholamines (epinephrine). Thus, the model incorporates the effects of both glucocorticoids and epinephrine.

In order to mathematically describe the secretion of stress hormones, the principles of IDR modeling are employed. To simulate conditions both with and without exogenous hydrocortisone infusion, the binary decision variable w_{Fex} is introduced. In the absence of exogenous steroid ($w_{\text{Fex}} = 0$), the plasma concentration of cortisol (F) is assumed to be the manifestation of a zero-order production rate ($K_{\text{in,Fen}}$) and a first-order degradation term ($K_{\text{out},F}$) (10.3a). Upon inflammatory stimulation, the proinflammatory response (P) indirectly stimulates the HPA axis through the linear function ($H_{\text{Fen},P}$) (10.3b). Further, the administration of exogenous steroid ($w_{\text{Fex}} = 1$) is simulated via the time invariant parameter (R_{in}) that reflects the constant infusion rate that persists as long as the infusion lasts. Therefore, total cortisol concentrations (F) are defined as the joint effect between endogenous and exogenous steroid as outlined in [120]. It is expected that under conditions of hypercortisolemia, which can be induced by constant steroid infusion, plasma cortisol levels will eventually increase and potentiate the anti-inflammatory arm of the host defense system [68, 121, 122].

To mathematically approximate the afferent transit mechanism describing the propagation of the local proinflammatory signal (P) to the SNS, the secretion of epinephrine is represented by a zero-order production rate ($K_{in,EPI}$) and a first-order degradation rate ($K_{out,EPI}$) (10.3b) and the proinflammatory response (P) stimulates epinephrine production through the linear function $H_{EPI,P}$.

As a result of the activation of the SNS, the released endogenous catecholamines (EPI) interact with α- and β-adrenergic receptors in immune cells, thus modulating the inflammatory response [123]. Short-term preexposure of healthy human subjects to epinephrine attenuates the proinflammatory components that are typically manifested in human endotoxemia, as shown by reduced TNF levels [124]. β-adrenergic stimulation mediates the anti-inflammatory effects of epinephrine, resulting in an increase in intracellular cAMP levels [125]. Additionally, experimental evidence further documents that increased cAMP signaling under conditions of acute epinephrine infusion decreases the production rate of IL-10 (A) [123, 124]. However, long-term preexposure of human subjects to epinephrine infusion before the main endotoxin challenge diminishes the anti-inflammatory capacity of epinephrine [124]. This tolerance effect is simulated with a precursor-dependent IDR model, as discussed in [126]. In particular, the precursor reflects a diminished anti-inflammatory effect in terms of changes in the duration of epinephrine infusion that can be simulated due to alterations in the concentration of β-adrenergic receptor (R_{EPI}).

The dynamic changes of epinephrine's receptor (R_{EPI}) depend on an apparent zero-order production rate k^0_{REPI}, while $k_{1,REPI}$ and $k_{2,REPI}$ represent first-order rate constants for the loss of the receptor (10.3c). The response to epinephrine is triggered in response to the binding of EPI to its receptor and the formation of an active signaling complex, so EPIR represents the formed signaling complex that decays with a first-order rate $k_{3,EPI}$ (10.3d). Further, as described in [127, 128], the stimulatory postadrenergic effect of sympathetic activity favors the production of cAMP signaling. This is described by the principles of a signal transduction model, where the production and loss of the cAMP signaling depends on first-order rate constants, which are equivalent to the reciprocal of the transit times (τ), consistent with the transit compartment model, while n is the shaping (scaling) factor (10.3e) [126]. The scaling factor is used to amplify the signal transduction cascade associated with the postadrenergic effect of epinephrine on the host.

Therefore, with this quantification of cortisol and epinephrine dynamics, the influence of neuroendocrine hormones on the host response to endotoxin can be further detailed. It has been demonstrated that hypercortisolemia increases plasma IL-10 (A) concentration during human endotoxemia [122]. As such, it is assumed that cortisol modulates the host response to endotoxin primarily via potentiation of IL-10 signaling (A). This steroid-dependent immunomodulatory effect is quantified via the stimulatory function $H_{A,FRN}$. In addition, epinephrine potentiates anti-inflammatory signaling through a cAMP-dependent mechanism that is quantified through the linear stimulatory function $H_{A,cAMP}$ (10.3f).

In addition to the neuroendocrine response evoked by endotoxin, the model incorporates autonomic dysfunction as assessed by HR. Recent studies imply that disordered neuroendocrine functions are associated with diminished HRV in stressed patients [16]. To quantify disturbances in autonomic activities, clinical measurements of HRV, at the systemic level, will be modeled.

10.3.1.5 Modeling the Effect of Endotoxin Injury on Heart Rate Variability

In human endotoxemia, the host undergoes a variety of responses, including a decrease in physiologic variability [59, 60] as evidenced in recent clinical data measuring HRV [43]. These data establish that the host response to endotoxin causes depressions in cardiac-vagal tone and in overall HRV. HRV analysis has been extensively used to evaluate autonomic modulation of sinus node in normal subjects. In addition, HRV analysis has been applied to critically ill patients with the goal of identifying those who are at increased risk with respect to mortality. Along these lines, several studies have described the use of HRV as a readily available vital sign [129] in the assessment of critically ill patients with the hope of earlier intervention for those patients deemed at higher risk [130–133]. These studies highlight how the prognostic significance of HRV has made it a critical enabler in the detection of physiologic deterioration and the response to treatment [134].

In our injury model, the overall HRV was assessed by the SDNN parameter, which is the standard deviation of normal to normal interbeat intervals; this parameter was significantly attenuated in response to a low dose of endotoxin [69]. In the context of modeling inflammation, a critical question arises that involves the relationship between proinflammatory markers and autonomic dysfunction. There is considerable evidence indicating that systemic low-grade proinflammatory activity is associated with reduced HRV in humans [135–137]. It is therefore reasonable to assume that proinflammation serves as a surrogate activating signal that indirectly modulates autonomic activity, as assessed by HRV. In order to establish quantifiable relationships between proinflammatory signaling (P) and autonomic control (HRV), the possibility of nonlinear effects must be addressed. Assuming a linear relationship between proinflammation and HRV would imply that any modulations in the peripheral immune response will subsequently drive changes in hemodynamic parameters. However, it cannot be taken for granted that factors modifying the immune response should affect all circulatory parameters equally [138].

Such nonlinearities in sinus node transduction processes may arise from the sensitivity of pacemaker discharge to the timing of pulsatile neural activity and from functional in-homogeneity within the sinus node tissue [139]. In addition, the concept of nonlinearity in cardiovascular variability has been stressed in the hemodynamic parameters of endotoxin-induced systemic inflammation under conditions of prior endocrine stress hormone infusion [69, 140]. In this context, it was observed that the anti-inflammatory influence of endocrine hormones, including cortisol and epinephrine, does not affect the changes in HRV induced by LPS. Thus, a reduction in endotoxin-induced proinflammation does not influence autonomic dysfunction (HRV).

On the other hand, this relationship was explored in a large study of trauma patients exhibiting diminished HRV [132]. It was suggested that among injured patients with diminished HRV, there exists a subset of patients with established adrenal insufficiency who, upon treatment with exogenous steroid, exhibit both increased HRV and improved clinical outcome.

In order to quantify such nonlinear interactions, the effect of endotoxin-induced inflammation on HRV will be mathematically approximated by employing appropriate sigmoidal activation functions as outlined in [141]. The physiological background for heart rate variation involves the activation of signal transduction mechanisms in the sinus node of the heart associated with the modulation of neuromediator concentrations [142]. The possible nonlinear modulatory effect of proinflammation (P) upon HRV is described by the dynamics of f_P (10.3g), where the switch-like behavior is determined by the sigmoid function ($\tanh(P - w)$) and w is a parameter greater than the proinflammatory response (P) elicited upon endotoxin-induced inflammation. This nonlinear gain modulatory function should be active during inflammation and inactive when the system is at homeostasis. We therefore model this event based on the function $\tanh(P^\phi - 1)^\phi$, which takes a value of zero when proinflammation (P) lies at its baseline value (homeostasis) and one otherwise. The underlying rationale for this function is predicated upon a neuro-computational model [143], which aims at simulating the firing rate of neuronal activity. In our model, f_P qualitatively reflects the activation of efferent nerve activity on the heart, eventually leading to the upregulation of intracellular mediators (S_f). The loss and production of such mediators (10.3h) is thereby described by the principles of a time-dependent transduction system [126] depending on two parameters including a first-order parameter (τ_S) and a scaling factor (n_S). The dynamics of HRV (10.3i) are described by a zero-order production rate ($K_{in,HRV}$) and a first-order degradation rate ($K_{out,HRV}$) stimulated by the effector biological signal (S_f). Taken together, the integrated module that describes critical aspects of the neuroendocrine-immune system interactions is presented in Eq. (10.3). This module can be used in conjunction with Eq. (10.2) to simulate a wide variety of conditions,

$$\frac{dF}{dt} = w_{F_{ex}} \cdot R_{in} + K_{in,F_{en}} \cdot (1 + H_{F_{en},P}) - K_{out,F} \cdot F \tag{10.3a}$$

$$\frac{dEPI}{dt} = w_{EPI_{ex}} \cdot R_{in,EPI} + K_{in,EPI} \cdot (1 + H_{EPI,P}) - K_{out,EPI} \cdot EPI \tag{10.3b}$$

$$\frac{dR_{EPI}}{dt} = k^0_{R_{EPI}} - \left[k_{1,R_{EPI}} \cdot (1 + H_{R_{EPI},EPI}) + k_{2,R_{EPI}}\right] \cdot R_{EPI} \tag{10.3c}$$

$$\frac{dEPIR}{dt} = k_{1,R_{EPI}} \cdot (1 + H_{R_{EPI},EPI}) \cdot R_{EPI} - k_{3,EPIR} \cdot EPIR \tag{10.3d}$$

$$\frac{dcAMP}{dt} = \frac{1}{\tau} \cdot \left(EPIR^n - cAMP\right) \tag{10.3e}$$

$$\frac{dA}{dt} = K_{in,A} \cdot (1 + H_{A,cAMP}) \cdot (1 + H_{A,E}) \cdot (1 + H_{A,FRN})$$
$$- K_{out,A} \cdot A \tag{10.3f}$$

$$\frac{df_P}{dt} = \left(1 + \tanh(P-w) - f_P\right) \cdot \tanh\left(P^j - 1\right)^\phi \tag{10.3g}$$

$$\frac{dS_f}{dt} = \frac{1}{\tau_S} \cdot \left(\left(\tanh\left(P^j - 1\right)^j \times f_P\right)^{ns} - S_f\right) \tag{10.3h}$$

$$\frac{d\text{HRV}}{dt} = K_{\text{in,HRV}} - K_{\text{out,HRV}} \cdot (1 + k_{\text{HRV},P} \times S_f) \cdot \text{HRV} \tag{10.3i}$$

$$H_{i,j} = k_{i,j} \times J$$

$$w_{i_{\text{ex}}} = \begin{cases} 1, & \text{infusion,} \\ 0, & \text{elsewhere,} \end{cases} \qquad i = \begin{cases} F \\ \text{EPI} \end{cases} \tag{10.3j}$$

10.4
Results

10.4.1
Transcriptional Analysis and Major Response Elements

The symbolic transformation of the expression motifs and the subsequent assignment of hash values to each expression profile [77] produces a distribution of motif values for all the available probes, Fig. 10.1. In order to estimate a p-value for each expression motif, we generate random data with the same dimensions as the original dataset (5093 probe sets and 6 time points). Genes that hash to the same integer value for the random data are characterized by a distribution that approximately follows an exponential decay and subsequently we can estimate the cumulative distribution for an exponential model. Thus, an appropriate p-value $= 1/($total number of expression motifs$)$, which equals 0.0045, is used to evaluate highly nonrandom expression motifs (clusters). The analysis generates a subset of 16 transcriptional motifs, which are considered to be statistically significant in terms of their population size, Fig. 10.2. Therefore, this family of expression motifs, and the associated probe sets, is most characteristic of the exposure of the host to LPS. In order to evaluate the essential basis set, subsets of the 16 profiles that exhibit the largest deviation from homeostasis, we solve the combinatorial selection problem which predicts that the maximum deviation from homeostasis would require three elementary motifs. As seen in Fig. 10.3 the maximum deviation from homeostasis is observed for three motifs, whereas further addition of motifs reduced the deviation, indicating the addition of less critical responses. Therefore, this result illustrates the existence of distinct critical sets of temporal responses that capture the intrinsic dynamics of the host response to endotoxin administration, Fig. 10.4. The first response is characterized by an early increase in the gene expression level during the first 2 h after the endotoxin challenge, whereas the second essential response shows an increase in the gene expression level at a later time (4–6 h). The third response is characterized by a downregulation during the time course of the exper-

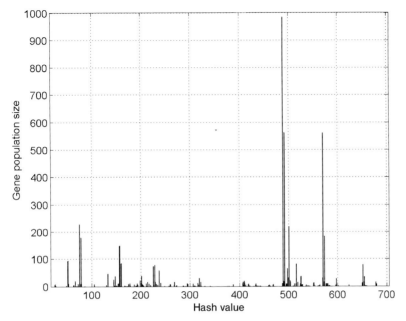

Fig. 10.1 Number of probe sets in each expression motif. A histogram of hash values is used to visualize the number of probe sets in each microcluster – 5093 probe sets are microclustered to 224 expression motifs.

iment and an eventual return to baseline at 24 h. All responses resolve, i.e., return to base line 24 h postexposure, which is in agreement with the overall design of the study and the reversible nature of the elicited response. Upon identification of the probe sets composing these three essential transcriptional responses, we identify significant localization in relevant biological pathways.

The early upregulation response, denoted as the proinflammatory component (P), is enriched in genes involved in cytokine–cytokine receptor interactions (C-X-C motifs and cytokines: CXCL1, CXCL2, CCL20) as well as in Toll-like receptor signaling pathway (CCL4, IL1B, IL8). Moreover, we identified genes (*IL1A, IL1B, IL1R1, IL1R2*) that participate in the MAPK signaling pathway, which is crucial in activating transcription factors that act synergistically with proinflammatory transcription factors such as members of NF-κB/RelA family. The late upregulation response, denoted as the anti-inflammatory component (A), is enriched in genes functionally participating in the JAK-STAT signaling pathway (*IL10RB, JAK3, STAT2, STAT5B*) which is essential to regulate the expression of target genes that counter-react the inflammatory response. In addition to this, it is emphasized [144] that a STAT pathway from a receptor signaling system is a major determinant of key regulatory systems, including feedback loops such as SOCS induction which subsequently suppresses the early induced cytokine signaling. In this class, we also capture the increased expression of the gene (*SP1*) that encodes

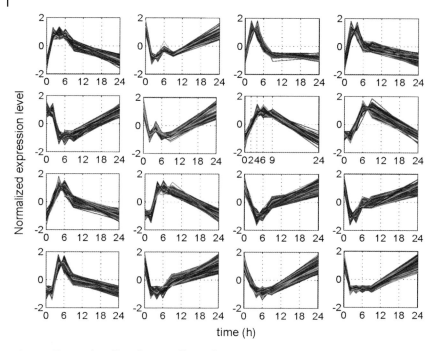

Fig. 10.2 Temporal profiles of statistically significant expression motifs – normalized expression values of motifs with p-value < 0.0045 (with respect to cluster size) vs. time.

a protein which acts as an essential activator for *IL10* signaling [145]. Moreover, we identified the late increased expression of *IL10RB*, which is assumed to be indicative of the *IL10* signaling cascade. Finally, the downregulation response, denoted as the energetic component (E), is characterized by a set of genes, which are mainly involved in the cellular bioenergetic processes. In addition we find genes essential to ribosome biogenesis and assembly (RPL/RPS family) as well as genes participating in the protein synthesis machinery, oxidative phosphorylation (ATP5A, COX11, NDUFA11), and pyruvate metabolism (PDHB, PDHX, MDH1). Endotoxin-induced inflammation causes the dysregulation of leukocyte bioenergetics and a persistent decrease in mitochondrial activity can lead to reduced cellular metabolism [146]. Organ function has been associated with changes in bioenergetics and metabolic activity [147]. These transcriptional responses effectively decompose the overall dynamic and potentially define the constitutive elements of the overall response, thus defining the transcriptional signatures in response to LPS administration. We will further explore the possibility of using the elementary responses as the surrogates for predicting the complex dynamic behavior of the system through an appropriately constructed mathematical model.

In order to reproduce the experimental data, we select the transcriptional signature of specific genes representative of each essential response. *IL1B* is selected to

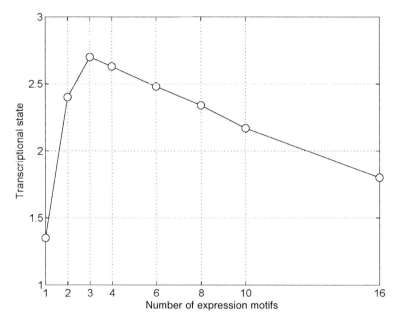

Fig. 10.3 Deviation of the transcriptional state of the system vs. number of expression motifs – the maximum perturbation in the intrinsic dynamics of the system occurs for three distinct expression motifs.

serve as the representative biomarker of the proinflammatory response. The gene transcript of *IL10RB* is considered to be indicative of the immune-regulatory signal of the anti-inflammatory response. Finally, a subunit of NADH ubiquinone dehydrogenase complex (mitochondrial component) NDUFC2 is considered as the proxy for the energetic component. These essential transcriptional signatures are normalized by taking the ratio of the measured mRNA level at each time point with respect to the control time point ($t = 0$ h). Selecting any other gene that belongs to the aforementioned essential inflammatory responses can very well be used as a surrogate for a representative of the response and will not alter the qualitative characteristic of our semimechanistic mathematical model.

10.4.2
Elements of a Multilevel Human Inflammation Model

We have previously demonstrated that the transcriptional dynamics of human leukocytes exposed to bacterial endotoxin can be decomposed into three elementary responses [87]. Unlike previous approaches [19, 25, 26, 31–33, 35] that focus on specific biomarkers, these elementary responses capture the functional dynamics in three categories: proinflammatory (P), anti-inflammatory (A), and energetic (E) transcriptional events. These responses are triggered by the activation of the

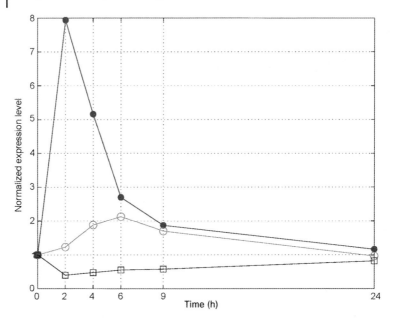

Fig. 10.4 Essential leukocyte transcriptional elements – (solid markers) pro-inflammatory response (P), (open circles) anti-inflammatory response (A), and (square markers) energetic response (E).

NF-κB signaling pathway as a result of the binding of LPS to its receptor (R). In order to introduce higher level biological information, we further consider essential aspects of the crosstalk between the peripheral immune response and the CNS. A schematic illustration of the network architecture that constitutes the multilevel host response model is presented in Fig. 10.5. At the cellular level, elementary signaling pathways propagate extracellular signals to the emergent transcriptional response level. At the autonomic (CNS) level, essential modules associated with the release of endocrine stress hormones are incorporated. These stress hormones are integral components of the bidirectional communication pathway between peripheral inflammation (cellular level) and the neuroendocrine axis (HPA, SNS). As such, the essential proinflammatory response (P) acts as the peripheral intercellular messenger that stimulates the hardwired brain network, particularly at the neuroendocrine level. Activation of the neuroendocrine-immune axis releases stress hormones, including cortisol and epinephrine, which interact with appropriate receptors in immune cells, thereby potentiating the production rate of humoral anti-inflammatory cytokines (A). Finally, at the systemic level, clinical measurements of HRV are also incorporated to assess disruptions in autonomic activities, which reflect the manifestation of systemic decomplexification and correlate with the severity of the host response. Thus, the proposed modeling effort associates acquired endocrine dysfunction with reduced HRV, enabling assessments of the influence of neuroendocrine activity at both the cellular and systemic levels.

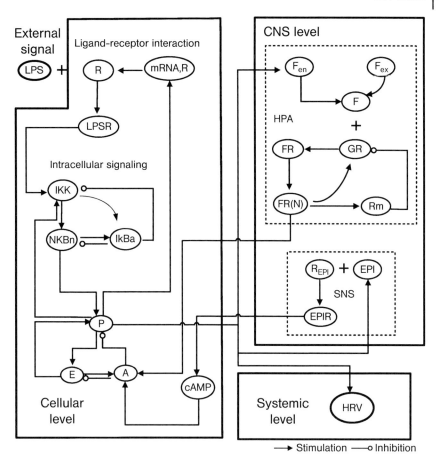

Fig. 10.5 Topological interactions composing the multiscale model of inflammation.

10.4.3
Estimation of Relevant Model Parameters

Given the available human experimental data, standard parameter estimation techniques are applied in order to evaluate appropriate model parameters [148]. In particular, we estimate those parameters that are involved in the dynamics of proinflammation (P), anti-inflammation (A), energetics (E), NF-κB inhibitor (mRNA$_{I\kappa Ba}$), TLR4 (mRNA$_R$), cAMP mediator (cAMP), epinephrine (EPI), cortisol (F), and heart rate response (HRV). Further, parameter estimation is performed to estimate the parameter ($R_{in,F}$) under conditions of hydrocortisone infusion (exogenous cortisol), reproducing human plasma cortisol levels in subjects exposed to cortisol from 6 h before to 6 h after LPS treatment. All the parameters associated with the propagation of LPS signaling on the transcriptional and neuroendocrine levels are shown in Table 10.1 while the performance of the multilevel human in-

Table 10.1 Estimated values of the parameters in the multilevel human inflammation model.

Parameter	Value	Parameter	Value	Parameter	Value
$k_{LPS,1}$	4.500	$K_{in,A}$	0.461	$K_{EPI,P}$	0.231
$k_{LPS,2}$	6.790	$K_{out,A}$	0.809	$k_{1,REPI}$	2.657
k_{syn}	0.020	$k_{A,cAMP}$	0.145	$k_{R,EPI}$	0.649
k_1	3.000	$k_{A,E}$	0.534	$k_{2,REPI}$	2.213
k_2	0.040	$k_{A,FRN}$	0.401	$k_{3,REPI}$	2.500
k_3	5.000	$K_{in,E}$	0.080	T	0.053
k_4	2.240	$K_{out,E}$	0.280	N	5.509
$k_{in,mRNA,R}$	0.090	$k_{P,R}$	1.740	$K_{out,EPI}$	7.286
$k_{out,mRNA,R}$	0.250	$k_{P,1}$	29.740	k_{REPI}^0	6.594
$k_{NF\kappa B,1}$	16.290	$k_{P,2}$	9.050	W	10
$k_{NF\kappa B,2}$	1.180	$k_{E,P}$	2.210	$\tau_{HRV,SNS}$	1.174
$K_{in,I\kappa Ba}$	0.460	$R_{in}\ (w_{Fex}=0)$	0	n_S	0.924
$k_{I\kappa Ba,1}$	13.270	$K_{in,Fen}$	0.796	$K_{in,HRV}$	0.101
$k_{I,1}$	1.400	$k_{Fen,P}$	0.203	$K_{out,HRV}$	0.101
$k_{I,2}$	0.870	$K_{out,F}$	0.957	$k_{HRV,SNS}$	9.984
$K_{in,P}$	0.030	$R_{in}\ (w_{Fex}=1)$	2.055		
$K_{out,P}$	0.330	$K_{in,EPI}$	5.921		

flammation model is shown in Fig. 10.6. In addition to this, the reconstruction of plasma cortisol levels under conditions of prior cortisol infusion is shown in Fig. 10.7.

In our computational model, the host restores homeostasis without any external perturbation. A self-limited inflammatory response to the endotoxin stimulus corresponds to resolved dynamic profiles for all the elements that constitute our model. In essence, a self-limited inflammatory response involves the successful elimination of the inflammatory stimulus within the first 2 h postendotoxin administration while followed by a subsequent resolution within 24 h. Although, the kinetic parameters associated with the epinephrine-receptor interactions are not calibrated, the dynamic profile of β-adrenergic receptor (R_{EPI}) lies in qualitative agreement with the basis of receptor occupancy theory [128] in that the concentration of free adrenergic receptors decreases in the presence of the ligand (EPI). Regarding the gain modulatory effect of peripheral proinflammation in the heart (f_P), such an exponential decrease would biologically reflect the decay rate of cardiac neuronal activity [149], which is eventually "translated" to the upregulation of neuro-ediator concentrations ($S_{HRV,P}$) in the heart. While comparing the top and bottom panel of Fig. 10.7, both plasma cortisol levels (F_{Fex}) and the steroid active signal ($FR(N)_{Fex}$) are expected to be greater under conditions of hypercortisolemia (Figs. 10.7(C) and (D)) relative to the baseline cortisol profiles (Figs. 10.7(A) and (B)).

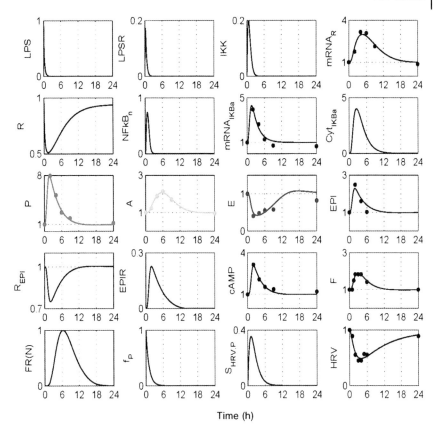

Fig. 10.6 Estimation of relevant model parameters to reproduce experimental results: solid lines (—) correspond to model predictions under LPS administration (LPS($t = 0$ h) $= 1$) while the symbols (•) refer to experimental data.

10.4.4
Qualitative Assessment of the Model

Ultimately, we are interested in applying mathematical models toward predicting relevant biological implications to the host response to endotoxin, particularly with respect to controlling and modulating the phenomenon. Thus, the correctness of the model is tested based on its ability to not only reproduce available data, but also on its ability to qualitatively predict and modulate uncontrolled responses. In the following sections, this ability is demonstrated to enable such predictions and provide further evidence of the appropriateness of the assumptions invoked in the development of the model. First, we explore the implications of increasing the level of initial insult (LPS), since this constitutes the most obvious irreversible disturbance. Second, we explore possible mechanistic dysregulations, which may reflect secondary effects that perturb the system toward sustained inflammation.

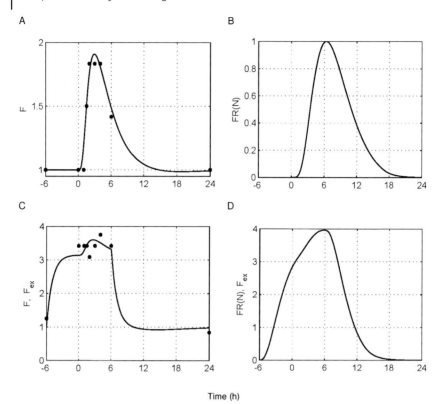

Fig. 10.7 (A) Plasma cortisol levels (F) and (B) steroid active signal ($FR(N)$) under conditions of acute endotoxin injury ($LPS(t=0\,h)=1$); (C) simulated plasma cortisol levels and (D) steroid active signals $FR(N)_{Fex}$ under conditions of low-dose steroid infusion ($w_{Fex}=1$), which is initiated at $t=-6\,h$ before LPS and continued for 6 h after LPS. Solid lines correspond to model predictions while solid markers represent experimental data.

Then, we explore the emergence of memory effects and evaluate the implication of priming the host with an initial endotoxin stimulus before applying the main endotoxin stimulus. Finally, we evaluate possible modulations in the dynamics of the host in response to an acute stress hormone (glucocorticoid, catecholamine) infusion under conditions of either high or low infectious challenge.

10.4.4.1 Implications of Increased Insult

While low levels of endotoxin stimulus invoke a self-limited inflammatory response, the system undergoes a bifurcation at some critical level of LPS. When the system is exposed to high levels of LPS, the response does not abate. This situation is simulated in Fig. 10.8. When the concentration of LPS is high enough, inflammation persists. Interestingly, the inflammatory stimulus itself is eventually cleared from the system, so the unconstrained inflammatory response is ultimately due to positive feedback within the host response. The uncontrolled (overwhelming) production of proinflammatory signaling contributes to derangements in au-

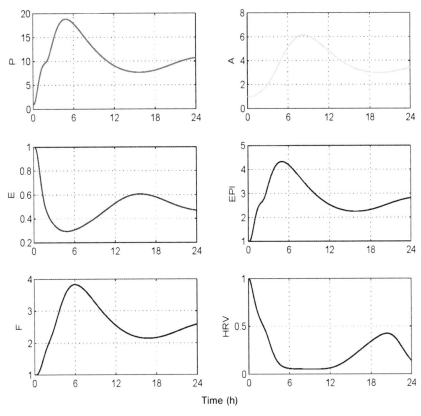

Fig. 10.8 Simulation of an unresolved inflammatory response due to high endotoxin concentration (LPS($t = 0$ h) $= 4$): a high concentration of LPS dysregulates the NF-κB signaling module, giving rise to an unconstrained immune response followed by abnormal hormonal responses that macroscopically are translated into diminished physiologic variability.

tonomic activity characterized by endocrine dysfunction and a loss of adrenergic responsiveness that leads to diminished HRV [136].

10.4.4.2 Modes of Dysregulation of the Inflammatory Response

The dynamics of the host response to an inflammatory stimulus are highly complex and nonlinear, suggesting that a dysregulation in the dynamics of the inflammatory response can be caused by many different factors. One such possibility is associated with a malfunction in the clearance rate of endotoxin, which corresponds to a higher exposure of the host response to the stimulus followed by increased persistence in the concentration of LPS. Secondly, a dysregulation in the intracellular dynamics can result in an aberrant response, given that improper regulation of NF-κB signaling might also be involved in the inflammatory component of a disease like sepsis [150]; such a mode of dysregulation isolates the host response dynamics

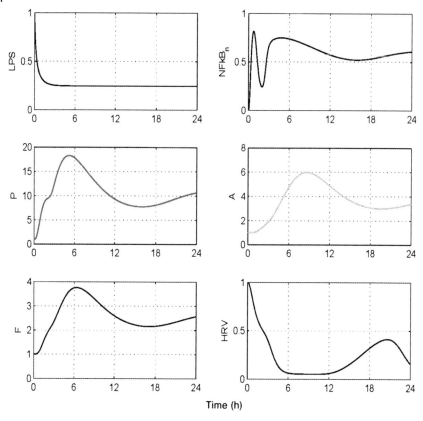

Fig. 10.9 Temporal responses of inflammatory components in a persistent infectious inflammatory response: the inflammatory stimulus cannot be eliminated, resulting in the observed persistence in the dynamic profiles of the inflammatory constituents. When the degradation rate of LPS is reduced to half of its initial value, the inflammatory stimulus cannot be cleared.

from the inflammatory insult. The model was probed by appropriately manipulating parameters to simulate the following scenarios: (i) a reduction in the first-order degradation rate of LPS, (ii) a knock-out of primary inhibitor of NF-κB (IκBa), and (iii) a potentiation in the production rate of proinflammatory mediators.

The first scenario is simulated in Fig. 10.9. A persistent inflammatory response is induced as a result of increased exposure of the host response to the inflammatory stimulus (LPS). This increased exposure is due to a reduction in the first-order degradation rate of LPS, causing LPS concentration to remain elevated longer. In this simulation, NF-κB activity can be characterized as a two-wave response; initially it increases due to the inflammatory stimulus while trying to adapt its regulatory activity at 2–3 h postendotoxin administration. However, at $t > 3$ h the activity of NF-κB cannot be regulated successfully and it moves to a sustained elevated state

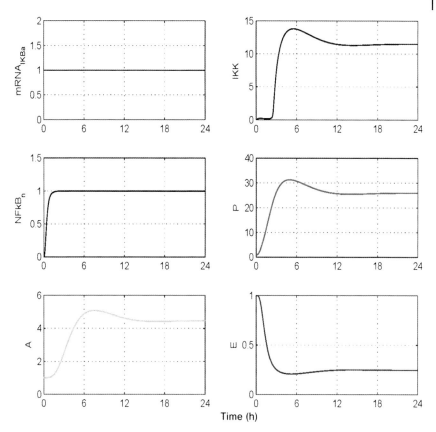

Fig. 10.10 Simulation of a knock-out *in silico* experiment (IκBa$^{-/-}$): the model is manipulated so that there is no *de novo* transcriptional synthesis of NF-κB's inhibitor (IκBa), which is responsible for the absence of the NF-κB autoregulatory feedback loop. Such a scenario accounts for maladapted activity of NF-κBn that triggers an uncompensated inflammatory response.

that drives downstream the over-excitation of both pro- and anti-inflammatory mediators, leading to an unconstrained inflammatory response. Interestingly, in [151] Klinke et al. aim at exploring the possibility of modulating the temporal control of NF-κB activation. Macrophages are exposed to a persistent inflammatory stimulus (LPS) and the available experimental data show the presence of damped oscillatory behavior in NF-κB activity.

The protein inhibitor of NF-κB (IκBa) retrieves nuclear NF-κB by forming an inactive complex with NF-κB in the cytoplasm, thus regulating the expression of various inflammatory genes. The transcription factor NF-κB upregulates the gene transcript of IκBa (mRNA$_{I\kappa Ba}$) so that the translated protein IκBa serves as the major component in regulating the transcriptional activity of NF-κB. Thus, the case in which there is no transcriptional activity of NF-κB in the promoter region

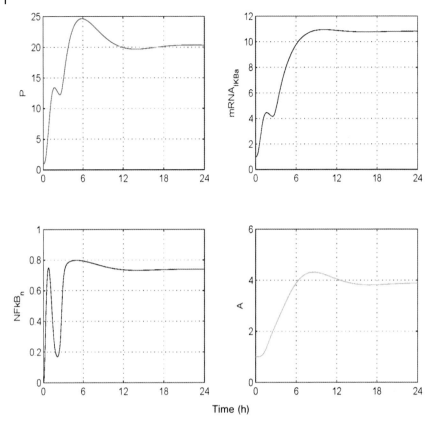

Fig. 10.11 Pre-existence of proinflammatory mediators abnormally enhancing the intracellular signaling through IKK: such a response leads to an unconstrained activation of NF-κB$_n$ that drives a persistent proinflammatory response downstream. This abnormal response cannot be counter-regulated by the anti-inflammatory arm of the host defense system. Such a mode of dysregulation is simulated by manipulating the zero production rate of proinflammation (Kin, P) so that Kin, P (unhealthy response) $\sim 2 *$ Kin, P (healthy response).

of IκBa is simulated in Fig. 10.10. In the absence of the NF-κB inhibitor (IκBa$^{-/-}$), the activity of NF-κB monotonically increases because there is no negative feedback loop to restore homeostasis. Such an *in silico* result has been experimentally tested, illustrating the impact of such a knock-out in inducing a chronic inflammatory response [93].

An additional mode of dysregulation concerns the production rate of proinflammatory mediators, which may disturb the bistable behavior of the system. In Fig. 10.11, the production rate of proinflammatory mediators is doubled. NFκB mediates a small decrease in P after ~ 3 h, but it is unable to overcome the increased production rate of P, so the system goes into unconstrained inflammation. Clinically, such an increased rate in the production of proinflammatory mediators might be the outcome of surgical trauma followed by bacterial infection, which is known as a two-hit scenario [152].

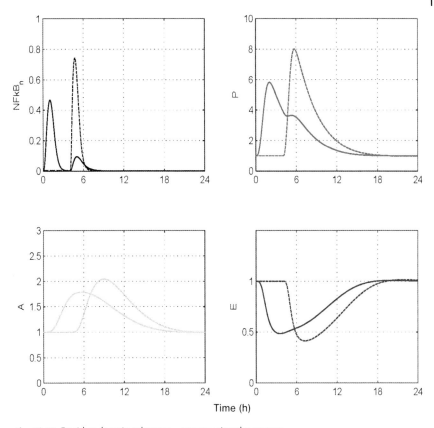

Fig. 10.12 Rapid endotoxin tolerance – preexposing the system into a small inflammatory insult results in a reduction in the cells' capacity to respond to the main endotoxin challenge. This is characterized as a short-time attenuation scenario. Solid line: $LPS(t = 0\,h) = 0.2$ and $LPS(t = 4\,h) = 1$; dashed line: $LPS(t = 0\,h) = 0$ and $LPS(t = 4\,h) = 1$.

10.4.4.3 The Emergence of Memory Effects

Repeated doses of endotoxin stimulus often lead to a less vigorous immune response; this phenomenon is known as endotoxin tolerance [153]. Even though endotoxin tolerance involves the administration of low, repeated doses of endotoxin over periods of time ranging from one day to a week [154], in the context of our model, we only consider the response of the system being preexposed to low doses of LPS for less than a day. This is because in our proposed model, all the interacting components do resolve within the first 24 h. Published studies [155, 156] report that rapid endotoxin tolerance can be induced when the system is preexposed to a low endotoxin challenge for between 3 and 6 h. This scenario is simulated in Fig. 10.12. When the system is preexposed to a low level of inflammatory stimulus for about 4 h before the main endotoxin challenge, the model predicts a much less vigorous inflammatory response. Such an event, which can be characterized as either a short-time attenuation effect or a rapid tolerance, is experimentally observed by decreased concentrations of various proinflammatory mediators, such as TNF-a and IL1B, in response to secondary *ex vivo* whole blood stimulation with LPS [155]. In addition, in an experimental study [156], concentrations of TNF-a were decreased profoundly *ex vivo* 3–6 h after *in vivo* endotoxin administration. However, by 24 h, the endotoxin tolerance had completely resolved. Such preconditioning results in an attenuation of the inflammatory response due to less vigorous intracellular signaling coupled with the decreased peak level of the proinflammatory response. Yet depending on the timing of the repeated doses, endotoxin tolerance may not be the outcome; rather, a deleterious effect on the system may be observed.

Repeated administration of low doses of LPS may perturb the system from its healthy homeostasis toward an unresolved inflammatory state. Thus, if the repeated doses separated by a short period of time, the dynamics of the system may lead to an overwhelming inflammatory response. Therefore, the successive administration of two inflammatory insults that individually account for constrained (self-limited) inflammatory responses might be detrimental to the outcome of sepsis (unresolved inflammatory response). Such an event can occur in the absence of a protective memory in the system. In this case, unlike the endotoxin tolerance phenomenon described above, the system has not yet activated its regulatory mechanism to compensate for the cumulative result of two successive doses. Such an abrupt secondary insult might dysregulate the dynamics of the host response to infection and lead to deterioration in the physiological state of the system, as seen in Fig. 10.13.

These results indicate that the cellular response is critically affected by the timing of exposure, thus demonstrating the need for an appropriate, quantifiable model to account for and integrate the various components constituting the response. Lacking a model-based approach, it is difficult to study these emergent timing effects. Furthermore, our results clearly indicate that the dynamics of the response are affected by the parameters defining the exposure to the inflammatory agent.

10.4.4.4 Evaluation of Stress Hormone Infusion in Modulating the Inflammatory Response

We have demonstrated the ability of our model to simulate an unconstrained inflammatory response, yet the true potential of the proposed model lies in its capability to respond to systematic perturbations designed to modulate the dynamics in favor of a balanced immune response and a restoration in autonomic balance. Considerable attention has been given to the effectiveness of pharmacological agents such as ligands of adrenergic receptors in influencing the production rate of both pro- and anti-inflammatory cytokines [157, 158]. In particular, significant modulations in the cytokine network were observed in human subjects exposed to epinephrine infusion [124], illustrating the role of neuroendocrine activity in dampening excessive proinflammatory effects. Therefore, the behavior of the model is studied under the conditions of epinephrine infusion prior to the inflammatory insult. Such an intervention strategy results in the potentiation of the total plasma concentration of epinephrine (EPI), which further increases intracellular cAMP

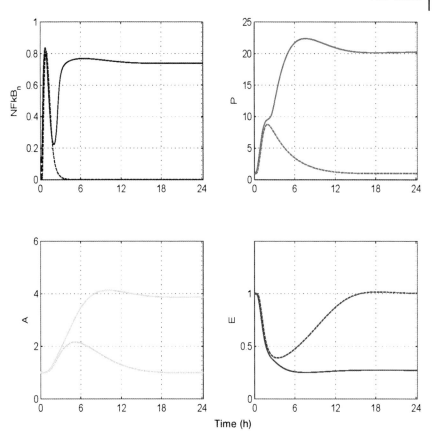

Fig. 10.13 Lethal potentiation: successive administration of small doses of endotoxin can lead to an unresolved inflammatory response due to the loss of regulatory memory. Solid line: LPS($t = 0$ h) $= 1$ and LPS($t = 0.2$ h) $= 2$; dashed line: LPS($t = 0$ h) $= 0$ and LPS($t = 0.2$ h) $= 2$.

signaling (dashed lines) as shown in Fig. 10.14. Based on the anti-inflammatory effect of acute EPI infusion via the cAMP-dependent mechanism, it is expected that an increase in intracellular cAMP levels will attenuate the proinflammatory response (P) and ultimately restore autonomic activity (HRV), which serves as a proxy indicator of improved survival [159]. This dramatic improvement in autonomic activity underscores the role of epinephrine in improving cardiac index under severe conditions (in this case, low-output septic shock) as supported by Court et al. [160].

In addition, the autonomic nervous system also controls inflammation by regulating the central secretion of glucocorticoids. During the progression of an SIRS, adrenal insufficiency is common in intensive-care units (ICU) [161]. Recently, the possible association between reduced HRV and adrenal insufficiency in trauma patients was explored in [132], where an increase in HRV is observed in patients

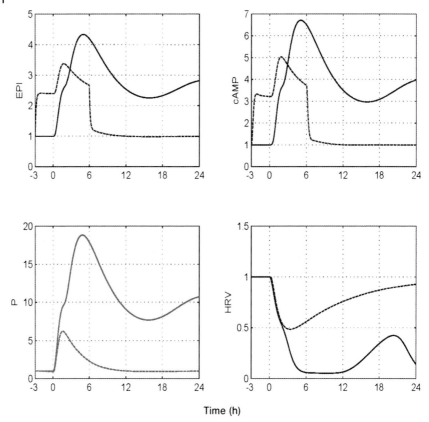

Fig. 10.14 Epinephrine preexposure: the effect of acute epinephrine infusion, which is initiated 3 h prior to the main endotoxin challenge (LPS($t = 0$ h) $= 4$) and continued for 6 h after LPS, is simulated. Dashed and solid lines represent the progression of balanced (due to the system's preexposure to epinephrine) and unconstrained (due to high levels of infectious challenge, LPS($t = 0$ h) $= 4$) inflammatory responses, respectively.

responding to steroid therapy. We opt to simulate this scenario in Fig. 10.15 by assuming that the trajectory of an unconstrained response (high LPS concentration) would qualitatively reflect critically ill patients that are diagnosed with adrenal insufficiency. This intervention strategy results in an increase in the total amount of endogenous cortisol (F), which subsequently potentiates the active steroid signal ($FR(N)$). Increased cortisol levels potentiate the anti-inflammatory arm of the system (A) immediately after the administration of LPS, thereby attenuating the proinflammatory response (P). Thus, the indirect decrease in proinflammatory signaling (P), as stimulated by anti-inflammatory signaling (A), reverses the inflammatory dynamics and restores autonomic balance.

Although the immunosuppressive effects of corticosteroids upon the systemic inflammatory effects of human endotoxemia have been well described, the influence of corticosteroids on overall autonomic dysfunction is not well understood.

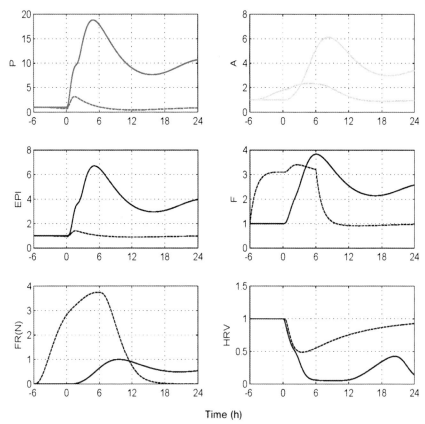

Fig. 10.15 Low-dose steroid preexposure: steroid administration is started 6 h prior to endotoxin challenge (dashed lines) while continued for another 6 h after LPS($w_{\text{Fex}} = 1$). Solid lines simulate the progression of an SIRS (high LPS concentration, LPS($t = 0$ h) $= 4$), while dashed lines reflect the protective effect that can be exerted by such hormonal (steroid) replacement therapy by augmenting the production of humoral anti-inflammatory mediators (A).

Predicated upon this, the influence of steroid administration on a self-limited endotoxin-induced inflammatory response is simulated in Fig. 10.16. Potentiation in IL-10 signaling (A) is observed under hypercortisolemia, followed by attenuation in the proinflammatory response (P). This leads to further hormonal changes, including a reduction in plasma epinephrine concentration. Experimentally, a decrease in endogenous epinephrine secretion under acute hypercortisolemia has been demonstrated [68], thus validating the assumptions of the proposed integrated model. In addition to cytokine and hormonal measurements under hypercortisolemia, HRV measurements are obtained in [69]. Remarkably, although acute hypercortisolemia significantly attenuated proinflammatory cytokines, such attenuation does not contribute to any alterations in HRV indices.

Another treatment that may attenuate the proinflammatory response is prior epinephrine (EPI) exposure; however, the influence of EPI treatment does not alter

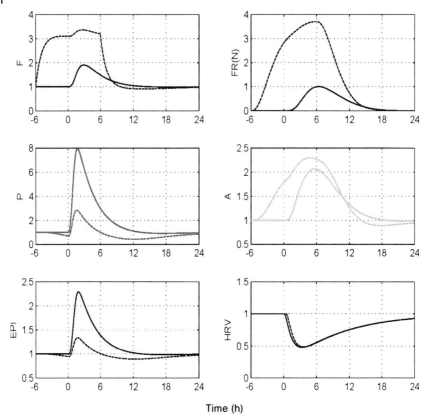

Fig. 10.16 The effect of hypercortisolemia on autonomic dysfunction under the systemic inflammatory manifestations of human endotoxemia: solid lines simulate a self-limited inflammatory response (low-dose endotoxin, LPS($t = 0$ h) $= 1$), while dashed lines reflect the potentiation of IL-10 signaling due to continuous steroid infusion initiated 6 h prior to LPS administration and continued for 6 h after endotoxin treatment ($w_{\text{Fex}} = 1$).

the system's overall adaptability (HRV) [140]. Since increased catecholamine secretion accompanies modest infections and EPI inhibits the LPS-induced proinflammatory response [162], in Fig. 10.17 we tested whether antecedent EPI infusion would alter the response to endotoxin. In particular, increasing plasma EPI levels modulate the activation of the innate immune system, particularly by attenuating the proinflammatory response through increased cAMP signaling. Such attenuation in the progression of the inflammatory response does not contribute to any changes in HRV, which is consistent with the aforementioned results concerning steroid administration before LPS treatment. From a modeling standpoint, such responses are captured due to the nonlinear crosstalk between peripheral proinflammation and the heart rate response, thus enabling the evaluation of the influence of steroid intervention strategies at the cellular and systemic levels.

The goal of this modeling approach is to demonstrate the feasibility of a multiscale model of human endotoxemia as a prototype model for systemic inflam-

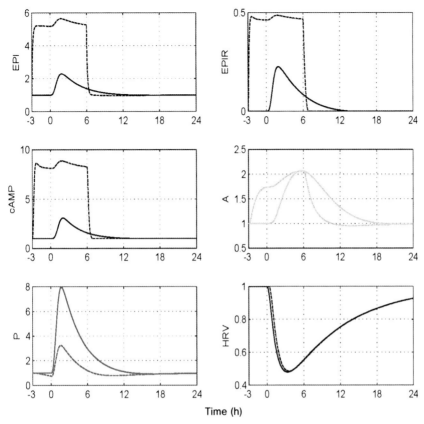

Fig. 10.17 Modulation in the progression of the inflammatory reaction due to short-term epinephrine infusion: epinephrine infusion is initiated 3 h before LPS and continued for 6 h after LPS at a normalized rate $R_{in,EPI} = 20\ h^{-1}$. Such intervention potentiates the secretion of epinephrine from SNS that through cAMP anti-inflammatory signaling can protect in part the host response by attenuating the proinflammatory response (P). Solid lines simulate a self-limited endotoxin-induced inflammatory reaction, while dashed lines reflect the scenario of prior epinephrine infusion ($w_{EPIex} = 1$).

mation in humans; specifically, the model currently focuses on critical aspects of the complex bidirectional relationship between the neuroendocrine axis and the immune system. This is accomplished by exploring the interaction networks at the (low) level of intercellular signaling in conjunction with the (high) level of interacting hormonal and physiological components. These multiscale interactions give rise to an overall systemic response. At the cellular level, elementary signaling pathways propagate extracellular signals to the transcriptional response level. At the autonomic level, the secretion of endocrine stress hormones (cortisol and epinephrine, specifically) are taken into account. Finally, at the systemic level, HRV is used to assess autonomic dysfunction, which is indicative of the severity of illness. Since both corticosteroids and catecholamines are used clinically in the treatment of systemic inflammation, the proposed model has the potential for direct clinical relevance. Thus, this modeling effort lays the foundation for a translational systems-based model of inflammation to clarify the clinical contexts in which vagal autonomic dysfunction contributes to morbidity and mortality in severely stressed patients.

10.5
Conclusions

A critical goal of translational research is to convert novel insights from basic science to clinically relevant actions related to disease prevention and diagnosis, eventually enabling physicians to identify and evaluate treatment strategies. Integrated initiatives are identified as valuable in uncovering the mechanisms underpinning the progression of human diseases. The advent of high-throughput technologies has enabled the generation of massive amounts of biological data at an unprecedented rate, facilitating a dramatic increase in the degree of quantification applied to modern biological research [163]. Further, the current definition of NIH's Road Map to Medical Research[1] clearly identifies and states the importance of a mechanistic systems approach to biological sciences. Thus, in this chapter we discussed the potential role of systems-based approaches in the quest to better understand critical physiological responses. We demonstrated how quantitative models of inflammation can be used as minimal representations of biological reality to formulate and test hypotheses, reconcile observations, and guide future experimental design. We also demonstrated the possibility of the generalization of this framework in a wide range of disease progression models. It is important to realize that *in silico* models will never replace either biological or clinical research. They could, however, rationalize the decision-making process by establishing the range of validity and predictability of intervention strategies, thus enabling the use of systems biology in translational research.

Disclaimer The Inflammation and the Host Response to Injury "Glue Grant" program is supported by the National Institute of General Medical Sciences. This Manuscript was prepared using a dataset obtained from the Glue Grant program and does not necessarily reflect the opinions or views of the Inflammation and the Host Response to Injury Investigators or the NIGMS.

Acknowledgements

We wish to acknowledge the invaluable input of our collaborators R.R. Almon and W.J. Jusko (Biological Sciences, SUNY Buffalo). J.D.S., P.T.F., and I.P.A. acknowledge financial support from the NIH under grant GM082974, the NSF under grant 0519563, the EPA under grant GAD R 832721-010, and a Busch Biomed-

---

1) http://nihroadmap.nih.gov.

ical Research Award. S.E.C. and S.F.L. are supported, in part, from USPHS Grant GM34695. The investigators acknowledge the contribution of the Inflammation and the Host Response to Injury Large-Scale Collaborative Project Award # 2-U54-GM062119 from the National Institute of General Medical Sciences.

References

1 WATSON, J. D., CRICK, F. H., Molecular structure of nucleic acids; a structure for deoxyribose nucleic acid, *Nature* 171(4356) (**1953**), pp. 737–738.
2 ADEREM, A., Systems biology: its practice and challenges, *Cell* 121(4) (**2005**), pp. 511–513.
3 MESAROVIC, M. D., Systems theory and biology – view of a theoretician, in: *Systems Theory and Biology*, Mesarovic, M. D. (ed.), Springer, New York, **1968**, pp. 59–87.
4 CSETE, M. E., DOYLE, J. C., Reverse engineering of biological complexity, *Science* 295(5560) (**2002**), pp. 1664–1669.
5 BARABASI, A. L., OLTVAI, Z. N., Network biology: understanding the cell's functional organization, *Nature Reviews Genetics* 5(2) (**2004**), pp. 101–113.
6 OLTVAI, Z. N., BARABASI, A. L., Systems biology: life's complexity pyramid, *Science* 298(5594) (**2002**), pp. 763–764.
7 VAZQUEZ, A., DOBRIN, R., et al., The topological relationship between the large-scale attributes and local interaction patterns of complex networks, *Proceedings of the National Academy of Sciences of the United States of America* 101(52) (**2004**), pp. 17940–17945.
8 BABU, M. M., LUSCOMBE, N. M., et al., Structure and evolution of transcriptional regulatory networks, *Current Opinion in Structural Biology* 14(3) (**2004**), pp. 283–291.
9 CALVANO, S. E., XIAO, W., et al., A network-based analysis of systemic inflammation in humans, *Nature* 437(7061) (**2005**), pp. 1032–1037.
10 VOGEL, T. R., DOMBROVSKIY, V. Y., et al., Trends in postoperative sepsis: are we improving outcomes?, *Surgical Infection (Larchmt)* 10(1) (**2009**), pp. 71–78.
11 BRUCE, J., RUSSELL, E. M., et al., The measurement and monitoring of surgical adverse events, *Health Technology Assessment* 5(22) (**2001**), pp. 1–194.
12 ANGUS, D. C., LINDE-ZWIRBLE, W. T., et al., Epidemiology of severe sepsis in the United States: analysis of incidence, outcome, and associated costs of care, *Critical Care Medicine* 29(7) (**2001**), pp. 1303–1310.
13 LOWRY, S. F., The stressed host response to infection: the disruptive signals and rhythms of systemic inflammation, *Surgical Clinics of North America* 89(2) (**2009**), pp. 311–326.
14 DECKER, T., Sepsis: avoiding its deadly toll, *The Journal of Clinical Investigation* 113(10) (**2004**), pp. 1387–1389.
15 CAVAILLON, J. M., ANNANE, D., Compartmentalization of the inflammatory response in sepsis and SIRS, *Journal of Endotoxin Research* 12(3) (**2006**), pp. 151–170.
16 LOWRY, S. F., CALVANO, S. E., Challenges for modeling and interpreting the complex biology of severe injury and inflammation, *Journal of Leukocyte Biology* 83(3) (**2008**), pp. 553–557.
17 MARSHALL, J. C., Modeling MODS: what can be learned from animal models of the multiple-organ dysfunction syndrome?, *Intensive Care Medicine* 31(5) (**2005**), pp. 605–608.
18 MARSHALL, J. C., DEITCH, E., et al., Preclinical models of shock and sepsis: what can they tell us?, *Shock* 24(1) (**2005**), pp. 1–6.
19 CLERMONT, G., BARTELS, J., et al., In silico design of clinical trials: a method coming of age, *Critical Care Medicine* 32(10) (**2004**), pp. 2061–2070.
20 VODOVOTZ, Y., CLERMONT, G., et al., Mathematical models of the acute inflammatory response, *Current Opinion in Critical Care* 10(5) (**2004**), pp. 383–390.
21 VODOVOTZ, Y., CONSTANTINE, G., et al., Mechanistic simulations of inflamma-

tion: current state and future prospects, *Mathematical Bioscience* 217(1) (**2009**), pp. 1–10.

22 FOTEINOU, P. T., CALVANO, S. E., et al., In silico simulation of corticosteroids effect on an NFκB-dependent physicochemical model of systemic inflammation, *PLoS One* 4(3) (**2009**), pp. e4706.

23 FOTEINOU, P. T., CALVANO, S. E., et al., Modeling endotoxin-induced systemic inflammation using an indirect response approach, *Mathematical Bioscience* 217(1) (**2009**), pp. 27–42.

24 FOTEINOU, P. T., CALVANO, S. E., et al., Translational potential of systems-based models of inflammation, *Clinical and Translational Science* 2(1) (**2009**), pp. 85–89.

25 DAY, J., RUBIN, J., et al., A reduced mathematical model of the acute inflammatory response II. Capturing scenarios of repeated endotoxin administration, *Journal of Theoretical Biology* 242(1) (**2006**), pp. 237–256.

26 REYNOLDS, A., RUBIN, J., et al., A reduced mathematical model of the acute inflammatory response: I. Derivation of model and analysis of anti-inflammation, *Journal of Theoretical Biology* 242(1) (**2006**), pp. 220–236.

27 LAUFFENBURGER, D. A., KENNEDY, C. R., Analysis of a lumped model for tissue inflammation dynamics, *Mathematical Biosciences* 53 (**1980**), pp. 189–221.

28 FISHER, E. S., LAUFFENBURGER, D. A., Analysis of the effects of immune cell motility and chemotaxis on target elimination dynamics, *Mathematical Biosciences* 98(1) (**1990**), pp. 73–102.

29 MAYER, H., ZAENKER, K. S., et al., A basic mathematical model of the immune response, *Chaos* 5(1) (**1995**), pp. 155–161.

30 DETILLEUX, J., VANGROENWEGHE, F., et al., Mathematical model of the acute inflammatory response to *Escherichia coli* in intramammary challenge, *Journal of Dairy Science* 89(9) (**2006**), pp. 3455–3465.

31 KUMAR, R., CLERMONT, G., et al., The dynamics of acute inflammation, *Journal of Theoretical Biology* 230(2) (**2004**), pp. 145–155.

32 CHOW, C. C., CLERMONT, G., et al., The acute inflammatory response in diverse shock states, *Shock* 24(1) (**2005**), pp. 74–84.

33 PRINCE, J. M., LEVY, R. M., et al., In silico and *in vivo* approach to elucidate the inflammatory complexity of CD14-deficient mice, *Molecular Medicine* 12(4–6) (**2006**), pp. 88–96.

34 HANCIOGLU, B., SWIGON, D., et al., A dynamical model of human immune response to influenza A virus infection, *Journal of Theoretical Biology* 246(1) (**2007**), pp. 70–86.

35 LAGOA, C. E., BARTELS, J., et al., The role of initial trauma in the host's response to injury and hemorrhage: insights from a correlation of mathematical simulations and hepatic transcriptomic analysis, *Shock* 26(6) (**2006**), pp. 592–600.

36 MI, Q., RIVIERE, B., et al., Agent-based model of inflammation and wound healing: insights into diabetic foot ulcer pathology and the role of transforming growth factor-β1, *Wound Repair and Regeneration* 15(5) (**2007**), pp. 671–682.

37 LI, N. Y. K., VERDOLINI, K., et al., An agent-based model of acute phonotrauma, *Journal of Critical Care* 20(4) (**2005**), pp. 393–394.

38 WIEDER, K. J., KING, K. R., et al., Optimization of reporter cells for expression profiling in a microfluidic device, *Biomedical Microdevices* 7(3) (**2005**), pp. 213–222.

39 KING, K. R., WANG, S., et al., A high-throughput microfluidic real-time gene expression living cell array, *Lab on a Chip* 7(1) (**2007**), pp. 77–85.

40 KING, K. R., WANG, S., et al., Large-scale profiling of gene expression dynamics in a microfabricated living cell array, **2006**, submitted for publication.

41 KALMAN, R. E., New development in systems theory relevant to biology, in: *Systems Theory and Biology*, Mesarovic, M. D. (ed.), Springer, New York, **1968**, pp. 222–245.

42 FANNIN, R. D., AUMAN, J. T., et al., Differential gene expression profiling in whole blood during acute systemic inflammation in lipopolysaccharide-treated rats, *Physiological Genomics* 21(1) (**2005**), pp. 92–104.

43 LOWRY, S. F., Human endotoxemia: a model for mechanistic insight and

therapeutic targeting, *Shock* 24(1) (**2005**), pp. 94–100.
44 TALWAR, S., MUNSON, P. J., et al., Gene expression profiles of peripheral blood leukocytes after endotoxin challenge in humans, *Physiological Genomics* 25(2) (**2006**), pp. 203–215.
45 WITTEBOLE, X., HAHM, S., et al., Nicotine exposure alters *in vivo* human responses to endotoxin, *Clinical and Experimental Immunology* 147(1) (**2007**), pp. 28–34.
46 VAN DEVENTER, S. J., BULLER, H. R., et al., Experimental endotoxemia in humans: analysis of cytokine release and coagulation, fibrinolytic, and complement pathways, *Blood* 76(12) (**1990**), pp. 2520–2526.
47 VAN ZEE, K. J., COYLE, S. M., et al., Influence of IL-1 receptor blockade on the human response to endotoxemia, *Journal of Immunology* 154(3) (**1995**), pp. 1499–1507.
48 COPELAND, S., WARREN, H. S., et al., Acute inflammatory response to endotoxin in mice and humans, *Clinical and Diagnostic Laboratory Immunology* 12(1) (**2005**), pp. 60–67.
49 BLALOCK, J. E., Harnessing a neural-immune circuit to control inflammation and shock, *Journal of Experimental Medicine* 195(6) (**2002**), pp. F25–F28.
50 ELENKOV, I. J., IEZZONI, D. G., et al., Cytokine dysregulation, inflammation and well-being, *Neuroimmunomodulation* 12(5) (**2005**), pp. 255–269.
51 JARA, L. J., NAVARRO, C., et al., Immune-neuroendocrine interactions and autoimmune diseases, *Clinical and Developmental Immunology* 13(2–4) (**2006**), pp. 109–123.
52 ELENKOV, I. J., Neurohormonal-cytokine interactions: implications for inflammation, common human diseases and well-being, *Neurochemistry International* 52(1–2) (**2008**), pp. 40–51.
53 TAUB, D. D., Neuroendocrine interactions in the immune system, *Cellular Immunology* 252(1–2) (**2008**), pp. 1–6.
54 ELENKOV, I. J., WILDER, R. L., et al., The sympathetic nerve – an integrative interface between two supersystems: the brain and the immune system, *Pharmacological Review* 52(4) (**2000**), pp. 595–638.
55 PAVLOV, V. A., TRACEY, K. J., Neural regulators of innate immune responses and inflammation, *Cellular and Molecular Life Sciences* 61(18) (**2004**), pp. 2322–2331.
56 SHARSHAR, T., HOPKINSON, N. S., et al., Science review: the brain in sepsis– culprit and victim, *Critical Care* 9(1) (**2005**), pp. 37–44.
57 LOMBARDI, F., Clinical implications of present physiological understanding of HRV components, *Cardiac Electrophysiology Review* 6(3) (**2002**), pp. 245–249.
58 NORRIS, P. R., MORRIS, J. A., JR., et al., Heart rate variability predicts trauma patient outcome as early as 12 h: implications for military and civilian triage, *Journal of Surgical Research* 129(1) (**2005**), pp. 122–128.
59 GODIN, P. J., FLEISHER, L. A., et al., Experimental human endotoxemia increases cardiac regularity: results from a prospective, randomized, crossover trial, *Critical Care Medicine* 24(7) (**1996**), pp. 1117–1124.
60 RASSIAS, A. J., HOLZBERGER, P. T., et al., Decreased physiologic variability as a generalized response to human endotoxemia, *Critical Care Medicine* 33(3) (**2005**), pp. 512–519.
61 SEELY, A. J., CHRISTOU, N. V., Multiple organ dysfunction syndrome: exploring the paradigm of complex nonlinear systems, *Critical Care Medicine* 28(7) (**2000**), pp. 2193–2200.
62 GODIN, P. J., BUCHMAN, T. G., Uncoupling of biological oscillators: a complementary hypothesis concerning the pathogenesis of multiple organ dysfunction syndrome, *Critical Care Medicine* 24(7) (**1996**), pp. 1107–1116.
63 BUCHMAN, T. G., Nonlinear dynamics, complex systems, and the pathobiology of critical illness, *Current Opinion in Critical Care* 10(5) (**2004**), pp. 378–382.
64 GOLUB, T. R., SLONIM, D. K., et al., Molecular classification of cancer: class discovery and class prediction by gene expression monitoring, *Science* 286(5439) (**1999**), pp. 531–537.
65 JIN, J. Y., ALMON, R. R., et al., Modeling of corticosteroid pharmacogenomics in rat liver using gene microarrays, *Journal of Pharmacology and Experimental Therapeutics* 307(1) (**2003**), pp. 93–109.

66 Cobb, J. P., Mindrinos, M. N., et al., Application of genome-wide expression analysis to human health and disease, *Proceedings of the National Academy of Sciences of the United States of America* 102(13) (2005), pp. 4801–4806.

67 Storey, J. D., Xiao, W., et al., Significance analysis of time course microarray experiments, *Proceedings of the National Academy of Sciences of the United States of America* 102(36) (2005), pp. 12837–12842.

68 Barber, A. E., Coyle, S. M., et al., Glucocorticoid therapy alters hormonal and cytokine responses to endotoxin in man, *Journal of Immunology* 150(5) (1993), pp. 1999–2006.

69 Alvarez, S. M., Katsamanis Karavidas, M., et al., Low-dose steroid alters *in vivo* endotoxin-induced systemic inflammation but does not influence autonomic dysfunction, *Journal of Endotoxin Research* 13(6) (2007), pp. 358–368.

70 Zamir, E., Bastiaens, P. I., Reverse engineering intracellular biochemical networks, *Nature Chemical Biology* 4(11) (2008), pp. 643–647.

71 Aldridge, B. B., Burke, J. M., et al., Physicochemical modelling of cell signalling pathways, *Nature Cell Biology* 8(11) (2006), pp. 1195–1203.

72 Jusko, W. J., Ko, H. C., Physiologic indirect response models characterize diverse types of pharmacodynamic effects, *Clinical Pharmacology and Therapeutics* 56(4) (1994), pp. 406–419.

73 Fong, Y. M., Marano, M. A., et al., The acute splanchnic and peripheral tissue metabolic response to endotoxin in humans, *The Journal of Clinical Investigation* 85(6) (1990), pp. 1896–1904.

74 Cross, A. S., Opal, S. M., A new paradigm for the treatment of sepsis: is it time to consider combination therapy, *Annals of Internal Medicine* 138(6) (2003), pp. 502–505.

75 Vodovotz, Y., Chow, C. C., et al., In silico models of acute inflammation in animals, *Shock* 26(3) (2006), pp. 235–244.

76 Huang, S., Eichler, G., et al., Cell fates as high-dimensional attractor states of a complex gene regulatory network, *Physical Review Letters* 94(12) (2005), p. 128701.

77 Yang, E., Maguire, T., et al., Bioinformatics analysis of the early inflammatory response in a rat thermal injury model, *BMC Bioinformatics* 8(1) (2007), p. 10.

78 Lin, J., Keogh, E., et al., Experiencing SAX: a novel symbolic representation of time series, *Data Mining and Knowledge Discovery* 15(2) (2007), pp. 107–144.

79 Storey, J. D., Dai, J. Y., et al., The optimal discovery procedure for large-scale significance testing, with applications to comparative microarray experiments, *Biostatistics* 8(2) (2007), p. 414.

80 Lampariello, F., On the use of the Kolmogorov–Smirnov statistical test for immunofluorescence histogram comparison, *Cytometry* 39(3) (2000), pp. 179–188.

81 Kirkpatrick, S., Gelatt, C. D., Jr., et al., Optimization by simulated annealing, *Science* 220(4598) (1983), pp. 671–680.

82 Vemula, M., Berthiaume, F., et al., Expression profiling analysis of the metabolic and inflammatory changes following burn injury in rats, *Physiological Genomics* 18(1) (2004), pp. 87–98.

83 Wells, C. A., Ravasi, T., et al., Inflammation suppressor genes: please switch out all the lights, *Journal of Leukocyte Biology* 78(1) (2005), pp. 9–13.

84 Dayneka, N. L., Garg, V., et al., Comparison of four basic models of indirect pharmacodynamic responses, *Journal of Pharmacokinetic Biopharmacy* 21(4) (1993), pp. 457–478.

85 Sharma, A., Jusko, W. J., Characteristics of indirect pharmacodynamic models and applications to clinical drug responses, *British Journal of Clinical Pharmacology* 45(3) (1998), pp. 229–239.

86 Krzyzanski, W., Jusko, W. J., Mathematical formalism for the properties of four basic models of indirect pharmacodynamic responses, *Journal of Pharmacokinetic Biopharmacy* 25(1) (1997), pp. 107–123.

87 Foteinou, P. T., Calvano, S. E., et al., An indirect response model of endotoxin-induced systemic inflammation, *Journal of Critical Care* 22(4) (2007), pp. 337–338.

88 Aderem, A., Smith, K. D., A systems approach to dissecting immunity and inflammation, *Seminar in Immunology* 16(1) (2004), pp. 55–67.

89 Protti, A., Singer, M., Strategies to modulate cellular energetic metabolism during sepsis, *Novartis Foundation Symposium* 280 (**2007**), pp. 7–16; discussion pp. 16–20, 160–164.

90 Zwietering, M. H., Jongenburger, I., et al., Modeling of the bacterial growth curve, *Applied and Environment Microbiology* 56(6) (**1990**), pp. 1875–1881.

91 Greisman, S. E., Hornick, R. B., et al., The role of endotoxin during typhoid fever and tularemia in man. IV. The integrity of the endotoxin tolerance mechanisms during infection, *The Journal of Clinical Investigation* 48(4) (**1969**), pp. 613–629.

92 Lauffenburger, D. A., Linderman, J. J., Receptors: models for binding, trafficking, and signalling, *The International Journal of Biochemistry and Cell Biology* 28 (**1996**), pp. 1418–1418.

93 Hoffmann, A., Levchenko A., et al., The IκB-NF-κB signaling module: temporal control and selective gene activation, *Science* 298(5596) (**2002**), pp. 1241–1245.

94 Saklatvala, J., Dean, J., et al., Control of the expression of inflammatory response genes, *Biochemical Society Symposium* 70 (**2003**), pp. 95–106.

95 Shin, H. J., Lee, H., et al., Kinetics of binding of LPS to recombinant CD14, TLR4, and MD-2 proteins, *Molecular Cells* 24(1) (**2007**), pp. 119–124.

96 Ihekwaba, A. E., Broomhead, D. S., et al., Sensitivity analysis of parameters controlling oscillatory signalling in the NF-κB pathway: the roles of IKK and IκBα, *Systematic Biology (Stevenage)* 1(1) (**2004**), pp. 93–103.

97 Krishna, S., Jensen, M. H., et al., Minimal model of spiky oscillations in NF-κB signaling, *Proceedings of the National Academy of Sciences of the United States of America* 103(29) (**2006**), pp. 10840–10845.

98 Rifkind, D., Prevention by polymyxin B of endotoxin lethality in mice, *Journal of Bacteriology* 93(4) (**1967**), pp. 1463–1464.

99 Lehmann, V., Freudenberg, M. A., et al., Lethal toxicity of lipopolysaccharide and tumor necrosis factor in normal and D-galactosamine-treated mice, *Journal of Experimental Medicine* 165(3) (**1987**), pp. 657–663.

100 Tschaikowsky, K., Schmidt, J., et al., Modulation of mouse endotoxin shock by inhibition of phosphatidylcholine-specific phospholipase C, *Journal of Pharmacological and Toxicological Methods* 285(2) (**1998**), pp. 800–804.

101 Kerschen, E. J., Fernandez, J. A., et al., Endotoxemia and sepsis mortality reduction by non-anticoagulant activated protein C, *Journal of Experimental Medicine* 204(10) (**2007**), pp. 2439–2448.

102 Wang, H., Bloom, O., et al., HMG-1 as a late mediator of endotoxin lethality in mice, *Science* 285(5425) (**1999**), pp. 248–251.

103 Barnes, P. J., Karin, M., Nuclear factor-κB: a pivotal transcription factor in chronic inflammatory diseases, *New England Journal of Medicine* 336(15) (**1997**), pp. 1066–1071.

104 Carmody, R. J., Chen, Y. H., Nuclear factor-κB: activation and regulation during toll-like receptor signaling, *Cellular and Molecular Immunology* 4(1) (**2007**), pp. 31–41.

105 Jusko, W. J., Receptor-mediated pharmacodynamics of corticosteroids, *Progress in Clinical Biological Research* 387 (**1994**), pp. 261–270.

106 DuBois, D. C., Xu, Z. X., et al., Differential dynamics of receptor down-regulation and tyrosine aminotransferase induction following glucocorticoid treatment, *Journal of Steroid Biochemistry and Molecular Biology* 54(5–6) (**1995**), pp. 237–243.

107 Xu, Z. X., Sun, Y. N., et al., Third-generation model for corticosteroid pharmacodynamics: roles of glucocorticoid receptor mRNA and tyrosine aminotransferase mRNA in rat liver, *Journal of Pharmacokinetics and Biopharmaceutics* 23(2) (**1995**), pp. 163–181.

108 Sun, Y. N., DuBois, D. C., et al., Fourth-generation model for corticosteroid pharmacodynamics: a model for methylprednisolone effects on receptor/gene-mediated glucocorticoid receptor down-regulation and tyrosine aminotransferase induction in rat liver, *Journal of Pharmacokinetics and Biopharmaceutics* 26(3) (**1998**), pp. 289–317.

109 Almon, R. R., DuBois, D. C., et al., Pharmacodynamics and pharmacogenomics

of diverse receptor-mediated effects of methylprednisolone in rats using microarray analysis, *Journal of Pharmacokinetics and Pharmacodynamics* 29(2) (2002), pp. 103–129.

110 ALMON, R. R., DUBOIS, D. C., et al., Pharmacogenomic responses of rat liver to methylprednisolone: an approach to mining a rich microarray time series, *AAPS Journal* 7(1) (2005), pp. E156–E194.

111 ALMON, R. R., DUBOIS, D. C., et al., A microarray analysis of the temporal response of liver to methylprednisolone: a comparative analysis of two dosing regimens, *Endocrinology* 148(5) (2007), pp. 2209–2225.

112 ALMON, R. R., LAI, W., et al., Corticosteroid-regulated genes in rat kidney: mining time series array data, *American Journal of Physiology-Endocrinology and Metabolism* 289(5) (2005), pp. E870–E882.

113 RAMAKRISHNAN, R., DUBOIS, D. C., et al., Fifth-generation model for corticosteroid pharmacodynamics: application to steady-state receptor down-regulation and enzyme induction patterns during seven-day continuous infusion of methylprednisolone in rats, *Journal of Pharmacokinetic and Pharmacodynamics* 29(1) (2002), pp. 1–24.

114 JUSKO, W. J., DUBOIS, D., et al., Sixth-generation model for corticosteroid pharmacodynamics: multi-hormonal regulation of tyrosine aminotransferase in rat liver, *Journal of Pharmacokinetic and Pharmacodynamics* (2005).

115 LIN, E., LOWRY, S. F., The human response to endotoxin, *Sepsis* 2 (1998), pp. 255–262.

116 CHROUSOS, G. P., The hypothalamic-pituitary-adrenal axis and immune-mediated inflammation, *New England Journal of Medicine* 332(20) (1995), pp. 1351–1362.

117 WEBSTER, J. I., TONELLI, L., et al., Neuroendocrine regulation of immunity, *Annual Review of Immunology* 20 (2002), pp. 125–163.

118 PADGETT, D. A., GLASER, R., How stress influences the immune response, *Trends in Immunology* 24(8) (2003), pp. 444–448.

119 BRIEGEL, J., JOCHUM, M., et al., Immunomodulation in septic shock: hydrocortisone differentially regulates cytokine responses, *Journal of American Society of Nephrology* 12(17) (2001), pp. S70–S74.

120 MAGER, D. E., LIN, S. X., et al., Dose equivalency evaluation of major corticosteroids: pharmacokinetics and cell trafficking and cortisol dynamics, *Journal of Clinical Pharmacology* 43(11) (2003), pp. 1216–1227.

121 HAWES, A. S., ROCK, C. S., et al., In vivo effects of the antiglucocorticoid RU 486 on glucocorticoid and cytokine responses to *Escherichia coli* endotoxin, *Infection and Immunology* 60(7) (1992), pp. 2641–2647.

122 VAN DER POLL, T., BARBER, A. E., et al., Hypercortisolemia increases plasma interleukin-10 concentrations during human endotoxemia – a clinical research center study, *Journal of Clinical Endocrinology Metabolism* 81(10) (1996), pp. 3604–3606.

123 VAN DER POLL, T., Effects of catecholamines on the inflammatory response, *Sepsis* 4 (2000), pp. 159–167.

124 VAN DER POLL, T., COYLE, S. M., et al., Epinephrine inhibits tumor necrosis factor-α and potentiates interleukin 10 production during human endotoxemia, *The Journal of Clinical Investigation* 97(3) (1996), pp. 713–719.

125 BARNES, P. J., Beta-adrenergic receptors and their regulation, *American Journal of Respiratory Critical Care Medicine* 152(3) (1995), pp. 838–860.

126 MAGER, D. E., WYSKA, E., et al., Diversity of mechanism-based pharmacodynamic models, *Drug Metabolism and Disposal* 31(5) (2003), pp. 510–518.

127 SUN, Y. N., JUSKO, W. J., Transit compartments versus gamma distribution function to model signal transduction processes in pharmacodynamics, *Journal of Pharmaceutical Science* 87(6) (1998), pp. 732–737.

128 MAGER, D. E., JUSKO, W. J., Pharmacodynamic modeling of time-dependent transduction systems, *Clinical Pharmacology and Therapeutics* 70(3) (2001), pp. 210–216.

129 HUIKURI, H. V., MAKIKALLIO, T., et al., Measurement of heart rate variability: a clinical tool or a research toy?, *Journal of the American College of Cardiology* 34(7) (1999), pp. 1878–1883.

130 WINCHELL, R. J., HOYT, D. B., Analysis of heart-rate variability: a noninvasive predictor of death and poor outcome in patients with severe head injury, *Journal of Trauma* 43(6) (**1997**), pp. 927–933.

131 MORRIS, J. A., JR., NORRIS, P. R., et al., Reduced heart rate variability: an indicator of cardiac uncoupling and diminished physiologic reserve in 1,425 trauma patients, *Journal of Trauma* 60(6) (**2006**), pp. 1165–1173; discussion pp. 1173–1174.

132 MORRIS, J. A., JR., NORRIS, P. R., et al., Adrenal insufficiency, heart rate variability, and complex biologic systems: a study of 1,871 critically ill trauma patients, *Journal of the American College of Surgery* 204(5) (**2007**), pp. 885–892; discussion pp. 892–893.

133 NORRIS, P. R., OZDAS, A., et al., Cardiac uncoupling and heart rate variability stratify ICU patients by mortality: a study of 2088 trauma patients, *Annals of Surgery* 243(6) (**2006**), pp. 804–812; discussion pp. 812–814.

134 WINCHELL, R. J., HOYT, D. B., Spectral analysis of heart rate variability in the ICU: a measure of autonomic function, *Journal of Surgical Research* 63(1) (**1996**), pp. 11–16.

135 ARONSON, D., MITTLEMAN, M. A., et al., Interleukin-6 levels are inversely correlated with heart rate variability in patients with decompensated heart failure, *Journal of Cardiovascular Electrophysiology* 12(3) (**2001**), pp. 294–300.

136 MALAVE, H. A., TAYLOR, A. A., et al., Circulating levels of tumor necrosis factor correlate with indexes of depressed heart rate variability: a study in patients with mild-to-moderate heart failure, *Chest* 123(3) (**2003**), pp. 716–724.

137 MARSLAND, A. L., GIANAROS, P. J., et al., Stimulated production of proinflammatory cytokines covaries inversely with heart rate variability, *Psychosomatic Medicine* 69(8) (**2007**), pp. 709–716.

138 BENDIXEN, H. H., OSGOOD, P. F., et al., Dose-dependent differences in catecholamine action on heart and periphery, *Journal of Pharmacology and Experimental Therapeutics* 145 (**1964**), pp. 299–306.

139 BROWN, G. L., ECCLES, J. C., The action of a single vagal volley on the rhythm of the heart beat, *Journal of Physiology* 82(2) (**1934**), pp. 211–241.

140 JAN, B. U., COYLE, S. M., et al., Influence of acute epinephrine infusion on endotoxin-induced parameters of heart rate variability: a randomized controlled trial, *Annals of Surgery* 249(5) (**2009**), pp. 750–756.

141 BERNTSON, G. G., CACIOPPO, J. T., et al., Autonomic determinism: the modes of autonomic control, the doctrine of autonomic space, and the laws of autonomic constraint, *Psychological Review* 98(4) (**1991**), pp. 459–487.

142 ZAZA, A., LOMBARDI, F., Autonomic indexes based on the analysis of heart rate variability: a view from the sinus node, *Cardiovascular Research* 50(3) (**2001**), pp. 434–442.

143 GUTKIN, B. S., DEHAENE, S., et al., A neurocomputational hypothesis for nicotine addiction, *Proceedings of the National Academy of Sciences of the United States of America* 103(4) (**2006**), pp. 1106–1111.

144 MURRAY, P. J., The JAK-STAT signaling pathway: input and output integration, *Journal of Immunology* 178(5) (**2007**), pp. 2623–2629.

145 BRIGHTBILL, H. D., PLEVY, S. E., et al., A prominent role for Sp1 during lipopolysaccharide-mediated induction of the IL-10 promoter in macrophages, *Journal of Immunology* 164(4) (**2000**), pp. 1940–1951.

146 SINGER, M., DE SANTIS, V., et al., Multiorgan failure is an adaptive, endocrine-mediated, metabolic response to overwhelming systemic inflammation, *Lancet* 364(9433) (**2004**), pp. 545–548.

147 BREALEY, D., BRAND, M., et al., Association between mitochondrial dysfunction and severity and outcome of septic shock, *Lancet* 360(9328) (**2002**), pp. 219–223.

148 CONTRERAS, M., RYAN, L. M., Fitting nonlinear and constrained generalized estimating equations with optimization software, *Biometrics* 56(4) (**2000**), pp. 1268–1271.

149 LAMERIS, T. W., DE ZEEUW, S., et al., Epinephrine in the heart: uptake and release, but no facilitation of norepinephrine release, *Circulation* 106(7) (**2002**), pp. 860–865.

150 Arnalich, F., Garcia-Palomero, E., et al., Predictive value of nuclear factor κB activity and plasma cytokine levels in patients with sepsis, *Infection and Immunity* 68(4) (2000), pp. 1942–1945.

151 Klinke, D. J., Ustyugova, I. V., et al., Modulating temporal control of NF-κB activation implications for therapeutic and assay selection (un-edited manuscript), *Biophysics Journal of Bio-FAST* (2008).

152 Romascin, A. D., Foster, D. M., et al., Let the cells speak: neutrophils as biologic markers of the inflammatory response, *Sepsis* 2 (1998), pp. 119–125.

153 Fan, H., Cook, J. A., Molecular mechanisms of endotoxin tolerance, *Journal of Endotoxin Research* 10(2) (2004), pp. 71–84.

154 Wysocka, M., Robertson, S., et al., IL-12 suppression during experimental endotoxin tolerance: dendritic cell loss and macrophage hyporesponsiveness, *Journal of Immunology* 166(12) (2001), pp. 7504–7513.

155 McCall, C. E., Grosso-Wilmoth, L. M., et al., Tolerance to endotoxin-induced expression of the interleukin-1 β gene in blood neutrophils of humans with the sepsis syndrome, *The Journal of Clinical Investigation* 91(3) (1993), pp. 853–861.

156 Poll, V. D., *Journal of Infectious Disease* 174 (1996), pp. 1356–1360.

157 Hasko, G., Elenkov, I. J., et al., Differential effect of selective block of α 2-adrenoreceptors on plasma levels of tumour necrosis factor-α, interleukin-6 and corticosterone induced by bacterial lipopolysaccharide in mice, *Journal of Endocrinology* 144(3) (1995), pp. 457–462.

158 Ignatowski, T. A., Spengler, R. N., Regulation of macrophage-derived tumor necrosis factor production by modification of adrenergic receptor sensitivity, *Journal of Neuroimmunology* 61(1) (1995), pp. 61–70.

159 Stein, P. K., Kleiger, R. E., Insights from the study of heart rate variability, *Annual Review of Medicine* 50 (1999), pp. 249–261.

160 Court, O., Kumar, A., et al., Clinical review: myocardial depression in sepsis and septic shock, *Critical Care* 6(6) (2002), pp. 500–508.

161 Keh, D., Sprung, C. L., Use of corticosteroid therapy in patients with sepsis and septic shock: an evidence-based review, *Critical Care Medicine* 32(11) (2004), pp. S527–S533.

162 Van der Poll, T., Lowry, S. F., Epinephrine inhibits endotoxin-induced IL-1 β production: roles of tumor necrosis factor-α and IL-10, *American Journal of Physiology* 273(6 Pt 2) (1997), pp. R1885–R1890.

163 Beard, D. A., Bassingthwaighte, J. B., et al., Computational modeling of physiological systems, *Physiological Genomics* 23(1) (2005), pp. 1–3; discussion p. 4.

164 An, G., Agent-based computer simulation and sirs: building a bridge between basic science and clinical trials, *Shock* 16(4) (2001), pp. 266–273.

165 An, G., In silico experiments of existing and hypothetical cytokine-directed clinical trials using agent-based modeling, *Critical Care Medicine* 32(10) (2004), pp. 2050–2060.

166 An, G., Concepts for developing a collaborative *in silico* model of the acute inflammatory response using agent-based modeling, *Journal of Critical Care* 21(1) (2006), pp. 105–110; discussion pp. 110–111.

11
Dynamic Modeling and Simulation for Robust Control of Distributed Processes and Bioprocesses

Antonio A. Alonso, Míriam R. García, and Carlos Vilas

Keywords

bioprocesses, model predictive control (MPC), convection–diffusion–reaction processes, distributed process systems (DPSs), tubular reactors, proper orthogonal decomposition (POD)

11.1
Introduction

The dynamic behavior of chemical and biochemical processes is usually the result of a diversity of physicochemical mechanisms acting on a wide range of time and length scales. Typical phenomena include mass, energy, and momentum transport, by convection or diffusion, phase equilibrium or chemical or biochemical transformations, for instance.

Depending on the spatial and temporal scale we might be interested in, there will be some mathematical formulations more appropriate than others to describe the system. In this way, if one is only interested in changes taking place on the whole system's domain during long time horizons, an inventory balance would suffice to describe the total amount of property (e.g., mass or energy) in the system.

On the other hand, it may happen that for such a spatial domain and observational "time window," diffusion-like transport mechanisms, active in a much more short scale, are relaxed so that the values of the states (densities of mass and energy, for instance) are homogeneously distributed in space [49]. We then say that the system is well-mixed, thus being the most appropriate description the one given by a set of ordinary differential equations representing macroscopic balances. If in contrast, such spatial domain or regions into that domain are inspected at much shorter periods of time, namely those corresponding to the time scale at which diffusion occurs, one would have to accept the densities also being a function of space. The appropriate description in that case should be cast in the context of partial differential equations.

This zooming approach to the system's dynamics could proceed into shorter time and length scales until the microscopic world of constantly colliding particles is reached. A mathematical description of such a world should encode the huge-dimensional space of particle positions and velocities into a reduced order formulation which usually takes the form of the Boltzmann equation. Finally, coarse graining methods will zoom out the view point of the system and take the description back to the mesoscopic and macroscopic world of continuum [29].

One conclusion to be drawn from this picture is that for any given time and length scales there will be a number of states dynamically active, coupled with others either relaxed (those states active at shorter scales) or frozen (as they are associated with much slower dynamics). Mathematically, such description translates into a highly coupled stiff system obeying a nonlinear set of algebraic, partial, and ordinary differential equations, which is usually difficult to solve. In this way, the stiffness of the resulting problem is the effect of the diversity of time scales coexisting in the system. From a control point of view or even from the more general perspective of "decision making," stiffness must be avoided, what calls for model reduction methods able to provide the simplest robust dynamic description representative of the system at the scales of interest.

The application to process control should be considered not just in the restricted sense of a PID regulator or any other such device equipped with a "compensation" rule which responds to deviations between the prescribed operation condition and the current one. From a broader perspective a controller consists of a virtual representation of the system one wants to operate (the plant), and a sort of agent which at regular time instants inspects the current state of the process and employs the system representation to screen future scenarios by asking whether a given decision should be made or not. Once the best possible action is found, the agent implements it in the real plant and waits for the response given by the sensors to initiate further corrective actions. This logic, properly repeated during the process operation, is what is known as a feedback control scheme. In its simple version a PID contains one such representation of the linear dynamic range, which in this case is explicitly inverted in some smart way as the internal model control (IMC) theory clearly highlights [39].

More sophisticated control algorithms will also share the same construction logic, namely a virtual representation of the system dynamics, an agent devising future operation, and a comparison rule which evaluates the real effect of the control actions on the process. Such control methods would include adaptive control [41], feedback linearization control [34], slide mode control [48], or model predictive control (MPC) in all its flavors (linear, nonlinear, constrained, etc.) [23, 32, 42].

In constructing a reliable dynamic representation for the system to be controlled, and depending on the class of description employed, there are a wide variety of methods at hand. However, they all rely on the very same basic principle, namely the existence of the so-called spectral gap (see [52] and references therein) which will establish a neat separation between the slow and fast dynamics.

Although each class of method has its own peculiarities, to a much extent connected with the mathematical formulation employed, in essence they share a common goal. This consists of finding a low-dimensional subspace in the state space of the system where the slow dynamics evolve. Those modes that collect the dynamics beyond the spectral gap, i.e., dynamics faster than the selected ones, would relax. Furthermore, such modes will be intrinsically stable while the slow ones (with eigenvalues near zero in the complex plane) will be the ones that might become unstable.[1] In this way, and from a control design perspective, the objective will be that of constructing feedback mechanisms to preserve the stability of the slow dynamics in the event of disturbances or under changes in the operation conditions. Model reduction methods will supply reliable but simple mathematical descriptions in the time scale of interest to be employed as part of the control scheme.

Although in this chapter the focus is on control applications, it must be remarked that reduced order models have been widely employed in different fields of engineering in the context of process analysis and design. In this way, systematic model reduction techniques have proved their value in chemical kinetics to produce reduced order reaction mechanisms which in fields such as combustion science helped improving the efficiency of reactive CFD (computational fluid dynamic) problems [30, 31, 52]. Other methods such as the so-called equation-free methods or recursive projection methods (RPM) have been successfully applied to general microscopic or mesoscopic systems representing complex spatially varying physicochemical processes to solve dynamic optimization problems or to do bifurcation analysis (see for instance [50, 51] and references therein). An exhaustive revision of model reduction methods applied to Boltzmann equation-based descriptions can be found in [29].

On the other hand, there are a number of general algebraic methods aimed at database compression which are capable of extracting the low-dimensional subspaces (in the space of data) which better represent the coherence of the full data set structure. Among those one of the most popular is the so-called Karhunen–Loève, also known as proper orthogonal decomposition, widely employed in the study of turbulence [33]. These methods are in one way or another related to SVD (singular value decomposition) or its extensions to tensor algebra (termed these methods as HSVD or High Order SVD methods) [14, 36] and have been employed in areas as diverse as computer vision [20, 40] or aerodynamics [37].

In this work we concentrate on two efficient model reduction techniques which are of particular interest in distributed process systems. The techniques take advantage of the dissipative nature of convection–diffusion–reaction processes to identify low-dimensional subspaces capturing most of the relevant features of the system dynamics (implicit in the slow modes of the system) [1, 4].

Depending on the method employed, the low-dimensional subspace will be defined either by the eigenfunctions of the Laplacian operator or through the so-called POD expansion as described in [3]. Projection of the original set of partial differential equations on these subspace will lead to a set of ordinary differential equations describing the slow (and possibly unstable) dynamics of the system. The low dimension of the resulting dynamic systems makes this approach specially attractive in dynamic reconstruction of the underlying infinite-dimensional field [25, 26] as well as in robust and nonlinear control design [2, 54, 55].

---

1) Some theoretical arguments and results in the context of distributed process systems are presented in [2, 12].

The structure of the chapter is as follows. The theoretical basis for model reduction as well as a self-contained description of the methods proposed, together with their implementation aspects in the framework of the finite element method (FEM), is presented in Section 11.2. Section 11.3 is devoted to discuss some applications of the resulting reduced order models in the context of system identification and optimal field reconstruction. The use of model reduction in robust control design and real-time optimization will be presented in Section 11.4. Finally, in Section 11.5 some conclusions and future research directions will be drawn.

11.2
Model Reduction of DPS: Theoretical Background

The class of distributed process systems (DPSs) we will consider are those representative of transport-reaction processes where the field $x(t, \vec{\xi})$ is usually described by the following quasilinear partial differential equation (PDE):

$$\frac{\partial x}{\partial t} = L(x) + f(x), \quad \forall \vec{\xi} \in \Omega \tag{11.1a}$$

where $x(t, \vec{\xi}) \in \mathbb{R}^{n_x}$ represents the state vector field as a function of time $t \in [0, \infty)$ and spatial coordinates $\vec{\xi} \in \mathbb{R}^D$. The spatial parabolic operator $L(\cdot)$ in (11.1a) has the following general representation:

$$L(\cdot) = \alpha \Delta - \vec{\beta} \cdot \vec{\nabla} + \gamma = \left(\alpha \sum_{j,i=1}^{D} \frac{\partial}{\partial \xi_j} \left(\frac{\partial}{\partial \xi_i} \right) - \sum_{i=1}^{D} \beta_i \frac{\partial}{\partial \xi_i} + \gamma \right) \tag{11.1b}$$

where $\alpha \in \mathbb{R}^+$ and $\gamma \in \mathbb{R}$ in (11.1b) are the parameters denoting, for example, dispersion and exchange with the environment and $\vec{\beta} \in \mathbb{R}^D$ is the vector of fluid velocities. Diffusive and convective terms are represented by the Laplacian operator $\Delta \in \mathbb{R}$ and the nabla operator $\vec{\nabla} \in \mathbb{R}^D$, respectively. Nonlinearities induced by the chemical reaction are accommodated into the system (11.1a) through the term $f(x)$ which is assumed to be Lipschitz continuous. The description of the system is completed with initial and boundary conditions of the form

$$x(0, \vec{\xi}) = x_0 \tag{11.1c}$$

$$\vec{n} \cdot \vec{\nabla} x + qx = g(x), \quad \forall t \in [0, +\infty), \vec{\xi} \in \Gamma \tag{11.1d}$$

The parameter q and function $g(x)$ are used in (11.1d) to settle either first- or second-order boundary conditions [25].

Solutions of the PDE system (11.1a)–(11.1d) can be represented on a Hilbert space equipped with an inner product, in terms of the infinite-dimensional basis $\{\phi_i^x(\vec{\xi})\}_{i=1}^{\infty}$ as follows:

$$x(t,\vec{\xi}) = \sum_{i=1}^{\infty} m_i^x(t)\phi_i^x(\vec{\xi}) \tag{11.2}$$

As pointed out elsewhere [11, 15], approaches based on spatial discretization present a number of disadvantages for the purpose of control applications: they are usually computationally involved and some essential control-theoretical properties, such as controllability or observability, may be lost by the discretization scheme or the degree of refinement.

Alternatively, one can take advantage of the dissipative nature of systems of the form (11.1) to approximate the field $x(t,\vec{\xi})$ by a truncated series expansion as follows [2]:

$$x(t,\vec{\xi}) \cong \tilde{x}(t,\vec{\xi}) = \sum_{i=1}^{k_x} m_i^x(t)\phi_i^x(\vec{\xi}) \tag{11.3}$$

where each element in the set of spatial functions $\{\phi_i^x(\vec{\xi})\}_{i=1}^{k_x}$ is computed by solving the following eigenvalue problem:

$$\int_{\Omega} R(\vec{\xi},\vec{\xi}')\phi_i^x(\vec{\xi}')\,d\vec{\xi}' = \lambda_i^x \phi_i^x(\vec{\xi}) \tag{11.4}$$

with λ_i^x being the eigenvalue associated with each eigenfunction ϕ_i^x of the field x. The ordered structure of the eigenspectrum $\{\lambda_i^x\}_{i=1}^{\infty}$ ensures that the set of time-dependent modes $\{m_i^x(t)\}_{i=1}^{k_x}$ will be those capturing the most relevant features of the solution, coincident with the slow dynamics. A geometric representation of this idea is depicted in Fig. 11.1 illustrating the fact that the field will evolve to a low-dimensional hyperplane and remain there in the future.

Depending on the nature of the kernel R different sets of basis functions, and therefore model reduction methods, will emerge [3]. In particular the following will be considered:

1. Proper orthogonal decomposition (POD), where R is a two-point correlation matrix constructed from empirical or simulation data (snapshots) [10].
2. Laplacian spectral decomposition (LSD), where R is the Green's function associated with the Laplacian [13].

The reduced order dynamic representation of the system is then obtained by projecting equation (11.1) on each element ϕ_i^x corresponding to the ith slowest eigenvalue λ_i^x:

$$\int_{\Omega} \left[\frac{\partial x}{\partial t} - L(x) - f(x)\right]\phi_i^x \, d\vec{\xi} = 0 \tag{11.5}$$

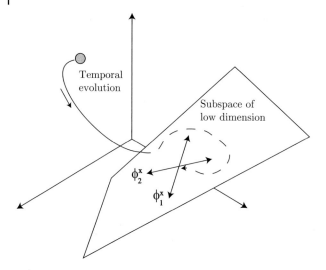

Fig. 11.1 Dynamic evolution of a dissipative process. After a transient, the trajectory of any point in the state space will converge to a low-dimensional manifold (in this case represented by a plane) where it will remain in the future.

This results into a set of ordinary differential equations of the form

$$\frac{dm^x}{dt} = Am^x + f(m^x)$$

$$\tilde{x} = \left[\phi_1^x, \ldots, \phi_{k_x}^x\right] m^x \tag{11.6}$$

Note that the dimension of the vector state m^x in (11.6) coincides with the dimension of the low-dimensional basis, while A and $f(m^x)$ correspond to the projection of the spatial operator and nonlinear terms $f(x)$, respectively, on the selected basis.

11.2.1
Model Reduction in the Context of the Finite Element Method

In order to numerically solve the eigenvalue problem (11.4) and obtain the set of basis functions needed to project the original PDE, some sort of discretization scheme must be employed. The approach we suggest benefits from the properties of the finite element method and its versatility to deal with arbitrary geometries (see [43] for a general description). As we will show next, the FEM framework will serve to approximate spatial integrals or derivatives in a straightforward manner, which considerably will simplify the numerical aspects of projection (i.e., (11.5)) [25].

Table 11.1 Algebraic relations to numerically compute integrals and derivatives using the FEM structure.

Continuous		Discrete
$\int_\Omega f(\vec{\xi}) h(\vec{\xi}) d\xi$	$\rightarrow$	$F^T(\mathcal{D}\mathcal{A})H$
$\int_\Omega f(\vec{\xi}) \frac{\partial h(\vec{\xi})}{\partial \xi_k} d\xi$	$\rightarrow$	$F^T(\mathcal{B}\mathcal{E}^k)H, \quad k = 1, \ldots, \mathcal{D}$
$\int_\Omega f(\vec{\xi}) \Delta h(\vec{\xi}) d\xi$	$\rightarrow$	$F^T(\mathcal{Q})G - F^T(\mathcal{C} + q\mathcal{Q})H$
$\frac{\partial f(\vec{\xi})}{\partial \xi_k}$	$\rightarrow$	$(\mathcal{D}\mathcal{A}^{-1})(\mathcal{B}\mathcal{E}^k)F, \quad k = 1, \ldots, \mathcal{D}$
$\Delta f(\vec{\xi})$	$\rightarrow$	$(\mathcal{D}\mathcal{A}^{-1})(\mathcal{Q})G - (\mathcal{D}\mathcal{A}^{-1})(\mathcal{C} + q\mathcal{Q})F$

To start with, let us first note that the FEM formulation also involves the approximation of the solution by a truncated series expansion of the form

$$x(t, \vec{\xi}) \simeq \tilde{x}(t, \vec{\xi}) = \sum_{i=1}^{n} x_i(t) \psi_i(\vec{\xi}) \quad (11.7)$$

where $\{\psi_i\}_{i=1}^{n}$ represents the FEM basis, and $X(t) = (x_1, x_2, \ldots, x_n)$ is the coordinate vector with respect to such basis. Projection of the original PDE (11.1) on the elements of the basis set produces a set of ODEs with the following structure:

$$\mathcal{D}\mathcal{A} \frac{dX(t)}{dt} + \left(\alpha(q\mathcal{Q} + \mathcal{C}) + \sum_{k=1}^{\mathcal{D}} \beta_k \mathcal{B}\mathcal{E}^k - \gamma \mathcal{D}\mathcal{A} \right) X(t) = \alpha \mathcal{G} + \mathcal{F} \quad (11.8)$$

the components i, j of the FEM matrices being computed as follows:

$$\mathcal{D}\mathcal{A}_{i,j} = \int_\Omega \psi_i \psi_j d\xi, \quad \mathcal{C}_{i,j} = \int_\Omega \vec{\nabla}\psi_i \cdot \vec{\nabla}\psi_j d\xi$$

$$\mathcal{B}\mathcal{E}^k_{i,j} = \int_\Omega \frac{\partial(\psi_i \psi_j)}{\partial \xi_k} d\xi_k \quad (11.9)$$

$$\mathcal{Q}_{i,j} = \int_{\partial\Omega} \psi_i \psi_j d\xi, \quad \mathcal{G}_i = \int_{\partial\Omega} \psi_i g \, d\xi, \quad \mathcal{F}_i = \int_\Omega \psi_i f \, d\xi \quad (11.10)$$

It should be noted that now the state becomes the vector $X(t)$ containing the n coordinates. Due to the nature of the FEM basis, n is usually a large number which translates into a large-dimensional ODE set, this being particularly critical when working in 2D or 3D domains. However, the FEM structure in (11.8) offers operators capable of efficient mappings between the original infinite-dimensional space and its finite counterpart. Table 11.1 summarizes the relationships between projection operators in the continuous and discrete FEM counterpart. Using these equivalences, we next provide a brief summary of the discrete versions of POD and LSD basis, respectively.

11.2.1.1 Proper Orthogonal Decomposition

In the POD method the kernel associated with the eigenvalue problem (11.4) corresponds with the two-point spatial correlation function [3, 7]:

$$R(\vec{\xi}, \vec{\xi}') = \lim_{T \to \infty} \frac{1}{T} \int_0^T x(t, \vec{\xi}) x(t, \vec{\xi}') \, dt \tag{11.11}$$

where $x(t, \vec{\xi})$ represents the value of the field on the time interval $[0, T]$. Since usually only a finite number of snapshots of the field is available over the interval $[0, T]$, the kernel in (11.11) is approximated as

$$R(\vec{\xi}, \vec{\xi}') = \frac{1}{\ell} \sum_{j=1}^{\ell} x(t_j, \vec{\xi}) x(t_j, \vec{\xi}') \tag{11.12}$$

where now $x(t_j, \vec{\xi})$ in (11.12) corresponds with the value of the field at each instant t_j and the summation extends over a sufficiently rich collection of uncorrelated snapshots at $j = 1, \ldots, \ell$. These can be obtained either from experiments or by direct numerical simulation of the original PDE [47].

The discrete counterpart of (11.4) is obtained by first combining it with (11.12) to get

$$\frac{1}{\ell} \sum_{j=1}^{\ell} \left(x(t_j, \vec{\xi}) \int_\Omega x(t_j, \vec{\xi}') \phi_i^x(\vec{\xi}') \, d\xi' \right) = \lambda_i^x \phi_i^x(\vec{\xi}) \tag{11.13}$$

and applying the relations given in Table 11.1, to transform the previous integral into

$$\int_\Omega x(t_j, \vec{\xi}') \phi_i^x(\vec{\xi}') \, d\xi' \to X^T \mathcal{D} \mathcal{A} \Phi_i^x$$

where $X \in \mathbb{R}^n$ and $\Phi_i^x \in \mathbb{R}^n$ are the discrete versions of the field and the ith eigenfunction, respectively. Applying the same relationships to the other terms in (11.13), the kernel (11.12) becomes

$$R = \frac{1}{\ell} \sum_{j=1}^{\ell} [X(t_j) X(t_j)^T] \tag{11.14}$$

and the FEM equivalent of the original eigenvalue problem takes the form

$$\frac{1}{\ell} \sum_{j=1}^{\ell} [X(t_j) X(t_j)^T] \mathcal{D} \mathcal{A} \Phi_i^x = \lambda_i^x \Phi_i^x \tag{11.15}$$

Standard algebraic methods to solve eigenvalue problems [28] can now be applied to (11.15) to extract the corresponding POD set in the FEM domain.

11.2.1.2 Laplacian Spectral Decomposition

It has been shown elsewhere [13] that (11.4) with the kernel $R(\vec{\xi}, \vec{\xi}')$ being the Green's function associated with the Laplacian is equivalent to the following eigenvalue problem:

$$\triangle\left(\phi_i^x(\vec{\xi})\right) = \overline{\lambda_i^x} \phi_i^x(\vec{\xi}') \tag{11.16}$$

where $\overline{\lambda_i^x}$ in (11.16) relates to the inverse of λ_i^x in the original formulation (11.4). In the same way as we did with PODs, we apply the equivalences in Table 11.1 to obtain a discrete version of the eigenvalue problem (11.16) which for homogeneous boundary conditions, takes the form

$$\mathcal{D}\mathcal{A}^{-1}(\mathcal{C} + q\mathcal{Q})\Phi_i^x = -\lambda_i^x \Phi_i^x \tag{11.17}$$

where Φ_i^x is the discrete version in the FEM mesh of the LSD eigenfunctions. It must be noted that systems with nonhomogeneous boundary conditions can be cast into this framework by first transforming them into the equivalent homogeneous ones [8, 19, 21].

11.3
Model Reduction in Identification of Bioprocesses

As was pointed out by [17], online measurements such as dissolved oxygen concentration, temperature or flow rates in the case of bioprocesses are usually available, whereas concentrations of biomass and some products and reactants require the use of state observers due to the lack of cheap or reliable online sensors. In addition, the reaction rates in bioprocesses are usually unknown and classical state estimators such as Luemberger or Kalman observers based on the perfect knowledge of the model structure cannot be applied. An observer design that circumvents such limitations was developed by Dochain and co-workers [9, 17, 18] for stirred tank reactors with asymptotic convergence depending on the dilution rate. In the work by [26] this theory was adapted to a general representation of the spatially distributed tubular reactors of the form

$$\frac{\partial x}{\partial t} = D \frac{\partial^2 x}{\partial \xi^2} - v \frac{\partial x}{\partial \xi} + K\varphi(x) + Q(x^* - x) \tag{11.18}$$

where $x(t, \xi) \in \mathbb{R}^{n_x}$ represents the state vector field as a function of time $t \in [0, \infty)$ and spatial coordinates $\xi \in \mathbb{R}$. $v \in \mathbb{R}^+$ denotes flow velocity and $D, Q \in \mathbb{R}^{n_x \times n_x}$ are positive and semipositive definite diagonal matrices which contain the parameters describing dispersion and exchange with the environment, property included in vector $x^* \in \mathbb{R}^{n_x}$. The kinetic part is described by a vector of nonlinear functions representing the reaction rates $\varphi(x) \in \mathbb{R}^{n_r}$ and a full column rank matrix $K \in \mathbb{R}^{n_x \times n_r}$ of yield coefficients [9]. Finally, the description is completed with initial and Danckwerts boundary conditions.

The transformation used in [26], which coincides with the one proposed by [9] and [16], determines which states $x_e \in \mathbb{R}^{n_e}$ can be observed from measurements $x_m \in \mathbb{R}^{n_r}$ without knowledge of the reaction rates (as shown next, the evolution of z is independent of them), provided that a reliable estimation z is at hand:

$$z = x_e + A_o x_m \quad \text{with } A_o = -K_e K_m^{-1} \tag{11.19}$$

Assuming that transport terms are known, each of the observed entities $\{z_k\}_{k=1}^{n_e}$ in the vector z is described by

$$\frac{\partial \hat{z}_k}{\partial t} = d_k \frac{\partial^2 \hat{z}_k}{\partial \xi^2} - v \frac{\partial \hat{z}_k}{\partial \xi} - q_k \hat{z}_k + h_k(x_m, x^*) \tag{11.20a}$$

$$d_k \frac{\partial \hat{z}_k}{\partial \xi} = v \hat{z}_k + g_k(x_m, x^{\text{in}}), \quad \forall t \in \mathbb{R}^+, \, \xi = 0 \tag{11.20b}$$

$$\frac{\partial \hat{z}_k}{\partial \xi} = 0, \quad \forall t \in \mathbb{R}^+, \, \xi = L \tag{11.20c}$$

$$\hat{z}_k = \hat{z}_k^0 = z_k - e_k^0, \quad \forall \xi \in [0, L], \, t = 0 \tag{11.20d}$$

where x_m represents the measurable states and $e_k^0 = x_k^0 - \hat{x}_k^0$ are the errors due to the unknown transformed initial condition. Finally, each of the observed states x_k that belong to the vector x_e is recovered by inverting the transformation.

In [26], the exponential convergence of the reduced order approximation of the observer (11.20) using the LSD method is demonstrated. In addition, the rate of convergence is also studied when the observed state is fed with an estimation of the measurable states from a few number of sensors optimally located in the spatial domain. Details for such optimal field reconstruction technique which uses the POD method can be seen elsewhere [3, 25].

11.3.1
Illustrative Example: Production of Gluconic Acid in a Tubular Reactor

The process takes place in a tubular reactor fed with glucose and oxygen. For the derivation of the mathematical representation, the system can be considered as a plug flow reactor with axial dispersion. In addition, the supply of oxygen can be introduced at several locations in the axial direction, resulting in a well-mixed and aerated medium inside any section of the reactor.

In accordance with the general dynamic structure proposed in (11.18) the state vector field, yield matrix, and reaction kinetic vector are formally written as

$$x = \begin{bmatrix} X \\ GA \\ G \\ O_2 \end{bmatrix}, \quad D = \begin{bmatrix} D_X & 0 & 0 & 0 \\ 0 & D_{GA} & 0 & 0 \\ 0 & 0 & D_G & 0 \\ 0 & 0 & 0 & D_{O_2} \end{bmatrix}$$

$$K = \begin{bmatrix} 1 & 0 \\ 0 & 1 \\ -1 & -1 \\ 0 & -0.5 \end{bmatrix}, \quad \varphi = [R_X] \tag{11.21}$$

where D_X, D_{GA}, D_G, and D_{O_2} represent the diffusion coefficients and R_X are considered to be equal to that obtained from the batch experiments from [38] in [24]:

$$R_X = \mu_{max} \frac{G}{K_G + G} \frac{O_2}{K_{O_2} + O_2} X$$

In addition, Q is a null matrix except for the last element $Q(4, 4) = kla$. As discussed in [26], the observation scheme for continuous gluconic acid production is designed to produce estimates of glucose, biomass, and gluconic acid from the limited number of measurements for oxygen and glucose, so we partition the system into

$$x_e = \begin{bmatrix} X \\ GA \\ G \end{bmatrix}, \quad x_m = [O_2]$$

$$K_e = \begin{bmatrix} 1 \\ Y_{GA} \\ Y_G \end{bmatrix}, \quad K_m = [Y_{O_2}] \tag{11.22a}$$

The rest of the matrices are null except the one that includes the oxygen exchange parameter $Q_m = [kla]$.

11.3.2
Observer Validation

The dynamic observer is of the form (11.19) for the glucose, biomass, and gluconic acid, respectively:

$$\frac{\partial Z_G}{\partial t} = D_G \frac{\partial^2 Z_G}{\partial \xi^2} + (D_G - D_{O_2}) \frac{Y_G}{Y_{O_2}} \frac{\partial^2 O_2}{\partial \xi^2}$$

$$- v \frac{\partial Z_G}{\partial \xi} - \frac{Y_G}{Y_{O_2}} K_{La}(O_2^* - O_2) \tag{11.23a}$$

$$\frac{\partial Z_X}{\partial t} = D_X \frac{\partial^2 Z_X}{\partial \xi^2} + (D_X - D_{O_2}) \frac{Y_X}{Y_{O_2}} \frac{\partial^2 O_2}{\partial \xi^2}$$

$$- v \frac{\partial Z_X}{\partial \xi} - \frac{Y_X}{Y_{O_2}} K_{La}(O_2^* - O_2) \tag{11.23b}$$

$$\frac{\partial Z_{GA}}{\partial t} = D_{GA} \frac{\partial^2 Z_{GA}}{\partial \xi^2} + (D_{GA} - D_{O_2}) \frac{Y_{GA}}{Y_{O_2}} \frac{\partial^2 O_2}{\partial \xi^2}$$

$$- v \frac{\partial Z_{GA}}{\partial \xi} - \frac{Y_{GA}}{Y_{O_2}} K_{La}(O_2^* - O_2) \tag{11.23c}$$

Table 11.2 Design parameters for the horizontal tubular fermenter.

Design parameter	Symbol value and units
Oxygen transfer rate	$K_{La} = 300\,\text{h}^{-1}$
Dispersion coefficients	$D_{GA} = D_G = 0.01\,\text{m}^2\,\text{h}^{-1}$, $D_X = 0.10\,\text{m}^2\,\text{h}^{-1}$, $D_{O_2} = 0.04\,\text{m}^2\,\text{h}^{-1}$
Flow velocity	$v = 0.01\,\text{mh}^{-1}$
Yield coefficients	$Y_G = -51.0$, $Y_{GA} = 44.9$, $Y_{O_2} = -2.6$
Input streams	$O_2^{\text{in}} = 0.0084\,\text{gl}^{-1}$, $X^{\text{in}} = 0.01\,\text{gl}^{-1}$, $GA^{\text{in}} = 0\,\text{gl}^{-1}$
Reaction rate parameters	$\mu_{\max} = 0.22\,\text{h}^{-1}$, $k_G = 9.9\,\text{gl}^{-1}$, $k_{O_2} = 0.00137\,\text{gl}^{-1}$

with the states in (11.23) being independent of the kinetic reactions and K_{La} denoting the maximum oxygen transfer rate. The concentration of the chemical compounds is recovered as follows:

$$G = Z_G + \frac{Y_G}{Y_{O_2}}O_2, \qquad X = Z_X + \frac{Y_X}{Y_{O_2}}O_2, \qquad GA = Z_{GA} + \frac{Y_{GA}}{Y_{O_2}}O_2$$

For the set of parameters presented in Table 11.2 (some of them estimated in [24]) and the glucose inlet of Fig. 11.2, the evolution and distribution of the concentrations are as the ones presented in Fig. 11.3. Partial differential equations were solved using the FEM with a mesh of 61 nodes (further discretizations will not alter the solution).

In order to illustrate the theoretical performance of the observer developed in [26] without any approximation, let us consider a 100% error in the initial concentration of the products, disturb the glucose inlet as illustrated in Fig. 11.2 and measure the input variables (oxygen) at any time and point of the spatial domain. The dynamic observer (11.20) is implemented using the FEM with a mesh of 61 nodes. In Fig. 11.4, the gluconic acid and biomass concentration errors are depicted at any

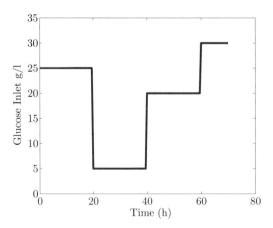

Fig. 11.2 Perturbations in the glucose inlet.

11.3 Model Reduction in Identification of Bioprocesses

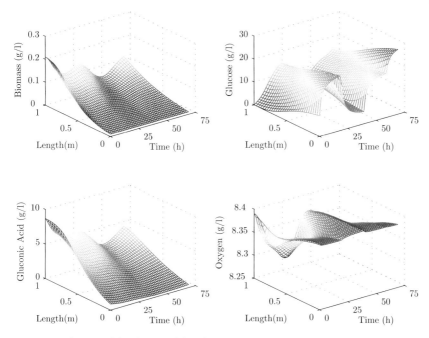

Fig. 11.3 Evolution and distribution of the relevant states.

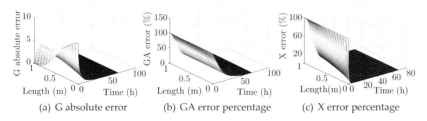

(a) G absolute error   (b) GA error percentage   (c) X error percentage

Fig. 11.4 Error convergence using the FEM with measurements at any time and point of the reactor.

time along the reactor, showing the exponential-type error convergence property discussed in [26].

At this point, let us illustrate how the observed variables (glucose, biomass, and gluconic acid) are affected by the model reduction. The POD technique was chosen for the derivation of the reduced order model. Following the steps indicated in Section 11.2.1, a version of Eq. (11.6) can be obtained for each particular observed variable of this example. A detailed derivation of the ROM is presented in [26]. It should be noted that biomass is the less diffusive compound, and a large number of ODEs will be required to get the same truncation error. Therefore, to obtain the error evolution and distribution shown in Fig. 11.5, we used six ODEs for the glucose and the gluconic acid and eight for the biomass (the less diffusive). It should be noted that the performance with the ROM in the cases of gluconic

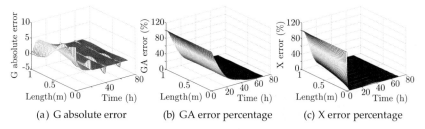

Fig. 11.5 Error convergence using the ROM with measurements at any time and point in the reactor.

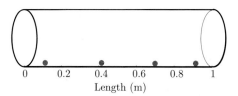

Fig. 11.6 Optimal sensor placement in the reactor for the reconstruction of oxygen concentration.

acid and biomass (Figs. 11.5(b) and (c)) seems to be the same as in the FEM case (Figs. 11.4(b) and (c)). However, when one compares the glucose absolute error evolution obtained with the ROM (Fig. 11.5(a)) and that obtained with the FEM (Fig. 11.4(a)) small differences can be appreciated when the reactor is highly perturbed.

So far we have considered oxygen measurements continuously taken along the axial dimension of the reactor. Now let us assume linear behavior between successive sampling times, and consider measurements taken every half an hour. In addition, let us make use of the method in [25] to select, from the POD bases set, the most appropriate number of oxygen sensors and their locations in the reactor. After obtaining by simulation a sufficiently rich number of dynamic snapshots, the POD set can be derived as proposed in Section 11.2.1.

In this particular case, four basis functions for the oxygen concentration are enough to capture almost the 100% of the energy. As pointed out in [25], the number of sensors required to ensure observability must be at least equal to the number of POD basis employed in the projection. The locations along the reactor for the four oxygen sensors are depicted in Fig. 11.6. Measurements are employed to reconstruct the whole oxygen field, as shown in Fig. 11.7. The reconstruction error remains extremely low in the event of perturbations thus showing good reconstruction capabilities.

Finally, the error obtained by combining all the approximations of the proposed methodology can be seen in Fig. 11.8 where the differences between the observation and the state after the transient remain below 3% in both cases (see Fig. 11.9).

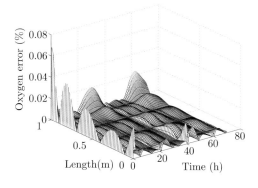

Fig. 11.7 Error (in percentage) in the reconstruction using a limited number of sensors for oxygen.

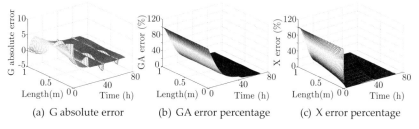

(a) G absolute error   (b) GA error percentage   (c) X error percentage

Fig. 11.8 Evolution of the ROM observation error in the estimation from partial measurements.

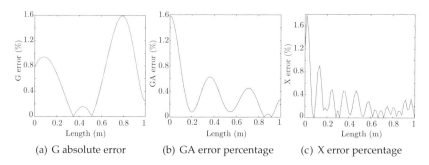

(a) G absolute error   (b) GA error percentage   (c) X error percentage

Fig. 11.9 Distribution of the ROM observation error in the estimation from partial measurements at final time.

11.4
Model Reduction in Control Applications

In this section we give an outline of some robust control design techniques for distributed process systems that make use of reduced order representations of the system's dynamics based on projections methods. In particular, we will show how this approach to modeling the system dynamics will provide the designer with some efficient tools to handle process uncertainties either through Lyapunov redesign techniques [35, 55] or by using a maximum allowed mismatch condition as the one proposed in [45] and [27] to update the model of the process. The application of these methodologies will be described in the context of tubular reactors.

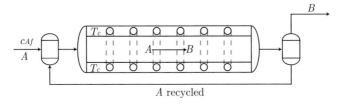

Fig. 11.10 General representation of a tubular reactor with recycle where a reaction $A \to B$ takes place. The recycle ratio is termed as r in the sequel.

Tubular reactors are widely employed in the process industry as they present a number of potential advantages over the well-mixed alternatives in terms of conversion and selectivity. However and despite their superior performance, they are difficult to operate. In fact, for exothermic reactions they may exhibit unstable behavior and hot spot phenomena which in some cases would lead to structural damage or even reactor explosion [22]. This makes tubular reactors good candidates for stabilizing robust controllers. A schematic representation of a typical tubular reactor is presented in Fig. 11.10. Fresh reactant A with concentration C_{Af} is pumped into the reactor at a given flowrate. As the flow crosses the reactor it produces the desired product B according to the irreversible reaction $A \to B$. At the reactor exit, a separation system recovers all the B produced while the remaining A is recycled to the reactor. The reactor temperature is controlled by means of a heating/cooling device at temperature T_c (see Fig. 11.10) which might consist of a piping system coiled around the reactor vessel.

11.4.0.1 Model Equations

Using the standard modeling assumptions (see for instance [6, 22]), the mass and energy balances result in the following dimensionless PDE set [5]:

$$\frac{\partial x(\xi, t)}{\partial t} + \nabla x(\xi, t) = D\Delta x(\xi, t) + \Sigma(x) + u(\xi, t) \tag{11.24}$$

$$[\nabla x = Pe_x(rx(1, t) - x(t, 0))]_{\xi=0}; \quad [\nabla x = 0]_{\xi=1} \tag{11.25}$$

where $x = [C, T]^T$, with C being the concentration of the component A, and T is the reactor temperature. For a more compact notation, let us rename the fields C and T as $x_1 = C$, $x_2 = T$, respectively. The diffusion, and Peclet matrices, nonlinear terms, and control inputs are, respectively,

$$D = \begin{bmatrix} \frac{1}{Pe_{x_1}} & 0 \\ 0 & \frac{1}{Pe_{x_2}} \end{bmatrix}; \quad Pe_x = \begin{bmatrix} Pe_{x_1} & 0 \\ 0 & Pe_{x_2} \end{bmatrix}$$

$$\Sigma(x) = \begin{bmatrix} -B_{c_A} f(x_1, x_2) \\ B_T B_{c_A} f(x_1, x_2) - \beta_T x_2 \end{bmatrix}; \quad u = \begin{bmatrix} 0 \\ \beta_T T_c \end{bmatrix}$$

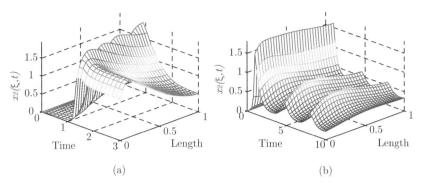

Fig. 11.11 Solution of system (11.24)–(11.25) for x_2 (a) without recycle ($r=0$) and (b) with recycle relation $r=0.5$. The following parameter values have been used in the simulations: $Pe_{x_1} = Pe_{x_2} = 7$; $B_{c_A} = 0.1$; $\gamma = 10$; $B_T = 2.5$; $\beta_T = 2$.

The nonlinear term associated with the chemical reaction term takes the form

$$f(x_1, x_2) = \exp\left(\frac{\gamma x_1}{x_2 + 1}\right)(1 + x_1) \tag{11.26}$$

The parameters γ, Pe_{x_1}, Pe_{x_2}, B_{c_A}, B_T, or β_T are related to the activation energy, the effect of diffusion–convection (the Peclet numbers) or heat transfer coefficients. For details the reader is referred to [5, 55].

The dynamic evolution of the system we just described is highly conditioned by the value of the recycle ratio. In order to illustrate this point, system (11.24)–(11.25) has been numerically solved using the FEM for different recycle ratios. The results for x_2 with $r = 0$ are represented in Fig. 11.11(a) showing the evolution of the field to a steady state, reached after $t = 2$. In contrast, a recycle ratio $r = 0.5$ induces oscillations on both fields (Fig. 11.11(b)). Such oscillatory behavior corresponds to a limit cycle as shown in Fig. 11.12, where the $\mathcal{L}_2$ norms of the fields x_1 and x_2 are plotted.

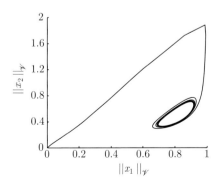

Fig. 11.12 Representation of the limit cycle reached when $r = 0.5$ in terms of the norm of the states.

11.4.1
Robust Control of Tubular Reactors

As a first step in the synthesis of the robust control law, a reduced order dynamic representation must be obtained. To this end, we make use of the methods described in Section 11.2, with emphasis on the implementation into the FEM framework (Section 11.2.1).

The Laplacian Spectral Decomposition Method As pointed out in Section 11.2, in order to apply this methodology, homogeneous boundary conditions are required. Unfortunately, this is not the case of Eqs. (11.24) and (11.25), thus a state transformation leading to an equivalent representation with homogeneous boundary conditions must be performed (see [8] for details). Projection of the homogeneous version of Eqs. (11.24) and (11.25) over the LSD eigenfunctions obtained from the solution of (11.17) leads to the following set of ODEs:

$$\frac{d\mathbf{m}^{z_1}}{dt} = \mathcal{A}\mathbf{m}^{z_1} - B_{C_A}\mathcal{F} - r\mathbf{\Phi}^T \mathcal{D}\mathcal{A}\frac{dX_1(1,t)}{dt} \tag{11.27}$$

$$\frac{d\mathbf{m}^{z_2}}{dt} = \mathcal{A}\mathbf{m}^{z_2} + B_T B_{C_A}\mathcal{F} + \beta_T(\mathcal{U} - \mathbf{m}^{z_2}) - r\mathbf{\Phi}^T \mathcal{D}\mathcal{A}\frac{dX_2(1,t)}{dt} \tag{11.28}$$

where

$$\mathcal{A} = -\mathbf{\Phi}^T\left(\frac{1}{Pe_{x_1}}\mathcal{C} + \mathcal{B}\mathcal{E} + \mathcal{Q}\right)\mathbf{\Phi}; \qquad \mathcal{F} = \mathbf{\Phi}^T \mathcal{D}\mathcal{A} F(\mathcal{Z}_1, \mathcal{Z}_2)$$

$$\mathcal{U} = \mathbf{\Phi}^T \mathcal{D}\mathcal{A} T_c$$

In this representation, z_1 and z_2 correspond to the transformed version of x_1 and x_2, respectively.

For $r = 0$, the first eight LSD eigenfunctions are enough to capture the relevant dynamics of the system, the maximum relative error between the FEM and the LSD being lower than the 1%. As shown in Fig. 11.13, the LSD basis is able to reproduce the tubular reactor behavior as accurately as the FEM approximation with almost four times less equations.

For a recycle ratio of $r = 0.5$ the LSD does not work as accurately as with $r = 0$ for the same number of eigenfunctions thus leading to relative errors between the LSD and the FEM greater than the 50% when using eight eigenfunctions. In order to recover a comparable accuracy, the number of eigenfunctions must be increased to 15 (the maximum relative error is reduced to the 5%). A comparison between the FEM and the LSD (with 15 eigenfunctions) solutions is presented in Fig. 11.14, showing slight differences between both solution methods.

The Proper Orthogonal Decomposition Method In order to obtain the POD basis set, a number of snapshots representative of the behavior of system (11.24)–(11.25)

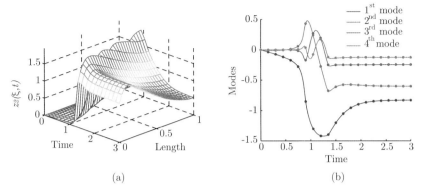

Fig. 11.13 Solution of system (11.24)–(11.25) with $r = 0$ using the LSD with eight eigenfunctions. (a) z_2 field representation, (b) the z_2 modes evolution. Lines indicate the real ones while dots correspond to the reduced order approximation.

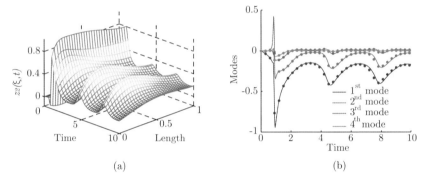

Fig. 11.14 Solution of system (11.24)–(11.25) with $r = 0.5$ using the LSD with 15 eigenfunctions. (a) z_2 field representation, (b) z_2 modes evolution. Lines indicate the real ones while dots correspond to the reduced order approximation.

in the range of operation conditions are needed. These data will then be used to compute the Kernel (11.14) required to solve the corresponding eigenvalue problem (11.15). In our case, the reduced order model should be capable of reproducing the dynamics of the original system for both $r = 0$ and $r = 0.5$. It should also be able to represent transition periods as well as steady states or limit cycles. For that purpose a number of snapshots were collected from direct numerical simulation covering mostly unsteady state scenarios with $r = 0$ and $r = 0.5$. For $r = 0.5$, the measurement sampling period was reduced to $\delta t = 0.01$ in order to accurately capture the limit cycle. Fig. 11.15(a) shows the shape of the first three PODs for the fields x_1 (continuous lines) and x_2 (dashed lines). Note that the higher the number of PODs, the larger the frequency of the spatial oscillations. The energy captured by the first nine PODs is depicted in Fig. 11.15(b). As shown in the figures, using

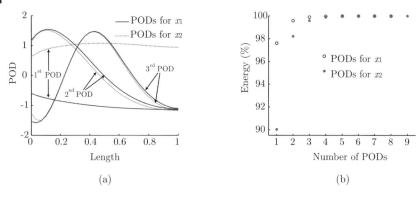

Fig. 11.15 (a) Shape of the first three PODs for the fields x_1 (continuous lines) and x_2 (dashed lines). (b) Energy captured by the PODs.

eight and nine PODs for the fields x_1 and x_2, respectively, is enough to capture the 99.999% of the energy which should be enough to explain most of the dynamic behavior.[2]

In this case, the projection of Eqs. (11.24) and (11.25) over the FEM version of the PODs results in the following ODE set:

$$\frac{d\mathbf{m}^{x_1}}{dt} = \mathcal{A}_{\mathbf{m}^{x_1}} \mathbf{m}^{x_1} - B_{C_A} \mathcal{F}_{x_1} + \mathcal{G}_{x_1} \tag{11.29}$$

$$\frac{d\mathbf{m}^{x_2}}{dt} = \mathcal{A}_{\mathbf{m}^{x_2}} \mathbf{m}^{x_2} + B_T B_{C_A} \mathcal{F}_{x_2} + \beta_T \left(\mathcal{T}_c - \mathbf{m}^{x_2} \right) + \mathcal{G}_{x_2} \tag{11.30}$$

Collecting the representative PODs ($\phi_i^{x_j}$, with $j = 1, 2$) into the matrix Φ^{x_j}, the different terms of the previous relations can be expressed as

$$\mathcal{A}_{\mathbf{m}^{x_j}} = -\left(\Phi^{x_j}\right)^T (1/Pe_{x_j} \mathcal{C} + \mathcal{Q} + \mathcal{B}\mathcal{E}) \Phi^{x_j}; \qquad \mathcal{F}_{x_j} = \left(\Phi^{x_j}\right)^T \mathcal{D} \mathcal{A} \mathcal{F}$$

$$\mathcal{G}_{x_j} = \left(\Phi^{x_j}\right)^T \mathcal{G}_j; \qquad \mathcal{T}_c = \left(\Phi^{x_2}\right)^T \mathcal{D} \mathcal{A} \mathcal{T}_c$$

Once these equations are solved, the field is recovered by applying $X_j = \Phi^{x_j} \mathbf{m}^{x_j}$. For $r = 0$, the field x_2 computed using the POD technique is plotted in Fig. 11.16(a). The field corresponding to $r = 0.5$ is presented in Fig. 11.16(c). The relative error between the FEM and the POD solution remains below the 0.5% with the exception of a few points for the case $r = 0$, which coincide with the sharp regions of the transient period. At these points the relative error increases to 5%. The main reason for this is that in the case of $r = 0.5$, a better set of snapshots were considered for the computation of the PODs. From the point of view of mode

---

2) A physical interpretation of energy in the context of fluid dynamics is presented [33]. For the relationships between energy and distance to the low-dimensional subspace the reader is referred to [3].

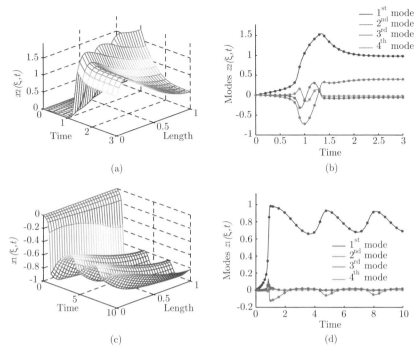

Fig. 11.16 (a) and (c) Evolution and distribution of the fields x_2 with $r = 0$ and x_1 with $r = 0.5$, respectively, computed with the POD technique. (b) and (d) Modes evolution. Lines indicate the true ones while dots correspond to the reduced order approximation.

evolution (Figs. 11.16(b) and (d)), the POD technique (dots) also shows good agreement with the FEM results (continuous lines). Note that, with the POD technique, the reduction resulted into a lower-dimensional system than the one obtained by the LSD method for an equivalent accuracy.

11.4.1.1 Controller Synthesis

Next we present a robust nonlinear controller for distributed process systems which is able to track either constant or variable (limit cycle-type) set points. The discussion will be centered in the control of tubular reactors, although it can be extended to other classes of distributed process systems of interest in biotechnology or bioengineering. For a complete discussion of theoretical and implementation aspects the reader is referred to [54–56].

The objective is to design a control law able to force a reactor as the one described in the previous section, to follow the limit cycle (reference trajectory) even for $r = 0$. The control variable corresponds with the temperature of the heating/cooling media T_c. In designing the controller it is assumed that the nonlinear reaction term $f(x_1, x_2)$ is unknown which calls for a robust design.

As discussed previously, the POD technique was shown to be more adequate than LSD to provide a reliable dynamic representation of the system. Note that in Figs. 11.16(c) and (d) one can distinguish two different behaviors: a transient period ($t < 2$) and a limit cycle ($t > 2$). In the previous section the objective was to construct a ROM able to represent both behaviors. Now, since the reference trajectory corresponds only with the limit cycle, the dynamics of the transient period can be ignored. For this reason, in this section a new POD basis set will be constructed using only data from the limit cycle dynamics[3] with $T_c = 0$. In this case, four PODs (per field) are enough to capture almost the 100% of the energy. The relative errors between the FEM and the ROM solution remain always below the 0.3%. The system in deviation form with respect to the reference ($\bar{x} = x - x^*$) becomes of the form

$$\frac{\partial \bar{x}}{\partial t} + \nabla \bar{x} = D \Delta \bar{x} + \bar{\Sigma} + \bar{u} \tag{11.31}$$

$$[\nabla \bar{x} = Pe_x (r\bar{x}(1,t) - \bar{x}(t,0))]_{\xi=0}; \qquad [\nabla \bar{x} = 0]_{\xi=1} \tag{11.32}$$

The construction of the control law then proceeds as follows: first, one chooses the sets $(\mathcal{E}_a, \mathcal{L}_a, \mathcal{N}_a)$ and $(\mathcal{E}_b, \mathcal{L}_b, \mathcal{N}_b)$ as the ones collecting the modes one wants to keep active (in fact one will force them to follow a given reference trajectory) and those that need to be stabilized, respectively. In our case, the sets $\mathcal{E}_a$ and $\mathcal{E}_b$ are those composed of the first four PODs and the remaining PODs, respectively.

Secondly, one projects Eqs. (11.31) and (11.32) over the sets $\mathcal{E}_a$ and $\mathcal{E}_b$ so as to obtain the following set of ODEs:

$$\frac{d\bar{m}^{x2a}}{dt} = \frac{1}{Pe_{x2}} \langle \Phi_a, \Delta \bar{x}_2 - \nabla \bar{x}_2 \rangle_\forall + B_T B_{cA} \langle \Phi_a, \bar{f} \rangle_\forall$$

$$+ \beta_T \langle \Phi_a, \bar{T}_c \rangle_\forall - \beta_T \langle \Phi_a, \bar{x}_2 \rangle_\forall \tag{11.33}$$

$$\frac{d\bar{m}^{x2b}}{dt} = \frac{1}{Pe_{x2}} \langle \Phi_b, \Delta \bar{x}_2 - \nabla \bar{x}_2 \rangle_\forall + B_T B_{cA} \langle \Phi_b, \bar{f} \rangle_\forall$$

$$+ \beta_T \langle \Phi_b, \bar{T}_c \rangle_\forall - \beta_T \langle \Phi_b, \bar{x}_2 \rangle_\forall \tag{11.34}$$

Note that the control action $\bar{T}_c$ only applies to the term x_2. Let us now split x_2 into two contributions using the sets $(\mathcal{E}_a, \mathcal{L}_a, \mathcal{N}_a)$ and $(\mathcal{E}_b, \mathcal{L}_b, \mathcal{N}_b)$ so that $x_{2a} = \sum_{i \in \mathcal{N}_a} \phi_i m^{x2i}$ and $x_{2b} = \sum_{i \in \mathcal{N}_b} \phi_i m^{x2i}$. On this basis, let us choose two Lyapunov functions of the form $\mathcal{B}_a(\bar{x}_2) = 1/2 \bar{x}_{2a}^2$ and $\mathcal{B}_b(\bar{x}_2) = 1/2 \bar{x}_{2b}^2$. The derivatives of $\mathcal{B}_a$ and $\mathcal{B}_b$ along the trajectories (11.33) and (11.34) lead, respectively, to the following relations (see [2] and [55]):

---

[3] The PODs of the previous section were obtained from data corresponding to limit cycle ($r = 0.5$), steady state ($r = 0$), and transient dynamics.

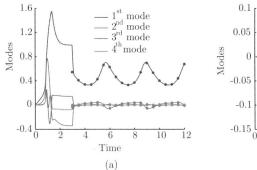
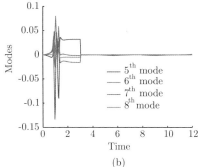

Fig. 11.17 Evolution of the modes corresponding to the sets (a) $(\mathcal{E}_a, \mathcal{L}_a, \mathcal{N}_a)$ and (b) $(\mathcal{E}_b, \mathcal{L}_b, \mathcal{N}_b)$, under the control law.

$$\dot{\mathcal{B}}_a \leqslant -\frac{1}{\beta_T}\left(\frac{1}{Pe_{x2}}\lambda_a + \beta_T\right)\frac{\delta_{0a}}{q_{1a}}\mathcal{B}_a + \beta_T B_{cA}\langle\bar{x}_{2a}, \bar{f}\rangle_{\mathcal{V}} + \beta_T\langle\bar{x}_{2a}, \bar{T}_{ca}\rangle_{\mathcal{V}}$$

$$\dot{\mathcal{B}}_b \leqslant -\frac{1}{\beta_T}\left(\frac{1}{Pe_{x2}}\lambda_b + \beta_T\right)\frac{\delta_{0b}}{q_{1b}}\mathcal{B}_b + \beta_T B_{cA}\langle\bar{x}_{2b}, \bar{f}\rangle_{\mathcal{V}} + \beta_T\langle\bar{x}_{2b}, \bar{T}_{cb}\rangle_{\mathcal{V}}$$

where λ_a and λ_b are the minimum eigenvalues of $\mathcal{A}_a = \langle\Phi_a, \Delta\bar{x}_2\rangle_{\mathcal{V}}$ and $\mathcal{A}_b = \langle\Phi_b, \Delta\bar{x}_2\rangle_{\mathcal{V}}$, respectively. One must realize first that the control objective is attained provided that the control law is able to drive $\mathcal{B}_a, \mathcal{B}_b \to 0$ [55]. As shown there, the first term on the right-hand side is negative so the control law only must compensate for the effects of the reaction term $\bar{f}$. As mentioned before, the exact form of $\bar{f}$ is unknown although a bound ζ so that $\zeta > \bar{f}$ is given. The following control law is shown to satisfy the stability condition described above [55]:

$$\bar{T}_{ca} = \begin{cases} -\frac{1}{\beta_T}\eta_a\frac{\bar{A}_a}{\|\bar{A}_a\|_{\mathcal{V}}} & \text{if } \eta_a\|\bar{A}_a\|_{\mathcal{V}} \geqslant \theta_a \\ -\frac{1}{\beta_T}(\eta_a)^2\frac{\bar{A}_a}{\theta_a} & \text{if } \eta_a\|\bar{A}_a\|_{\mathcal{V}} < \theta_a \end{cases} \quad (11.35)$$

$$\bar{T}_{cb} = \begin{cases} -\frac{1}{\beta_T}\eta_b\frac{\bar{A}_b}{\|\bar{A}_b\|_{\mathcal{V}}} & \text{if } \eta_b\|\bar{A}_b\|_{\mathcal{V}} \geqslant \theta_b \\ -\frac{1}{\beta_T}(\eta_b)^2\frac{\bar{A}_b}{\theta_b} & \text{if } \eta_b\|\bar{A}_b\|_{\mathcal{V}} < \theta_b \end{cases} \quad (11.36)$$

where $\bar{A}_i$, referred to as the dual of the field [2, 55], is computed as $\bar{A}_i = \partial\mathcal{B}_i/\partial x_{2i} = x_{2i}$. The function η_i with $i = a, b$ in Eqs. (11.35) and (11.36) is of the form $\eta_i > \|\zeta\|_{\mathcal{V}}$. It must be mentioned that the price to pay for robustness is that convergence to the exact reference cannot be assured. Nevertheless, convergence to an arbitrarily small region, defined by the parameters θ_a and θ_b, can be proved. The lower the values of such parameters the smaller the size of the region (and the better the control performance) although at the expenses of larger control efforts.

A simulation experiment was carried out to illustrate the behavior of the tubular reactor under this control law acting after $t = 3$. The results of this experiment are presented in Figs. 11.17 and 11.18. On the one hand, Figs. 11.17(a) and (b) show the

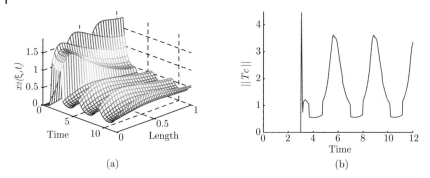

Fig. 11.18 (a) Evolution of the field x_2 under the control law. (b) Control effort.

evolution of the modes corresponding to the sets $(\mathscr{E}_a, \mathscr{L}_a, \mathcal{N}_a)$ and $(\mathscr{E}_b, \mathscr{L}_b, \mathcal{N}_b)$, respectively. As expected the modes $\mathbf{m}^{x2b}$ and $\mathbf{m}^{x2a}$ start to behave like in the case of $r = 0$; when the control law enters in action at $t = 3$, the modes $\mathbf{m}^{x2b}$ are stabilized while the modes $\mathbf{m}^{x2a}$ (continuous lines in Fig. 11.17(a)) are forced to follow the reference trajectory given by $\mathbf{m}^{x_{2a}^*}$ (dots in Fig. 11.17(a)). On the other hand, the effects of the control law on the field are depicted in Fig. 11.18(a). This figure illustrates that after entering the control at $t = 3$ the field is forced to follow the desired reference. The control effort is depicted in Fig. 11.18(b).

11.4.1.2 Robust Control with a Finite Number of Actuators

In this section, the problem of controlling the tubular reactor using a finite number of actuators is considered. The objective is the same as in the previous section, i.e., to actuate on the tubular reactor with a recycle relation $r = 0$ so as to recover the limit cycle dynamics exhibited when $r = 0.5$.

As shown before, the projection of the system (11.24)–(11.25) over the four more representative PODs resulted into a ROM able to reproduce the limit cycle dynamics. Furthermore, the evolution of the representative modes (first four modes) in deviation form with respect to the reference trajectory is described by Eq. (11.33). Let us denote by $\mathscr{M}_a$ the set containing the representative modes, i.e., $\mathscr{M}_a = \{\overline{m}_{x2i}\}_{i \in \mathcal{N}_a}$.

The control device would consist of a small number of pipes around the reactor. The temperature of the fluid inside the pipes can be manipulated and is employed to control the temperature inside the reactor. Consider that only seven pipes of diameter $d = L/30$, with L being the reactor length, are available. As shown in [3, 57], a number of actuators at least equal to the number of modes are needed to ensure stabilization. In our case this means that in order to stabilize the first four modes at least four "zone actuators" [55] are needed. It should be noted that if the control law is able to stabilize the modes of $\mathscr{M}_a$ (that is $\overline{m}_{x_{2a}} \to 0$), the modes $\mathbf{m}^{x2a}$ will follow the reference given by $\mathbf{m}^{x_{2a}^*}$ since $\overline{m}_{x_{2a}} = \mathbf{m}^{x2a} - \mathbf{m}^{x_{2a}^*}$. The other three pipes will be employed to stabilize the first three modes of the set $\mathscr{M}_b$. It must be stressed that the remaining infinite modes of the set $\mathscr{M}_b$ are stable as discussed in Section 11.2.

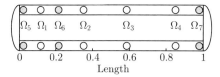

Fig. 11.19 Longitudinal section of the tubular reactor showing the optimal position of the zone actuators (pipes). Blank and gray pipes are employed to stabilize the modes belonging to the sets $\mathcal{M}_a$ and $\mathcal{M}_b$, respectively.

Actuator placement has been computed by methods similar to the ones employed in [3] and [25]. The selection criteria aim at minimizing the control effort by finding the position which maximizes the minimum eigenvalue of a given matrix **P** constructed from integrals of the PODs over the geometric support of the controller. A detailed description on the construction of the matrix **P** and the derivation of the control law is given in [53]. In this case, the optimal locations for the four and three actuators employed for stabilizing the modes of the set $\mathcal{M}_a$ and $\mathcal{M}_b$ are, respectively,

$$\mathcal{M}_a: \quad \Omega_1 = [0.1, 0.133]; \quad \Omega_2 = [0.333, 0.367]; \quad \Omega_3 = [0.567, 0.6]$$

$$\Omega_4 = [0.833, 0.867]$$

$$\mathcal{M}_b: \quad \Omega_5 = [0, 0.033]; \quad \Omega_6 = [0.2, 0.233]; \quad \Omega_7 = [0.967, 1]$$

Figure 11.19 schematically represents such locations. The blank pipes are employed to control the modes of the set $\mathcal{M}_a$ while the objective of the gray pipes is to stabilize the first three modes of the set $\mathcal{M}_b$. Using these locations, the eigenvalues of the matrix **P** are $\underline{\lambda}_a = 0.243$ for the set $\mathcal{M}_a$ and $\underline{\lambda}_b = 0.219$ for the set $\mathcal{M}_b$. Essentially, the steps to construct the control law are similar to those taken when the input is a function spatially distributed. In this case, the expression for the control law is $\overline{T}_c = \overline{T}_{ca} + \overline{T}_{cb}$, with

$$\overline{T}_{ca} = \begin{cases} -\frac{1}{\beta_T} \eta_a \frac{\overline{A}_a}{\|\overline{A}_a\|_\gamma} & \text{if } \eta_a \|\overline{A}_a\|_\gamma \geq \theta_a \\ -\frac{1}{\beta_T} (\eta_a)^2 \frac{\overline{A}_a}{\theta_a} & \text{if } \eta_a \|\overline{A}_a\|_\gamma < \theta_a \end{cases}$$

$$\overline{T}_{cb} = \begin{cases} -\frac{1}{\beta_T} \eta_b \frac{\overline{A}_b}{\|\overline{A}_b\|_\gamma} & \text{if } \eta_b \|\overline{A}_b\|_\gamma \geq \theta_b \\ -\frac{1}{\beta_T} \eta_b^2 \frac{\overline{A}_b}{\theta_b} & \text{if } \eta_b \|\overline{A}_b\|_\gamma < \theta_b \end{cases}$$

where $\overline{A}_i = \partial \mathcal{B}_i / \partial x_{2i} = x_{2i}$ and $\eta_i > \frac{\|\zeta\|_\gamma}{\underline{\lambda}_i}$ with $i = a, b$. The effects of this control over the modes of the system are depicted in Fig. 11.20. Before entering the control law ($t < 3$) the evolution of the modes is that corresponding to the tubular reactor with recycle $r = 0$. Once the control law enters in action and after a short transition period, the first four modes (lines in Fig. 11.20(a)) are forced to

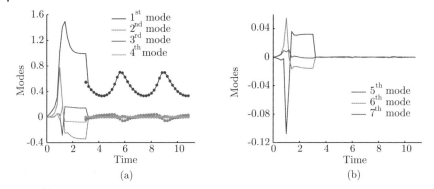

Fig. 11.20 Evolution of the modes corresponding to the sets (a) ($\mathscr{E}_a, \mathscr{L}_a, \mathcal{N}_a$) and (b) ($\mathscr{E}_b, \mathscr{L}_b, \mathcal{N}_b$), under the control law.

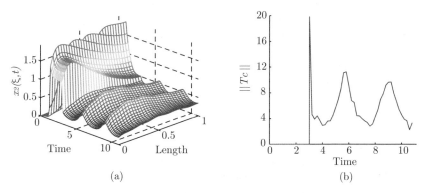

Fig. 11.21 (a) Evolution of the field x_2 under the control law using a finite number of actuators. (b) Control effort.

follow the reference (marks in Fig. 11.20(a)) while the first three modes of the set $\mathscr{M}_b$ are stabilized (Fig. 11.20(b)). If the field evolution is recovered – Fig. 11.21(a) – one can see that the objective of reproducing the limit cycle is reached. Note that, as expected, in this case the control effort – Fig. 11.21(b) – is larger than when an infinite number of actuators were available.

11.4.2
Real-Time Optimization: Multimodel Predictive Control

The use of a ROM methodology is specially advantageous in real time optimization, such as in the classical model predictive control (MPC). To be precise, the use of the POD might be the solution to the usually high dimensionality associated with the discretization of the PDEs by classical methods such as the FEM. However, the local nature of the POD basis involves two major problems: rich and numerous snapshots should be available and a large perturbation can lead the system to a region not represented by the POD basis.

In order to solve both problems, we propose the use of a multimodel predictive control (MMPC), which ensures stability under model/plant mismatch by updating the process of the model from the last snapshots available. Let us denote by $J_i^j(\mathbf{y}_i)$ the value of the optimum cost function in the time interval $[t_i, t_j]$, with $\mathbf{y}_i$ being the measurement at time t_i (initial condition). In addition, those states estimated from the ROM (initialized with previous measurements) will be distinguished by adding an upper hat to the symbol. In this way, $\hat{\mathbf{y}}_i$ will represent the estimation at time t_i of the states initializing the reduced model at t_{i-1} with $\mathbf{y}_{i-1}$. With this notation, the sufficient condition for robust stability to be used as follows:

$$J_{i-1}^i(\mathbf{y}_{i-1}) - [J_i^{i+p}(\mathbf{y}_i) - J_i^{i+p}(\hat{\mathbf{y}}_i)] \geqslant W \tag{11.37}$$

where $W \geqslant 0$. The difference $J_i^{i+p}(\mathbf{y}_i) - J_i^{i+p}(\hat{\mathbf{y}}_i)$ is a measure of the process/model mismatch and henceforth it will be referred as the *mismatch term*. Stability of the system under control is ensured whenever (11.37) is satisfied; otherwise the plant may become unstable. This observation motivates a POD and model updating method which whenever the criterion is violated makes use of the last snapshots to readapt the model to the new operation regime and thus to recover stability by reducing the *mismatch term*. The sufficient condition for robust stability is based on the work in [44–46] and the details for the MMPC technique can be seen in [24, 27].

11.4.2.1 Optimization Problem

The objective of the optimization problem is to enforce stability of system (11.24)–(11.25) and to achieve a concentration in the output of $x_1(t, L) = C_{\text{out}} = -0.9$. A weighted quadratic function is selected for the output concentration (now denoted by C_{out}) and control T_c, to avoid large deviations from the control reference T_c^r. Its expression reads

$$L(C_{\text{out}}, T_c) = 100\big(C_{\text{out}}(t) - C_{\text{out}}^r\big)^2 + \big(T_c(t) - T_c^r\big)^2 \tag{11.38}$$

where $C_{\text{out}}^r = -0.9$ and $T_c^r = -0.030$ denote the references to be achieved by the optimizer. The latter is computed by solving the nonlinear system (11.24)–(11.25) at a steady state when $C_{\text{out}} = -0.9$. Therefore the cost functional to be solved at each sampling time i becomes

$$\min_{T_c} J_i^{i+p}$$

with
$$J_i^{i+p} = \int_{t_i}^{t_i+p} \big[100\big(C_{\text{out}}(t) - C_{\text{out}}^r\big)^2 + \big(T_c(t) - T_c^r\big)^2\big] dt \tag{11.39}$$

subject to the dynamics of the POD model (11.29) and (11.30).

11.4.2.2 The Online Strategy

For the case study considered and after some preliminary trials, the predictive horizon is fixed to $p = 6$ and the manipulated horizon to $m = 2$. It should be noted that the conditions to assure stability without mismatch should be satisfied since the manipulated horizon is much smaller than the predictive horizon and this has been chosen in the order of the period of the oscillation. The cooling water temperature is approximated by a piecewise function with steps of length one unit of time, so that it can be represented as

$$T_c^i(t) = \sum_{c=1}^{3} u_c^i \ell_c^i(t)$$

where each $u_c^i \in \mathbb{R}$ is the temperature value in the control step c and $\{\ell_c^i(t)\}_{c=1}^{3}$ are the following unitary step functions:

$$\ell_1^i(t) = \begin{cases} 1, & \text{if } t \in [t_i, t_{i+1}), \\ 0, & \text{otherwise,} \end{cases} \qquad \ell_2^i(t) = \begin{cases} 1, & \text{if } t \in [t_{i+1}, t_{i+2}) \\ 0, & \text{otherwise} \end{cases}$$

$$\ell_3^i(t) = \begin{cases} 1, & \text{if } t \in [t_{i+2}, t_{i+6}) \\ 0, & \text{otherwise} \end{cases}$$

The measurements are taken at each unit of time including the initial conditions at the reactor startup ($t_0 = 0, t_1 = 1, t_2 = 2, \ldots, t_i = i$). In order to illustrate the MMPC method, the reactor is simulated by means of the FEM with a grid of 31 nodes. Since no previous off-line snapshots are taken into account, the reactor is operated in open-loop until the first two measurement vectors are available. At this moment, a POD that captures the 99.9% of the energy is computed, which allows the construction of a first ROM model to be used for optimization. Before each new optimization, violation of condition (11.37) will be checked to decide whether the POD set must be updated or not. If that is so, the most recent measurements will be employed to compute a new POD basis. In this case, the number of PODs will be that which captures at least the 99.5% which being sufficient, always resulted in reduced representations with less than five ODEs.

Also, and in order to avoid unnecessary model updating near the references, a stopping criterion is imposed through the following inequality that involves a minimum acceptable upper bound on the cost function:

$$\min_{T_c} J_i^{i+p} < 0.01$$

meaning that whenever the states and controls are so near the references that the cost function is less than 0.01, the performance is considered good enough and model updating stops.

In order to check the performance of the MMPC not only under mismatch, but also in the presence of disturbances, a perturbation is introduced at 10 units of time. The closed loop response obtained is presented in Fig. 11.22 showing a quite

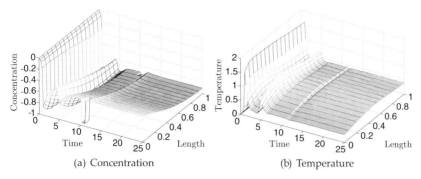

Fig. 11.22 Reactor behavior under MMPC. The evolution and spatial distribution of (a) concentration and (b) temperature is shown under the MMPC policy.

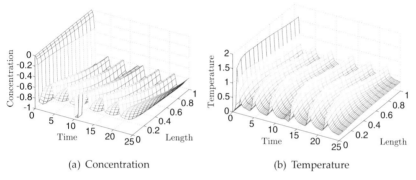

Fig. 11.23 Reactor behavior under the classical MPC. Evolution and spatial distribution of (a) concentration and (b) temperature with classical MPC.

reasonable performance, specially if one compares these results with those presented in Fig. 11.23 which were obtained by classical MPC (i.e., without model updating). In this case, the MPC controller is neither able to reach the objective nor able to stabilize the oscillations. This fact is also illustrated in Fig. 11.24 where the performance of MPC and MMPC is compared in terms of the output concentration and cooling temperature (the input).

Finally, the convergence of the objective function for both approaches, with and without updating, is presented in Fig. 11.25(a) showing once again the inability of MPC to achieve the desired reference.

11.5
Conclusions

In this chapter, we have presented a class of model reduction techniques for nonlinear distributed process systems representative of convection–diffusion–reaction

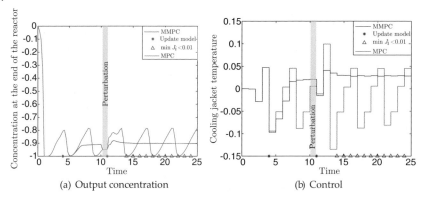

Fig. 11.24 A comparison between MPC and MMPC: (a) output concentration and (b) control action. The asterisks represent the times when the PODs are updated whenever the stability criterion (11.37) is not satisfied and triangles those instants when the output concentration is so close to the optimum that no updating is required.

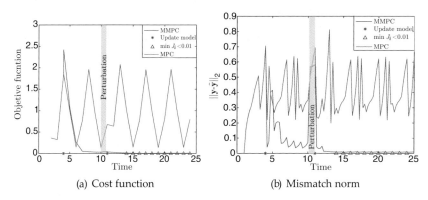

Fig. 11.25 A comparison between MPC and MMPC: (a) cost function evolution and (b) mismatch evolution. The mismatch is represented as the L_2 norm of the differences between the measured states y and the estimated states $\hat{y}$.

phenomena. The methodology exploits the dissipative nature of these processes which guarantees the existence of a time scale separation property associated with the dynamics of the original infinite-dimensional distributed field. This allows the identification of a low-dimensional subspace representing the slow (and possibly unstable) dynamics of the system. Projection of the original PDE equation set on such subspaces will produce a reduced order description of the relevant dynamics which is found to be particularly useful in the context of state estimation from partial measurements and robust process control.

Some obstacles however must be surmounted in producing a reliable representation suitable for identification and control: one of them is of numerical nature and refers to the lack of efficient projection formulations for arbitrary geometries. This has been overcome by taking advantage of the underlying algebraic structure associated with the finite element method which allows building simple yet accurate equivalences between discrete and continuous operators. The other obstacle is related to the presence of parametric and structural uncertainty, the latter being manifested in the form of changes in the dimension required to properly describe the slow dynamics as the process moves into different operation regimes. The solutions we just have discussed in the context of control applications include some extensions of the so-called Lyapunov redesign to cope with functional uncertainty or some model re-adaptation protocols as the ones described in the context of real time optimization and model predictive control of tubular reactors. It is our intention for future work to explore further these approaches in the context of complex phenomena taking place in fluid dynamics as well as that observed in physico-chemical systems at a microscopic or mesoscopic scales.

References

1 Alonso, A. A., Banga, J. R., Sánchez, I., Passive control design for distributed process systems: theory and applications, *AIChE Journal* 46(8) (**2000**), pp. 1593–1606.

2 Alonso, A. A., Fernández, C. V., Banga, J. R., Dissipative systems: from physics to robust nonlinear control, *International Journal of Robust Nonlinear Control* 14(2) (**2004**), pp. 157–179.

3 Alonso, A. A., Frouzakis, C. E., Kevrekidis, I. G., Optimal sensor placement for state reconstruction of distributed process systems, *AIChE Journal* 50(7) (**2004**), pp. 1438–1452.

4 Alonso, A. A., Ydstie, B. E., Banga, J. R., From irreversible thermodynamics to a robust control theory for distributed process systems, *Journal of Process Control* 12(4) (**2002**), pp. 507–517.

5 Antoniades, C., Christofides, P. D., Studies on nonlinear dynamics and control of a tubular reactor with recycle, *Nonlinear Analysis* 47(9) (**2001**), pp. 5933–5944.

6 Aris, R., *Análisis de Reactores*, Alhambra, Madrid, Spain, **1973**.

7 Balsa-Canto, E., Alonso, A. A., Banga, J. R., A novel, efficient and reliable method for thermal process design and optimization. Part I: Theory, *Journal of Food Engineering* 52(3) (**2002**), pp. 227–234.

8 Balsa-Canto, E., Alonso, A. A., Banga, J. R., Reduced-order models for nonlinear distributed process systems and their application in dynamic optimization, *Industrial & Engineering Chemistry Research* 43(13) (**2004**), pp. 3353–3363.

9 Bastin, G., Dochain, D., *On-Line Estimation and Adaptive Control Bioreactors*, Elsevier, Amsterdam, **1990**.

10 Berkooz, G., Holmes, P., Lumley, L., The proper orthogonal decomposition in the analysis of turbulent flows, *Annual Review of Fluid Mechanics* 25 (**1993**), pp. 539–575.

11 Christofides, P. D., *Nonlinear and Robust Control of PDE Systems: Methods and Applications to Transport-Reaction Processes*, Birkhäuser, Boston, **2001**.

12 Christofides, P. D., Daoutidis, P., Finite-dimensional control of parabolic PDE systems using approximate inertial manifolds, *Journal of Mathematical Analysis and Applications* 216 (**1997**), pp. 398–420.

13 Courant, R., Hilbert, D., *Methods of Mathematical Physics*, 1st edn., John Wiley & Sons, Inc., New York, **1989**.

14 De Lathauwer, L., De Moor, B., Vandewalle, J., A multilinear singular value decomposition, *SIAM Journal on Matrix Analysis and Applications* 21(4) (**2000**), pp. 1253–1278.

15 Delattre, C., Dochain, D., Winkin, J., Observability analysis of nonlinear tubular (bio)reactor models: a case study, *Journal of Process Control* 14 (**2004**), pp. 661–669.

16 Dochain, D., State observers for tubular reactors with unknown kinetics, *Journal of Process Control* 10 (**2000**), pp. 259–268.

17 Dochain, D., State and parameter estimation in chemical and biochemical processes: a tutorial, *Journal of Process Control*, 13 (**2003**), pp. 801–818.

18 Dochain, D., Perrier, M., Ydstie, B. E., Asymptotic observers for stirred tank reactors, *Chemical Engineering Science* 47(15/16) (**1992**), pp. 4167–4177.

19 Emirsjlow, Z., Townley, S., From PDEs with boundary control to the abstract state equation with an unbounded input operator: a tutorial, *European Journal of Control* 6 (**2000**), pp. 27–49.

20 Everson, R., Sirovich, L., Karhunen–Loève procedure for gappy data, *Journal of the Optical Society of America* 12(8) (**1995**), pp. 1657–1664.

21 Fattorini, H. O., Boundary control systems, *SIAM Journal of Control* 6(3) (**1968**), pp. 349–385.

22 Fogler, H. S., *Elements of Chemical Reaction Engineering*, 2nd edn., Prentice-Hall, Englewood Cliffs, NJ, **1992**.

23 Garcia, C. E., Prett, D. M., Morari, M., Model predictive control – theory and practice – a survey, *Automatica* 25(3) (**1989**), pp. 335–348.

24 García, M. R., Identification and real time optimisation in the food processing and biotechnology industries, PhD thesis, Universidad de Vigo, Spain, 2001, http://digital.csic.es/bitstream/10261/4662/1/MiriamRGarcia_thesis.pdf.

25 García, M. R., Vilas, C., Banga, J. R., Alonso, A. A., Optimal field reconstruction of distributed process systems from partial measurements, *Industrial & Engineering Chemistry Research* 46(2) (**2007**), pp. 530–539.

26 García, M. R., Vilas, C., Banga, J. R., Alonso, A. A., Exponential observers for distributed tubular (bio)reactors, *AIChE Journal* 54(11) (**2008**), pp. 2943–2956.

27 García, M. R., Vilas, C., Santos, L. O., Alonso, A. A., A robust and stabilizing multi-model predictive control approach to command the operation of distributed process systems, in: *The 2006 Annual Meeting*, **2006**.

28 Golub, G., Van Loan, C., *Matrix Computations*, John Hopkins University Press, Baltimore, MD, **1996**.

29 Gorban, A. N., Karlin, I. V., Zinovyev, A. Y., Constructive methods of invariant manifolds for kinetic problems, *Physics Reports* 396(4–6) (**2004**), pp. 197–403.

30 Goussis, D. A., On the construction and use of reduced chemical kinetic mechanisms produced on the basis of given algebraic relations, *Journal of Computational Physics* 128(2) (**1996**), pp. 261–273.

31 Goussis, D. A., Valorani, M., An efficient iterative algorithm for the approximation of the fast and slow dynamics of stiff systems, *Journal of Computational Physics* 214(1) (**2006**), pp. 316–346.

32 Henson, M. A., Nonlinear model predictive control: current status and future directions, *Computers and Chemical Engineering* 23(2) (**1998**), pp. 187–202.

33 Holmes, P., Lumley, J. L., Berkooz, G., *Turbulence, Coherent Structures, Dynamical Systems and Symmetry*, Cambridge University Press, Cambridge, **1996**.

34 Isidori, A., *Nonlinear Control Systems*, Springer, Berlin, **1995**.

35 Khalil, H. K., *Nonlinear Systems*, 2nd edn., Prentice-Hall, Upper Saddle River, NJ, **1996**.

36 Kolda, T. G., Orthogonal tensor decompositions, *SIAM Journal on Matrix Analysis and Applications* 23(1) (**2001**), pp. 243–255.

37 Lorente, L. S., Vega, J. M., Velazquez, A., Generation of aerodynamic databases using high-order singular value decomposition, *Journal of Aircraft* 45(5) (**2008**), pp. 1779–1787.

38 Mirón, J., González, M. P., Pastrana, L., Murado, M. A., Diauxic production of glucose oxidase by *Aspergillus Niger* in submerged culture. A dynamic model, *Enzime and Microbial Technology* 31 (**2002**), pp. 615–620.

39 Morari, M., Zafiriou, E., *Robust Process Control*, Prentice-Hall, Upper Saddle River, NJ, **1989**.

40 Muller, N., Magaia, L., Herbst, B. M., Singular value decomposition, eigenfaces, and 3D reconstructions, *SIAM Review* 46(3) (**2004**), pp. 518–545.

41 Narendra, K. S., Annaswamy, A. M., *Stable Adaptive Control*, Prentice-Hall, Englewood Cliffs, NJ, **1989**.

42 Qin, S. J., Badgwell, T. A., A survey of industrial model predictive control technology, *Control Engineering Practice* 11(7) (**2003**), pp. 733–764.

43 Reddy, J. N., *An Introduction to the Finite Element Method*, 2nd edn., McGraw-Hill, New York, **1993**.

44 Santos, L. O., Multivariable predictive control of chemical processes, PhD thesis, Universidade de Coimbra, **2001**.

45 Santos, L. O., Biegler, L. T., A tool to analyze robust stability for model predictive controllers, *Journal of Process Control* 9(11) (**1999**), pp. 233–246.

46 Santos, L. O., Biegler, L. T., Castro, J. A. A. M., A tool to analyze robust stability for constrained nonlinear MPC, in: *Process Optimisation, ADCHEM 2003, International Symposium on Advanced Control of Chemical Processes*, **2004**.

47 Sirovich, L., Turbulence and the dynamics of coherent structures. Part I: Coherent structures, *Quarterly of Applied Mathematics* 45(3) (**1987**), pp. 561–571.

48 Slotine, J. J. E., Li, W., *Applied Nonlinear Control*, Prentice-Hall, Englewood Cliffs, NJ, **1991**.

49 Smoller, J., *Shock Waves and Redaction–Diffusion Equations*, Springer, New York, **1983**.

50 Theodoropoulos, C., Luna-Ortiz, E., A reduced input/output dynamic optimization method for macroscopic and microscopic systems, in: *Workshop on Model Reduction and Coarse Graining Approaches for Multiscale Phenomena*, Leicester, England, Springer, Berlin, **2006**, pp. 535–560.

51 Theodoropoulos, C., Qian, Y. H., Kevrekidis, I. G., "Coarse" stability and bifurcation analysis using time-steppers: a reaction-diffusion example, *Proceedings of the National Academy of Sciences of the USA* 97 (**2000**), pp. 9840–9843.

52 Valorani, M., Creta, F., Goussis, D. A., Lee, J. C., Najm, H. N., An automatic procedure for the simplification of chemical kinetic mechanisms based on CSP, *Combustion and Flame* 146(1–2) (**2006**), pp. 29–51.

53 Vilas, C., Modelling, simulation and robust control of distributed processes: application to chemical and biological systems, PhD thesis, University of Vigo, Spain, May **2008**. Available online at http://digital.csic.es/handle/10261/4236.

54 Vilas, C., García, M. R., Banga, J. R., Alonso, A. A., Stabilization of inhomogeneous patterns in a diffusion–reaction system under structural and parametric uncertainties, *Journal of Theoretical Biology* 241(2) (**2006**), pp. 295–306.

55 Vilas, C., García, M. R., Banga, J. R., Alonso, A. A., Robust feed-back control of distributed chemical reaction systems, *Chemical Engineering Science* 62(11) (**2007**), pp. 2941–2957.

56 Vilas, C., García, M. R., Banga, J. R., Alonso, A. A., Robust feedback control of travelling waves in a class of reaction–diffusion distributed biological systems, *Physica D: Nonlinear Phenomena* 237(18) (**2008**), pp. 2353–2364.

57 Zerrik, E., Boutoulout, A., Bourray, H., Boundary strategic actuators, *Sensors and Actuators A – Physical* 94(3) (**2001**), pp. 197–203.

12
Model Development and Analysis of Mammalian Cell Culture Systems

Alexandros Kiparissides, Michalis Koutinas, Efstratios N. Pistikopoulos, and Athanasios Mantalaris

Keywords

single cell model (SCM), population balance modeling (PBM), generalized simulated annealing (GSA), sensitivity index (SI), mammalian cell culture systems, monoclonal antibody (MAb)

12.1
Introduction

The advancements in molecular biology and analytical techniques over the last century have significantly elevated the biological industry in the economical scale. Monoclonal antibodies (MAb) alone have a projected market of $49 bn by 2013, according to "Monoclonal antibodies Report, 2007." Considering that MAb industry is a mere fraction of the applications that utilize mammalian cell culture systems, one can appreciate the size of the biological industry. However as initially Bailey [1] and later Sidoli *et al.* [43] argue, the development of mathematically and computationally orientated research has failed to catch up with the recent developments in biology. Moreover, the little credit that mathematical modeling of biological systems receives from experimentalists may be the offspring of the lack of effective communication of the benefits of making a mathematical model [1].

In all aspects of science where modeling is involved, the first step, before making the model, is to determine its use and define *a priori* the problem the said model intends to address. Even the simplest possible bacterial strain, or the most exhaustively studied for that matter, is a complex network of a myriad of interconnected processes occurring on diverse time scales within a confined volume. To add to the complexity, the cell regulates its activities on multiple levels, deploying an elaborate control network, which to a large extent still remains gray territory. Moreover when cells grow in the neighborhood of other cells, an intricate communication network of signals and interactions mediates the macroscopic behavior of the culture. Therefore, any attempt to elaborately model the function of even a single cell will encounter numerous challenges. First of all, the amount of delicate intracellular measurements required to validate such a model is exhaustive both in

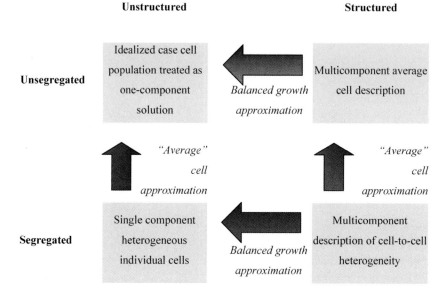

Fig. 12.1 Classification of biological models according to Fredrickson [16].

terms of labor as well as cost, in addition to numerical identifiability issues arising due to the large number of parameters involved [43]. Furthermore, not all aspects of biology are thoroughly understood, while many have been studied under very specific conditions. Therefore, one will reasonably wonder what role mathematical modeling can play when studying biological systems.

Borrowing research principles from the Chemical and Process Engineering paradigms, mathematical modeling of biological systems can provide a systematic means to quantitatively study the characteristics of the complex and multilevel interactions that occur in cell bioprocessing. In a way, it can be viewed as an effective way to organize in a meaningful way the vast plethora of available biological information. Mathematical models have successfully been used to design optimal media [53], identify previously ignored growth-limiting factors [10], optimize culture growth and productivity [8, 41, 11], and apply control principles to cell culture processes [14]. Thus the potential of modeling as a scientific and engineering tool has proven its worth; however in order to maximize the gains from the ever increasing influx of information from biology, especially with the development of the omics techniques, our view of modeling needs to be shifted toward a closed-loop framework from conception to optimization.

The way biochemical engineers conceive of and mathematically describe biological processes is still defined by the framework presented by Tsuchiya et al. [50] and Fredrickson et al. [16]. According to the framework shown in Fig. 12.1, a model can be *structured* (or unstructured) and/or *segregated* (or unsegregated). Most models fall in one of the four subcategories formed, depending on whether they possess one, two, or none of the above properties. Structured models enable the identification of a cell by assigning structure to it. The use of the term "structure" does not necessarily refer solely to the physical meaning of the term. Appart from physical structure which can be incorporated through the creation of intracellular compartments representing the various organelles, it can be incorporated by distinction amongst the various biochemical species hence giving rise to biochemical structure. Therefore, a structured metabolic model would consist of the reactions of at least two intracellular species. On the other hand, if cells are treated individually so that they differ from each other in some distinct way, we have a segregated model. The other extreme is to treat the entire population as a sum of averaged cellular behavior, leading to unsegregated models. Segregated models can account for cells in different cycle phases and generally depict the inherent heterogeneity of a cell population, whereas unsegregated models describe a homogeneous culture composed by a number of "average" cells.

Discriminations that exist for mathematical models in general, such as stochastic or deterministic, static or dynamic, are also applicable to biological models. Stochastic models account for the uncertainty inherent in all systems and implement this through some probabilistic based variation of the input variables. Deterministic modeling is usually based on experimental observations, accounting in a straightforward manner for the most common behavior of the system under observation. The drawback of deterministic modeling is that it cannot account for any possible set of inputs but only for the most probable ones. Dynamic models observe the evolution of the modeled system over a predetermined time horizon, whereas static models focus on a specific instance of the population. Dynamic models usually consist of sets of differential algebraic equations (DAE) and are computationally more demanding than static models, which usually contain algebraic equations and can be used for more detailed modeling of a system while remaining tractable.

A typical example of the above can be found in metabolic models, which can be either stoichiometric or kinetic. A kinetic model is represented usually by a set of DAEs, which are integrated over a time domain of interest and result in well-defined time trajectories of all the variables involved. A key drawback of kinetic modeling is that their additional predictive capability is associated with the incorporation of complex dynamic expressions, which usually result in nonlinearities both in the parameters and the variables. On the other hand, stoichiometric models are represented by a system of flux balance equations based on reaction stoichiometry of a metabolic network with accompanying constraints on flux values and are solved as a constrained optimization problem using some *a priori* assumed cellular objective. A key advantage of stoichiometric modeling is that it can take into account competing reactions, which enables the study of the relative activity of certain pathways under different culture conditions. However, the main drawback of stoichiometric models is that they are not dynamic and hence they cannot provide information on the temporal evolution of the variables under study [43].

Two extremes exist on the dimension scale on which cell models are considered. The first one is the single cell model (SCM) approach, first presented by Shuler *et al.* [42], according to which a single cell is modeled exhaustively incorporating as much information possible. SCMs are detailed descriptions of the functions occurring within a single cell ignoring any interactions with other cells. The more holistic modeling counterpart to single cell modeling is population balance modeling (PBM), where multiple populations with varying parameters can be studied. This type of modeling can account for cells being in different phases of the cell cycle and, therefore, displaying different behavior and different protein production rates. In the core of every PBM lies a simple model that describes cell metabolism, growth kinetics, and when division occurs. The drawback of PB models is that they tend to be computationally demanding and include large numbers of parameters.

Notable studies that have shaped new subcategories of biological systems modeling include, but are not limited to, cybernetic modeling presented by Ramkrishna [37] and the introduction of structure as defined by Fredrickson *et al.* [16] at the genetic level by Lee and Bailey [27, 28]. In brief, the concept behind cybernetic modeling is the adaptation of a mathematically simple description of a complex organism, which is compensated for the oversimplification by assigning an optimal control motive to its response [23]. For example, microbial cells growing in the presence of multiple substrates are assumed to be following an invariant strategy to optimize a certain goal by choosing which substrate to consume first. So assuming, for example, a multisubstrate environment containing cells that follow different strategies of substrate consumption, cells that will, somehow, choose to grow first on the fastest substrate available will proliferate much faster than cells that respond differently. After some time, all the cells that remain in the environment will be those that have responded in the optimal manner [23]. It is, therefore, reasonable to assume that over the many years of evolution cells have acquired the ability to respond optimally to environmental conditions.

Lee and Bailey [27, 28] extended the concept of structure as presented above to the level of nucleotide sequences. Lee introduced an explicit connection between a particular nucleotide sequence and the affinity of a particular protein for that sequence which in turn will influence the corresponding transcription event, thus deriving a quantitative mapping from nucleotide sequence to overall phenotype. Even though in his detailed review Bailey [1] predicted that this new "genetically structured model" would be widely embraced in the future, supported by the advancement of the omics techniques, little work has yet been done in that direction.

12.2
Review of Mathematical Models of Mammalian Cell Culture Systems

Mathematical biology, and biotechnology for that matter, can be subdivided in two broad categories depending on the type of cells studied. The oldest and more exhaustively studied category deals with microbial systems mostly and prokaryotic cells in general. On the other hand, we have the more recently emerged field of eukaryotic (or mammalian) cell modeling. In light of recent advancements, both scientific and regulatory, it is worthwhile mentioning the prominent advancement of a third category, namely that of stem cell modeling. Even though all three types of cells share a lot of common elements with respect to their core metabolism, they have distinct differences in their behavior in culture necessitating

Table 12.1 Enzyme and microbial growth kinetic expressions.

Name	Expression	Function
Michaelis–Menten	$v_0 = \frac{V_{MAX}[S]}{K_M + [S]}$	Describes the kinetics of the simple enzyme catalyzed reaction: $E + S \underset{k_{-1}}{\overset{k_1}{\longleftrightarrow}} ES \overset{k_2}{\longrightarrow} P$
Hill	$\theta = \frac{[L]^n}{(K_A)^n + [L]^n}$	Describes the fraction of the macromolecule saturated by ligand as a function of the ligand concentration
Monod	$\mu = \mu_{MAX} \frac{[S]}{K_S + [S]}$	Describes microbial growth based on the consumption of one substrate

their study within these broad categories. Undoubtedly, developments and research in prokaryotic cell modeling lead by a fair margin the respective developments in both mammalian and stem cell modeling. This can be attributed to a number of reasons, the main being that prokaryotic cells have to a certain degree simpler metabolic characteristics. The literature around mathematical modeling of biological systems, be they prokaryotic or eukaryotic, is arguably too vast to summarize within the limited space of a book chapter. Therefore, we will attempt to review contributions that have either shaped or can successfully highlight a new way of approaching dynamical modeling of mammalian systems.

The earliest reference and possibly the most significant one is the mathematical formulation that describes enzyme kinetics, presented by Michaelis and Menten [56]. Although the hypothetical system studied was the simplest possible, the conversion of one molecule of a given substrate to a product via a single enzymatic reaction, in many ways it shaped the way we conceive of kinetic rates in biology. Since then, the theory provided by Michaelis and Menten has evolved, now being used as a starting point when attempting to describe much more complex enzyme kinetics. Around the same time that Michaelis and Menten presented their work, Archibald Vivian Hill, in his effort to describe the sigmoidal binding curve of oxygen to hemoglobin, derives what is now know as the Hill function (1910). In essence, the Hill function describes the binding of a given ligand to a macromolecule when the latter is already saturated with ligands. Finally in 1948, Jacques Monod presented a function identical to the Michaelis–Menten rate equation, which successfully described microbial growth. The basic concept behind Monod's work was that the kinetics observed in every metabolic pathway are largely shaped by its rate limiting step, ultimately an enzyme catalyzed reaction. All these kinetic equations are summarized in Table 12.1.

Shifting our focus toward the area of mammalian cell metabolism, most mathematical models examine glucose and glutamine as the primary nutrients and lactate and ammonia as the main metabolites. A typical layout of an unstructured model for cell metabolism consists of mass balances on glucose, glutamine, ammonia, and lactate around the bioreactor. These account for the uptake of glucose and glutamine from viable cells for cell growth, as well as glucose consumption by glucokinase, glucose maintenance energy, and spontaneous degradation of glutamine in the medium. Lactate and ammonia production are described as functions of glucose and glutamine consumption, respectively [18]. Monod-type kinetics are used for most metabolic models [18, 49, 51].

One of the first attempts at developing a structured model for mammalian cells was that of Batt and Kompala [2] who adapt ideas presented by Fredrickson et al. [16] and Shuler et al. [42] to mammalian cell culture systems. Cell mass is divided in four intracellular metabolic pools accounting for amino acids, nucleotides proteins, and lipids. These are derived from the extracellular substrates, glucose glutamine, and amino acids found in the culture media, while the secreted products include lactate, ammonia, and monoclonal antibody (MAb). Borrowing experimental data from the extensive work of Miller et al. [34], Batt and Kompala [2] show that the model successfully describes experimental data but more importantly can be used in order to study the effects of various feeding strategies.

Bibila and Flickinger [3] presented one of the most significant structured models describing MAb synthesis in hybridoma cells. Based on the mechanism proposed by Percy [36] for the covalent assembly of MAb, the authors present in detail the derivation of the structured model that successfully describes experimental data of MAb synthesis and secretion. On subsequent studies [4, 5], they move on to use the proposed model for both steady-state and dynamic optimization of the culture conditions, suggesting strategies that increase final antibody titer. Moreover they perform a parameter sensitivity analysis (SA) [4] through factorial design in the steady-state version of their model and draw conclusions on the parameters that affect antibody secretion positively. Finally, they suggest the assembly step of antibodies within the endoplasmic reticulum (ER) as the most probable candidate for a rate-limiting step of the secretion process, based on perturbation studies conducted with their model [5].

Xie and Wang [53] presented a detailed stoichiometric model for animal cell growth and utilized it to optimize culture media composition. Their stoichiometric analysis covers various aspects of cellular metabolism including energy requirements, lipid, carbohydrate, nucleotide, and protein synthesis. Moreover they provide formulae for the derivation of stoichiometric coefficients both for nutrients and products by studying their roles in animal cell metabolism. Later work [54, 55] by the same authors has provided valuable insight on mammalian cell metabolism. Utilizing the devised model [53], the authors reach a number of valuable conclusions including the necessity to control glucose feed at low concentrations in order to shift mammalian cell metabolism toward more energy-efficient pathways. Finally [55], they were among the first to exhaustively study energy metabolism in mammalian cell culture systems by studying the stoichiometry of the simplified metabolic reaction network they devised.

De Tremblay et al. [8] showcase the potential of dynamic programming for the optimization of fed-batch hybridoma cultures. Having verified the applicability of dynamic programming in [9], they went one step forward and examined the benefits of using and optimal control approach versus a closed loop strategy on fed-batch hybridoma cultures, also presenting experimental data to support their results. Frahm et al. [58, 14] presented a novel open-loop-feedback-optimal controller for the fed-batch cultivation of hybridomas. The utilized unstructured model accounts for MAb production and culture growth based on the consumption of glucose and glutamine and the production of lactate and ammonia as basic by-products of metabolism.

DiMasi *et al.* [12] present a mechanistic structured kinetic model of mammalian cell culture dynamics. The developed model specifically addresses the dynamics of substrate consumption and energy metabolism in mammalian cell culture. Borrowing experimental data from Miller *et al.* [34], the authors compare their model to the unstructured model of Batt and Kompala [2] and reach the conclusion that a structured model that successfully predicts specific growth rates and utilization rates of the major substrates (glutamine, glucose, essential amino acids, and oxygen) is a more suitable candidate for model-based optimization and control studies. Their work provides a solid framework for the development of structured dynamic models that capture the dynamics of mammalian energy metabolism; however, parameters have been estimated from literature data, therefore leading to low confidence levels in the model output.

Even though cell growth is a well-studied area of animal cell cultures, there appear to be many differences between the mathematical models that describe it. These differences mainly involve its dependency on nutrients, metabolites, and oxygen. Cell growth has been mathematically related to glucose concentration alone [15], glucose and glutamine [8], glutamine, ammonia, and lactate [57], glucose and lactate [25], and to all four nutrients and metabolites [35, 18]. All the above models assume Monod-type kinetics. Tatiraju *et al.* [49] have also suggested an equation for oxygen consumption and its relation to cell growth, which is of little use as it is decoupled from the other nutrients with which it has been proven to be associated. Similar models have been obtained for cell death, which relate the rate of cell death to glutamine, lactate, and ammonia concentrations [8, 7], or glucose [15], or glutamine [7], or ammonia and lactate [2], or ammonia [18].

Pörtner and Schäfer [38] compared a selection of models and model parameters that existed in the literature at that time and carried out an analytic error and range of validity analysis. They found significant variations in the values of maximum growth rate, yields, and nutrient Monod constants that were used by other researchers. They came to the conclusion that the models' predictions involved significant errors, particularly due to the lack of understanding of cellular metabolism and the limited data ranges within which the model was valid. They further suggested that static batch cultures could be used, for example, for the determination of maximum specific growth rate, but not for establishing a relationship between the growth rate and substrate concentration, whereas continuous cultures could yield reliable data due to the steady-state operation conditions. For very low substrate concentrations, they suggest using fed-batch cultures. Finally, they recognized that for significant improvements, parameter identification techniques and control strategies need to be applied to mammalian cell cultures as has previously been the case in other biotechnological processes.

Significant efforts, lead by Fredrickson and co-workers [13, 50], have been made to introduce PBM in biological systems modeling. Even though PBMs have the unique ability to account for the inherent heterogeneity in all cell cultures, unfortunately they are difficult to solve and usually lead to intractable models. Despite their promising characteristics, their limited usage in mathematical biology is mainly due to two major drawbacks [48, 52]. They are complicated to handle and solve and accurate determination of model parameters is not possible due to the lack of distribution data.

One of the main contributors in the field of PBM has been Mantzaris and coworkers, who have presented a series of papers [29–33] covering a variety of different PBM cases, some of which were compared to models that could be analytically solved. The combination of SCMs and PBMs represents the next logical challenge. Due to the additional model fidelity, such a hybrid model is extremely computationally intensive; hence the solution of even the PBM component of the overall model becomes intractable. To overcome this problem, investigators have used finite-representation techniques to discretize populations avoiding the problems of continuous distributions and the integral differential features they bring. Sidoli *et al.* [44] presented a coupled SCM-PBM model that uses a highly structured SCM to characterize single-cell growth and death rates in each stage of a multistage PBM. The model was validated against batch and fed-batch experimental data achieving a satisfactory agreement with some but not all of the modeled variables.

This section was intended as a brief overview of some, but not all, key contributions in the field of mammalian cell culture systems' modeling. The following section will pose the key questions that need to be answered in the near future and intends to motivate the reader to follow through the remainder of this chapter.

12.3
Motivation

A thorough overview of the previous sections reveals that the optimal point on the scale between tractability and fidelity does not lie near the boundaries. On the contrary, an approach that would attempt to exploit the advantages of structured models while maintaining tractability could result in a robust yet computationally flexible hybrid model. Such an approach has successfully been followed by Mantalaris and coworkers [17, 22, 26] for secreting mammalian cell culture systems. The idea was to maintain structure for the protein formation and secretion process while using an unstructured model to describe growth and proliferation. All models successfully predict final antibody titers have gone a step further having used the mathematical model for the derivation of an optimal feeding profile.

What ultimately discriminates a good model from a bad model is its ability to successfully describe the modeled process while minimizing the uncertainty of its output variables. The majority of studies presented either utilize literature data to validate the models or generate their own experimental data without any form of systematic design of experiments (DOE). One of the challenges in biochemical engineering is the development of high fidelity models that are able to capture the required biological functions involved in the generation of the end-product while remaining computationally tractable in order to be viable candidates for model-based control and optimization. However, high fidelity models, inherently, contain a large number of parameters. Therefore, use of a systematic framework that designs experiments in a way that minimizes the necessary experimentation while simultaneously maximizing information obtainable from the data, is the first step toward achieving a uniquely validated model [43]. The increasingly available biological information, both theoretical and analytical, necessitates the use of such a framework from model conception to validation in order to avoid unnecessary experimentation and poorly informative experiments.

The work presented by Asprey and Machietto [59, 60], Sidoli et al. [43], Ho et al. [17], Kontoravdi et al. [22], Lam et al. [26], and Kiparissides et al. [19] defines a unique and systematic approach to modeling biological systems which is depicted in Fig. 12.2.

Let us assume a first principle mathematical model, $g(x, \bar{P})$, formulated to describe a real-life process, where x denotes the input vector, and $\bar{P}_i$ (where $i = 1, \ldots, \nu$) denotes the parameter vector. The first step of the model development algorithm is to determine, before actually designing or performing any experiments, whether model parameters can be uniquely identified from the mathematical structure of the model. Failure to pass the identifiability test implies mathematical singularity with respect to the model parameters; therefore, there is no need to perform any further analysis or experiments for a model whose parameters are know a priori to be unidentifiable [60, 43]. Models that fail the identifiability test should either be reformulated in a way that avoids singularity for the problematic parameters or discarded.

For models that satisfy the criteria of the identifiability test, the next step is to apportion the uncertainty in the model output to the sources of variation. Model analysis techniques and SA, in particular, can provide valuable insight regarding the dependence of the model output to its parameters. The output of SA will be a vector of size ν, containing the sensitivity indices (SI) of the model parameters. As a rule of thumb if ν larger than 20, the use of parameter grouping will become necessary [19]. This merely affects the notation of Fig. 12.2 and in no way the algorithm itself. Therefore in the case of parameter grouping, $\bar{P}_I$ is the parameter vector and ν is the number of partitions it contains, corresponding to the number of parameter groups formed.

Consequently an empirical criterion, determined by the modeler, is applied in order to discriminate the significant from the insignificant model parameters. The criterion is a threshold value for the normalized SI, usually set between 0 and 0.2 [40]. Any parameters with values below the set threshold are considered insignificant to the model output and are allocated in a partition of the parameter vector termed $\bar{p}_j^1$ ($j = 1, \ldots, \nu'$). The remaining parameters whose SI is above the threshold value are allocated in a second partition of the parameter vector termed $\bar{p}_k^2$ ($k = 1, \ldots, \nu''$). The sum of ν' and ν'' should of course equal the size of the parameter vector $\bar{P}_I$, at all times. The values of the parameters in partition $\bar{p}_j^1$ are

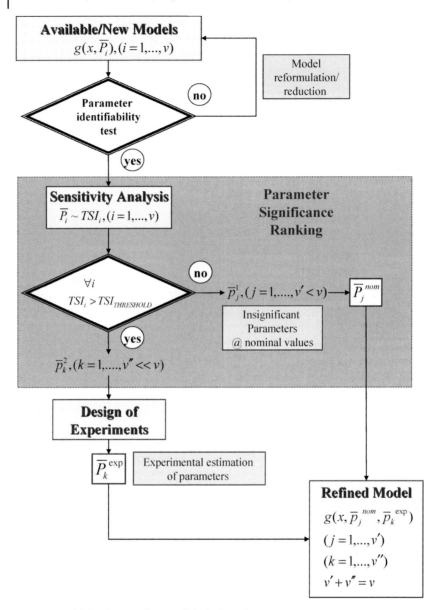

Fig. 12.2 Model development framework for biological systems.

set to the nominal values, which can be derived either from existing literature or from a parameter estimation algorithm, hence yielding the parameter vector $\bar{p}_j^{nom}$.

The values of the model affecting parameters in partition $\bar{p}_k^2$ need to be determined experimentally with accuracy in order to reduce the uncertainty in the model output. Therefore, experiments are specifically designed [59, 60] for the determination of the parameters in vector $\bar{p}_k^2$ and once the experimental data are available the values for parameters $\bar{p}_k^2$ are determined explicitly, yielding vector $\bar{p}_k^{\exp}$. Finally by substituting the initial parameter vector $\bar{P}_I$, with the newly derived $\bar{p}_j^{\text{nom}}$ and $\bar{p}_k^{\exp}$, we derive a refined version of the original model, $g(x, \bar{p}_j^{\text{nom}}, \bar{p}_k^{\exp})$.

The above presented framework successfully minimizes model uncertainty but more significantly minimizes experimental costs and labor. Moreover, it sets a scientific platform of communication between modeler and experimentalist, thus bridging the communication gap between engineers and biologists. In the sections that follow, the individual steps of the model development framework will be discussed in more detail, illustrated by the presentation of relative research examples.

12.4
Dynamic Modeling of Biological Systems – An Illustrative Example

The biological systems model building framework described in the previous section will be explained in detail through a "real-life" illustrative example in this section. Let us utilize, as an example, an industrial process for the production of MAbs harvested from cultures of hybridoma cells. The reason we wish to model this process is the maximization of final antibody titer in our culture through *in silico* experimentation. Batch and fed-batch cultures are currently the cultivation methods of choice from the biologics industry for the large-scale production of MAb, due to their operational simplicity, reliability, and flexibility for implementation in multipurpose facilities [61]. Therefore for the purposes of our example, a model capable of describing both batch and fed-batch cultures of antibody secreting hybridomas is required.

Bearing in mind that the model will ultimately be utilized for optimization studies, which are inherently computationally expensive, renders structured models a less attractive idea. Moreover, as Sidoli *et al.* [44] have proved, overparametrized models lead to parameter identifiability issues, which in turn reduce confidence in the model output. However, the model should contain adequate level of information regarding the antibody formation process and how its various steps are affected by the growth characteristics of the culture and the availability of nutrients in order to yield meaningful results. Balancing the trade-off between tractability and fidelity is the first challenge we need to address. Hybridoma growth kinetics have been widely studied and unstructured models have proven to be capable of capturing their dynamics. Therefore, structure can be avoided at a low cost and an unstructured model can be used to describe cell proliferation and nutrient uptake in batch and fed-batch cultures. The model has been adapted by Kontoravdi *et al.* (2005) based on the work of Jang and Barford (2000) and describes cellular growth based on the consumption of two basic nutrients (glucose and glutamine) and the inhibition by the two corresponding byproducts of the cell's metabolism (lactate and ammonia).

The formation and secretion of MAb is an inherently complex process (Fig. 12.3). MAbs are Y-shaped proteins formed from two identical heavy and two identical

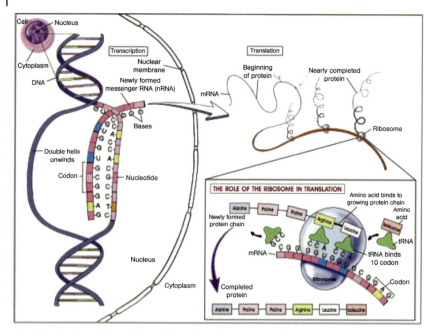

Fig. 12.3 Protein synthesis (adapted from: http://www.nih.gov/).

light polypeptide chains. The heavy and light chains are encoded from different genes and, therefore, a situation might easily arise where there is an abundance of one type of chains yet a shortage in the other resulting in a small, and even zero, production rate. The formation of these proteins starts from the nucleus of the cell where the chain-specific DNA sequence is copied on an mRNA molecule in a process known as the transcription. The mRNA molecule will migrate to the ER where it will bind to a ribosome and start the process of transcription. Once an antibody molecule has been formed by two heavy and two light chains, it will be transferred to the Golgi apparatus where it will undergo posttranslational modifications in order to become a biologically active molecule prior to its secretion to the extracellular environment.

Since a significant number of processes, each occurring at a separate site, are involved in antibody formation, utilizing an unstructured model to describe the rate of antibody accumulation in the media would result in a significant loss of information. Therefore, the structured model, presented by Bibila and Flickinger [4, 5], is an ideal candidate for the description of the antibody formation process. Kontoravdi *et al.* [21] successfully managed to couple this structured model to the unstructured model (mentioned above) describing cell growth and proliferation. Following this brief discussion around the conceptual formulation of the model, a first principle derivation of the hybrid model is presented below.

12.4.1
First Principles Model Derivation

A material balance for viable cells within the bioreactor is given by the following equation:

$$\frac{dVX_u}{dt} = \mu V X_u - \mu_d V X_u \quad (12.1)$$

where X_u is the concentration of viable cells in the bioreactor measured in cell per liter and μ, μ_d are the specific growth and death rates, respectively, measured in h^{-1}. Detailed formulas for the estimation of the specific growth and death rates will be presented at a later stage. The material balance for the total cell concentration (the sum of both dead and viable cells within the bioreactor) is

$$\frac{dVX_t}{dt} = \mu V X_u - K_{\text{lysis}} V (x_t - x_v) \quad (12.2)$$

where X_t denotes the total cell concentration and is measured in cells per liter.

The specific growth rate that appears in Eqs. (12.1) and (12.2) is estimated through the following formula:

$$\mu = \mu_{\max} f_{\lim} f_{\inh} \quad (12.3)$$

where $\mu_{\max}$ is the maximum possible growth rate for the specific cell line (h^{-1}) and the terms $f_{\lim}$ and $f_{\inh}$ represent, respectively, the nutrient limitation and product inhibition. These can be defined through the following equations:

$$f_{\lim} = \left(\frac{[GLC]}{K_{\text{glc}} + [GLC]}\right)\left(\frac{[GLN]}{K_{\text{gln}} + [GLN]}\right)$$

$$f_{\inh} = \left(\frac{KI_{\text{amm}}}{KI_{\text{amm}} + [AMM]}\right)\left(\frac{KI_{\text{lac}}}{KI_{\text{lac}} + [LAC]}\right) \quad (12.4)$$

where the K_i parameters are the Monod constants for the primary nutrients, namely glucose and glutamine. Similarly, the KI_i parameters are the inhibition constants of the primary products of metabolism, namely lactate and ammonia. [GLC], [GLN], [LAC], and [AMM] represent the extracellular concentrations of the aforementioned nutrients and products and are measured in mM.

The term μ_d represents the specific death rate of the cells within the bioreactor and can be defined in a way similar to the specific growth rate

$$\mu_d = \frac{\mu_{d,\max}}{1 + \left(\frac{K_{d,\text{amm}}}{[AMM]}\right)^n}, \quad \text{with } n > 1 \quad (12.5)$$

where, $\mu_{d,\max}$ represents the maximum specific death rate (h^{-1}) and $K_{d,\text{amm}}$ describes the rate of cell death by ammonia.

The presented differential equations along with the accompanying algebraic equations describe the growth and proliferation of the mammalian cell culture within the bioreactor. Since the model is unsegregated, it only represents the overall concentrations of nutrients and by-products of cellular metabolism within the bioreactor. Therefore, by performing material balances on each biological compound, four ordinary differential equations yielding the temporal evolution of the concentration of nutrients/metabolites are obtained. Specifically, the material balance for the concentration of glucose can be formulated as shown:

$$\frac{d(V[GLC])}{dt} = -Q_{glc} V X_u \tag{12.6}$$

where Q_{glc} is the specific glucose consumption rate (mmol/cell/hr) and is defined as:

$$Q_{glc} = \frac{\mu}{Y_{x,glc}} + m_{glc} \tag{12.7}$$

The parameters $Y_{x,glc}$ and m_{glc} that appear in Eq. (12.7) are the cell yield on glucose (cell/mmol) and maintenance energy of glucose (mmol/cell/h), respectively. Equation (12.6) was originally presented [18] with an additional term for glucose consumption by glucokinase, which as Kontoravdi et al. [21] later argued, based on evidence by Tatiraju et al. [49], has negligible effects. The material balance for glutamine similarly is described by the following equation:

$$\frac{d(V[GLN])}{dt} = -Q_{gln} V X_u - K_{d,gln} V[GLN] \tag{12.8}$$

The only difference is the term containing glutamine degradation. Glutamine is known to be spontaneously converted into pyrolidonecarboxylic acid at high temperatures and when in weakly acidic or alkaline solutions [62]. Bray et al. [6] showed that even in medium temperatures, around 37 °C, glutamine degrades in the presence of weakly acidic or alkaline solutions. The degradation is more pronounced when the solution contains phosphate buffer, which is often the case with media used for mammalian cell cultures. The specific consumption rate for glutamine is calculated through a formulation containing the cell yield on glutamine, $Y_{x,gln}$, and the maintenance energy of glutamine, m_{gln}.

$$Q_{gln} = \frac{\mu}{Y_{x,gln}} + m_{gln} \tag{12.9}$$

where

$$m_{gln} = \frac{a_1[GLN]}{a_2 + [GLN]} \tag{12.10}$$

with a_1 and a_2 being the relevant kinetic constants. Equation (12.9) is presented in the updated version [22] and not as originally presented [49].

Similarly, mass balances can be formulated to describe the temporal evolution of the concentrations of the primary by-products of cell metabolism. More specifically, the mass balance for ammonia is given by

$$\frac{d(V[AMM])}{dt} = Q_{amm} V X_u + K_{d,gln} V[GLN] \quad (12.11)$$

with

$$Q_{amm} = Y_{amm,gln} Q_{gln} \quad (12.12)$$

Similarly for lactate

$$\frac{d(V[LAC])}{dt} = Q_{lac} V X_u \quad (12.13)$$

with

$$Q_{lac} = Y_{lac,glc} Q_{glc} \quad (12.14)$$

where Q_{lac} and Q_{amm} represent the specific production rate (mmol/cell/h), while $Y_{lac,glc}$ and $Y_{amm,gln}$ represent the yield of the particular product on its primary nutrient (mmol of metabolite/mmol of nutrient).

The structured model describing antibody formation and secretion, as presented by Kontoravdi et al. [22], consists of an intracellular heavy- and light-chain mRNA balance:

$$\frac{dm_H}{dt} = N_H S_H - K m_H \quad (12.15)$$

and

$$\frac{dm_L}{dt} = N_L S_L - K m_L \quad (12.16)$$

where m_H and m_L are the intracellular heavy- and light-chain mRNA concentrations (mRNAs/cell), N_H and N are the heavy- and light-chain gene copy numbers (gene/cell), S_H and S_L are the heavy- and light-chain gene-specific transcription rates (mRNAs/gene/h), and, finally, K is the heavy- and light-chain mRNA decay rate (h^{-1}).

The intra-ER heavy- and light-chain balances are

$$\frac{d[H]}{dt} = T_H m_H - R_H \quad (12.17)$$

and

$$\frac{d[L]}{dt} = T_L m_L - R_L \quad (12.18)$$

where $[H]$ and $[L]$ are the free heavy- and light-chain concentrations in the ER (chain/cell), T_H and T_L are the heavy- and light-chain specific translation rates (chain/mRNA/h), and R_H and R_L are the rates of heavy- and light-chain consumption in assembly (chain/cell/h). MAbs consist of two heavy (H) and two light (L) amino acid chains. Each molecule is synthesized in the ER according to the following mechanism [36]:

$$H + H \longleftrightarrow H_2$$

$$H_2 + L \longleftrightarrow H_2L$$

$$H_2L + L \longleftrightarrow H_2L_2 \tag{12.19}$$

Assuming that the rates of heavy- and light-chain consumption in the assembly stage are given by

$$R_H = \frac{2}{3}K_A[H]^2$$

$$R_L = 2K_A[H_2][L] + K_A[H_2L][L] \tag{12.20}$$

an intra-ER balance can be performed for each of the assembly intermediates:

$$\frac{d[H_2]}{dt} = \frac{1}{3}K_A[H]^2 - 2K_A[H_2][L] \tag{12.21}$$

$$\frac{d[H_2L]}{dt} = 2K_A[H_2][L] - K_A[H_2L][L] \tag{12.22}$$

where $[H_2]$ and $[H_2L]$ are the concentrations of the assembly intermediates in the ER (molecule/cell), and K_A is the assembly rate constant ((molecule/cell) h^{-1}).

A balance can then be performed on the assembled MAb structure ($[H_2L_2]_{ER}$) in the ER:

$$\frac{d[H_2L_2]_{ER}}{dt} = K_A[H_2L][L] - K_{ER}[H_2L_2]_{ER} \tag{12.23}$$

where $[H_2L_2]_{ER}$ is the MAb concentration in the ER (molecule/cell) and K_{ER} is the rate constant for ER-to-Golgi antibody transport (h^{-1}). Once the MAb is assembled in the ER, it proceeds to the Golgi apparatus, where the main part of its glycosylation process takes place. An intra-Golgi MAb balance yields

$$\frac{d[H_2L_2]_G}{dt} = \varepsilon_1 K_{ER}[H_2L_2]_{ER} - K_G[H_2L_2]_G \tag{12.24}$$

where $[H_2L_2]_G$ is the MAb concentration in the Golgi (molecule/cell), ε_1 is the ER glycosylation efficiency factor, and K_G is the rate constant for Golgi-to-extracellular medium antibody transport (h^{-1}). Finally, the expression for antibody secretion (production) is

$$\frac{d(V[\text{MAb}])}{dt} = (\gamma_2 - \gamma_1 \mu) Q_{\text{MAb}} V X_V \tag{12.25}$$

where

$$Q_{\text{MAb}} = \varepsilon_2 \lambda K_G [H_2 L_2]_G \tag{12.26}$$

where Q_{MAb} is the specific MAb production rate (mg/cell/h), λ is the molecular weight of IgG$_1$ (g/mol), and ε_2 is the Golgi glycosylation efficiency factor. In Eq. (12.25), [MAb] is the MAb concentration in the culture, and γ_1, γ_2 are constants.

Equations (12.1)–(12.26) form a first principles model consisting of a total of 16 differential equations and 30 model parameters. In order to save time and effort from performing tedious computations manually, models are usually implemented in a computer aided design (CAD) tool of choice. Common choices among biochemical engineers include, but are not limited to, Fortran, C^{2+}, the Mathworks Matlab® suite, Mathematica®, Mathcad®, and gPROMS®. For the purposes of our example, we choose gPROMS® as the software of choice due to its superior solvers and seamless integration of experimental data [39].

The next step is the derivation of initial estimates for the model parameters from relevant experimental data. In the case that the utilized model already exists, this step can utilize parameter values obtained from relevant literature. For the derivation of estimates for the presented model's parameters, we will borrow experimental data from batch hybridoma cultures from the work of Kontoravdi et al. (2006). All parameter estimation experiments and model simulations were implemented in the advanced process modeling environment gPROMS® (Process Systems Enterprise, 2010).

gPROMS is an equation-oriented modeling system used for building, validating, and executing first-principles models within a flow sheeting framework. Parameter estimation in gPROMS is based on the maximum likelihood formulation, which provides simultaneous estimation of parameters in both the physical model of the process as well as the variance model of the measuring instruments. gPROMS attempts to determine values for the uncertain physical and variance model parameters, θ, that maximize the probability that the mathematical model will predict the measurement values obtained from the experiments. Assuming independent, normally distributed measurement errors, ε_{ijk}, with zero means and standard deviations, σ_{ijk}, this maximum likelihood goal can be captured through the following objective function:

$$\Phi = \frac{N}{2} \ln(2\pi) + \frac{1}{2} \min_{\theta} \left\{ \sum_{i=1}^{NE} \sum_{j=1}^{NV_i} \sum_{k=1}^{NM_{ij}} \left[\ln(\sigma_{ihk}^2) + \frac{(\bar{z}_{ijk} - z_{ijk})^2}{\sigma_{ihk}^2} \right] \right\} \tag{12.27}$$

where N stands for the total number of measurements taken during all the experiments, θ is the set of model parameters to be estimated, NE is the number of

Table 12.2 Model parameter estimates derived from batch hybridoma culture data.

Symbol	Units	Nominal value
μ_{max}	h^{-1}	5.8×10^{-3}
$K_{I,Amm}$	mM	28.484
$K_{I,Lac}$	mM	171.756
K_{Glc}	mM	0.75
K_{Gln}	mM	0.075
$m_{d,max}$	h^{-1}	0.03
$K_{d,Amm}$	mM	0.1386
N	Real integer	0.995
$Y_{Lac,Glc}$	Dimensionless	1.399
m_{Glc}	Mmol/cell/h	4.853×10^{-14}
$Y_{x,Glc}$	Cell/mmol	1.061×10^{8}
$Y_{x,Gln}$	Cell/mmol	5.565×10^{8}
$K_{d,gln}$	h^{-1}	9.6×10^{-3}
A_1	mM × L/cell/h	3.4×10^{-13}
A_2	mM	4
$Y_{Amm,Gln}$	Dimensionless	0.4269
K	h^{-1}	0.1
N_H	Gene/cell	139.8
S_H	mRNAs/gene/h	300
N_L	Gene/cell	117.5
S_L	mRNAs/gene/h	4500
T_H	Chain/mRNA/h	17
T_L	Chain/mRNA/h	11.5
K_A	(Molecule/cell) h^{-1}	10^{-6}
K_{ER}	h^{-1}	0.693
K_G	h^{-1}	0.1386
E_1	Dimensionless	0.995
γ_1	Dimensionless	0.1
γ_2	Dimensionless	2
E_2	Dimensionless	1
λ	g/mol	146 000
K_{lysis}	h^{-1}	0.03014

experiments performed, NV_i is the number of variables measured in the ith experiment, and NM_{ij} is the number of measurements of the jth variable in the ith experiment. The variance of the kth measurement of variable j in experiment i is denoted by σ^2_{ijk}, while $\bar{z}_{ijk}$ is the kth measured value of variable j in experiment i and z_{ijk} is the kth (model-) predicted value of variable j in experiment i. The above formulation can be reduced to a recursive least squares parameter estimation if no variance model for the sensor is selected.

Table 12.2 summarizes the list of model parameter estimates obtained from the parameter estimation algorithm while Fig. 12.4 presents an overview of experimental data and model simulations. The model is in good agreement with the experimental data and successfully captures the trends of nutrient consumption and

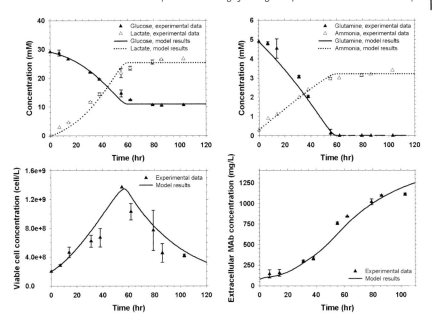

Fig. 12.4 Experimental data from batch hybridoma cultures and model predictions.

metabolite accumulation. This indicates a well-posed model, capable of describing the process under study even with initial parameter estimates. Having obtained initial estimates for the values of the model parameters we can proceed to the next step of the algorithm, namely model analysis.

12.4.2
Model Analysis

Model analysis techniques and SA in particular, can provide valuable insight regarding the dependence of the model output to its parameters. Allocating model uncertainty to the various sources of uncertainty (i.e., model parameters) facilitates the targeted reduction of output uncertainty by accurately estimating model parameters through tailor-made experiments indicated by a model-based DOE algorithm. On the other hand, parameters indicated as insignificant, with respect to the model output, can be fixed at their literature values (if available) or approximated, hence reducing unnecessary experimentation. There is a vast plethora of available model analysis techniques, enough to confuse even the experienced modeler. Below follows a rational discussion, leading to the proposal of the most suitable group of model analysis techniques, in the context of dynamical modeling of biological systems. A detailed study on the performance and applicability of SA techniques in the context of biological models can be found in the work of Kiparissides et al. [19].

Dynamic models describing complex biological functions involve highly nonlinear terms and include a large number of parameters with varying orders of magnitude. Thus, commonly used SA techniques are not able to provide results with any practical value for such models. Sensitivity analysis methods are commonly grouped in three main categories, namely screening, local, and global methods.

Screening methods are randomized, one-at-a-time numerical experiments, which aim to indicate the most important factors among the totality of model parameters. While screening methods involve computationally efficient algorithms, their use is limited to only preliminary results due to calculation of only first-order effects (i.e., effects the input factors have on the model output, without including their mutual interactions) and inherently lack precision, especially when used on nonlinear models [40]. Efforts to calculate higher-order effects, through screening methods, have been recorded in the literature [63, 64], though these methods fall short either in terms of accuracy or computational time.

Local methods derive measures of importance by estimating the effects infinitesimal variations of each factor have on the model output, in the area of a predetermined nominal point. Local methods are commonly used on steady-state models, or on studies dealing with the stability of a nominal point. Consequently, local methods fail to capture large variations in the parameter set and can only account for small variations from the parameter nominal values.

Global methods have the unique advantage of performing a full search of the parameter space, hence providing data independent of nominal points and are applicable to the whole range of the model's existence. Moreover global methods apportion the total uncertainty in the model output to the various sources of variation, while all parameters are varied at the same time. Generalized simulated annealing (GSA) provides the most complete set of results and mapping of the system, being able to cope with nonlinearities and identify parameter interaction effects [40]. The main drawback of GSA methods is their extensive computational requirements for large models. GSA methods are commonly grouped in two categories, namely methods that utilize a model approximation in order to generate measures of importance, and methods that study the total output variance of the model. Model approximation methods, such as regression analysis, correlation ratios, and rank transformation cannot account for higher-order effects.

Variance-based methods provide measures of importance, i.e., SI that apportion the total output variance to its contributors, namely the model parameters. In order to estimate the total output variance and its fractions, model parameters are treated as random variables within the parameter space. In the present context, randomness refers to the statistical independence of the generated samples. Since the models' parameters are treated as random variables, the resulting model output will be a random variable itself. The model output can thus be decomposed into summands of increasing dimensionality, a procedure also known as analysis of variance (ANOVA) decomposition:

$$f(X_1, \ldots, X_n) = f_0 + \sum_{i=1}^{n} f_i(X_i) + \sum_{1 \leq i < j \leq n} f_{ij}(X_i, X_j) + \cdots$$

$$+ f_{1,2,\ldots,n}(X_1, \ldots, X_n) \qquad (12.28)$$

Under the assumption that each of the terms in (12.28) is orthogonal [46, 65], the decomposition is unique and, therefore, integration of any term over any of the variables it may contain results to zero. This unique decomposition enables variance-based methods to discriminate between the first-order and higher-order effects. First-order information refers to the significance of merely the first summand with respect to the model output, while higher-order information explore the effect of parameter interactions and their contribution to the total output variance.

Main effects can be used to generate a significance – with respect to the model output – ranking of the model parameters. While rankings based solely on main effects are quite efficient in the case of linear models, the effect of the remaining summands cannot be neglected for nonlinear models. Chan *et al.* [66] have illustrated the significance of higher-order SI in understanding the behavior of the model parameters and how the uncertainty associated with them propagates through the model. Significance ranking for the model parameters should be based on the calculation of the total sensitivity index (TSI) [45, 65, 66]. The TSI for parameter i is estimated as the sum of all higher-order terms in (12.28), which include parameter i. The vast majority of SA techniques do not include a decomposition similar to the one presented in Eq. (12.28); therefore, it is not possible to discriminate whether the measure of importance they estimate refers to first- or higher-order information. Therefore, in order to obtain a realistic insight into the model's affecting parameters, obtaining information in the form of TSI is required. For a more comprehensive description of ANOVA decomposition and the TSI, refer to the work of Sobol' [46], Saltelli [40], and Chan *et al.* [66]. The most commonly used variance-based methods include the Sobol' global SI and the Fourier Amplitude Sensitivity Test (FAST). As reported in Kiparissides *et al.* [19], such methods can be computationally exhaustive; however, robust and less cumbersome alternatives exist [24].

For the purposes of our example, we have chosen to use the global SA. The dimensionality of the SA problem is ultimately defined by the number of model parameters; therefore, a feasibility constraint regarding the maximum possible number of individually scanned parameters is imposed implicitly in terms of computational time. This constraint is unavoidable due to the – increasing with dimension – number of model evaluations required for the Monte Carlo integrals to converge. It is common practice to resolve to the parameter grouping in order to reduce the dimensionality of the problem, thus solving a more tractable version of the original problem. A detailed discussion on parameter grouping and various methods for grouping can be found in the work of Kiparissides *et al.* [19], and is summarized in Table 12.3, adapted from the same work.

An overview of Table 12.3 will help us decide upon the most beneficial for our analysis method of grouping. Having already tested our model's agreement with experimental data (Fig. 12.4), we have concluded that our model is well posed and therefore there is no need for reduction or structural reformulation. Therefore, grouping according to biological significance seems to be the most suitable method of grouping our parameters prior to the application of SA.

Table 12.3 Grouping parameters in GSA (adapted from [19]).

Consideration	Scope of the analysis
Number of groups	
1. Few groups	Computationally efficient, low resolution
2. Many groups	Computationally expensive, high resolution
Grouping method	
1. Random/arbitrary	Parameter significance ranking
2. Biological significance related	Model analysis leading to DOE
3. Functional	Model analysis/reduction

As stated earlier, a batch operation mode was considered, the model was simulated for 120 h of culture time, and the SA was performed at three characteristic time points (20, 50, and 120 h). SIs have been known to change dynamically along the time trajectory of the model output. For example, as the culture progresses and nutrients start being depleted, the model output will become more sensitive toward the parameters affecting nutrient uptake and metabolism. This is a valuable property as it can provide information regarding the time point that would yield the most informative experiments. SA was conducted at different phases of the cell culture in order to capture the dynamics of the various growth phases of a batch cell culture. Specifically time points from the lag phase, the exponential growth phase, and the decline phase were evaluated. The output variables of interest, from a process point of view, are viable cell concentration and MAb concentration as these ultimately define the final amount of MAb titer available. The simulations involved scanning of all model parameters with respect to the output variables of interest. The uncertainty range associated with each of the 30 model parameters was set to $\pm 100\%$ from the parameter nominal value. Following the discussion in the previous section, using parameter grouping was imposed by the dimensionality of the problem and the model's 30 parameters were grouped into four unequal groups, which were formulated based on their biological function and can be seen Table 12.4.

When the goal of SA is to indicate candidates for model-based DOE, a "first-layer" analysis of grouped parameters does not suffice. More specifically, considering the formulation of our problem, the first analysis will indicate one – or more – group of parameters as significant instead of identifying individual parameters. A group of parameters with a high sensitivity index (SI) does not necessarily translate in all parameters within that group being significant with respect to the model output. Moreover, it is often the case that different groups will have high SI at different points of the predetermined time horizon. Therefore, a second analysis, this time within the significant groups, is required in order to identify individual parameters that affect the model output.

From a computational point of view, the "first-layer" analysis is computationally more demanding than the "second layer" as the entirety of the parameter space is

Table 12.4 Model parameters: biological significance grouping.

Group 1 – Growth/death related	Group 2 – Metabolism related
μ_{max}	$Y_{Lac,Glc}$
$K_{I,Amm}$	m_{Glc}
$K_{I,Lac}$	$Y_{x,Glc}$
K_{Glc}	$Y_{x,Gln}$
K_{Gln}	$K_{d,gln}$
$m_{d,max}$	a_1
$K_{d,Amm}$	a_2
N	$Y_{Amm,Gln}$
Group 3 – MAb synthesis related	**Group 4 – MAb secretion related**
K	γ_1
N_H	γ_2
S_H	ε_2
N_L	
S_L	
T_H	
T_L	
K_A	
K_{ER}	
K_G	
ε_1	

sampled. Furthermore, since the "first layer" ultimately is a stepping stone toward model-based DOE, the acquisition of SIs of first order is somewhat of a luxury. Taking this into account, we have chosen to use derivative based global sensitivity measures (DGSM) [24] for the "first layer" of GSA. DGSM is a global screening method, proved [19] to provide results similar to the variance-based methods' TSI, and infact has been shown to have a direct correlation with the Sobol' TSI in most cases [47]. The main benefit of using DGSM over a variance-based method is the significant gain in terms of computational time. DGSM provides only TSI information, and not first or higher-order information, but as discussed previously this is not an issue for this stage of the analysis. Figure 12.5 presents the results of the analysis when studying parameters in the groups defined in Table 12.4.

Observing the results of Fig. 12.5, one can conclude that different parameter groups affect different model outputs. This is both expected and logical. The unstructured model as seen through its output variable, namely viable cell concentration, is affected mainly by parameter groups 1 and 2, which contain the parameters of the unstructured model. In the early stages (20, 50 h), group 1 is the more significant by a fair margin while group 2 becomes increasingly significant as the culture progresses. Group 2 contains the parameters associated with nutrient uptake and metabolism, and as discussed earlier is expected to become more significant as the depletion of nutrients starts being an increasingly crucial factor for the culture. On the other hand, the structured MAb formation model is affected mainly by param-

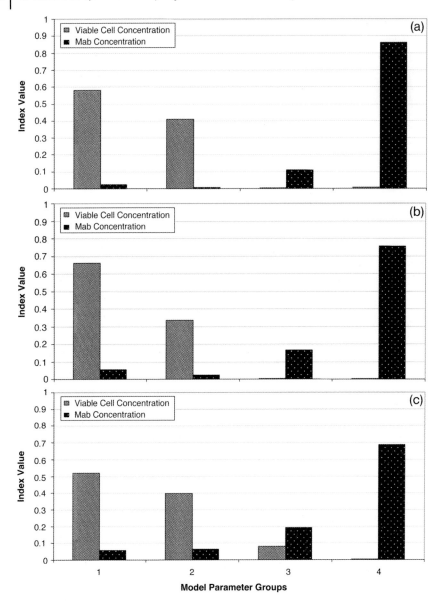

Fig. 12.5 "First layer" GSA: (a) 20 h, (b) 50 h, and (c) 120 h of culture time.

eter groups 3 and 4, which contain the respective model parameters, but as the culture progresses starts to be slightly affected by groups 1 and 2.

One could arguably discuss that this "first layer" of GSA has yielded no useful results as it appears that all parameter groups show some significance throughout the culture, indicating possibly a poor choice in parameter grouping. However, a closer observation of the results of Fig. 12.5 will yield valuable information. That is, group 4, for example, is constantly significant throughout the culture with its SI varying slightly. Therefore a "second layer" analysis within group 4 at only one time point would be sufficient to indicate which parameters of that group are responsible for the high SI of that group. Due to the nature of the grouping we have chosen, the significant individual parameters of group 4 are expected to maintain their significance, in accordance to the behavior of the SI of group 4, throughout the culture. Therefore, we opt to perform a "second-layer" analysis of the parameters within group 4 at the first time point (20 h) for two reasons. First and foremost, the earlier time points are less computationally demanding, even if marginally, than latter time points due to the required integration time. Moreover, group 4 has the highest SI at 20 h of culture time.

Group 1 has the highest SI at 50 h of culture time, which makes sense from a biological point of view, since it is the time point closest to the peak in viable cell concentration. Therefore, the best time point to scan the parameters of group 1 is 50 h of culture time. Using the same rationale, group 2 will be studied at 20 h while group 3 will be studied at 120 h. Therefore this "first layer" of GSA has indeed provided considerable amount of information, serving its purpose to guide us through the "second layer," which will indicate the actual model parameters and that affect the model output the most, and the point in time where they have the highest SI associated with them, leading to suggestions for model-based DOE.

Since the "second layer" of GSA will involve smaller problems as only partitions of the parameter space will be sampled and individual parameters will be studied, we have chosen to use the Sobol' global SI for this part of the analysis. The Sobol' SI, even though more computationally cumbersome, will provide information on parameter interactions and can, therefore, be used to exclude parameters with a high level of nonlinear interactions from DOE on the basis of singularity. Figures 12.6–12.9 present the results of the "second layer" of GSA, scanning individual parameters as discussed above.

The largest contributor to the high SI of group 1 is the maximum growth rate (μ_{max}) as can be seen from Fig. 12.6, while from the remaining parameters, only the maximum death rate ($\mu_{d,max}$) seems to affect the output of viable cell concentration. Another important conclusion drawn from Fig. 12.6 is that all parameters have a very small contribution of parameter interaction toward their overall SI. As discussed earlier, the difference in value between the total index and the individual index is an indication of parameter interactions. Low level of interactions is a desirable property as it allows for more accurate parameter estimations.

The SIs for the parameters of group 2 are quite similar between the two studied model outputs. Taking a closer look at the model formulation, this result is both reasonable and expected. The parameters of group 2 are associated with the nutrient uptake rates and the metabolite accumulation rates. Therefore they ultimately define the overall growth rate, which in turn affects the viable cell concentration profile. Moreover, the final antibody titer is a function of both the overall growth rate and the viable cell concentration. Figure 12.7 is a confirmation of the above, indicating the same parameters as significant for both model outputs. Furthermore, a slight increase in the difference between the total and individual indices is notice-

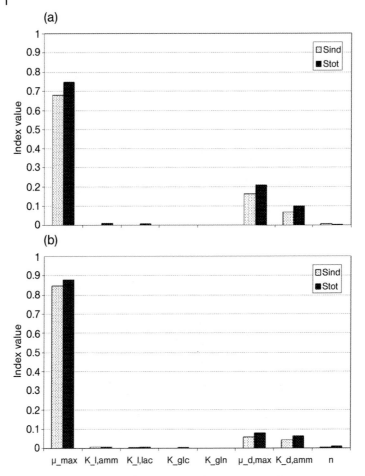

Fig. 12.6 Sensitivity indices for parameters of group 1 at 50 h of culture time. (a) Viable cell concentration as the output. (b) MAb concentration as the output.

able when MAb concentration is the studied output, indicating the indirect effect these parameters have on the said output variable. The significant parameters are the yields of cell mass on both substrates ($Y_{x,\text{glc}}$, $Y_{x,\text{gln}}$) and the yield of ammonia from glutamine consumption ($Y_{\text{amm,gln}}$). The threshold level below which a parameter is considered insignificant is arbitrarily chosen by the modeler as stated earlier. In the present work, parameters with SI smaller than 0.1 are considered insignificant, while others [40] favor a higher cut-off point such as 0.2 or even 0.3.

According to Fig. 12.5, groups 3 and 4 have SI equal to zero with respect to viable cell concentration as the model output. Referring back to the model equations, this is readily justifiable since parameters from both groups are the parameters of the structured model describing MAb formation and secretion. The structured model is coupled with the unstructured model in a one-way manner. That is, the unstruc-

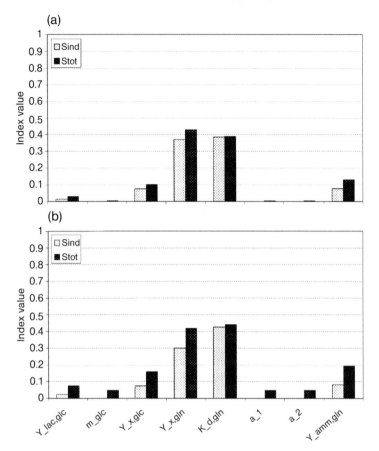

Fig. 12.7 Sensitivity indices for parameters of group 2 at 20 h of culture time. (a) Viable cell concentration as the output. (b) MAb concentration as the output.

tured growth model affects the output of the structured model but the structured part of the model has no effect on the output of the growth model. Therefore, for groups 3 and 4 only the SI with respect to MAb concentration as the model output will be considered.

Figure 12.8 seems difficult to interpret at a first glance. Almost all the model parameters have a nonnegligible TSI, yet only one (ε_1) has an individual index higher than 0.1. As mentioned earlier, parameters whose total index is strongly influenced by nonlinear interactions are poor candidates for parameter estimation as they cannot be uniquely identified [44]. However this does not necessarily mean that none of the parameters of group 3 can be uniquely estimated. It is highly probable that the time point of the analysis was a poor choice and performing the analysis at a different time point might yield more informative results. Few other conclusions can be drawn from Fig. 12.7 alone. Therefore, we have chosen to repeat the analysis of the parameters of group 3 at a different time point, namely at 20 h of

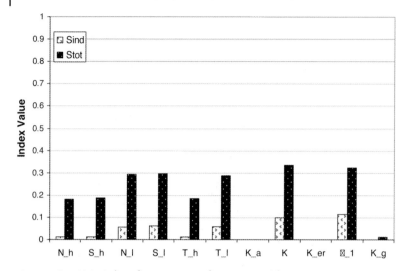

Fig. 12.8 Sensitivity indices for parameters of group 3 at 120 h of culture time; MAb concentration as the output.

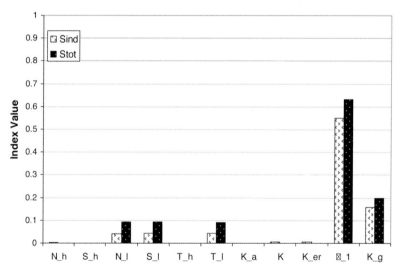

Fig. 12.9 Sensitivity indices for parameters of group 3 at 20 h of culture time; MAb concentration as the output.

culture time. Should the results of the new analysis resemble the results shown in Fig. 12.7, this would be an indication of an over-parameterized and ill-posed model.

Figure 12.9 contains the SI for the parameters of group 3 as calculated after 20 h of culture time. The results at this time point are indeed much more informative than the ambiguous results of Fig. 12.8. The glycosylation efficiency factor (ε_1) and the rate constant for Golgi to extracellular media transport (K_g) are the only parameters with a TSI greater than 0.1. Both parameters also have quite high indi-

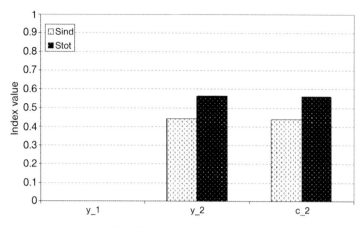

Fig. 12.10 Sensitivity indices for parameters of group 4 at 20 h of culture time; MAb concentration as the output.

vidual indices, which suggest that they could be uniquely identified from a suitably designed experiment. It is worthwhile identifying the reasons behind the differences between Figs. 12.8 and 12.9. After 120 h of culture time, glutamine has been completely depleted and the culture is well in its decline phase. A near zero concentration for glutamine would yield a near zero value for the specific growth rate as given in Eqs. (12.3) and (12.4). This in turn would affect the right-hand side of Eq. (12.25), which in fact yields the output variable. Since GSA is a numerical tool, it shares the same limitations as the numerical solvers it utilizes for the integration of the DAE system. Therefore, when the right-hand side of Eq. (12.25) is known to lie at a near zero (if not exactly zero) value, any variance from parameter value alterations is difficult to be quantified at the specific time point.

The analysis of the parameters of group 4 (Fig. 12.10) highlighted both the glycosylation efficiency factor (ε_2) and one of the constants for MAb secretion (γ_2) as significant parameters. Both parameters display a high individual index value which, as mentioned earlier, is a desirable property from a parameter estimation point of view. However, even though a parameter may be significant to the model output and can mathematically guarantee unique identification, it may still not be feasible to conduct the necessary experimental measurements that would allow the precise estimation of its value.

Glycosylation efficiency measurements, for example, are rather complex and cumbersome and require equipment not readily available in every analytical laboratory. Therefore, while aware of the uncertainty associated with these parameters ($\varepsilon_1, \varepsilon_2$), we have no choice but to omit them from the DOE algorithm. Similarly, parameter (γ_2) is closely linked to the cells' position in the cell cycle, making its experimental estimation particularly difficult and, therefore, was excluded from the DOE algorithm. The difficulty of obtaining experimental measurements for certain parameters is often a "real-life" problem. Some might argue that on the basis of experimental estimation, such parameters could be excluded from model analysis since they cannot be estimated. However, the fact that a parameter cannot be

Table 12.5 Summary of GSA results.

Parameters to be input to the DOE algorithm	
Parameter	Time point of GSA (h)
μ_{max}	50
$\mu_{d,max}$	50
$Y_{amm,gln}$	20
$Y_{x,glc}$	20
$Y_{x,gln}$	20
$K_{d,gln}$	20
K_g	20

experimentally estimated does not invalidate the SA and on the contrary raises the awareness of the modeler to possible weaknesses of the developed model.

The gain from the detailed analysis in this section is the reduction of the number of parameters that need to be experimentally validated in order to increase the fidelity of our model. Having started from a total of 30 parameters, we have successfully narrowed down the parameters that need to be experimentally validated to a mere 7, summarized in Table 12.5. Moreover, we have gained valuable information regarding the time points that would yield the most informative experiments leading to a more accurate estimation of our model parameters. Therefore, model analysis is a key step toward the development of a robust and well-posed dynamic model and should always be performed prior to experimentation in order to avoid unnecessary experimental costs and labor. The next step of the model development algorithm is DOE and finally model validation through an independent set of experiments.

12.4.3
Design of Experiments and Model Validation

Thus far, we have presented the derivation of a model that can accurately describe batch cultures of MAb secreting hybridoma cultures as it flows through the model development algorithm of Fig. 12.2. Using as an example the model and experimental results presented in the work of Kontoravdi *et al.* [22], we have successfully created the partitions of the parameter vector containing the significant (Table 12.5) and the insignificant parameters that will be set at their nominal values. The next step of the model development algorithm is to design tailor-made experiments for the significant parameters in order to facilitate their accurate estimation. In order to guide the reader through this step of the algorithm, we will again use as an example the relevant work of Kontoravdi *et al.* [22] in an attempt to provide a "closed-loop" overview of the model development framework presented in Fig. 12.2.

However, from a process engineering point of view, fed-batch operation is the most important for industrial applications as it can prolong culture longevity, thus

Table 12.6 Optimal experiment schedule (adapted from [22]).

Time (h)	Feed volume (mL)	Time (h)	Feed volume (mL)
12		90	
F–12.1	1.25	96	
18		108	
24		F–108.1	1.25
36		114	
F–36.1	1.25	120	
42		132	
48		F–132.1	1.25
60		138	
F–60.1	1.25	144	
66		156	
72		F–156.1	1.25
84		162	
F–84.1	1.25	168	

increasing MAb productivity and final titer. Therefore, the goal set out is to extend the model's predictive capabilities to fed-batch conditions so that it can be used for the application of model-based optimization. Assuming that the model is valid under such conditions, the specific objective is to accurately estimate the significant model parameters from fed-batch experimental data. As previous studies have discussed [67–69], optimal experimental design uses the model to design sufficiently informative experiments for this purpose. Borrowing experimental data from the work of Kontoravdi et al. [22], Table 12.6 has been presented.

Following the detailed analysis of the model and its parameters in the previous section, we have already identified the most significant model parameters that can readily be input to the optimal experimental design algorithm. Reaping the benefits of using an advanced CAD tool like gPROMS, an optimal experimental design utility is already implemented trivializing the application of DOE. The concentrations of glucose and glutamine in the feed were set at 500 and 100 mM, respectively. The maximum total volume of feed was fixed at 8.75 ml, which represents nearly 5% of the total culture volume (200 ml), so as to avoid dilution effects. Sampling times, at which measurements were conducted were determined (indicated from the work of [22]) a priori and can be found in Table 12.6. The output of the algorithm (Table 12.7) provided us with the optimal amount of feed supplied at each feeding interval as well as the optimal timing of the intervals and the optimal duration of the experiment (168 h) [22].

The "refined" version of the model is simulated for fed-batch operation and is plotted against relevant experimental data. The model is found to be in good agreement with the experimental data and can successfully capture the dynamics of a fed-batch culture as shown in Fig. 12.11. Table 12.7 contains the values of the "refined" model parameters.

Table 12.7 Results of the parameter estimation algorithm.

Parameter	New estimated value
μ_{max}	0.0554067
$\mu_{d,max}$	0.0345536
$Y_{amm,gln}$	0.399768
$Y_{x,glc}$	2.6×10^8
$Y_{x,gln}$	8×10^8
$K_{d,gln}$	0.00905775
K_g	12.4518

Having verified the validity of the model structure for the simulation of fed-batch processes, its applicability should be examined against an independent set of experimental data. This will prove that model parameters were not just fitted to experimental data but were properly estimated and the model is valid under various operating conditions. It is beyond the subject of this chapter, however, to provide such a validation and for a proof of concept the reader can refer to the work of Kontoravdi [21], where such data are available.

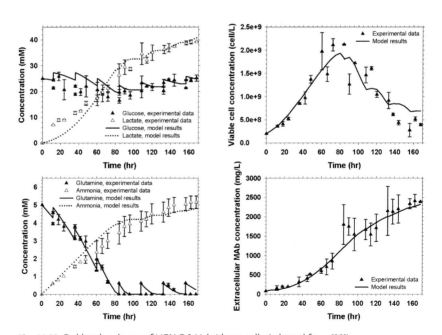

Fig. 12.11 Fed-batch cultures of HFN 7.1 Hybridoma cells (adapted from [22]).

12.5
Concluding Remarks

The balancing point for the trade-off between fidelity and tractability is constantly shifting with the advancements both in numerical tools and raw computational power. Most of the models describing mammalian cell culture presented thus far in the literature are based on the consumption of up to two basic nutrients and the toxic effects from the accumulation of the corresponding end-products of metabolism. However, in order to truly capture the dynamics and behavior of a culture and achieve truly optimal feeding strategies, we need to start paying attention to a number of other components that have so far being ignored. The work of Xie and Wang [55] has already taught us that excessive feed of glucose is not always the best means toward higher titers of product, as they proved it shifts metabolism toward energy-inefficient pathways. Moreover, deZengotita *et al.* [10] have shown that apart from the established growth-limiting nutrients (namely glucose and glutamine), there are many other components that might be limiting the growth of a culture. The question that naturally arises is whether an optimal feeding profile should be derived on the basis of availability of nutrients while disregarding the energy requirements of the cell or on the provision of adequate yet not excessive amounts of energy through the provision of controlled quantities of nutrients. Energy metabolism is a significant element of cell culture that has thus far been ignored from a modeling point of view.

Optimization of cultures secreting a valuable end-product has always been centered around the balance between prolonged culture life and increased specific productivity. Usually these two goals are reached through competing paths. Conditions that seem to prolong culture viability reduce specific productivity, while conditions that increase specific productivity seem to affect culture longevity. Even though a lot of published studies have shown the potential of *in silico* experimentation, a lot of work still remains to be done. In order to achieve the global optimum between prolonged culture life and increased productivity, we must first understand and incorporate the significant elements of metabolism in our models. For example, there has been to the extent of our knowledge no model that takes into account the availability of amino acids in the culture medium as a factor affecting either growth or productivity apart from the work of Kontoravdi *et al.* [70]. However, it is well known from biology that the building blocks for the synthesis of biological macromolecules are amino acids and that not all the amino acids can be produced by the cell. Therefore, seeking optimal feeding profiles based on the provision of glucose alone, and in the best case glucose and glutamine, might yield an increase in final titer on the one hand; however, it is still quite far from the global optimum.

The visionary remarks of Bailey [1] predicted the need to shift modeling focus upstream toward the gene level in order to truly understand the dynamics of cellular metabolism. Little – to no – work has been published since containing structure on the gene level where the kernel of the cell's control mechanism lies. The advancements in analytical and theoretical biology will increasingly provide more information in the future, especially with the increasing popularity and availability of the omics techniques. However from an engineering point of view, this information can only be utilized through a systematic and rigorous framework that will organize and prioritize necessary measurements and experiments.

Paving the way toward a "closed-loop" holistic framework for bioprocess automation, this chapter covers the development of dynamical models of biological systems. The biological model development framework presented in Fig. 12.2 is explained in a step-by-step fashion, highlighting scientific concerns, challenges, and "real-life" problems associated with each step of the framework. Adapting a "real-life" example from the work of Kontoravdi *et al.* [22], we present the logical and systematic evolution of a model from conception to validation as it flows through the various steps of the model development framework. The key conclusion of this chapter is that by utilizing a systematic way of organizing available information, one can avoid conducting experiments for the sake of experimentation and develop models with an *a priori* set aim.

References

1 Bailey, J. E., Mathematical modeling and analysis in biochemical engineering: past accomplishments and future opportunities, *Biotechnology Progress* 14 (**1998**), pp. 8–20.

2 Batt, B. C., Kompala, D. S., A structured kinetic modelling framework for the dynamics of hybridoma growth and monoclonal antibody production in continuous suspension cultures, *Biotechnology and Bioengineering* 34 (**1989**), pp. 515–531.

3 Bibila, T. A., Flickenger, M. C., A structured model for monoclonal antibody synthesis in exponentially growing and stationary phase hybridoma cells, *Biotechnology and Bioengineering* 37 (**1991**), pp. 210–226.

4 Bibila, T. A., Flickenger, M. C., Use of a structured kinetic model of antibody synthesis and secretion for the optimisation of antibody product synthesis. Part 1. Steady-state analysis, *Biotechnology and Bioengineering* 39 (**1992**), pp. 251–261.

5 Bibila, T. A., Flickenger, M. C., Use of a structured kinetic model of antibody synthesis and secretion for the optimisation of antibody product synthesis. Part 2. Transient analysis, *Biotechnology and Bioengineering* 39 (**1992**), pp. 262–272.

6 Bray, H. G., James, S. P., Raffan, I. M., Thorpe, W. V., The enzymic hydrolysis of glutamine and its spontaneous decomposition in buffer solutions, *The Biochemistry Journal* 44 (**1948**), pp. 625–627.

7 Dalili, M., Sayles, G. D., Ollis, D. F., Glutamine-limited batch hybridoma cell growth and antibody production: experiment and model, *Biotechnology and Bioengineering* 36 (**1990**), pp. 74–82.

8 De Tremblay, M., Perrier, M., Chavarie, C., Archambault, J., Optimization of fed-batch culture of hybridoma cells using dynamic programming: single and multi feed cases, *Bioprocess Engineering* 7 (**1992**), pp. 229–234.

9 De Tremblay, M., Perrier, M., Chavarie, C., Achambault, J., Fed-batch culture of hybridoma cells: comparison of optimal control approach and closed loop strategies, *Bioprocess Engineering* 9 (**1993**), pp. 13–21.

10 deZengotita, M. V., Miller, W. M., Aunins, J. G., Phosphate feeding improves high-cell-concentration NS0 myeloma culture performance for monoclonal antibody production, *Biotechnology and Bioengineering* 69 (**2000**), pp. 566–676.

11 Dhir, S., Morrow, K. J., Rhinehart, R. R., Wiesner, T., Dynamic optimisation of hybridoma growth in a fed-batch bioreactor, *Biotechnology and Bioengineering* 67(2) (**1999**), pp. 197–205.

12 DiMasi, D., Swartz, R. W., An energetically structured model of mammalian cell metabolism. 1. Model development and application to steady-state hybridoma cell growth in continuous culture, *Biotechnology Progress* 11(6) (1995), pp. 664–676.

13 Eakman, J. M., Fredrickson, A. G., Tscuchiya, H. M., Statistics and dynamics of microbial cell populations, *Chemical Engineering Progress* 62 (1966), pp. 37–49.

14 Frahm, B., Lane, P., Märkl, H., Pörtner, R., Improvement of a mammalian cell culture process by adaptive, model-based dialysis fed-batch cultivation and suspension of apoptosis, *Bioprocess and Biosystem Engineering* 26 (2003), pp. 1–10.

15 Frame, K. K., Hu, W. S., Kinetic study of hybridoma cell growth in continuous culture: I. A model for nonproducing cells, *Biotechnology and Bioengineering* 37 (1991), pp. 55–64.

16 Fredrickson, A. G., Mcgee, R. D., III, Tsuchiya, H. M., Mathematical models in fermentation processes, *Advanced Applied Microbiology* 23 (1970), p. 419.

17 Ho, Y., Varley, J., Mantalaris, A., Development and analysis of a mathematical model for antibody-producing GS-NS0 cells under normal and hyperosmotic culture conditions, *Biotechnology Progress* 22(6) (2006), 1560–1569.

18 Jang, J. D., Barford, J. P., An unstructured kinetic model of macromolecular metabolism in batch and fed-batch cultures of hybridoma cells producing monoclonal antibodies, *Biochemical Engineering Journal* 4 (2000), pp. 153–168.

19 Kiparissides, A., Kucherenko, S., Mantalaris, A., Pistikopoulos, E. N., Global sensitivity analysis challenges in biological systems modelling, *Industrial and Engineering Chemistry Research* 48(15) (2009), pp. 7168–7180.

20 Kontoravdi, C., Asprey, S. P., Pistikopoulos, E. N., Mantalaris, A., Application of global sensitivity analysis to determine goals for design of experiments: an example study on antibody-producing cell cultures, *Biotechnology Progress* 21(4) (2005), pp. 1128–1135.

21 Kontoravdi, C., Development of a combined mathematical and experimental framework model for modelling mammalian cell cultures, PhD Thesis, Department of Chemical Engineering and Chemical Technology Imperial College, London, 2006.

22 Kontoravdi, C., Asprey, S. P., Pistikopoulos, E. N., Mantalaris, A., Development of a dynamic model of monoclonal antibody production and glycosylation for product quality monitoring, *Computers and Chemical Engineering* 31 (2007), pp. 392–400.

23 Kompala, D. S., Ramkrishna, D., Tsao, G. T., Cybernetic modeling of microbial growth on multiple substrates, *Biotechnology and Bioengineering* 26 (1984), pp. 1272–1281.

24 Kucherenko, S., Rodriguez-Fernandez, M., Pantelides, C., Shah, N., Monte Carlo evaluation of derivative based global sensitivity indices, *Reliability Engineering and System Safety* 94(7) (2007), pp. 1135–1148.

25 Kurokawa, H., Park, Y. S., Iijima, S., Kobayashi, T., Growth characteristics in fed-batch culture of hybridoma cells with control of glucose and glutamine concentration, *Biotechnology and Bioengineering* 44 (1994), pp. 95–103.

26 Lam, M. C., Sriyudthsak, K., Kontoravdi, C., Kothari, K., Park, H. H., Pistikopoulos, E. N., Mantalaris, A., Cell cycle modelling for off-line dynamic optimisation of mammalian culture, *Computer Aided Chemical Engineering* 25 (2008), pp. 109–114.

27 Lee, S. B., Bailey, J. E., A mathematical model for λdv plasmid replication: analysis of wild-type plasmid, *Plasmid* 11(2) (1984a), pp. 151–165.

28 Lee, S. B., Bailey, J. E., A mathematical model for λdv plasmid replication: analysis of copy number mutants, *Plasmid* 11(2) (1984b), pp. 166–177.

29 Mantzaris, N., Liou, N. J., Daoutidis, P., Srienc, F., Numerical solution of a mass structured cell population balance model in an environment of changing substrate concentration, *Journal of Biotechnology* 71 (1999), pp. 157–174.

30 Mantzaris, N., Daoutidis, P., Srienc, F., Nonlinear productivity control using a multi-staged cell population balance model, *Chemical Engineering Science* 57 (2002), pp. 1–14.

31 Mantzaris, N., Daoutidis, P., Srienc, F., Numerical solution of multi-variable cell population balance models: I. Finite difference methods, *Computers and Chemical Engineering* 25 (**2001a**), pp. 1411–1440.

32 Mantzaris, N., Daoutidis, P., Srienc, F., Numerical solution of multi-variable cell population balance models: II. Spectral methods, *Computers and Chemical Engineering*, 25 (**2001b**), pp. 1441–1462.

33 Mantzaris, N., Daoutidis, P., Srienc, F., Numerical solution of multi-variable cell population balance models: III. Finite element methods, *Computers and Chemical Engineering* 25 (**2001c**), pp. 1463–1481.

34 Miller, W. M., Blanch, H., Wilke, C., Kinetic analysis of hybridoma growth in continuous suspension culture, in: *ACS National Meeting, September 11*, **1986**.

35 Miller, W. M., Blanch, H., Wilke, C., A kinetic analysis of hybridoma growth and metabolism in batch and continuous suspension culture: effect of nutrient concentration, dilution rate and ph, *Biotechnology and Bioengineering* 67 (**2000**), pp. 852–871.

36 Percy, J. R., A theoretical model for the covalent assembly of immunoglobulins, *Journal of Biological Chemistry* 250(6) (**1975**), pp. 2398–2400.

37 Ramkrishna, D., A cybernetic perspective of microbial growth, in: *Foundations of Biochemical Engineering: Kinetics and Thermodynamics in Biological Systems*, Blanch, H. W., Papoutsakis, E. T., Stephanopoulos, G. N. (eds.), American Chemical Society, Washington, DC, **1982**, pp. 161–178.

38 Portner, R., Schafer, T., Modelling hybridoma cell growth and metabolism – a comparison of selected models and data, *Journal of Biotechnology* 49 (**1996**), pp. 119–135.

39 Process Systems Enterprise, gPROMS, http://www.psenterprise.com/gproms, **1997–2009**.

40 Saltelli, A., Chan, K., Scott, E. M., *Sensitivity Analysis*, Wiley Press, New York, **2000**.

41 San, K. Y., Stephanopoulos, G., Optimization of fed-batch penicillin fermentation: a case of singular optimal control with state constraints, *Biotechnology and Bioengineering* 34(1) (**1989**), pp. 72–78.

42 Shuler, M. L., Leung, S., Dick, C. C., A mathematical model for the growth of a single bacterial cell, *Annals of the New York Academy of Sciences* 326 (**1979**), pp. 35–52.

43 Sidoli, F. R., Mantalaris, A., Asprey, S. P., Modelling of mammalian cells and cell culture processes, *Cytotechnology* 44 (**2004**), pp. 27–46.

44 Sidoli, F. R., Mantalaris, A., Asprey, S. P., Toward global parametric estimability of a large-scale kinetic single-cell model for mammalian cell culture, *Industrial and Engineering Chemistry Research* 44(4) (**2006**), pp. 868–878.

45 Sobol', I. M., On sensitivity estimation for nonlinear mathematical models, *Mathematics Modeling* 2 (**1990**), pp. 112–118.

46 Sobol', I. M., Global sensitivity indices for nonlinear mathematical models and their Monte Carlo estimates, *Mathematics and Computers in Simulation* 55 (**2001**), pp. 271–280.

47 Sobol', I. M., Kucherenko, S., Derivative based global sensitivity measures and their link with global sensitivity indices, *Mathematics and Computers in Simulation* 79(10) (**2009**), pp. 3009–3017.

48 Srienc, F., Short communication: cytometric data as the basis for rigorous models of cell population dynamics, *Journal of Biotechnology* 71 (**1999**), pp. 233–238.

49 Tatiraju, S., Soroush, M., Mutharassan, R., Multi-rate nonlinear state and parameter estimation in a bioreactor, *Biotechnology and Bioengineering* 63(1) (**1999**), p. 22.

50 Tsuchiya, H. M., Fredrickson, A. G., Aris, R. D., Dynamics of microbial cell populations, *Advanced Chemical Engineering* 6 (**1966**), pp. 125–206.

51 Tziampazis, E., Sambanis, A., Modelling of cell culture processes, *Cytotechnology* 14 (**1994**), pp. 191–204.

52 Villadsen, J., Short communication: on the use of population balances, *Journal of Biotechnology* 71 (**1999**), pp. 251–253.

53 Xie, L., Wang, D. I. C., Applications of improved stoichiometric model in medium design and fed-batch cultivation of animal cells in bioreactor, *Growth Factors* 1 (**1994**), pp. 17–29.

54 Xie, L., Wang, D. I. C., Material balance studies on animal cell metabolism using a stoichiometrically based reaction network, *Biotechnology and Bioengineering* 52(5) (**1996**), pp. 578–590.

55 Xie, L., Wang, D. I. C., Energy metabolism and ATP balance in animal cell cultivation using a stoichiometrically based reaction network, *Biotechnology and Bioengineering* 52(5) (**1996**), pp. 591–601.

56 Michaelis, L., Menten, M. L., Die Kinetik der Invertinwirkung, *Biochemische Zeitschrift* 49 (**1913**), pp. 334–336.

57 Bree, M. A., Dhurjati, P., Geoghegan, R. F., Robnett, B., Kinetic modelling of hybridoma cell growth and immunoglobulin production in a large-scale suspension culture, *Biotechnology and Bioengineering* 32 (**1988**), pp. 1067–1072.

58 Frahm, B., Lane, P., Atzert, H., Munack, A., Hoffmann, M., Hass, V. C., Portner, R., Adaptive, model-based control by the open-loop-feedback-optimal (OLFO) controller for the effective fed-batch cultivation of hybridoma cells, *Biotechnol. Progress* 18 (**2002**), pp. 1095–1103.

59 Asprey, S. P., Macchietto, S., Statistical tools for optimal dynamic model building, *Computers & Chemical Engineering* 24(2–7) (**2000**), pp. 1261–1267.

60 Asprey, S. P., Macchietto, S., Designing robust optimal dynamic experiments, *Journal of Process Control* 12(4) (**2002**), pp. 545–556.

61 Bibila, T. A., Robinson, D. K., In pursuit of the optimal fed-batch process for monoclonal antibody production, *Biotechnology Progress* 11 (**1995**), pp. 1–13.

62 Chibnall, A. C., Westall, R. G., The estimation of glutamine in the presence of aspargine, *Biochem. Journal* 26(1) (**1932**).

63 Box, G. E. P., Hunter, W. G., Hunter, J. S., *Statistics for Experimenters, An Introduction to Design, Data Analysis and Model Building*, Wiley, New York, **1978**.

64 Cotter, S. C., A screening design for factorial experiments with interactions, *Biometrika* 66 (**1979**), pp. 317–320.

65 Hommaa, T., Andrea Saltelli, A., Importance measures in global sensitivity analysis of nonlinear models, *Reliability Engineering & System Safety* 52(1) (**1996**), pp. 1–17.

66 Chan, K., Saltelli, A., Tarantola, S., Sensitivity analysis of the model output: variance based methods make the difference, in: *Proceedings of the 1997 Winter Simulation Conference*, **1997**

67 Versyck, K. J., Claes, J. E., Impe, J. F. V., Practical identification of unstructured growth kinetics by application of optimal experimental design, *Biotechnology Progress* 13 (**1997**), pp. 524–531.

68 Nathanson, M. H., Saidel, G. M., Multiple-objective criteria for optimal experimental design: application to ferrokinetics, *American Journal of Physiology* 248 (**1985**), pp. 378–386.

69 Munack, A., Posten, C., Design of optimal dynamical experiments for parameter estimation, *Proceedings of the American Control Conference ACC* 89 (**1989**), pp. 2011–2016.

70 Kontoravdi, C., Pistikopoulos, E. N., Mantalaris, A., Systematic development of predictive mathematical models for animal cell cultures, Computers & Chemical Engineering (**2010**), accepted for publication.

13
Dynamic Model Building Using Optimal Identification Strategies, with Applications in Bioprocess Engineering
Eva Balsa-Canto, Julio R. Banga, and Miriam R. García

Keywords
nonlinear dynamic models, nonlinear programming method (NLP), initial value problem (IVP), Fisher information matrix (FIM), optimal experimental design (OED), thermal death time (TDT) model

13.1
Introduction

The purpose of system identification is to build dynamic models from measured data [1]. Although identification and optimal experimental design (OED) are certainly not new research topics (see reviews by, e.g., Walter and Pronzato [2] and Ford *et al.* [3], for a list of classical papers), the last two decades have witnessed a significant increase in publications regarding both novel methodologies and applications, with emphasis in nonlinear dynamic models. Here, we will try to briefly review them, focusing on bioprocess engineering [4] and biological systems (systems biology; [5]).

Model building is an iterative process, usually represented as a cycle (Fig. 13.1), which must start from the definition of the purpose of the model. That is, the goal (purpose) and application of the model conditions the selection of the modeling framework, i.e., models must start with questions to be addressed. In the next step, using the *a priori* available knowledge and preliminary experimental data, a modeling framework is chosen and a first mathematical model structure is proposed. This first model usually contains unknown nonmeasurable parameters that may be estimated by means of experimental data fitting. In this regard, we need to know whether it is possible to uniquely determine their values (identifiability analysis) and if so, to estimate them with maximum precision and accuracy (parameter estimation step). This leads to a first working model that must be (in) validated with new experiments, revealing in most cases a number of deficiencies. In this case, a new model structure and/or a new (optimal) experimental design must be planned, and the process is repeated iteratively until the validation step is considered satisfactory. Regarding validation and invalidation of models, Anderson and

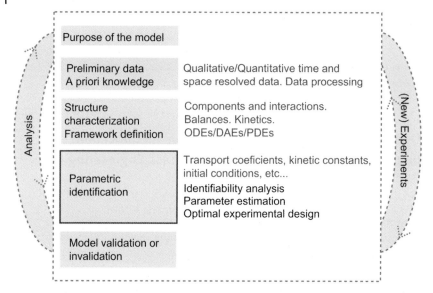

Fig. 13.1 Schematic representation of the model building loop.

Papachristodoulou [6] have recently presented a nicely written essay, which clarifies a number of myths.

In this chapter, we will consider nonlinear dynamic models, typically stated as (ordinary and/or partial) differential equations, and we will focus on parameter estimation, designed, and identifiability analysis, assuming the structure of the (nonlinear dynamic) model as given.

OED aims to devise the dynamic experiments that provide the maximum information content for subsequent nonlinear model identification, estimation, and/or discrimination. A related important problem is parameter estimation, which aims to find the unknown parameters of the model which give the best fit to a set of experimental data (ideally, obtained in the experiments designed with OED). Thus, a proper OED will facilitate proper parameter estimation. The identifiability analysis is intended to assess the quality of parameter estimates.

Iterative approaches for parametric identification in this domain have been studied by a number of authors [7–10]. Those approaches incorporate some or all of the following steps: identifiability and sensitivity analysis, experimental design, and parameter estimation.

In the context of OED, recent research directions have considered model-based design of experiments for parameter precision [11] and model discrimination [12, 13]. Others have explored the design of sequential and parallel optimal designs [14, 15].

These OED problems have been often stated as dynamic optimization statements (optimal control), resorting to numerical methods to obtain satisfactory solutions [17, 18]. However, these problems are usually nonconvex, so local gradient-based optimization methods can converge to local solutions. Certain global optimization methods have been used to surmount these difficulties [19, 20]. Development of new methodologies for OED continues to be a very active area of research, as it can be seen by the large number of relevant recent works [20–25]. A particularly interesting scenario is that of robust experimental design, that is, design under uncertainty [26–29].

On the other hand, estimating the parameters of a nonlinear dynamic model is much more difficult than for the linear case, as no general analytic result exists. Traditional methods for parameter estimation in dynamical systems rely on local optimization solvers; thus there is a risk of convergence to local optima. To deal with this difficulty, several authors have proposed the use of deterministic [30–32] and stochastic global optimization methods [33, 34]. Recently suggested, in addition, was the use of hybrid global–local methods [34–36] to enhance efficiency, particularly for highly multimodal systems.

Most systems of interest in bioprocess engineering and systems biology are highly nonlinear, so system identification in these areas has attracted many research efforts. The recent literature presents many applications of OED and parameter estimation in systems biology [33–44], food process engineering and predictive microbiology [45–50], and industrial biotechnology [51–53].

In the reminder of this chapter, we will introduce the parameter estimation problem and the associated OED problem, together with the different identifiability analyses. Numerical techniques will also be detailed with special emphasis on global optimization methods. All these concepts will be illustrated with two examples: the first, related to the modeling of microbial growth, and the second related to the modeling of gluconic acid production in a fed-batch reactor.

13.2
Parameter Estimation: Problem Formulation

Given a general set of differential equations explaining the dynamics of a system, as in Eqs. (13.1a) and (13.1b), the values assigned to the parameters will give rise to different system behaviors. The problem of parameter estimation may be formulated as follows:

> *Find model unknown parameters (e.g., transfer coefficients, kinetic coefficients, initial conditions, etc.) so as to minimize a measure of the distance among the model predictions and the available experimental data.*

The following elements are therefore necessary to formulate the parameter estimation problem: the model, the experimental scheme, the experimental data, and a measure of the distance among model predictions and experimental data, i.e., a cost function.

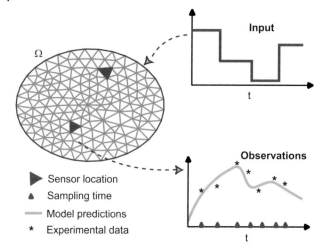

Fig. 13.2 Illustrative representation of the experimental scheme.

13.2.1
Mathematical Model Formulation

The mathematical model will consist of two essential elements – first, the set of differential equations describing the system dynamics (here we consider a general deterministic nonlinear dynamic model):

$$\mathbf{f}(\mathbf{x}, \dot{\mathbf{x}}, \mathbf{x}_\xi, \mathbf{x}_{\xi\xi}, \mathbf{u}, \boldsymbol{\theta}, \xi, t) = 0 \tag{13.1a}$$

Second, the observation function, describing the relationship among the states in the model and the available measured quantities:

$$\mathbf{y}^\varepsilon = \mathbf{g}^\varepsilon(\mathbf{x}, \mathbf{u}, \boldsymbol{\theta}, \xi, t) \tag{13.1b}$$

where $\mathbf{x} \in X \subset R^{n_x}$ are the state variables that may be distributed; $\mathbf{y}^\varepsilon \in Y \subset R^{n_o^\varepsilon \times n_s^\varepsilon \times n_l^\varepsilon}$ represents the vector of n_o^ε observables with n_s^ε discrete time measurements and n_l^ε sensor locations each for a given experiment ε, $\mathbf{u} \in U \subset R^{n_u}$ specifies the vector of inputs (i.e., all manipulable variables for a particular experiment), $\boldsymbol{\theta} \in \Theta \subset R^{n_\theta}$ is the vector of model parameters where Θ represents the set of admissible parameters that may be fixed by physical, chemical, or biological considerations, and $\xi \in \Omega \subset R^3$ corresponds to the spatial coordinates.

13.2.2
Experimental Scheme and Experimental Data

The experimental scheme (see Fig. 13.2) collects all information related to the way experimental data are obtained, i.e., the number of experiments, the observed or measured quantities (concentrations, temperatures, etc.), the (time-dependent) input profile, the spatial domain of interest and the location of the sensors, when relevant for the case under consideration, the experiment duration and the sampling times.

The experimental data consist on matrices of values corresponding to individual measurements obtained under the conditions specified by the experimental scheme ε. For the sake of clarity the experimental data and model predictions corresponding to an experimental scheme will be encoded in the following two vectors:

$$\mathbf{y}^{\varepsilon} = [y_1, y_2, \ldots, y_d, \ldots, y_{n_d}]^T; \qquad \tilde{\mathbf{y}}^{\varepsilon} = [\tilde{y}_1, \tilde{y}_2, \ldots, \tilde{y}_d, \ldots, \tilde{y}_{n_d}]^T \qquad (13.2)$$

where d represents a certain experimental condition defined by the subindexes ε (for the experiment), o (for the observables in the experiment ε), s (for the sampling times in the experiment ε) and l (for the sensor locations in the experiment ε), and n_d represents the total number of such conditions, that is, the total number of data. Note that the operators to be defined in the sequel can be then easily condensed as follows:

$$\sum_{d=1}^{n_d} (.) = \sum_{\varepsilon=1}^{n_\varepsilon} \left(\sum_{o=1}^{n_o^\varepsilon} \left(\sum_{s=1}^{n_s^\varepsilon} \left(\sum_{l=1}^{n_l^\varepsilon} (.) \right) \right) \right) \qquad (13.3)$$

It is also desirable to provide information about the type and quantity of noise in the experimental data. In this concern, replicates of the experiments are often required to determine the variance of the data, which may depend on what is being measured or may be different for every measurement.

Output-additive experimental noise is assumed as follows:

$$\tilde{y}_d = y_d + e_d, \quad d = 1, \ldots, n_d \qquad (13.4)$$

where e_d belongs to a sequence of independent random variables with a given probability density $\pi_{e_d}(e_d)$. In many practical examples e_d are independent random variables $N(0, \sigma_d^2)$, where the variance σ_d^2 of the noise is either constant or known for all ds in the so-called homoscedastic case, or unknown and dependent in the heteroscedastic case.

13.2.3
Cost Function

The definition of the scalar measure of the distance among the experimental data and the model predictions will depend on the available information for a particular example. The most well-known cost function is the generalized least squares, given by

$$J_{\text{lsq}}(\boldsymbol{\theta}) = \sum_{d=1}^{n_d} q_d (y_d(\boldsymbol{\theta}) - \tilde{y}_d)^2 \qquad (13.5)$$

The weighting coefficients $\{q_d\}_{d=1}^{n_d} \geq 0$ are fixed *a priori*. The choice of the weights will express the relative confidence in the various experimental data and the consequent importance attached to the model performance with regard to each data. It should be noted that nonprior information is required to use the least-squares function.

When information about the nature of the experimental noise is available, one may use the maximum (log-)likelihood function that looks for the value of the parameters that give the highest probability to the measured data:

$$J_{\text{llk}}(\boldsymbol{\theta}) = \ln(\pi(\tilde{\mathbf{y}}|\boldsymbol{\theta})) \tag{13.6}$$

The probability density function (π) selected will condition the type of cost function. In practice it is often assumed Gaussian additive noise. In this concern two situations may appear: (i) the homoscedastic case, for which the cost function results to be similar to the generalized least squares, with weights taken as the inverse of the variance of the experimental data:

$$J(\boldsymbol{\theta}) = \sum_{d=1}^{n_d} \frac{(\mathbf{y}_d(\boldsymbol{\theta}) - \tilde{\mathbf{y}}_d)^2}{\sigma_d^2} \tag{13.7}$$

Or (ii) the more general heteroscedastic case for which the following cost function is obtained:

$$J(\boldsymbol{\theta}) = \sum_{d=1}^{n_d} \log(\sigma_d^2) + \frac{(\mathbf{y}_d(\boldsymbol{\theta}) - \tilde{\mathbf{y}}_d)^2}{\sigma_d^2} \tag{13.8}$$

13.2.4
Numerical Methods: Single Shooting vs. Multiple Shooting

The parameter estimation problem is thus formulated as a nonlinear optimization problem subject to the system dynamics and possibly bounds on the parameter values. Therefore, its numerical solution involves an outer iterative procedure to generate values for the unknown parameters and initial conditions, the nonlinear programming method (NLP) and an iterative procedure to solve the differential equations, the initial value problem (IVP) solver. Note here that the boundary value problem may be transformed into an IVP by a spatial discretization technique such as, for example, the finite differences or the finite elements methods.

In the so-called single shooting approach, the IVP is solved from the initial conditions till the final time for all the iterates generated by the NLP solver. Alternatively, in the multiple shooting approach [54–56], the duration of the process is partitioned into a number of shooting intervals, in such a way that at least one experimental data may be found in each shooting, and the several IVPs are to be solved.

It should be noted that in the multiple shooting, the initial conditions for the different intervals are also to be computed during optimization. Therefore, the addition of further constraints to the parameter estimation problem is required so as to guarantee that at the optimum the solution is smooth. This leads to a constrained nonlinear optimization problem.

13.3 Identifiability

Parameter identifiability is concerned with the possibility of finding a unique value for the parameters. We can distinguish between structural and practical identifiability. Structural identifiability is a theoretical property of the model structure depending only on the observation and the input functions. The parameters of a model are structurally globally identifiable if, under ideal conditions of noise-free observations and error-free model structure, and independently of the particular values of the parameters, they can be uniquely estimated from the designed experiment [57].

Practical identifiability analysis is related to the question whether it is possible to find a unique solution for the parameters given a pair model–experimental data. At this point, it is important to note that, in fact, in the presence of experimental error, there are several equivalent solutions defining the confidence region of the parameters, but of course this does not necessarily mean that the model is not practically identifiable.

Although the questions seem pretty similar, there are several crucial differences:

- Structural analysis is performed in the absence of noise, whereas for the practical analysis the experimental error is crucial.
- In most of the cases, lack of structural identifiability will be unsolvable. In most examples, one should consider reformulating the model or aggregating measured quantities.
- Lack of practical identifiability will be in general terms solvable, provided the experimental constraints allow for designing sufficiently rich experiments.
- Performing a structural identifiability analysis is by far more complicated as complex symbolic manipulations are required, and this might make a full analysis impossible for highly nonlinear large-scale models. Practical identifiability analysis will be in general computationally intensive but it is extremely helpful to asses parameter estimates reliability and to compare possible experimental designs.

There are, at least, two obvious reasons to assess identifiability: first, most of the model parameters have a physical meaning, and we are interested in knowing whether it is at all possible to determine their values from experimental data; second, numerical optimization approach will find difficulties when trying to estimate the parameters of an unidentifiable model.

There are a few methods for testing the *structural identifiability* of nonlinear models in ordinary differential equations [58, 59]: the similarity transformation approach [60], differential algebra methods [61, 62], and power series approaches including the Taylor series [63], and the generating series methods [64].

Unfortunately there is no method amenable to every model; thus at some point, the selection of one of the possibilities has to be faced. All of them present limitations related to the nonlinearity and the size of the system under consideration, while size implying the number of state variables, the number of parameters, and the number of observables.

Practical identifiability may be evaluated by using observables' sensitivity with respect to the parameters evaluated at the optimum. If the sensitivity functions are linearly dependent, the model is not identifiable, and if they are nearly linearly dependent, parameters are highly correlated [1]. In fact, it is also possible to perform a so-called *global sensitivity analysis* [65] to analyze the relative influence of the parameters on model predictions over the region of interest ($\Theta \subset \Re^{n_\theta}$). Global sensitivities may be computed by somehow sampling the parameter space. An alternative that can yield precise estimates without requiring an excessively large number of samples is Latin hypercube sampling (LHS). Poor identifiability is expected when the model output is insensitive to some parameters. In this case, less influencing parameters may be fixed in order to improve the overall identifiability.

Additionally, one may compute the confidence or uncertainty region for the parameters since the shape and size of such regions will determine whether practical identifiability is or not guaranteed.

Confidence regions may be computed by using the Fisher information matrix (FIM) or a Monte Carlo based approach. Both methods are described in detail in Balsa-Canto et al. [20], and here they are only briefly discussed.

FIM provides a measure of the quantity and quality of the information of a given experiment for a particular value of parameters. It is mathematically defined as follows:

$$\text{FIM} = \underset{\tilde{y}|\mu}{E}\left\{\left[\frac{dJ(\theta)}{d\theta}\right]\left[\frac{dJ(\theta)}{d\theta}\right]^T\right\} \tag{13.9}$$

where E represents expected value and μ is a value of the parameters hopefully closed to their "real" value. It should be noted that J regards the cost function used for parameter estimation. Therefore when the log-likelihood is being used, the FIM will depend on the noise distribution and whether it is constant or not for all experimental data. A general formulation of the FIM for additive Gaussian noise is obtained by Garcia [66].

The Crammèr–Rao inequality provides a lower bound on the covariance of the estimators:

$$\mathbf{C} \geq \text{FIM}^{-1}(\mu) \tag{13.10}$$

The diagonal elements of the covariance matrix correspond to half the confidence intervals for the parameters. The correlation between parameters may be computed as follows:

$$Cr_{ij} = \frac{C_{ij}}{\sqrt{C_{ii}C_{jj}}} \tag{13.11}$$

in such a way that if $Cr_{ij} = \pm 1$, the parameters ij are highly correlated whereas if $Cr_{ij} = 0$ the parameters are fully uncorrelated.

The *Monte Carlo based approach* requires the solution of the parameter estimation problem under hundreds of realizations of the experimental data. This allows the generation of a cloud of solutions for the parameters that represents the confidence region. The numerical information about the parameter uncertainties is then obtained through the manipulation of the resulting matrix of estimated parameters [20].

13.4
Optimal Experimental Design

Performing experiments to obtain a rich enough set of experimental data is a costly and time-consuming activity. The main purpose of OED is to devise the necessary dynamic experiments in such a way that the parameters are estimated from the resulting experimental data with the best possible statistical quality, which is usually a measure of the accuracy and/or decorrelation of the estimated parameters. In other words, based on model candidates, we seek to design the best possible experiments in order to facilitate system identification. OED is, therefore, critical in the model building cycle, outlined in Fig. 13.1, which should also contain coupled elements like model discrimination and/or model selection.

The traditional way of designing an experiment requires two steps [2] – (i) selecting adequate inputs (control variables) and outputs (measured variables), and (ii) obtain the optimal input and sampling strategy, which optimizes a certain criterion (related with maximum information content of the experiment).

Mathematically, the OED problem can be formulated as a dynamic optimization problem where the objective is to find a set of inputs $\mathbf{u}$, usually time-varying variables together with sampling times, experiment durations, and possibly initial conditions, so as to minimize a cost function related to the FIM (Eq. (13.9)). This can be mathematically stated as follows:

$$\min_{\mathbf{u}^\varepsilon, t_s^\varepsilon, t_f^\varepsilon, \mathbf{x}_0^\varepsilon} \phi(\text{FIM}) \quad \text{for } \varepsilon = 1, \ldots, n_\varepsilon \tag{13.12}$$

subject to experimental constraints such as

$$\mathbf{u}_L^\varepsilon \leq \mathbf{u}^\varepsilon \leq \mathbf{u}_U^\varepsilon \quad \text{for } \varepsilon = 1, \ldots, n_\varepsilon \tag{13.13}$$

$$\mathbf{x}_{0,L}^\varepsilon \leq \mathbf{x}_0^\varepsilon \leq \mathbf{x}_{0,U}^\varepsilon \quad \text{for } \varepsilon = 1, \ldots, n_\varepsilon \tag{13.14}$$

$$t_{f,L}^\varepsilon \leq t_f^\varepsilon \leq t_{f,U}^\varepsilon \quad \text{for } \varepsilon = 1, \ldots, n_\varepsilon \tag{13.15}$$

$$0 \leq t_{s,i}^\varepsilon \leq t_{s,i+1}^\varepsilon \leq t_f^\varepsilon, \quad i = 1, \ldots, n_s^\varepsilon - 1; \; \varepsilon = 1, \ldots, n_\varepsilon \tag{13.16}$$

where $\mathbf{u}^\varepsilon$ represents the control variables, $\mathbf{x}_0^\varepsilon$ the initial conditions, t_f^ε the experiment duration, and $t_{s,i}^\varepsilon$ the sampling times for the experiment ε.

Since the FIM may be related to the confidence hyperellipsoid for the parameters, the different OED criteria provide information about its shape and size [17]. The D-criterion, that is, the maximization of the determinant of the FIM, minimizes the confidence ellipsoids volume; the E-criterion, that is, the maximization of the minimum eigenvalue of the FIM, minimizes the length of the largest axis, whereas the modified E-criterion, that is, the minimization of the condition number of the FIM, minimizes the ratio of the largest to the smallest axis, seeking to make those ellipsoids as spherical as possible.

13.4.1
Numerical Methods: The Control Vector Parameterization Approach

Numerical solutions for the dynamic optimization problem stated above (Eq. (13.12)) can be obtained using direct methods, which transform the original problem into an NLP problem via parameterizations of the inputs and/or the states. One such direct method is the control vector parameterization (CVP) method that proceeds by dividing the duration of the experiment(s) into a number of elements and approximating the inputs using low-order polynomials.

Once the CVP has been applied, the general OED problem is transformed into an NLP, being the decision variables the amount of stimulation, plus the experimental sampling times, the duration of the experiments and possibly some initial conditions. Note that the solution of this NLP requires suitable NLP and IVP solvers similar to the parameter estimation problem.

Regarding the IVP solver, several alternatives exist: (i) to use symbolical manipulation or automatic differentiation to obtain the parametric sensitivities, which can then be solved together with the systems dynamics using a standard IVP solver or (ii) to exploit a backward differentiation formulae based (BDF) method so as to exploit the fact that the original ODE and the sensitivities share the same Jacobian to compute states and sensitivities. Typical solvers are ODESSA [67] or the more recently developed CVODES, included in the SUNDIALS suite [68].

13.5
Nonlinear Programming Solvers

Both the parameter estimation and the OED problems are ultimately formulated as nonlinear optimization problems subject to the system dynamics and algebraic constraints.

Optimization methods are designed to generate a sequence of solutions that eventually converges to the minimum of the cost function. The way this sequence is generated gives rise to hundreds of different NLP solvers. Figure 13.3 summarizes some alternatives.

13.5 Nonlinear Programming Solvers

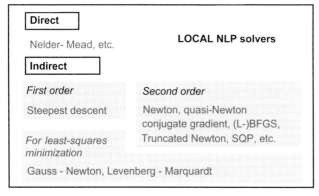

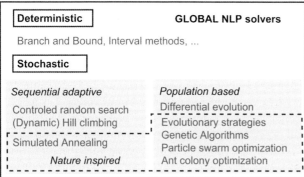

Fig. 13.3 Rough classification of NLP solvers and some typical examples.

Local methods use information about the cost function and possibly its gradient and its Hessian in the neighborhood of every iterate. Thus these methods are expected to converge to the closest minimum to the initial guess.

In the context of least-squares minimization, the most widely used local method is the Levenberg–Marquardt (see for example, Seber and Wild [69]), a combination of the steepest descent with the Newton method for a least-squares cost function. It should be noted that most of the methods for least-squares problems, such as the Levenberg–Marquardt, are based on the Gauss–Newton modification of the Newton method, that is, part of the Hessian of the objective with respect to the parameters is ignored so as to avoid computing second-order derivatives. If this approximation is not valid, the method may converge slowly or even fail. Schittkowski [70] describes how to combine the Gauss–Newton method with a sequential quadratic approach (SQP) for the specific case of minimizing the least squares function.

The two major advantages of this type of methods are as follows:

1. Convergence to a minimum ($\boldsymbol{\theta}^*$) is guaranteed by the fact that the gradient of the cost function evaluated at the optimum is zero, $\nabla J(\boldsymbol{\theta}^*) = 0$, and the Hessian is positive definite ($\nabla^2 J(\boldsymbol{\theta}^*) > 0$).
2. The methods are highly efficient when started close to the solution.

Local methods have been largely used in combination with the single and the multiple shooting approaches for the purpose of parameter estimation. However, the nonlinear character of the bioprocess models leads to the presence of several suboptimal solutions and thus local methods may end up in a suboptimal solution.

It has been argued that multiple shooting based approaches can circumvent some local minima by allowing for discontinuous trajectories while searching the global minimum. And even though this may be true for some cases, for example oscillatory systems, convergence to the global solution cannot be guaranteed [36].

Moreover, in the presence of a bad fit, there is no way of knowing if it is due to a wrong model formulation, or if it is simply a consequence of local convergence.

Global methods have emerged as the alternative to search the global optimum. One of the simplest global methods is a multistart method. Here, a large amount of initial guesses are drawn from a distribution and subjected to a parameter estimation algorithm based on a local optimization approach. The smallest minimum is then regarded as being the global optimum. In practice, however, there is no guarantee of arriving to the global solution and the computational effort can be quite large.

Over the last decade, more suitable techniques for the solution of multimodal optimization problems have been developed (see, e.g., Pardalos *et al.* [71], for a review). The successful methodologies combine effective mechanisms of exploration of the search space and exploitation of the previous knowledge obtained by the search. Depending on how the search is performed and the information they are exploiting, the alternatives may be classified in three major groups: deterministic, stochastic, and hybrid.

Global *deterministic* [72, 73] methods in general take advantage of the problem's structure and even guarantee convergence to the global minimum for some particular problems that verify specific conditions of smoothness and differentiability.

Several recent works propose the application of global deterministic methods for parameter estimation in the context of chemical processes, biochemical processes, and in the modeling of metabolic and signaling pathways [30–32, 38]. Although very promising and powerful, there are still limitations to their application, mainly due to rapid increase in computational cost with the size of the considered system and the number of its parameters.

As opposed to deterministic approaches, global *stochastic* methods do not require any assumptions about the problem's structure. Stochastic global optimization algorithms are making use of pseudorandom sequences to determine search directions toward the global optimum. This leads to an increasing probability of finding the global optimum during the runtime of the algorithm. The main advantage of these methods is that they rapidly arrive to the proximity of the solution.

The number of stochastic methods has rapidly increased in last decades. The most successful approaches lie in one (or more) of the following groups: pure random search and adaptive sequential methods, clustering methods, population-based methods, or nature-inspired methods [74].

Within the adaptive search algorithms, simulated annealing (*SA*, [75]) is probably the most popular algorithm. The main difference with respect to the other random search methods consists in the possibility to accept, in some conditions, a detrimental search step with a given probability, thus avoiding local optima.

Population-based strategies such as genetic algorithms (*GA*) or evolutionary search (*ES*) methods are based on the ideas of biological evolution [76], which is driven by the mechanisms of reproduction, mutation, and the principle of survival of the fittest.

Recent works present comparisons of different implementations of *SA*, *GA*, and *ES* strategies in the context of parameter estimation [33, 77] concluding that although all of them are usually reliable in solving complex multimodal problems, *SA* experiences a very low convergence rate as compared to population-based strategies.

In our group, we have compared a number of different stochastic population-based strategies for the solution of parameter estimation problems concluding that differential evolution [78] and stochastic ranking *ES* [79] are the most competitive in terms of robustness and efficiency in finding the global solution [33, 34, 36].

Despite the fact that many stochastic methods can locate the vicinity of global solutions very rapidly, the computational cost associated with the refinement of the solution is usually very large. In order to surmount this difficulty, *hybrid* methods and *metaheuristics* have been recently presented for the solution of parameter estimation problems [34–36] that speed up these methodologies while retaining their robustness.

In particular, the scatter search metaheuristic [34, 80] and the sequential hybrid with automatic switching [36] showed speeds up between one and two orders of magnitude with respect to the use of stochastic global methods.

13.6
Illustrative Examples

13.6.1
Modeling of the Microbial Growth

The mathematical modeling of microbial growth, survival, or death, usually regarded as predictive microbiology, has received major attention in previous years due to its enormous potential for improving processes in the food industry.

Predictive microbiology models can be classified in three major groups – primary, secondary, and tertiary [81]. Primary level models describe changes in microbial numbers with time; secondary level models show how the parameters of the primary models vary with environmental factors (temperature, water activity, pH, etc.) and the tertiary level combines the first two types of models with user-friendly application software or expert systems that calculate the microbial behavior under the specified conditions.

Growth models may be used to estimate values of kinetic parameters such as the lag-time, the maximum specific growth rate (μ_{max}), and the maximum cell concentration (N_{max}) from experimental data. For this purpose, an adequate experimental design is necessary.

Versyck et al. [82] presented a pioneering study on the use of OED for parameter estimation for the Arrhenius-type thermal inactivation model, the thermal death time (TDT) model. In their work, the authors confirmed the advantages of using time-varying processing temperatures for parameter estimation, in the case of "perfect" (continuous and noiseless) experiments. By the same time, it was shown that the use of optimal sampling could improve results [83]. More recently, it has been shown that parallel designs where the temperature profiles, initial conditions, the sampling times, and experiment durations are simultaneously optimized may considerably improve identifiability properties [50].

We consider here an example consisting of the combination of the primary model by Geeraerd et al. [84] (Eq. (13.13)) with the TDT secondary model (Eq. (13.18)):

$$\frac{dn}{dt} = -\mu_{max}(T) \frac{1}{1 + \exp(q_p)} (1 - \exp(n_{res} - n)) \quad (13.17a)$$

$$\frac{dq_p}{dt} = -\mu_{max}(T) \quad (13.17b)$$

$$\mu_{max}(t) = \frac{2.303}{D_{ref}} \exp\left[\frac{2.303}{z}(T - T_{ref})\right] \quad (13.18)$$

The objective is to estimate D_{ref} and z by measuring n. It has been demonstrated that the model is structurally nonidentifiable when using a constant-temperature profile for getting experimental data and that step-wise profiles may largely improve the identifiability for a perfect noise-free experiment [82].

First we define a qualitative experimental design (QED), which consists of a stepwise treatment of 20 min. The microorganisms are subject to $T = 50\,°C$ during first 10 min and then $T = 80\,°C$ till the end of the experiment. To obtain the pseudoexperimental data under such conditions, the following values for the nominal parameters have been selected: $D^*_{ref} = 1$ min and $z^* = 10\,°C$. The initial value for q_p was fixed to the value $\ln(10^{-5})$ and $n_{res} = \ln(10^2)$ was assumed. The resultant data are used for parameter fitting.

First, a multistart optimization using a local method (n2fb and adaptive nonlinear least-squares algorithm [85] with 200 different initial guesses) is used to solve the problem. Results reveal that different initial guesses end up in different suboptimal solutions. If we analyze in more detail the best 50 solutions by plotting the histograms of results (Fig. 13.4), we observe that several parameter values result in the same cost function.

This means that the log-likelihood function is almost flat (see Fig. 13.5), that is, the parameters are not practically identifiable. Therefore an OED is needed.

The experimental equipment allows for the following degrees of freedom for the OED:

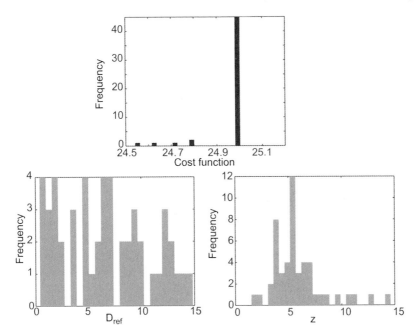

Fig. 13.4 Microbial growth: histograms for the best 50 solutions when performing parameter fitting with a 200 multistart of a local method under a QED.

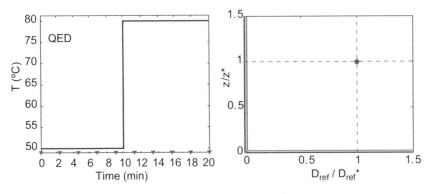

Fig. 13.5 Microbial growth: qualitative experimental design and the resultant log-likelihood contour plot.

- Step-wise experiments with: $50\,°\mathrm{C} \leq T(t) \leq 80\,°\mathrm{C}$.
- Initial cell density can vary within the following limits $\ln(10^6) \leq n(0) \leq \ln(10^9)$ cfu/ml.
- The duration of the experiment can also be modified: $12 \leq t_f \leq 20$ min.
- The number of sampling times is fixed to $n_s = 10$, being either equidistantly or optimally located.

Table 13.1 Microbial growth: summary of results for the optimal experimental design with different number of equidistant and optimally located sampling times.

	n_s	t_s type	E
OED1	20	Equidistant	4.5×10^7
OED2	10	Equidistant	3.7×10^7
OED3	10	Optimal	8.4×10^7

Several OEDs were computed under the constraints detailed above. Previous works [50] showed that D-optimal designs, even resulting in less uncertainty on the parameters, are rather difficult to implement in the laboratory, whereas modified E-designs are rather insensitive to the experimental constraints, that is, the global optimum (modified $E = 1$) is obtained regardless the constraints imposed in the optimization. Therefore we will consider here E-optimality criterion. Table 13.1 summarizes the results obtained for different number of equidistant or optimally located sampling times.

From the results, it becomes apparent that E-optimality is not as sensitive to the number of sampling times as to their location. Note that using 10 optimal sampling times results in almost twice the E-value as compared to the case of using 20 equidistant sampling times.

In order to compare with the QED, Fig. 13.6 presents the OED2 with 10 equidistant sampling times and the corresponding contour plot for the log-likelihood. Note that a suitable experimental design makes the parameters identifiable even though the uncertainty in their values may still be substantial.

The robust identifiability analysis is performed for all three optimal designs to estimate the expected uncertainty on the parameters for all cases. Figure 13.7 presents the OEDs together with a comparison of the corresponding uncertainty ellipses. Note that in any case, the expected uncertainty is within the 3%, which is reasonable considering the experimental error assumed.

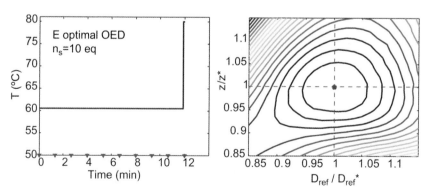

Fig. 13.6 Microbial growth: optimal experimental design OED1 and the resultant log-likelihood contour plot.

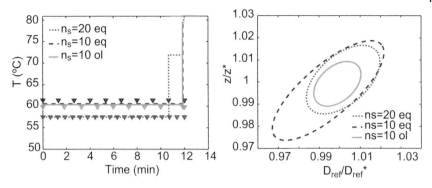

Fig. 13.7 Microbial growth: comparison of optimal experimental designs and the corresponding robust identifiability ellipses.

13.6.2
Modeling the Production of Gluconic Acid in a Fed-Batch Reactor

Most of the fermentation processes to obtain gluconic acid (GA) are carried out by *Aspergillus niger*. The objective here is that of building a model with good predictive capabilities to describe the dynamics of glucose (G), oxygen (O_2), gluconic acid (GA), and biomass (X) during the growth phase of *A. niger*. We consider a fed-batch fermenter with two valves to regulate the incoming flux of glucose and water mixture (u_1) and the oxygen transfer rate described by the Henry's law (u_2). The controls take the value 0 when they are completely closed and 1 when completely opened. Mathematically, the process may be described as follows:

$$\frac{dX}{dt} = R_X$$

$$\frac{dGA}{dt} = R_{GA}$$

$$\frac{dG}{dt} = R_G + u_1 \frac{F_{in}}{V}(G^{in} - G)$$

$$\frac{dO_2}{dt} = R_{O_2} + u_2 K_{La}(O_2^* - O_2)$$

$$\frac{dV}{dt} = u_1 F_{in} \tag{13.19}$$

where F_{in} and K_{La} represent the maximum incoming flux and oxygen transfer rate, respectively, O_2^* is the saturation of dissolved oxygen, G^{in} the concentration of glucose in the inlet, and R_X, R_{GA}, R_G, and R_{O_2} are the kinetic rates of the biomass, GA, glucose, and oxygen, respectively.

Several works have studied the mathematical modeling of the kinetic rates to obtain GA by means of A. niger. Those works conclude that either the Monod [86] or the Contois [87] law can take into account the dependence of the biomass growth with the limiting substrates (oxygen and glucose); however, since the Monod law is simpler, it is convenient to check whether this law is able to reproduce the data. A further simplification would be to represent the kinetic mechanisms as a unique reaction rate of the following form:

$$Y_G G + Y_{O_2} O_2 \xrightarrow{R_X} X + Y_{GA} GA$$

the model (13.15) can be then rewritten as follows:

$$\frac{dX}{dt} = R_X = \mu_{max} X \frac{G}{K_G + G} \frac{O_2}{K_{O_2} + O_2}$$

$$\frac{dGA}{dt} = Y_{GA} R_X$$

$$\frac{dG}{dt} = Y_G R_X + u_1 \frac{F_{in}}{V}(G^{in} - G)$$

$$\frac{dO_2}{dt} = Y_{O_2} R_X + u_2 K_{La}(O_2^* - O_2)$$

$$\frac{dV}{dt} = u_1 F_{in} \tag{13.20}$$

Next in model building loop is to compute model unknowns, in this case $\theta = [\mu_{max}, Y_G, Y_{O_2}, Y_{GA}, K_{La}]$ by measuring $\mathbf{y} = [X, GA, G, O_2]$. First step in the parametric identification loop is to check whether this is possible by performing a structural identifiability analysis.

In fact using the first terms of the generating or the Taylor series approaches for structural identifiability, it may be concluded that the parameters are globally structurally identifiable.

To estimate their values, we will first consider a QED. Basically two completely different experiments are designed: (1) where the incoming flux valve is almost closed $u_1 = 0.01$ and the oxygen transfer is completely open $u_2 = 1$ and (2) where the incoming flux valve is completely open $u_1 = 1$ and the oxygen transfer is almost closed $u_2 = 0.01$. Pseudoexperimental data are obtained by direct numerical simulation of the model (Eq. (13.20)) assuming the nominal values for the parameters given in Table 13.2.

Heterocesdastic experimental error is added to the model predictions as follows: $e = 0.05 y Rnd$, where Rnd is a random variable $N(0, 1)$, in such a way that the variance of the measurements becomes $\sigma^2 = (0.05y)^2$ corresponding to what is usually called power-in-the-mean variance [57]. Forty equidistant sampling times are used per experiment.

13.6 Illustrative Examples

Table 13.2 Parameters nominal values for the gluconic acid production in a fed-batch reactor.

Parameter	Nominal value	Parameter	Nominal value
K_{La}	600 h^{-1}	K_G	9.9222 gl^{-1}
O_2^*	0.0084 gl^{-1}	K_{O_2}	0.0137 gl^{-1}
G^{in}	250 gl^{-1}	Y_{GA}	44.8887
F_{in}	0.5 min^{-1}	Y_{O_2}	-2.5598
μ_{max}	0.2242 h^{-1}	Y_G	-51.0365

First a multistart of n2fb with 200 different initial guesses was used to check for the possible multimodal nature of the problem. Figure 13.8 presents the histogram of solutions achieved showing how the solution is highly dependent on the initial guess. The best objective function corresponds to $J = 159.1$ achieved only once.

Moreover, the histograms for the cost function and for the parameters reveal poor identifiability of some parameters, although this problem is not as obvious as in

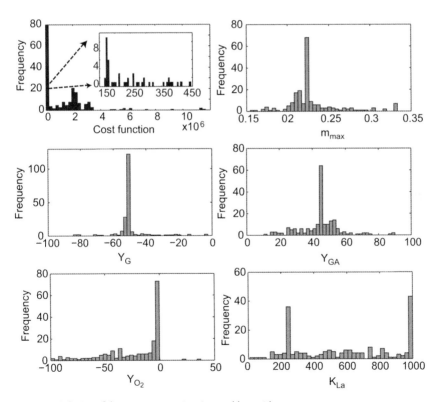

Fig. 13.8 Solution of the parameter estimation problem with a multistart of a local optimization method. The best objective function corresponds to $J = 159.1$.

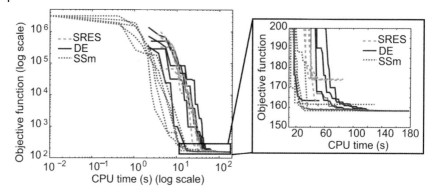

Fig. 13.9 Solution of the parameter estimation problem with global solvers: comparison of the curves of convergence.

previous example. We decided to solve the problem by using a global optimization method.

For the sake of illustrating the performance of several global optimization methods, DE [78], SRES [79], and SSm [80] were selected to solve the problem and guarantee convergence to the global solution. Five different runs were performed for all methods. Figure 13.9 presents the curves of convergence for all runs and methods and Table 13.3 summarizes the results. It should be noted that both DE and SSm were able to rapidly converge to the global solution $J = 158.35$, which in fact was never achieved by the multistart approach.

Figures 13.10(a) and (b) present the model predictions for the optimal solution $\hat{\theta} = [0.2241, -51.049, -2.1341, 44.923, 500.02]$.

It should be noted that even though the values obtained for μ_{max}, Y_{GA}, and Y_G are within the 1% of the global optimum, this is not the case for Y_{O_2} and K_{La} for which the difference achieved was 17%. Note, in addition that the Monte Carlo based identifiability analysis for the parameter estimates reveals uncertainties over the 20% but for μ_{max}.

In view of the results, a parallel-sequential OED was pursued in order to improve parameter estimates. The two qualitative designs are incorporated in the FIM and two new experiments are designed allowing for constant control profiles that are optimized together with the final time and the initial conditions of glucose and

Table 13.3 Parameter estimation for the qualitative experimental design using global optimization methods: summary of results.

	Best J	Tcpu (s)	Mean J	Mean Tcpu (s)
DE	158.35	138	159.65	126
SRES	159.01	76.2	171.06	76
SSm	158.35	129	159.27	126

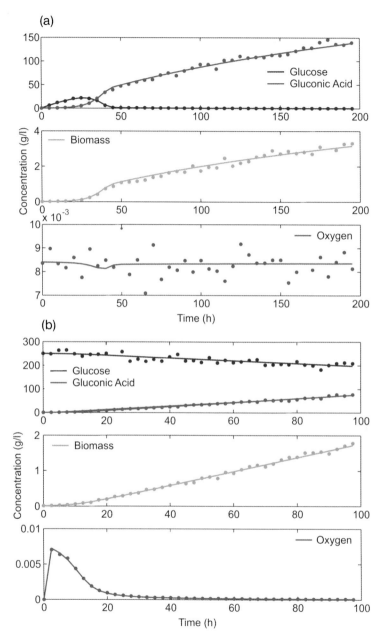

Fig. 13.10 (a) Calibration of the GA model by using qualitative experimental design. Fit of the experiment with $u_1 = 0.01$ and $u_2 = 1$. (b) Fit of the experiment with $u_1 = 1$ and $u_2 = 0.01$.

Table 13.4 Gluconic acid production: summary of results for the optimal experimental design with different FIM-based criteria[a].

Design	Expected uncertainty in %				
	μ_{max}	Y_G	Y_{O_2}	Y_{GA}	K_{La}
D	1.5	5.5	5.8	5.3	5.9
E	1.2	2.9	1.8	3.8	1.7
Modified E	0.8	1.0	0.8	1.5	1.1

a) Note that the parameter space is normalized to select that criterion whose largest relative parameter error is the smallest.

biomass. Note that an appropriate formulation of the FIM is necessary because of the presence of heteroscedastic noise [66].

The OED problem was solved for D-, E-, and E-modified criteria, and the Monte Carlo based practical identifiability analysis was performed for the resultant experimental schemes so as to compare the expected uncertainty in the parameter estimates. Table 13.4 summarizes the results showing that in this example, the best optimal design corresponds to the modified E-criterion that achieves the smallest expected parameter uncertainty. It is to be noted that the uncertainty has been reduced up to two orders of magnitude with respect to the uncertainties for the QED.

In order to graphically illustrate the differences among the different criteria, the worst case expected uncertainty by pairs of parameters are shown in Fig. 13.11.

This particular example reveals that in the experimental conditions considered, improving quantity of information results in smallest uncertainty hyperellipsoids. Therefore, why D gets the worse result? From our experience, these criteria may end up in highly noisy or multimodal log-likelihood functions; therefore even though the experiment is more informative, finding the global solution is almost impossible, thus resulting in largest Monte Carlo based hyperellipsoids. To illus-

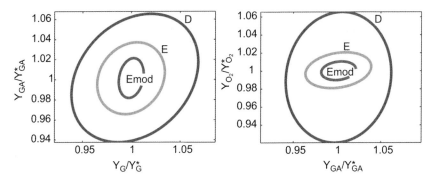

Fig. 13.11 Gluconic acid production: worst case expected parameter uncertainty ellipses by pairs of parameters after optimal experimental design.

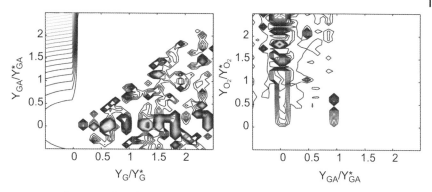

Fig. 13.12 Gluconic acid production: illustrative examples of the contour plots by pairs of parameters corresponding to the log-likelihood under the D optimum OED.

trate this point, Fig. 13.12 presents two examples of the contour plots by pairs of parameters corresponding to the log-likelihood under the D-optimum OED.

The parameter estimation problem is now solved by using the four experiments in the optimally modified E-experimental scheme. Figures 13.13(a) and (b) show the two optimally designed experiments together with the optimal fits obtained by the use of SSm that correspond to the following parameter set: $\hat{\hat{\theta}} = [0.2241, 44.908, -51.04, -2.5606, 600.04]$, which is within the 0.04% of the optimal value θ^*.

13.7
Overview

In this chapter, we have focused on three critical steps of the model-building loop: parameter estimation, identifiability analysis, and optimal experimental design. The parameter estimation and the optimal experimental design problems can be formulated as, or transformed to, nonlinear optimization problems that may be solved by using nonlinear programming methods. We have presented a brief overview of available optimization techniques with special emphasis on the necessity of using global optimization methods to ensure proper solutions for these problems. In addition, identifiability analysis was presented as a way to measure the quality of the parameter estimates or to anticipate the quality of a given experimental design.

A first example related to the modeling of microbial growth helped us to illustrate how strong practical identifiability problems can make it difficult to find suitable solutions for the parameters, and how a suitable experimental design may help to improve such identifiability. A second example, related to the modeling of production of GA in a fed-batch reactor, showed the importance of using global optimization methods to deal with multimodal problems, and how a sequential-

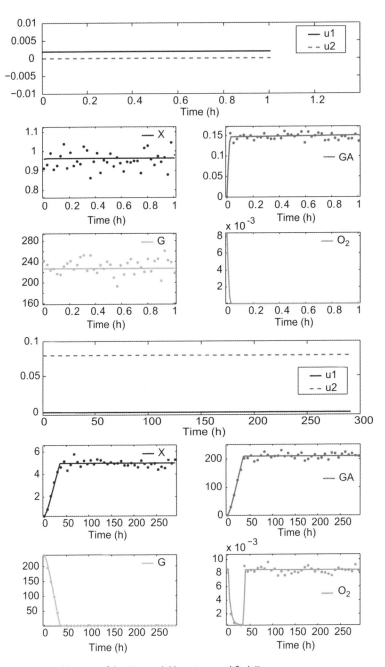

Fig. 13.13 Calibration of the GA model by using modified E optimal experimental design. (a) First and (b) second optimally designed experiments and the corresponding fit.

parallel OED may improve the quality of parameter estimates together with model predictive capabilities.

Acknowledgements

The authors acknowledge financial support from Spanish MICINN project "MultiSysBio" ref. DPI2008-06880-C03-02 and Xunta de Galicia project "IDECOP" ref. 08DPI007402PR.

References

1 Ljung, L., *System Identification – Theory for the User*, Prentice-Hall, Upper Saddle River, NJ, **1999**.
2 Walter, E., Pronzato, L., *Automatica* 26(2) (**1990**), pp. 195–213.
3 Ford, I., Titterington, D. M., Kitsos, C. P., *Technometrics* 31(1) (**1989**), pp. 49–60.
4 Shuler, M. L., Kargi F., *Bioprocess Engineering: Basic Concepts*, Prentice-Hall PTR, Upper Saddle River, NJ, **2001**.
5 Doyle, F. J., Stelling, J., *Journal of the Royal Society, Interface* 3(10) (**2006**), pp. 603–616.
6 Anderson, J., Papachristodoulou, A., *BMC Bioinformatics* 10 (**2009**), p. 132.
7 Feng, X., Rabitz, H., *Biophysics Journal* 6(3) (**2004**), pp. 1270–1281.
8 Gadkar, K. G., Gunawan, R., Doyle, F. J., *BMC Bioinformatics* 6 (**2005**), art. no. 155.
9 van Riel, N. A. W., *Briefings in Bioinformatics* 7(4) (**2006**), pp. 364–374.
10 Balsa-Canto, E., Alonso A. A., Banga J. R., in: *Foundations of System Biology in Engineering*, Allgöwer, F., Reuss, M. (eds.), Universität Stuttgart, Stuttgart, Germany, **2007**, p. 51.
11 Franceschini, G., Macchietto, S., *Chemical Engineering Science* 63(19) (**2008**), pp. 4846–4872.
12 Chen, B. H., Asprey, S. P., *Industrial and Engineering Chemistry Research* 42(7) (**2003**), pp. 1379–1390.
13 Donckels, B. M. R., De Pauw, D. J. W., De Baets, B., Maertens, J., Vanrolleghem, P. A., *Chemometric and Intelligent Laboratory System* 95(1) (**2009**), pp. 53–63.
14 Galvanin, F., Macchietto, S., Bezzo, F., *Industrial and Engineering Chemistry Research* 46(3) (**2007**), pp. 871–882.
15 Schwaab, M., Monteiro, J. L., Pinto, J. C., *Chemical Engineering Science* 63(9) (**2008**), pp. 2408–2419.
16 Galvanin, F., Macchietto, S., Bezzo, F., *Industrial and Engineering Chemistry Research* 46(3) (**1987**), pp. 871–882.
17 Asprey, S. P., Macchietto, S., *Computers and Chemical Engineering* 24(2–7) (**2000**), pp. 1261–1267.
18 Körkel, S., Kostina, E., Bock, H. G., Schlöder, J. P., *Optimization Methods and Software* 19(3–4) (**2004**), pp. 327–338.
19 Banga, J. R., Versyck, K. J., Van Impe, J. F., *Industrial and Engineering Chemistry Research* 41(10) (**2002**), pp. 2425–2430.
20 Balsa-Canto, E., Alonso A. A., Banga J. R., *IET Systems Biology* 2(4) (**2008**), pp. 163–172.
21 Schittkowski, K., *Industrial and Engineering Chemistry Research* 46(26) (**2007**), pp. 9137–9147.
22 Zhang, Y., Edgar, T. F., *Industrial and Engineering Chemistry Research* 47(20) (**2008**), pp. 7772–7783.
23 Franceschini, G., Macchietto, S., *AIChE Journal* 54(4) (**2008**), pp. 1009–1024.
24 Galvanin, F., Barolo, M., Bezzo, F., *Industrial and Engineering Chemistry Research* 48(9) (**2009**), pp. 4415–4427.
25 Rasch, A., Bucker, H. M., Bardow, A., *Computers and Chemical Engineering* 33(4) (**2009**), pp. 838–849.
26 Pronzato, L., Walter, E., *Mathematical Biosciences* 75(1) (**1985**), pp. 103–120.
27 Asprey, S. P., Macchietto, S., *Journal of Process Control* 12(4) (**2002**), pp. 545–556.

28 Yue, H., Brown, M., He, F., Jia, J. F., Kell, D. B., *International Journal of Chemical Kinetics* 40(11) (**2008**), pp. 730–741.
29 Chu, Y. F., Hahn, J., *AIChE Journal* 54(9) (**2008**), pp. 2310–2320.
30 Esposito, W. R., Floudas, C. A., *Industrial and Engineering Chemistry Research* 39 (**2000**), pp. 1291–1310.
31 Gau, C. Y., Stadtherr, M. A., *Computers and Chemical Engineering* 24 (**2000**), pp. 631–637.
32 Lin, Y., Stadtherr, M. A., *Industrial and Engineering Chemistry Research* 45 (**2006**), pp. 8438–8448.
33 Moles, C. G., Mendes P., Banga J. R., *Genome Research* 13(11) (**2003**), pp. 2467–2474.
34 Rodriguez-Fernandez, M., Egea, J. A., Banga, J. R., *BMC Bioinformatics* 7 (**2006**), p. 483.
35 Rodriguez-Fernandez, M., Mendes P., Banga J. R., *BioSystems* 83(2–3) (**2006**), pp. 248–265.
36 Balsa-Canto, E., Peifer, M., Banga, J., Timmer, J., Fleck, C., *BMC Systems Biology* 2 (**2008**), p. 26.
37 Zwolak, J., Tyson, J., Watson, L., *IEE Proceedings System Biology* 152 (**2005**), pp. 81–92.
38 Polisetty, P. K., Voit, E. O., Gatzke, E. P., *Theoretical Biology and Medical Modelling* 3 (**2006**), p. 4.
39 van Riel, N. A. W., *Briefings in Bioinformatics* 7(4) (**2006**), pp. 364–374.
40 Banga, J. R., Balsa-Canto, E., *Essays in Biochemistry* 45 (**2008**), pp. 195–210.
41 Kreutz, C., Timmer, J., *FEBS Journal* 276(4) (**2009**), pp. 923–942.
42 Galvanin, F., Barolo, M., Macchietto, S., Bezzo, F., *Industrial and Engineering Chemistry Research* 48(4) (**2009**), pp. 1989–2002.
43 Schenkendorf, R., Kremling, A., Mangold, M., *IET Systems Biology* 3(1) (**2009**), pp. 10–23.
44 Ashyraliyev, M., Fomekong-Nanfack, Y., Kaandorp, J. A., Blom, J. G., *FEBS Journal* 276(4) (**2009**), pp. 886–902.
45 Versyck, K. J., Bernaerts, K., Geeraerd, A. H., Van Impe, J. F., *International Journal of Food Microbiology* 51(1) (**1999**), pp. 39–51.
46 Bernaerts, K., Smets, I., Gysemans, K., Cappuyns, A., Van Impe, J. F., in: *FOOD-SIM 2004*, DeJong, P., Verschueren, M. (eds.), NIZO Food Research, Wageningen, The Netherlands, **2004**, p. 69–74.
47 Balsa-Canto, E., Rodriguez-Fernandez, M., Banga, J. R., *Journal of Food Engineering* 82(2) (**2007**), pp. 178–188.
48 Rodriguez-Fernandez, M., Balsa-Canto, E., Egea, J. A., Banga, J. R., *Journal of Food Engineering* 83(3) (**2007**), pp. 374–383.
49 Scheerlinck, N., Berhane, N. H., Moles, C. G., Banga, J. R., Nicolaï, B. M., *Journal of Food Engineering* 84(2) (**2008**), pp. 297–306.
50 Balsa-Canto, E., Alonso, A. A., Banga, J. R., *Journal of Food Process Engineering* 31(2) (**2008**), pp. 186–206.
51 Baltes, M., Schneider, R., Sturm, C., Reuss, M., *Biotechnology Progress* 10(5) (**1994**), pp. 480–488.
52 Franceschini, G., Macchietto, S., *Industrial and Engineering Chemistry Research* 46(1) (**2007**), pp. 220–232.
53 Franceschini, G., Macchietto, S., *Industrial and Engineering Chemistry Research* 47(7) (**2008**), pp. 2331–2348.
54 Bock, H., in: *Modelling of Chemical Reaction Systems*, Ebert, K. H., Deuflhard, P., Jäger, W. (eds.), Springer, Berlin, **1981**, pp. 102–125.
55 Bock, H., in: *Numerical Treatment of Inverse Problems in Differential and Integral Equations*, Deuflhard, P., Hairer, E., (eds.), Birkhäuser, Basel, **1983**, pp. 95–121.
56 Peifer, M., Timmer, J., *IET Systems Biology* 1(2) (**2007**), pp. 78–88.
57 Walter, E., Pronzato, L., *Identification of Parametric Models from Experimental Data*, Springer, Masson, **1997**.
58 Chapman, M. J., Godfrey, K., Chappell, M. J., Evans, N. D., *Mathematical Biosciences* 183 (**2003**), pp. 1–14.
59 Xia, X., Moog, C. H. I., *IEEE Transactions on Automatic Control* 48(2) (**2003**), pp. 330–336.
60 Vajda, S., Godfrey, K., Rabitz, H., *Mathematical Biosciences* 93 (**1989**), pp. 217–248.
61 Ljung, L., Glad, T., *Automatica* 30(2) (**1994**), pp. 265–276.
62 Bellu, G., Saccomani, M. P., Audoly, S., D'Angio, L., *Computer Methods and Programs in Biomedicine* 88 (**2007**), pp. 52–61.
63 Pohjanpalo, H., *Mathematical Biosciences* 41(1–2) (**1978**), pp. 21–33.

64 WALTER, E., LECOURTIER, Y., *Mathematics and Computers in Simulation* 24 (**1982**), pp. 472–482.
65 SALTELLI, A., RATTO, M., ANDRES, T., CAMPOLONGO, F., CARIBONI, J., GATELLI, D., SAISANA, M., TARANTOLA, S., *Global Sensitivity Analysis: The Primer*, Wiley, New York, **2008**.
66 GARCIA, M. R., Identification and real time optimisation in the food processing and biotechnology industries, PhD Thesis, Applied Mathematics Department, University of Vigo, Spain, **2007**.
67 LEIS, J. R., KRAMER, M. A., *ACM Transactions on Mathematical Software* 14 (**1988**), pp. 61–67.
68 HINDMARSH, A. C., BROWN, P. N., GRANT, K. E., LEE, S. L., SERBAN, R., SHUMAKER, D. E., WOODWARD, C. S., *ACM Transactions on Mathematical Software* 31(3) (**2005**), pp. 363–396.
69 SEBER, G. A. F., WILD, C. J., *Nonlinear Regression*, Wiley Series in Probability and Mathematical Statistics, John Wiley & Sons, USA, **1989**.
70 SCHITTKOWSKI, K., *Numerical Data Fitting in Dynamical Systems*, Kluwer, Dordrecht, **2002**.
71 PARDALOS, P., ROMEIJNA, H., TUYB, H., *Journal of Computational and Applied Mathematics* 124 (**2000**), pp. 209–228.
72 PINTER, J., *Global Optimization in Action. Continuous and Lipschitz Optimization: Algorithms, Implementations and Applications*, Kluwer Academics, The Netherlands, **1996**.
73 FLOUDAS, C. A., *Deterministic Global Optimization: Theory, Methods and Applications*, Kluwer Academics, The Netherlands, **2000**.
74 DRÉO, J., PETROWSKI, A., TAILLARD, E., SIARRY, P., *Metaheuristics for Hard Optimization. Methods and Case Studies*, Springer, Berlin, **2006**.
75 KIRCKPATRICK, S., GELLAT C. D., VECCHI M. P., *Science* 220 (**1983**), pp. 671–680.
76 FOGEL, D. B., *Evolutionary Computation: Toward a New Philosophy of Machine Intelligence*, IEEE Press, New York, **2000**.
77 MENDES, P., KELL, D. B., *Bioinformatics* 14 (**1998**), p. 869.
78 STORN, R., PRICE, K., *Journal of Global Optimization* 11 (**1997**), pp. 341–359.
79 RUNARSSON, T., YAO, X., *IEEE Transactions on Evolutionary Computation* 564 (**2000**), pp. 284–294.
80 EGEA, J. A., RODRIGUEZ-FERNANDEZ, M., BANGA, J. R., MARTÍ, R., *Journal of Global Optimization* 37(3) (**2007**), pp. 481–503.
81 WHITING, R., *Critical Reviews in Food Science and Nutrition* 35(6) (**1995**), pp. 467–494.
82 VERSYCK, K., BERNAERTS, K., GEERAERD, A. H., VAN IMPE, J. F., *International Journal of Food Microbiology* 51 (**1999**), pp. 39–51.
83 GRIJSPEERDT, K., VANROLLEGHEM, P., *Food Microbiology* 16(6) (**1999**), pp. 593–605.
84 GEERAERD, C., HERREMANS H., VAN IMPE, J. F., *International Journal of Food Microbiology* 59(3) (**2000**), pp. 185–209.
85 DENNIS, J. E., GAY, D. M., WELSCH, R. E., *ACM Transactions on Mathematical Software* 7(3) (**1981**), pp. 348–368.
86 RINAS, U., EL-ENSHASY, H., EMMLER, M., HILLE, A., HEMPEL, D. C., HORN, H., *Chemical Engineering Science* 60 (**2005**), pp. 2729–2739.
87 ZNAD, H., BLAZEJ, M., BALES, V., MARKOS, J., *Chemical Papers – Chemicke Zvesti* 58(1) (**2003**), pp. 23–28.
88 RINAS, U., EL-ENSHASY, H., EMMLER, M., HILLE, A., HEMPEL, D. C., HORN, H., *Chemical Engineering Science* 60 (**2005**), pp. 2729–2739.

14
Multiscale Modeling of Transport Phenomena in Plant-Based Foods

Quang Tri Ho, Pieter Verboven, Bert E. Verlinden, Els Herremans, and Bart M. Nicolaï

Keywords

mass transport, aquaporins, symplast, Fick's law of diffusion, multiscale modeling, Michaelis–Menten kinetics

14.1
Introduction

Plants or plant parts such as fruit are intrinsically multiscale assemblies, containing nanoscale and microscale features [1]. Most plant parts are alive until further processing, i.e., they maintain metabolic processes in an attempt to preserve their natural state. Nature has designed their structure through evolution in such a way as to facilitate transport of water, respiratory gasses, and other chemical substances which serve as building blocks, energy source, and signaling molecules. Modeling of transport phenomena is essential to understand and predict the response of the plants or plant parts – further called plant materials – to external changes of the environment, and to optimize metabolic processes during further handling and processing.

Continuum models have been used widely to describe the transport phenomena in bulky plant organs [2–4]. However, contrary to typical engineering materials such as steel or brick, plant materials cannot be considered as a continuum material because of their cellular nature. A continuum model is phenomenological, and the corresponding material properties such as the diffusion properties [5] should be considered as *apparent* parameters. These parameters cannot only incorporate both actual physical material constants such as the thermodynamic properties of different components but also the microscale structure of the material. The relationship between the macroscopic apparent properties and the microscopic features is not understood well to date. As a consequence, the available continuum models have a limited range of validity.

A *multiscale modeling approach* is more appropriate than a continuum modeling approach to understand the relative importance of microscopic features on the overall transport phenomena within plant materials [6–8]. Multiscale models are basically a hierarchy of submodels which describe the transport behavior at different spatial scales in such a way that the submodels are interconnected. Multiscale modeling is of particular importance in biological materials, engineering applications because measurements of microscale properties are indirect, require theoretical interpretation, and often the ideal experiment simply cannot be performed due to technological limitations [9].

This chapter is devoted to the multiscale modeling paradigm for describing transport phenomena in plant materials used for food applications. We will start with defining the different length scales of plant materials. We will then apply the multiscale modeling paradigm to transport processes that occur at the different length scales inside the plant materials. As the resulting models cannot be solved analytically, we will discuss some numerical aspects of the solution. As an illustration, we will subsequently develop a multiscale model of gas exchange in fruit tissue. Finally, some conclusions and future perspectives will be given.

14.2
Length Scales of Biological Materials

The typical multiscale structure of plant materials is illustrated in Fig. 14.1 for pome fruit. The following spatial scales can be defined:

- Macroscale (10^{-3}–10^{-1} m): The macroscale addresses the biological material as a whole. At this scale the plant material is considered as a continuum, and may consist of different connected tissues, all with homogeneous properties [10–12]. Tissue is an intermediate building block and can be considered as a conglomerate of cells with similar biological function, features, and physical properties.
- Microscale (10^{-6}–10^{-3} m): At the microscale level, the actual topology of the biological tissue is considered, incorporating the layout of the intercellular space and individual cells as basic building blocks. Cells are defined as the fundamental, structural, and functional units or subunits of living organisms or biological materials.
- Nanoscale (10^{-9}–10^{-6} m): The subcellular components of the cell include cell walls, cell membranes, and organelles. Because of their size, their structure cannot be observed using ordinary light microscopy.

Fruit consists of different types of tissue. The skin is made up of an epidermis and hypodermis cell layer and provides a natural barrier between the fruit and the environment. It contains specialized structures for water and gas transport such as lenticels and stomata. The epidermis is covered with a cuticle consisting of a cutine and epicuticular wax layer, which protects the fruit against fungal attack and excessive water loss. The bulk of the fruit is cortex tissue which consists of relatively large polyhedrical parenchyma cells. The cell size is not uniform throughout the fruit and this causes differences in gas diffusion properties [11, 13, 14] and probably also in moisture transport properties and mechanical strength. The cortex tissue contains intercellular space; its porosity can be up to 25% in some apple cultivars such as Jonagold [15]. The histology of pear is similar to that of apple with the

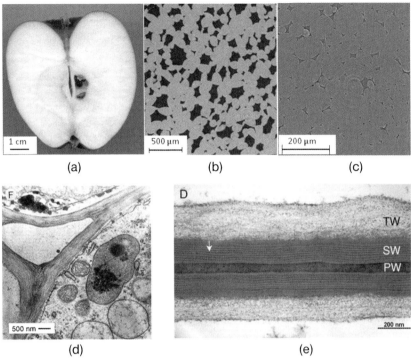

Fig. 14.1 Multiscale structure of plant tissue. (a) Cross-section image of intact apple fruit; (b) and (c). Tomographic images of the cortex of apple (b) and pear (c). Shown are slices obtained by absorption tomography at 5- and 1.4-μm pixel resolution for apple and pear. Dark regions are gas-filled intercellular spaces; light regions are cells [15] respectively. (d) Transmission electron micrograph of parenchymal cells and an interstitial gas space (from *Philodendron selloum*) [16]. (e) TEM micrograph of the suberized cell wall of wild-type peridermal cells (from potato tube *Solanum tuberosum 'Desirée'*). PW, Primary wall; SW, secondary suberin wall; TW, tertiary wall; white arrows, suberin lamellation [17]. (http://www.plantphysiol.org, Copyright American Society of Plant Biologists.)

exception of the presence of stone cells which are responsible for the often gritty texture of pear. Also, the porosity of some pear cultivars such as Conference (3–5%) is often much smaller compared to that of apple [14, 15]. Vascular bundles lead from the stem to the calyx around the core of the fruit and throughout the cortex and provide a major transport path for water loaded with minerals and metabolites. The core contains the seeds and may be empty in some cultivars.

The most basic structure and functional unit of fruit is the cell. Apple and pear parenchyma cells ($\varnothing$200–500 μm) [14] are essentially semipermeable phospholipid membranes ($d \cong 10$ nm) filled with cytoplasm. They are surrounded by a cell wall ($d \cong 1$ μm) thick [18] and consist of cellulose fibrils embedded in a pectin matrix. In between the cells there is a pectin layer ($d \cong 10$ nm) called the middle lamella, which glues the cells together. The most important cell organel is the vacuole that is surrounded by a phospholipid membrane and which typically covers almost the entire cell volume. The cytoplasm of neighboring cells is connected through *plas-*

modesmata – narrow channels that act as intercellular cytoplasmic bridges to facilitate communication and transport of nutrients between plant cells. Plasmodesmata (⌀25 nm) have a complicated ultrastructure, and the effective diameter for water transport is only about 3 nm [19]. The plasmatic membrane and also the membranes of all intracellular organels contain transporter proteins – the so-called *aquaporins* [20] – which serve as additional passive channels (⌀0.5 nm) for one-at-a-time transport of water molecules (⌀0.3 nm). Because of these transport structures, the cytoplasm of the apple cells form an interconnected spatial region called the *symplast*. The cell walls and the intercellular space form another region through which water can be transported and which is called the *apoplast* [21]. Epidermal cells are similar to parenchyma cells but are smaller, have a larger aspect ratio and more tightly packed. Vascular bundle cells are elongated and form a tubular network throughout the cortex.

14.3
Multiscale Modeling of Transport Phenomena

14.3.1
Mass Transport Fundamentals

Mass transport is associated with many biological processes. In biological systems, transport processes are complex and are driven by different mechanisms. For example, O_2 and CO_2 transport in plant tissue occurs mainly by diffusion through the intercellular space; transport of water and solutes from the intercellular space to the cell across the cell membrane is driven by passive diffusion, hydraulic, and osmotic transport of water, passive ion transport, and/or active ion transport. Transport processes are often multiphysics phenomena: for example, water transport is coupled to mechanical deformation of tissue.

From a physical and engineering point of view, transport phenomena follow the laws of thermodynamics and are driven by differences in thermodynamic potentials until equilibrium is reached. A simple example is that when a system contains two or more components whose concentration varies from point to point, there is a natural tendency for mass to be transferred, thereby minimizing the concentration difference within the system. Mass transport can be due to random molecular motion or by advection flow. The two distinct models of transport, molecular mass transport and advection mass transfer, are analogous to conduction and advection heat transfer [22].

Molecular diffusion can be defined as the transfer or movement of individual molecules by means of their random, individual movements and is described by Fick's law [23]. Fick's first law of diffusion states that the driving force leading to net molecular movement is the concentration gradient vector. This gradient indicates how a concentration C of a chemical species i changes with distance and is represented by ∇C_i. The diffusion flux $\mathbf{j}_d$ of a given species i is a vector quantity denoting the amount of species i crossing a certain area normal to the vector per

14.3 Multiscale Modeling of Transport Phenomena

unit time. The relation showing the dependence of the flux density on the driving force is given by [23]

$$\mathbf{j}_d = -D_i \nabla C_i \tag{14.1}$$

where D_i is the diffusivity of species i. For $\mathbf{j}_d$ in mol m^{-2} s^{-1} and C_i in mol m^{-3}, D_i has units of m^2 s^{-1}. The minus sign is due to the direction of net diffusion which is toward regions of lower concentration.

While Fick's law of diffusion is derived from the basic molecular mass transport, it has been applied often for describing mass transfer in biological and chemical processes. The diffusive medium is usually considered as a continuum [3, 4, 24–26] although the actual microscopic mass transfer processes are much more complex.

Gas exchange in tissue occurs mainly by means of intercellular spaces or pores. If the pore diameter is smaller than the mean free path of the diffusing gas, and the density of the gas is low, the gas molecules will collide with the pore walls more frequently than with each other. Then the diffusion is determined by the size of the pore instead of by the solvents or solutes. This process is known gas Knudsen flow or Knudsen diffusion. The Knudsen diffusion coefficient of gas i is a function of the pore diameter d and its properties [23]:

$$D_{K,i} = \frac{d}{3}\sqrt{\frac{8RT}{\pi M_i}} \tag{14.2}$$

where R and T are the gas constant (8.314 J mol^{-1} K^{-1}) and temperature (K), respectively, and M_i is the molar mass of gas i (kg mol^{-1}).

In macroscopic flow of a solvent characterized by a velocity vector $\mathbf{u}$, the total amount of species i crossing a certain area A normal to $\mathbf{u}$ is found by multiplying the concentration of species C_i by the fluid volume velocity ($A\mathbf{u}$). The corresponding flux is then

$$\mathbf{j}_c = C_i \mathbf{u} \tag{14.3}$$

The flux $\mathbf{j}_c$ (mol m^{-2} s^{-1}) is called the advection flux. Note that whereas diffusion flux is proportional to the concentration gradient, advection flux is proportional to concentration itself. For mass transfer in porous medium such as biological tissue, advection flux of macroscopic bulk flow through the tissue (permeation) by a pressure gradient can be described by Darcy's law for laminar flow in porous media

$$\mathbf{u} = -\frac{K}{\mu} \nabla P \tag{14.4}$$

with K (m^2) the permeation coefficient, P (Pa) the pressure, μ (Pa s) the viscosity of the gas, ∇ (m^{-1}) the gradient operator. For laminar flow in porous media such as tissue, permeation coefficients can be determined empirically and are usually considered to be independent of the gas passing the tissue [26].

14.3.2
Multiscale Transport Phenomena

Multiscale modeling is a fundamental paradigm to quantify complex biological phenomena by means of a mathematical description. This approach is aimed not only at providing comprehensive details of complex transport phenomena at the cellular level but also at describing the global transport at the level of the biological entity. We will consider both the macroscale and the microscale here; nanoscale transport requires atomistic transport approaches such as molecular dynamics and Monte Carlo methods which are beyond the scope of this chapter.

14.3.2.1 Macroscale Approach

In the macroscopic approach (the continuum model), it is assumed that the tissue is homogeneous and the modeling is carried out on the lumped properties of microstructure of biological materials including pores and cells [3, 4, 24–26]. Often volume-based conservation equations are derived [24–26], where the microscale description of gas transport for cells, pores can be replaced by averaging gas transport (representative elemental volume). For example, transport of a component i can be governed by the following equation:

$$\alpha_i \frac{\partial C_i}{\partial t} + \nabla \cdot (\mathbf{u} C_i) = \nabla \cdot D_i \nabla C_i + R_i \tag{14.5}$$

where C_i is the concentration of component i in macroscale (for example, gas concentration of tissue; water potential of plant tissue ...), α_i is the capacity of tissue, $\mathbf{u}$ is the effective permeation velocity vector, D_i is the effective diffusion coefficient, and R_i is the source term (production or consumption of component i due to metabolic processes). The capacity α_i represents a measure of the tissue's ability to store i in accordance to the external environment. For example, a typical representative elementary volume of tissue in biological materials contains cells (liquid) and pores (gas). The total amount of the gas i per unit volume (mol m^{-3}) of tissue in equilibrium with the gas phase of environment is always less than that of gas phase of the environment due to the low solubility of the gas in the liquid phase. The gas capacity is therefore the ratio of the equilibrium gas concentration of the tissue to the concentration of the gas phase derived from Henry's law [4, 26]. In the case of water transport, this term can be derived from the sorption isotherm curve which relates dry mass basic water content to water potential [29].

This continuum approach to mass transfer is a means to avoid the necessity of modeling the microscopic structure of the plant material. It constitutes a phenomenological approach as the parameters that appear in the macroscopic balances are effective or apparent properties which need to be determined experimentally or computed by the microscale model [10, 11, 13, 24, 30].

14.3.2.2 Microscale Approach

The microscopic approach recognizes the heterogeneous properties of the tissue, and the complex cellular structure is represented by a geometrical model [30, 31].

The microscale approach is often more complex due to different fluid phases and geometrical features in the usually quite heterogeneous medium.

Microscale modeling of transport requires geometrical information at the microscopic scale. In the traditional approach, microscale models can be constructed based on a distinct geometry model including pores and cells [30, 31]. The actual topology of the tissue incorporating geometrical details of the intercellular space and cell arrangement can be modeled and the diffusion equations can be discretized over this geometry. The transport of a component in the pore and cell can be modeled using diffusion laws while the relationship between the equilibrium concentrations in different phases can be assumed to be governed by the thermodynamic equilibrium laws such as Henry's law of gas exchange, and water potential equilibrium in water transport [29, 30]. The finite element or finite volume method can be used to solve the microscale model defined on the detailed geometry.

Another approach is the percolation network concept [32]. In this approach the model essentially consists of a network of nodes representing the pores that are connected by means of tubes of variable diameter and length which represent the intercellular channels. Every pore is also connected to a cell cluster via tubes representing the cellular membrane. Some tubes may be closed, and this naturally leads to clusters of interconnected cells without channels in between. It is further assumed that transport in every tube is partial pressure driven and governed by Fick's law. Each pipe is considered to have the same equivalent diameter D of the pore tube. If a flux quantity J_i of gas enters and leaves the pipe network, it is necessary to compute the gas nodal concentration and the volume flow rate in each pipe. The gas volume transport for an element is written as

$$J_i = -\frac{AD}{L}\Delta C \tag{14.6}$$

where L is the length of the pipe section; A the area of the pipe section, D (m^2 s^{-1}), the diffusion coefficient of component i in the intercellular channel, and ΔC (mol m^{-3}) is the concentration difference over the tube. In every node additional constraints are applied to satisfy the mass balances.

14.3.2.3 Kinetic Modeling

Many mass transport processes in plant materials are associated with (bio)chemical kinetics. For example, respiration and photosynthesis are associated with transport of respiratory gasses. The kinetic reactions must be modeled and lead to sources/sinks in the mass balance equations.

Enzyme kinetics are often used to model the complicated biochemical reactions taking place in the cell as a function of the concentration of the substrate S. The metabolic pathway is usually assumed to be linear and determined by one rate limited enzymatic reaction [33–35] so that Michaelis–Menten kinetics can be used to describe the reaction rate:

$$R_i = \frac{V_{\max} C_S}{K_M + C_S} \tag{14.7}$$

where V_{max} and K_M are the kinetic parameters which need to be determined experimentally; K_M is the substrate concentration required to obtain half of the maximum substrate conversion rate V_{max}. The effect of temperature on the maximum reaction rate is usually described by Arrhenius' law. Equation (14.7) describes many experimental results surprisingly well. The Michaelis–Menten model has been used as a semiempirical mode to describe enzyme kinetics at both the microscale and the macroscale.

14.3.2.4 Multiscale Model

In the multiscale approach, models that operate at the finer scale are incorporated into models which operate at the coarser scale through a procedure called *homogenization* or *upscaling*. Typically this is achieved by a numerical experiment that is carried out at a lower scale to derive apparent material properties. These apparent material properties are inserted into the coarser scale model. For example, a gradient in water potential or gas concentration can be applied to a virtual microstructure of tissue sample and the corresponding diffusion model is solved; the apparent diffusion coefficient of the sample is subsequently calculated from the computed flux. This diffusion coefficient can then be used in the macroscale diffusion model.

Homogenization essentially removes small scale details. On the other hand, *localization* is the *downscaling* procedure where the finer scale is analyzed from the computed results of coarser scale. For example, in the case of water or gas transport in fruit, first the gas or water distribution inside the fruit is calculated at the macroscale using the apparent transport properties derived from the homogenization procedure. Critical zones can then be identified as zones with high hygrostresses or extreme gas concentrations. In the subsequent localization procedure, the microscale model will be used to analyze the transport processes in these critical zones at cellular level.

14.4
Numerical Solution

Although it is possible to derive the governing partial differential equations and boundary conditions from first principles, it is difficult to obtain any form of analytical solution of multiscale models. The complexity is due to either the complexity of the geometry, or some other feature of the problem such as nonlinear parameters. The finite element/finite volume method is a numerical tool for determining approximate solutions to solve partial differential equations. These methods require a detailed geometrical description of the computational domain.

14.4.1
Geometrical Model

The construction of a macroscale geometrical model is usually based on the reconstruction of scanned images in the form of 3D points on the surface of pho-

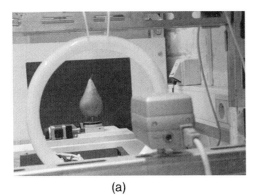 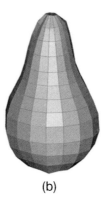

(a) (b)

Fig. 14.2 Construction of a geometrical model of an intact pear.
(a) Rotational table with pear fruit, lighting, and CCD camera.
(b) Reconstructured three-dimensional geometry of a pear.

tographs, video recordings, computed tomography (CT), or nuclear magnetic resonance (NMR) images [36]. In a typical photographic experiment of a plant material such as intact fruit, the object is placed on a rotating disk to get 2D snapshots differing by small angles [37, 38]. The contours of the fruit are then extracted from the images and used to reconstruct the three-dimensional shape of the fruit [38] (see Fig. 14.2). Internal features such as cavities and the core of the fruit can be constructed from X-ray CT or MRI scans. The reconstructed geometric models can easily be transferred to a finite element package for further numerical analysis [3, 4, 29].

The microscale geometry can be reconstructed from microscopy images (2D) [14, 39, 40] or X-ray computed tomography (3D) [18] of tissue samples. Mebatsion et al. [39, 40] digitized 2D microscopic images of apple cortex tissue using an image analysis software program. The geometrical cell structure was described by means of statistical distributions of the cell area, cell aspect ratio, and cell orientation. The cell orientation and cell aspect ratio were calculated based on the moments of inertia (area moments) and least-square ellipse fitting, respectively. Finally, geometrical models were generated by means of the Poisson–Voronoi tessellation algorithm or the ellipse tessellation algorithm to produce virtual cell tissue with the same statistical properties as the real tissue [39, 40] (see Fig. 14.3). The 3D microstructure and the connectivity of the pore space – important features that affect complex gas transport phenomena in plant materials – can be reconstructed using microfocus X-ray computed tomography (CT) technique [15, 41]. X-ray CT is a noninvasive image acquisition procedure that avoids cutting and fixation procedures and allows visualization and analysis of the architecture of cellular materials with an axial and lateral resolution down to a few micrometers [15, 41, 42]. The 3D pore space of plant tissue has been reconstructed successfully by means of X-ray CT [16, 41] (see Fig. 14.4). Reconstruction of 3D geometrical models for application in transport simulation based on X-ray images is challenging [36]. 3D skeletonization of the pore space in which the medial axis of an object is the skeleton of the void space of

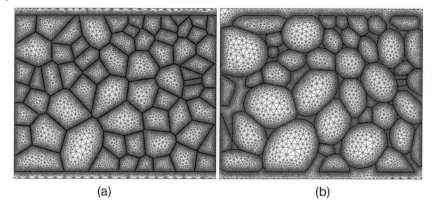

Fig. 14.3 Geometrical model and finite element meshes of Voronoi (a) and ellipse tessellations (b) generated from pear cortex tissue (adapted from [39, 40]).

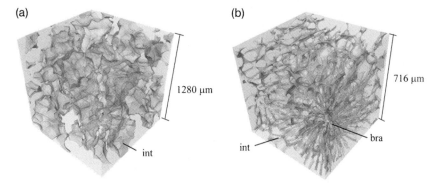

Fig. 14.4 3D rendering of the void network of apple (a) and pear (b) fruit cortex. The marked components are the intercellular void spaces (int) and the brachysclereids (bra). While the voids between apple parenchyma are large and form an incompletely connected network, those of pear are very small and form a complete network throughout the cortex sample without preferential direction [15]. (http://www.plantophysiol.org, Copyright American Society of Plant Biologists.)

an object running along its geometrical middle [39, 43, 44] has been beneficial for network modeling. Gas transport can be simulated through the skeleton of the connected pore network of the real samples or an artificially generated equivalent network having similar properties compared to the real samples.

14.4.2
Discretization

Various discretization methods have been used in the past for the numerical solution of heat and mass transfer problems arising in biological engineering. Among the most commonly used are the finite difference method, the finite element method, and the finite volume method. It must be emphasized that – particularly in the case of nonlinear mass transfer problems – the numerical solution must always be validated.

The finite element and finite volume method [45–47] require a geometrical description and is based on the discretization of the computational domain into many small, interconnected, subregions or elements and gives a piecewise approximation to the governing equations. That is, the complex partial differential equations are reduced to either linear or nonlinear simultaneous equations. Thus, discretization procedure reduces the continuum problem to one with a finite number of unknowns at specified points referred to as nodes. The number of nodes forms an element/control volume; the nature and number of unknowns at each node decide the variation of a field variable within the element/control volume. To solve the problem, the individual element equations are assembled into the overall system. That is the matrix equations of each element are combined in an appropriate way such that the resulting matrix represents the behavior of the entire solution region of the problem.

Several finite element packages are commercially available and they allow the user to focus on the geometry and physics of the problem to be solved (Comsol Multiphysics, http://www.comsol.com; Ansys Fluent and CFX, http://www.ansys.com). Several linear system solvers can be chosen to solve one or several systems of linear equations. Direct solvers in which the linear equation systems are solved by Gaussian elimination, are preferable for 1D and 2D models, and for 3D models with few degrees of freedom. For models with many degrees of freedom (roughly, more than 100,000), the direct solvers typically need too much memory. In those cases, the more memory-efficient iterative solvers perform better. However, iterative solvers are less stable than direct solvers in that they do not always converge. At the macroscale, the characteristics of lower scales are captured in an averaged sense, e.g., via homogenization procedures and the geometry is much simpler compared to the microscale model. Microscale approaches offer a more accurate representation, but need more details and, hence, a higher number of elements are typically required. Therefore, high-performance computation with large resources (memory, CPU time) is needed to solve such problems [48, 49].

Network models are based on nodes which are connected by means of tubes. Application of mass balances to such percolation networks results in algebraic equations, which can also be stacked to the linear system. Network models may be easier to solve than finite element/finite volume models since the number of unknown variables is usually less. However, they should be applied with caution since they are limited to transport in the network of the connected pores. When the transport phenomena occur simultaneously in both the intercellular space as well as in the cells, the finite element/finite volume modeling is more appropriate to get an accurate solution.

14.5
Case Study: Application of Multiscale Gas Exchange in Fruit

In this section, the multiscale paradigm will be illustrated based on a case study of gas exchange in fruit [4, 30, 50]. Pome fruits such as pears are typically stored under a controlled atmosphere with reduced O_2 and increased CO_2 levels to extend their commercial storage life, which can be as long as 9 months. The exact optimal gas conditions depend on factors such as cultivar, origin, growing conditions, and picking date of the fruit [51, 52]. At too low an oxygen concentration, anoxia may occur, eventually leading to cell death and loss of the product [52, 53]. Other fruit such as apples are considerably less sensitive to variations in low oxygen conditions. This is probably related to differences in gas concentration gradients resulting from differences in tissue diffusivity and respiratory activity [52], there is little information about such gas gradients in fruit. Knowledge on internal gas exchange would be, nevertheless, very valuable to guide commercial storage practices, since disorders under controlled atmosphere related to fermentation are a prime cause of concern. Multiscale modeling can help to better understand the processes of gas exchange and the kinetics of respiration associated with fruit storage potential and allows performing numerical experiments to determine optimal storage conditions [30, 50].

14.5.1
Macroscale Model

At the macroscopic spatial scale, the fruit tissue was considered as a continuum material. A permeation–diffusion–reaction model [4] was applied to study gas exchange due to respiration of intact fruit with the environment:

$$\alpha_i \frac{\partial C_i}{\partial t} + \nabla \cdot (\mathbf{u} C_i) = \nabla \cdot D_i \nabla C_i + R_i \tag{14.8}$$

and boundary conditions at the external surface of the pear:

$$C_i = C_{i,\infty} \tag{14.9}$$

In the equations, α_i is the gas capacity of the component i of the tissue, C_i (mol m^{-3}) is the concentration of i of tissue, D_i (m^2 s^{-1}) is the apparent diffusion coefficient, $\mathbf{u}$ (m s^{-1}) the apparent velocity vector, R_i (mol m^{-3} s^{-1}) the production term of the gas component i related to O_2 consumption or CO_2 production, ∇ (m^{-1}) the gradient operator, and t (s) the time. The index ∞ refers to the gas concentration of the ambient atmosphere. The first term in the equation represents the accumulation of gas i, the second term permeation transport driven by an overall pressure gradient, the third term molecular diffusion due to a partial pressure gradient, and the last term consumption or production of gas i because of respiration or fermentation. If, for example, oxygen is consumed in the center of the fruit, it creates a local partial pressure gradient which drives molecular diffusion. However, if the rates of transport of different gasses are different, overall pressure gradients may build up and cause permeation transport.

The gas capacity α_i of the component i of the tissue is defined from the following expression:

$$\alpha_i = \varepsilon + (1-\varepsilon) \cdot R \cdot T \cdot H_i = \frac{C_{i,\text{tissue}}}{C_{i,g}} \quad (14.10)$$

where ε is the porosity of tissue, $C_{i,g}$ (mol m^{-3}) and $C_{i,\text{tissue}}$ (mol m^{-3}) are the concentration of the gas component i in the gas phase and the tissue. The concentration of the compound in the liquid phase of fruit tissue normally follows Henry's law represented by constant H_i (mol m^{-3} kPa^{-1}). R (8.314 J mol^{-1} K^{-1}) is the universal gas constant and T (K) the temperature.

The extended Michaelis–Menten kinetics has been applied as a semiempirical model to describe the respiration characteristic of intact and tissue fruit [4, 33–35]. A noncompetitive inhibition model was used to describe O_2 consumption as formulated by Eq. (14.11)

$$R_{O_2} = -\frac{V_{m,O_2} \cdot C_{O_2}}{(K_{m,O_2} + C_{O_2}) \cdot (1 + \frac{C_{CO_2}}{K_{mn,CO_2}})} \quad (14.11)$$

with V_{m,O_2} (mol m^{-3} s^{-1}) is the maximum oxygen consumption rate, C_{O_2} (mol m^{-3}) the O_2 concentration, C_{CO_2} (mol m^{-3}) the CO_2 concentration, K_{m,O_2} (mol m^{-3}) the Michaelis–Menten constant for O_2 consumption, K_{mn,CO_2} (mol m^{-3}) the Michaelis–Menten constant for noncompetitive CO_2 inhibition, and R_{O_2} (mol m^{-3} s^{-1}) the O_2 consumption rate of the sample. The equation for production rate of CO_2 consists of an oxidative respiration part and a fermentative part:

$$R_{CO_2} = -r_{q,\text{ox}} \cdot R_{O_2} + \frac{V_{m,f,CO_2}}{(1 + \frac{C_{O_2}}{K_{m,f,O_2}})} \quad (14.12)$$

with V_{m,f,CO_2} (mol m^{-3} s^{-1}) the maximum fermentative CO_2 production rate, K_{m,f,O_2} (mol m^{-3}) the Michaelis–Menten constant of O_2 inhibition on fermentative CO_2 production, $r_{q,\text{ox}}$ the respiration quotient at high O_2 partial pressure, and R_{CO_2} (mol m^{-3} s^{-1}) the CO_2 production rate of the sample. The effect of temperature on the respiration rate was described by Arrhenius' law [34].

A typical simulation of macroscale gas exchange in pear tissue is shown in Fig. 14.5. Due to the respiration of the tissue, the O_2 partial pressure decreases from the surface to the center of the pear while the CO_2 partial pressure decreases in the opposite direction. The N_2 partial pressure increases from the surface to the center of the pear.

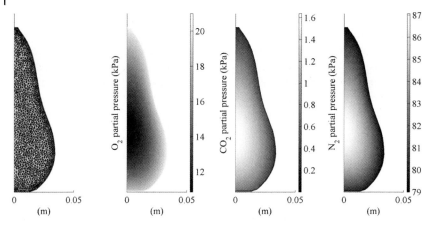

Fig. 14.5 Finite element mesh of pear geometry and simulated gas partial pressure distribution in pear intact fruit using permeation–diffusion–reaction model. Simulation was carried out at 0 °C, 21 kPa O_2, 0 kPa CO_2 at the ambient atmosphere and applied to an axisymmetrical pear shape. Parameters were taken from [4].

14.5.2
Microscale Model

The microscale model for gas exchange includes mass transport of the respiratory gases in the intercellular space (pore), the cell wall network, through the cell membrane into the cytoplasm. Respiration was taken into account. It was assumed that the size of the pores and channels connecting the pores (equivalent diameter of about 18 µm for pear, [15]) is large compared to the mean free path of molecular motions which is typically 0.07 µm for N_2 at 20 °C and 105 kPa [54]. Therefore, microscale diffusion was considered to be mainly through Fickian instead of Knudsen diffusion.

14.5.3
O_2 Transport Model

Microscale diffusion was assumed to dominate transport through each of the compartments and can be described by Fick's second law. Respiration is incorporated into the model as a source term:

$$\frac{\partial C_{O_2}}{\partial t} = \nabla \cdot D_{O_2} \nabla C_{O_2} + R_{O_2} \tag{14.13}$$

where C_{O_2} (mol m^{-3}) is the O_2 concentration in a certain phase, D_{O_2} (m^2 s^{-1}) is the O_2 diffusivity, and R_{O_2} (mol m^{-3} s^{-1}) is the O_2 consumption rate.

Redgwell et al. [55] assumed that the cell wall of fruit was a porous network of cellulose and hemicellulose, and swelling of the cell wall during fruit ripening was associated with movement of water into the voids of the cell wall network by solubilised pectin. In such case, the resistance would be high. However, no data are available yet to support this hypothesis. Here, the cell wall was assumed to be a porous material with air voids so that in the intercellular space and the cell wall network O_2 diffuses through the gas phase while in the cytoplasm through the liquid phase [30]. The relationship between the O_2 concentration in the gas phase $C_{O_2,g}$ and that in the liquid phase $C_{O_2,l}$ is given by Henry's law:

$$C_{O_2,l} = R \cdot T \cdot H_{O_2} \cdot C_{O_2,g} \tag{14.14}$$

with H_{O_2} (mol m^{-3} kPa^{-1}) the Henry's constant for O_2.

In the intercellular space and the cell wall network, the respiration R_{O_2} is zero while in the intracellular liquid phase this term is the consumption rate of O_2. Michaelis–Menten kinetics was used as a phenomenological model to describe the O_2 consumption rate of cell protoplasts [35]:

$$R_{O_2} = -\frac{V_{m,O_2} \cdot C_{O_2}}{K_{m,O_2} + C_{O_2}} \tag{14.15}$$

with V_{m,O_2} (mol m^{-3} s^{-1}) the maximal O_2 consumption rate in liquid phase, and K_{m,O_2} (mol m^{-3}) the Michaelis constant for O_2 consumption. The cell membrane is essentially a phospholipid bilayer. Passive gas transfer across the cell membrane occurs according to Fick's first law as a consequence of a concentration difference over the membrane. The flux J_{O_2} (mol m^{-2} s^{-1}) through the membrane was equal to

$$J_{O_2} = -h_{O_2,mem} \Delta C_{O_2} = -h_{O_2,mem}\left(C_{O_2} - C_{O_2}^*\right) \tag{14.16}$$

where $h_{O_2,mem}$ (m s^{-1}) is the O_2 permeability of the cell membrane, C_{O_2} (mol m^{-3}) is the equilibrium O_2 concentration in the liquid phase of the outer membrane.

14.5.4
CO$_2$ Transport Model (Lumped CO$_2$ Transport Model)

Transport of CO_2 was considered to be governed by diffusion through the intercellular space, the cell walls and the cytoplasm. For pores and cell walls, the following equation was used:

$$\frac{\partial C_{CO_2}}{\partial t} = \nabla \cdot D_{CO_2} \nabla C_{CO_2} \tag{14.17}$$

with C_{CO_2} (mol m^{-3}) the CO_2 concentration in a certain phase, D_{CO_2} (m^2 s^{-1}) the CO_2 diffusivity. Similar to O_2, transport of CO_2 in the cell wall was assumed to happen in the gas phase. The model for CO_2 transport in the cytoplasm was more complex because of the various equilibria of CO_2 in the liquid phase. The cytoplasmic pH of plant cells appears to be fairly constant and around 7 despite metabolic processes which generate or consume H$^+$ and despite the wide variation in the external pH [56, 57]. The pH in the vacuole is lower due to the large pool of organic acids. Therefore, dissociation of HCO_3^- to H^+ and CO_3^{2-} was neglected in both the cytoplasm and the vacuole. The diffusivity of H^+ is high (9.3×10^{-9} m² s⁻¹ [58]) so that within the cytoplasm or vacuole the pH can be assumed constant and uniform in this model. The model of CO_2 transport in the liquid phase was, therefore

$$\frac{\partial C_{CO_2}}{\partial t} = \nabla \cdot D_{CO_2} \nabla C_{CO_2} + R_{CO_2} - k_1 C_{CO_2} + k_2 \frac{[H]^+ C_{HCO_3^-}}{K} \tag{14.18}$$

$$\frac{\partial C_{HCO_3^-}}{\partial t} = \nabla \cdot D_{HCO_3^-} \nabla C_{HCO_3^-} + k_1 C_{CO_2} - k_2 \frac{[H]^+ C_{HCO_3^-}}{K} \tag{14.19}$$

with $C_{HCO_3^-}$ (mol m⁻³), $D_{HCO_3^-}$ (m² s⁻¹) the cytoplasmic concentration and diffusivity of HCO_3^-, R_{CO_2} (mol m⁻³ s⁻¹) the cytoplasmic CO_2 production rate, k_1 (s⁻¹) and k_2 (s⁻¹) the CO_2 hydration rate constant and H_2CO_3 dehydration rate constant, respectively. $[H]^+$ (mol L⁻¹) and K (mol L⁻¹) are the concentration of protons H^+ and the acid dissociation constant for H_2CO_3. The latter two terms of Eqs. (14.18) and (14.19) represent the forward and backward conversion rate of CO_2 to, HCO_3^- respectively. The equation for production rate of CO_2 in the cytoplasm accounts for both oxidative and fermentative respiration [33]:

$$R_{CO_2} = -r_{q,ox} \cdot R_{O_2} + \frac{V_{m,f,CO_2}}{(1 + \frac{C_{O_2}}{K_{m,f,O_2}})} \tag{14.20}$$

with V_{m,f,CO_2} (mol m⁻³ s⁻¹) the maximum fermentative CO_2 production rate of the intra-cellular liquid phase, K_{m,f,O_2} (mol m⁻³) the Michaelis–Menten constant of O_2 inhibition on fermentative CO_2 production, $r_{q,ox}$ the respiration quotient at high O_2 partial pressure (ratio between CO_2 production and O_2 consumption). The first term on the right-hand side indicates the oxidative CO_2 production rate due to consumption of O_2; the second term represents anoxic conditions in the cell where the oxidative respiration process is inhibited and replaced by a fermentation pathway.

Similar to O_2 transport, the flux J_{CO_2} (mol m⁻² s⁻¹) through the membrane was written as

$$J_{CO_2} = -h_{CO_2,mem} \Delta C_{CO_2} = -h_{CO_2,mem} (C_{CO_2} - C^*_{CO_2}) \tag{14.21}$$

where $h_{CO_2,mem}$ (m s⁻¹) is the CO_2 permeability of the cell membrane, $C^*_{CO_2}$ (mol m⁻³) is the equilibrium CO_2 concentration in the liquid phase of the outer membrane. The relationship between the equilibrium CO_2 concentration in the gas and liquid phase was again assumed to be described by Henry's law:

$$C_{CO_2,l} = R \cdot T \cdot H_{CO_2} \cdot C_{CO_2,g} \tag{14.22}$$

with H_{CO_2} (mol m⁻³ kPa⁻¹) Henry's constant for CO_2. Sources of the material properties and respiration parameters can be found in [30].

The microscale geometry model of tissue was constructed by the ellipse tessellation algorithm [39]. The generated geometric model was, then, meshed into 267,863 quadratic elements with triangular shape. The nonlinear coupled model equations from (14.13) to (14.22) were discretized over the mesh and solved using the finite element method (Comsol Multiphysics vs. 3.3, Comsol AB, Stockholm). The simultaneous gas exchange of O_2 and CO_2 coupled with respiration kinetics in a tissue sample of 4.19×10^{-4} m thickness [30] was shown in Fig. 14.6. In the simulation, the two opposite sides were applied with gas partial pressure differences ($\Delta O_2 = 20$ kPa, $\Delta CO_2 = 5$ kPa, respectively); while the other sides were assumed to be impermeable. The resulting 2D oxygen profile shows that the O_2 concentration is low inside the cells (Fig. 14.6(a)). There was a one order of magnitude difference in O_2 concentration between the gas and liquid phases. This is due to the fact that O_2 has a low solubility in the cell (Henry's constant for O_2 at $20\,°C$ is 1.37×10^{-2} mol m^{-3} kPa^{-1}) and the O_2 diffusivity in air is about 10^4 times that in water. The CO_2 concentration profiles are shown in Fig. 14.6(b). The results indicate a high CO_2 concentration in the cytoplasm, because CO_2 has high solubility in the cell (Henry's constant for CO_2 at $20\,°C$ is 3.876×10^{-1} mol m^{-3} kPa^{-1}). As a result, the CO_2 concentration in the gas and liquid phase has the same order of magnitude.

The apparent diffusivities of the microscopic samples were calculated. The mean values of apparent diffusivity and corresponding 95% confidence interval of nine different cellular structures in the pear cortex are given in Table 14.1. The resulting apparent O_2 and CO_2 diffusivities from microscale simulations are clearly very variable. The mean apparent diffusivity of O_2 at $20\,°C$ was $(3.54 \pm 0.68) \times 10^{-10}$ m^2 s^{-1}, while the experimental values were $(2.87 \pm 0.45) \times 10^{-10}$ m^2 s^{-1} as reported in [11]. The apparent CO_2 diffusivity of the tissue in both cases was almost the same. The estimated apparent diffusivity of CO_2 was $(3.13 \pm 0.59) \times 10^{-9}$ m^2 s^{-1} at $20\,°C$, while the experimental values were $(2.6 \pm 0.36) \times 10^{-9}$ m^2 s^{-1} as reported by [11]. The measured apparent CO_2 diffusivity was one order of magnitude larger than that of O_2. The microscale models confirm this observation. *In silico* analysis showed that temperature does not have much effect on the gas diffusivity. The mean apparent diffusivity of O_2 and CO_2 was $(3.54 \pm 0.68) \times 10^{-10}$ m^2 s^{-1} and $(3.13 \pm 0.59) \times 10^{-9}$ m^2 s^{-1} at $20\,°C$ while the tissue diffusivities were $(3.23 \pm 0.63) \times 10^{-10}$ m^2 s^{-1} and $(2.83 \pm 0.45) \times 10^{-9}$ m^2 s^{-1} at $0\,°C$. The predicted activation energy for O_2 and CO_2 diffusivity were (3.1 ± 7.4) kJ mol^{-1} and (3.2 ± 6.5) kJ mol^{-1} [30]. The effect of temperature on tissue diffusivity was not statistically significant due to the high variability of the tissue diffusivity.

14.6
Conclusions and Outlook

In this chapter the basic mathematical approaches for describing multiscale transport phenomena in plant materials have been outlined. Microscale models incorporating the microstructure geometry can provide detailed information about the

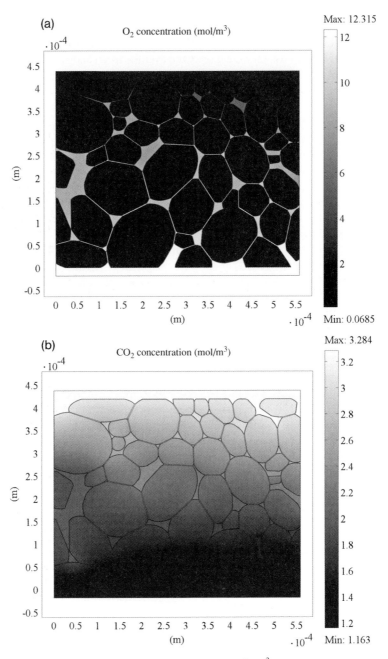

Fig. 14.6 Simulation respiratory gas concentration (mol m^{-3}) in pear cortex tissue at steady state, taking into account intracellular respiration ($\Delta O_2 = 20$ kPa, $\Delta CO_2 = 5$ kPa, $T = 20\,°C$) [28]. (a) O_2 concentration; (b) CO_2 concentration.

Table 14.1 Macroscopic apparent diffusivities of O_2 and CO_2 in pear parenchyma tissue[a].

	T (C)	$D_{O_2,\text{tissue}}$ (m² s⁻¹)	$D_{CO_2,\text{tissue}}$ (m² s⁻¹)
Microscale	20	$(3.54 \pm 0.68) \times 10^{-10}$ [30]	$(3.13 \pm 0.59) \times 10^{-9}$ [30]
	0	$(3.23 \pm 0.63) \times 10^{-10}$ [30]	$(2.83 \pm 0.45) \times 10^{-9}$ [30]
Measurement	20	$(2.87 \pm 0.45) \times 10^{-10}$ [11]	$(2.6 \pm 0.36) \times 10^{-9}$ [11]
	12	$(4.3 \pm 1.7) \times 10^{-10}$ [13]	$(1.73 \pm 1.15) \times 10^{-9}$ [13]
	20	$(5.63 \pm 3.09) \times 10^{-10}$ [30]	$(5.32 \pm 1.43) \times 10^{-9}$ [30]

±95% confidence limits.

a) The values reported for the microscale model are average values of nine different tissue geometries [30].

mechanisms of transport phenomena but require extensive computations; they typically cannot encompass the full computational domain as the resulting discretized model would be too large to solve. Macroscale models, on the other hand, provide macroscopic information about the transported quantities within the full computational domain but do not provide microscopic detail. The multiscale paradigm combines both approaches through homogenization (upscaling, or computation of the apparent material properties in the macroscale model from numerical experiments at the microscale) and localization (downscaling, or computation of microscale transport phenomena in regions of interested identified through the macroscale model), and, hence, bridges the gap between length scales.

While gas transport in plant materials is now understood relatively well, much more research is required with respect to modeling water transport in plant materials. Water transport in plant materials is a multiphysics problem as it is coupled to mechanical deformation due to hygrostresses. It is very well possible that the typical viscoelastic behavior at the macroscopic scale is due to more simple constitutive laws at the microscale through this coupling, but this remains to be shown.

The range of scales that are relevant for transport phenomena in plant materials extends from about 10^{-9} m (the *nanoscale*) to 10^{-1} m (the *macroscale*). Transport through nanoscale features such as cell membranes, plasmodesmata (tiny structures which control component transport from one cell to its neighbors), and aquaporins (water transporter proteins) is governed by atomistic models and has not been considered in this chapter. Such features certainly have an effect on microscale transport phenomena and pose a challenging problem yet to be solved.

Multiscale models are now increasingly being used to understand the fundamental processes in plant materials during food processing operations. More research is required before such models can actually be used for process design purposes, such as the design of fruit storage or drying processes.

Acknowledgements

The authors wish to thank the Research Council of the K.U. Leuven (OT 08/023), the Flanders Fund for Scientific Research (project G.0603.08), and the Institute for the Promotion of Innovation by Science and Technology in Flanders (project IWT-050633) for financial support. Quang Tri Ho is postdoctoral fellow of the Research Council of the K.U. Leuven.

List of Symbols

A – the area of the pipe section (m^2)
C_{CO_2} – CO$_2$ concentration (mol m^{-3})
$C^*_{CO_2}$ – equilibrium CO$_2$ concentration in liquid phase of outer membrane (mol m^{-3})
$C_{HCO_3^-}$ – HCO$_3^-$ concentration (mol m^{-3})
C_i – concentration of gas i (mol m^{-3})
C_{O_2} – O$_2$ concentration (mol m^{-3})
$C^*_{O_2}$ – equilibrium O$_2$ concentration in liquid phase of outer membrane (mol m^{-3})
C_s – the substrate concentration (mol m^{-3})
D_{CO_2} – CO$_2$ diffusivity (m^2 s^{-1})
$D_{CO_2,tissue}$ – CO$_2$ apparent diffusivity of tissue (m^2 s^{-1})
$D_{HCO_3^-}$ – HCO$_3^-$ diffusivity (m^2 s^{-1})
D_i – macroscopic apparent diffusivity of gas i (m^2 s^{-1})
$D_{K,i}$ – Knudsen diffusivity of gas i (m^2 s^{-1})
D_{O_2} – O$_2$ diffusivity (m^2 s^{-1})
$D_{O_2,mem}$ – O$_2$ diffusivity through cell membrane (m^2 s^{-1})
d – the pore diameter (m)
$h_{CO_2,mem}$ – CO$_2$ permeability through cell membrane (m s^{-1})
$h_{O_2,mem}$ – O$_2$ permeability through cell membrane (m s^{-1})
H_{CO_2} – Henry's constant for CO$_2$ (mol m^{-3} kPa^{-1})
H_{O_2} – Henry's constant for O$_2$ (mol m^{-3} kPa^{-1})
$[H^+]$ – H$^+$ concentration (mol L^{-1})
j_d – diffusion flux (mol m^{-2} s^{-1})
J_{CO_2} – the flux of CO$_2$ (mol m^{-2} s^{-1})
J_{O_2} – the flux of O$_2$ (mol m^{-2} s^{-1})
k_1 – CO$_2$ hydration velocity constant (s^{-1})
k_2 – CO$_2$ dehydration velocity constant (s^{-1})
K – permeation coefficient (m^2)
K_a – acid dissociation constant for H$_2$CO$_3$ (mol L^{-1})
K_{m,f,O_2} – Michaelis–Menten constant of O$_2$ inhibition on fermentative CO$_2$ production (mol m^{-3})
K_{mn,CO_2} – Michaelis–Menten constant for noncompetitive CO$_2$ inhibition (mol m^{-3})

K_{m,O_2} – Michaelis–Menten constant for O_2 consumption (mol m^{-3})
K_m – Michaelis–Menten constant (mol m^{-3})
L – the length of the pipe section (m)
M_i – the molar mass of gas i (kg mol^{-1})
P – total pressure (kPa)
$r_{q,\text{ox}}$ – respiration quotient
R – universal gas constant (8.314 J mol^{-1} K^{-1})
R_i – effective respiration term of the tissue (mol m^{-3} s^{-1})
R_{O_2} – O_2 consumption rate (mol m^{-3} s^{-1})
R_{CO_2} – CO_2 consumption rate (mol m^{-3} s^{-1})
S – source term
T – temperature (K)
T_{ref} – reference temperature (K)
t – time (s)
$\mathbf{u}$ – velocity vector (m s^{-1})
V_{m,f,CO_2} – maximal fermentative CO_2 production rate (mol m^{-3} s^{-1})
V_{m,O_2} – maximal O_2 consumption rate (mol m^{-3} s^{-1})

Greek symbols

α_i – gas capacity of the component i
μ – viscosity (Pa s)
ε – porosity of tissue
∇ – gradient operator (m^{-1})
Δ – difference

Subcripts

g – gas phase
i – O_2, CO_2, or N_2
l – liquid phase
∞ – ambient atmosphere

References

1 KLEIN, M. L., SHINODA, W., Large-scale molecular dynamics simulations of self-assembling systems, *Science* 321 (**2008**), pp. 798–800.
2 MANNAPPERUMA, J. D., SINGH, R. P., MONTERO, M. E., Simultaneous gas diffusion and chemical reaction in foods stored in modified atmospheres, *Journal of Food Engineering* 14 (**1991**), pp. 167–183.
3 LAMMERTYN, J., SCHEERLINCK, N., JANCSÓK, P., VERLINDEN, B. E., NICOLAÏ, B. M., A respiration–diffusion model for 'Conference' pears I: model development and validation, *Postharvest Biology and Technology* 30 (**2003**), pp. 29–42.
4 HO, Q. T., VERBOVEN, P., VERLINDEN, B. E., LAMMERTYN, J., VANDEWALLE, S., NICOLAÏ, B. M., A continuum model for gas ex-

change in pear fruit, *PLoS Computational Biology* 4(3) (**2008**), p. e1000023.

5 NESVADBA, P., HOUSKA, M., WOLF, W., GEKAS, V., JARVIS, D., SADD, P. A., JOHNS, A. I., Database of physical properties of agro-food materials, *Journal of Food Engineering* 61(4) (**2004**), pp. 497–503.

6 LÜDING, S., From microscopic simulations to macroscopic material behavior, *Computer Physics Communications* 147 (**2002**), pp. 134–140.

7 CHANG, C. S., WANG, T. K., SLUYS, L. J., van MIER, J. G. M., Fracture modelling using a micro-structural mechanics approach – I. Theory and formulation, *Engineering Fracture Mechanics* 69(17) (**2002**), pp. 1941–1958.

8 HAYES, D. M., BAXTER, J., TUZUN, U., QIN, R. S., Discrete-element method simulations: from micro to macro scales, *Philosophical Transactions – Royal Society, Mathematical, Physical and Engineering Sciences* 362(1822) (**2004**), pp. 1853–1865.

9 CARTER, E. A., Challenges in modelling materials properties without experimental input, *Science* 321 (**2008**), pp. 800–803.

10 LAMMERTYN, J., SCHEERLINCK, N., VERLINDEN, B. E., SCHOTSMANS, W., NICOLAÏ, B. M., Simultaneous determination of oxygen diffusivity and respiration in pear skin and tissue, *Postharvest Biology and Technology* 23 (**2001**), pp. 93–104.

11 HO, Q. T., VERLINDEN, B. E., VERBOVEN, P., NICOLAÏ, B. M., Gas diffusion properties at different positions in the pear, *Postharvest Biology and Technology* 41 (**2006**), pp. 113–120.

12 MEBATSION, H. K., VERBOVEN, P., HO, Q. T., VERLINDEN, B. E., NICOLAÏ, B. M., Modelling fruit microstructures, why and how?, *Trends in Food Science and Technology* 19(2) (**2008**), pp. 59–66.

13 SCHOTSMANS, W., VERLINDEN, B. E., LAMMERTYN, J., NICOLAÏ, B. M., Simultaneous measurement of oxygen and carbon dioxide diffusivity in pear fruit tissue, *Postharvest Biology and Technology* 29(2) (**2003**), pp. 155–166.

14 SCHOTSMANS, W., VERLINDEN, B. E., LAMMERTYN, J., NICOLAÏ, B. M., The relationship between gas transport properties and the histology of apple, *Journal of the Science of Food and Agriculture* 84 (**2004**), pp. 1131–1140.

15 VERBOVEN, P., KERCKHOFS, G., MEBATSION, H. K., HO, Q. T., TEMST, K., WEVERS, M., CLOETENS, P., NICOLAÏ, B. M., 3-D gas exchange pathways in pome fruit characterised by synchrotron X-ray computed tomography, *Plant Physiology* 147 (**2008**), pp. 518–527.

16 SEYMOUR, R. S., Diffusion pathway for oxygen into highly thermogenic florets of the arum lily *Philodendron selloum*, *Journal of Experimental Botany* 52 (**2001**), pp. 1465–1472.

17 SERRA, O., SOLER, M., HOHN, C., SAUVEPLANE, V., PINOT, F., FRANKE, R., SCHREIBER, L., PRAT, S., MOLINAS, M., FIGUERAS, M., CYP86A33 – targeted gene silencing in potato tuber alters suberin composition, distorts suberin lamellae, and impairs the periderm's water barrier function, *Plant Physiology* 149 (**2009**), pp. 1050–1060.

18 MEBATSION, H. K., VERBOVEN, P., ENDALEW, A. M., BILLEN, J., HO, Q. T., NICOLAÏ, B. M., A novel method for 3D microstructure modelling of pome fruit tissue using synchrotron radiation tomography images, *Journal of Food Engineering* 93 (**2009**), pp. 141–148.

19 ROBERTS, A. G., OPARKA, K. J., Plasmodesmata and the control of symplastic transport, *Plant, Cell and Environment* 26(1) (**2003**), pp. 103–124.

20 TYERMAN, S. D., NIEMIETZ, C. M., BRAMLEY, H., Plant aquaporins: multifunctional water and solute channels with expanding roles, *Plant, Cell and Environment* 25(2) (**2002**), pp. 173–194.

21 NOBEL, P. S. (ed.), *Physicochemical and Environmental Plant Physiology*, Academic Press, San Diego, **1991**.

22 WELTY, J. R., WICKS, C. E., WILSON, R. E., RORRER, G. (eds.), *Fundamentals of Momentum, Heat, and Mass Transfer*, John Wiley and Sons, Ltd., New York, **2001**.

23 GEANKOPLIS, J. C. (ed.), *Transport Processes and Unit Operations*, Prentice-Hall, Englewood Cliffs, NJ, **1993**.

24 NGUYEN, T. A., VERBOVEN, P., SCHEERLINCK, N., NICOLAÏ, B. M., Estimation of effective diffusivity in pear tissue and cuticle by means of a numerical water diffusion model, *Journal of Food Engineering* 72(1) (**2006**), 63–72.

25 Mannapperuma, J. D., Singh, R. P., Montero, M. E., Simultaneous gas diffusion and chemical reaction in foods stored in modified atmospheres, *Journal of Food Engineering* 14 **(1991)**, pp. 167–183.
26 Ho, Q. T., Verlinden, B. E., Verboven, P., Vandewalle, S., Nicolaï, B. M., A permeation–diffusion–reaction model of gas transport in cellular tissue of plant materials, *Journal of Experimental Botany* 57 **(2006)**, pp. 4215–4224.
27 Wood, B. D., Quintard, M., Whitaker, S., Calculation of effective diffusivities for biofilms and tissues, *Biotechnology and Bioengineering* 77(5) **(2002)**, pp. 495–516.
28 Helmis, R., Miller, C. T., Jakobs, H., Class, H., Hilpert, M., Kees, C., Niessner, J., Multiphase flow and transport modelling in heterogeneous porous media, in: *Progress in Industrial Mathematics at ECMI 2004*, Bucchianico, A. D., Mattheij, R. M. M., Peletier, M. A. (eds.), Springer, Berlin, Heidelberg, 2006.
29 Nguyen, T. A., Dresselaers, T., Verboven, P., D'hallewin, G., Culeddu, N., Van Hecke, P., Nicolaï, B. M., Finite element modelling and MRI validation of 3D transient water profiles in pears during postharvest storage, *Journal of the Science of Food and Agriculture* 86 **(2006)**, pp. 745–756.
30 Ho, Q. T., Verboven, P., Mebatsion, H. K., Verlinden, B. E., Vandewalle, S., Nicolaï, B. M., Microscale mechanisms of gas exchange in fruit tissue, *New Phytologist* 182 **(2009)**, pp. 163–174.
31 Aalto, T., Juurola, E., A three-dimensional model of CO_2 transport in airspaces and mesophyll cells of a silver birch leaf, *Plant, Cell and Environment* 25 **(2002)**, pp. 1399–1409.
32 Levitz, P., Toolbox for 3D imaging and modelling of porous media: relationship with transport properties, *Cement and Concrete Research* 37 **(2007)**, pp. 351–359.
33 Peppelenbos, H. W., van't Leven, J., Evaluation of four types of inhibition for modelling the influence of carbon dioxide on oxygen consumption of fruits and vegetables, *Postharvest Biology and Technology* 7 **(1996)**, pp. 27–40.
34 Hertog, M. L. A. T. M., Peppelenbos, H. W., Evelo, R. G., Tijskens, L. M. M., A dynamic and generic model on the gas exchange of respiring produce: the effects of oxygen, carbon dioxide and temperature, *Postharvest Biology and Technology* 14 **(1998)**, pp. 335–349.
35 Lammertyn, J., Franck, C., Verlinden, B. E., Nicolaï, B. M., Comparative study of the O_2, CO_2 and temperature effect on respiration between 'Conference' pear cells in suspension and intact pears, *Journal of Experimental Botany* 52 **(2001)**, pp. 1769–1777.
36 Mebatsion, H. K., Verboven, P., Ho, Q. T., Verlinden, B. E., Nicolaï, B. M., Modelling fruit micro structures, why and how?, *Trends in Food Science and Technology* 19 **(2008)**, pp. 59–66.
37 Moustakides, G., Briassoulis, D., Psarakis, E., Dimas, E., 3D image acquisition and NURBS based geometry modelling of natural objects, *Advances in Engineering Software* 31(12) **(2000)**, pp. 955–969.
38 Jancsók, P., Clijmans, L., Nicolaï, B. M., De Baerdemaeker, J., Investigation of the effect of shape on the acoustic response of Conference pears by finite element modelling, *Postharvest Biology and Technology* 23 **(2001)**, pp. 1–12.
39 Mebatsion, H. K., Verboven, P., Ho, Q. T., Mendoza, F., Verlinden, B. E., Nguyen, T. A., Nicolaï, B. M., Modelling fruit microstructure using novel ellipse tessellation algorithm, *Computer Modelling in Engineering and Science* 14(1) **(2006)**, pp. 1–14.
40 Mebatsion, H. K., Verboven, P., Verlinden, B. E., Ho, Q. T., Nguyen, T. A., Nicolaï, B. M., Microscale modelling of fruit tissue using Voronoi tessellations, *Computers and Electronics in Agriculture* 52 **(2006)**, pp. 36–48.
41 Mendoza, F., Verboven, P., Mebatsion, H. K., Kerckhofs, G., Wevers, M., Nicolaï, B. M., Three-dimensional pore space quantification of apple tissue using X-ray computed microtomography, *Planta* 226 **(2007)**, pp. 559–570.
42 Cloetens, P., Mache, R., Schlenker, M., Mach, S. L., Quantitative phase tomography of arabidopsis seeds reveals intercellular void network, *Proceedings of National Academy of Sciences of the United States of America* 103 **(2006)**, pp. 14626–14630.

43 Thovert, J., Salles, J., Adler, P., Computerised characterisation of the geometry of real porous media: their discretisation, analysis and interpretation, *Journal of Microscopy* 170 (**1993**), pp. 65–79.

44 Lindquist, W. D., Lee, S. M., Coker, D., Jones, K., Spanne, P., Medial axis analysis of three dimensional tomographic images of drill core samples, *Journal of Geophysical Research* 101B (**1996**), pp. 8297–8310.

45 Segerlind, L. (ed.), *Applied Finite Element Analysis*, John Wiley and Sons, Ltd., New York, **1984**.

46 Versteeg, H. K., Malalasekera, W. (eds.), *An Introduction to Computational Fluid Dynamics, the Finite Volume Method*, Longman Scientific & Technical, Harlow, **1995**.

47 Lewis, R. W., Nithiarasu, P., Seetharamu, K. N. (eds.), *Fundamentals of the Finite Element Method for Heat and Fluid Flow*, John Wiley and Sons, Ltd., Chichester, **2004**.

48 Sanz-Herrera, J. A., Garcia-Aznar, J. M., Doblaré, M., Micro–macro numerical modelling of bone regeneration in tissue engineering, *Computer Methods in Applied Mechanics and Engineering* 197 (**2008**), pp. 3092–3107.

49 Lacroix, D., Planell, J. A., Prendergast, P. J., Computer-aided design and finite-element modelling of biomaterial scaffolds for bone tissue engineering, *Philosophical Transactions of The Royal Society A – Mathematical Physical and Engineering Sciences* 367 (**2009**), pp. 1993–2009.

50 Ho, Q. T., Multiscale modelling of gas transport in fruit during controlled atmosphere storage, K.U. Leuven University Dissertation, **2009**.

51 Lammertyn, J., Aerts, M., Verlinden, B. E., Schotsmans, W., Nicolaï, B. M., Logistic regression analysis of factors influencing core breakdown in 'Conference' pears, *Postharvest Biology and Technology* 20 (**2000**), pp. 25–37.

52 Franck, C., Lammertyn, J., Ho, Q. T., Verboven, P., Verlinden, B. E., Nicolaï, B. M., Browning disorders in pear: a review, *Postharvest Biology and Technology* 43 (**2007**), pp. 1–13.

53 Veltman, R. H., Lenthric, I., Van der Plas, L. H. W., Peppelenbos, H. W., Browning in pear fruit (*Pyrus communis* L. cv Conference) may be a result of a limited availability of energy and antioxidants, *Postharvest Biology and Technology* 28 (**2003**), pp. 295–302.

54 Leuning, R., Transport of gas into leaves, *Plant, Cell and Environment* 6 (**1983**), pp. 181–194.

55 Redgwell, R. J., MacRae, E., Hallett, I., Fischer, M., Perry, J., Harker, R., In vivo and in vitro swelling of cell walls during fruit ripening, *Planta* 203 (**1997**), pp. 162–173.

56 Smith, F. A., Raven, J. A., Intracellular pH and its regulation, *Annual Review of Plant Physiology* 30 (**1979**), pp. 289–311.

57 Roberts, J. K. M., Wemmer, D., Ray, P. M., Jardetzky, O., Regulation of cytoplasmic and vacuolar pH in maise root tips under different experimental conditions, *Plant Physiology* 69 (**1982**), pp. 1344–1347.

58 Moore, W. J. (ed.), *Electrochemistry: Conductance and Ionic Reactions*, in: *Physical Chemistry*, Prentice-Hall, London, **1962**.

15
Synthetic Biology: Dynamic Modeling and Construction of Cell Systems
Tatiana T. Marquez-Lago and Mario Andrea Marchisio

Keywords
synthetic biology, biological parts and pools, deterministic and stochastic modeling

15.1
Introduction

In "traditional" biology, systems are generally studied using a descriptive approach by being illustrated in a rather qualitative way, such as cartoon models. In response to this, Systems Biology has introduced physical concepts and mathematical instruments into the modeling of biological systems, in order to provide for a quantitative description, aimed at a better understanding of the underlying fundamental processes. In this perspective, Synthetic Biology represents a step further from systems biology: it relies on the same mathematical framework but is focused on the *de novo* engineering of life systems.

So far, synthetic biology experimental applications have been implemented mostly in prokaryotic cells and have the structure of rather simple *genetic circuits* [1, 2] whose working relies on two main processes: DNA *transcription* into mRNA and mRNA *translation* into proteins. In the process of transcription, RNA polymerases bind to a specific DNA chain and copy it into an mRNA sequence. In turn, ribosomes bind to the mRNA, starting the process of translation, where they convert each mRNA nucleotide triplet into an amino acid, giving rise to a protein. Furthermore, synthetic gene circuits exploit various ways to regulate transcription and translation, in order to reproduce a well-defined behavior, such as oscillations in the concentration of their final products. For instance, some proteins, the *transcription factors*, can be used either to block the access to DNA or, on the contrary, to recruit directly RNA polymerases and start transcription in response to the presence or the absence of particular *chemicals*.

Potential benefits from synthetic biology might be achieved in a wide range of fields: from cancer treatment [3] and tissue engineering [4] to clean fuel production [5], and waste/mines recognition [6]. Hence, methods and tools for the computer-aided *design* of more complex gene networks and algorithms for their numerical *simulations* are essential for the progress of Synthetic Biology.

Process Systems Engineering: Vol. 7 Dynamic Process Modeling
Edited by Michael C. Georgiadis, Julio R. Banga, and Efstratios N. Pistikopoulos
Copyright © 2011 WILEY-VCH Verlag GmbH & Co. KGaA, Weinheim
ISBN: 978-3-527-31696-0

This chapter focuses on some recent achievements for *in silico* synthetic biology. Section 15.2 provides the reader with an introduction to a novel approach for the computational design of synthetic gene circuits based on the concepts of biological parts and pools. On the other hand, Section 15.3 provides the reader with guidelines for adopting certain modeling regimes, as well as computational strategies for the simulation of biological systems with an emphasis on available stochastic algorithms. Finally, Section 15.4 illustrates the design and modeling of the Repressilator circuit [7] by means of biological parts and pools, and highlight the difference of results stemming from using a deterministic or stochastic modeling and simulation approach.

15.2
Constructing a Model with Parts

In electronics, circuits are constructed from *standard* basic parts such as batteries, resistors, capacitors, and soleinoids. These parts can be assembled into more complex components – *devices* like diodes and transistors – and circuits (*abstraction hierarchy*) that represent the connections among all parts and devices. The exchange of information between different circuit components is mediated by the electrons that represent the only input/output of every part and device. The feature of sharing the same input and the same output (*composability*) allows for connecting, inside a circuit, any two parts through wires where the electrons are able to flow.

Inspired by electronic circuits, a general method for modeling and building synthetic gene circuits can be derived by the following three main concepts of electrical engineering:

- part standardization;
- abstraction hierarchy; and
- part composability.

For such a condition, synthetic biology needs (1) the definition of a set of standard circuit components, which can be put together into devices associated with new functionalities, and (2) *common signal carriers*, that is, molecules that lead the communication among the parts and represent the biological counterpart of the electrons.

15.2.1
General Nomenclature

15.2.1.1 Parts and Devices
We will adopt the Standard Biological Parts classification of the MIT Registry of standard biological parts (http://partsregistry.org), depicted in Fig. 15.1. This clas-

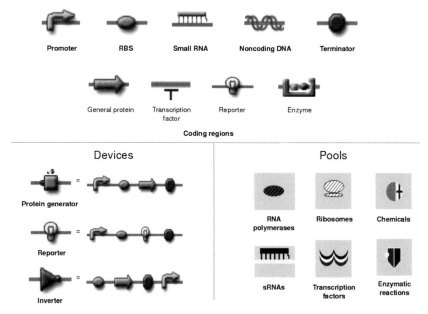

Fig. 15.1 Symbols for standard biological parts, pools, and devices. Each device is also accompanied by a scheme of its internal structure.

sification holds for bacterial cells, and different parts are defined as *DNA traits* characterized by specific functions. They are grouped into the following categories:

- promoters;
- ribosome-binding sites (RBS);
- coding regions;
- noncoding DNAs;
- small RNAs (sRNA), and
- terminators.

All the necessary information to synthesize (at least) a single protein is contained in a *transcription unit*. This portion of DNA is leaded by a *promoter*, the part of the sequence where RNA polymerases bind to initiate transcription. Specific sites (*operators*) recognized by transcription factors lie on the promoter as well, the occupancy of which can prevent or facilitate RNA polymerase binding.

Beside the promoter, we generally observe an *RBS* part, sitting on the DNA. Here, RNA polymerases start the DNA transcription and produce the initial, small piece of mRNA that contains, in bacteria, the so-called *Shine–Dalgarno* sequence. This very short sequence is detected by ribosomes and is essential for their binding to the mRNA.

An RBS is followed by a *coding region*, commonly referred to as gene. This is the longest part and carries all the essential information to produce a protein. Interestingly, the coding region includes "punctuation signs," such as the STOP codon, a sequence of three nucleotides that signals to the ribosomes to finish translating and to leave the mRNA. Similarly, on the DNA, a transcription unit shows a *terminator* sequence encountered by RNA polymerases, on which they release the mRNA product and detach from the DNA double chain.

However, it should be noted that bacteria can have *multicistronic sequences* that are delimited, again, by a promoter and a terminator, but encode for more than one protein. The encoding sequences (cistrons) are separated by *noncoding DNAs* that are transcribed, into the so-called spacers, but are not translated. Moreover, a transcription unit is able to produce an sRNA instead of a protein, in which case, the RBS and the coding region are replaced by an *sRNA* part, as is explained later.

As it is clear from the above description, the simplest device that can be constructed with standard biological parts is a transcription unit for the production of specific proteins, such as fluorescent reporters. Another commonly reported example is the *Inverter* (Fig. 15.1). This device is made of the same four parts that are normally present in a transcription unit. However, they are placed in a different order: RBS, coding region, terminator, and promoter. The coding region produces a *repressor*, which is a transcription factor that binds the Inverter promoter and blocks the access to the RNA polymerases. Hence, when the Inverter receives a high input signal, the Inverter promoter is fully inactivated and the output tends to zero. On the contrary, a low input is accompanied by a negligible repressor concentration and it is turned into a high output.

15.2.1.2 Common Signal Carriers

In electrical circuits, electrons are the sole responsible entity for the spread of information. Namely, they work as a "currency," where every component within the circuit gets it as an input and sends it as an output. However, genetic circuits have special requirements, as they are based on DNA transcription and mRNA translation that, as was explained before, are carried out by two different molecules: RNA polymerases and ribosomes, respectively. Both of them allow communication among basic parts and are necessary for the production of new proteins. Hence, they both represent biological common signal carriers [8].

For the sake of clarity, it should be mentioned that all standard biological parts are defined to be sitting on the DNA, even though their names could be misleadingly interpreted as sitting on the mRNA or else. As such, RNA polymerases act at DNA level and scan all the above-mentioned standard biological parts. Ribosomes, on the contrary, have access only to the parts that are transcribed into mRNA: RBS, coding regions, and noncoding DNA. However, other molecules play an important role in transcription and translation regulation.

Transcription factors, produced by circuit devices, can hinder (repressors) or favor (activators) the binding of RNA polymerases to the promoters. Normally, these proteins show two distinctive 3D configurations: the active and inactive, that is, when they are able or unable to bind to a promoter. Moreover, small chemical molecules in the cell environment can bind transcription factors causing a change in their structure. There are two types: *inducers* or *corepressors* [9]. Inducers aid transcription by inactivating repressors or by activating activators. On the other hand, corepressors activate repressors or inactivate activators, preventing in this way the synthesis of new proteins. However, control mechanisms are not limited to DNA as, at the mRNA level, controls for translation are also present. In bacteria, for instance, sRNA molecules (typically no longer than 20 nucleotides) can bind, by simple base pairing, to their complementary sequence on the RBS or along the coding region. In this way, they either form or remove a hurdle for ribosomes that prevents their binding or forces them to interrupt translation. Hence, sRNAs either activate or repress translation.

Chemicals do not interact with sRNAs but are able to modify the mRNA directly at *ribozymes* and *riboswitches*. These are mRNA sequences that contain an *aptamer* (a region recognized by environmental signals) and an *expression platform* that undergoes structural modifications (or cut, in the ribozymes' case), upon the arrival of chemicals. In this sense, both ribozymes and riboswitches can either turn on or turn off translation.

In a genetic circuit, transcription factors and sRNAs are produced by certain devices – inside their coding region or sRNA part – and are able to regulate the transcriptional or translational activity of other devices by acting on their promoters or their RBSs. Hence, they carry out communication among transcription units. Chemicals, on the contrary, are normally sent by a source external to the circuit and represent a channel for the information coming from the surrounding environment. Therefore, transcription factors, sRNAs, and chemicals must be also considered as common signal carriers [10, 11].

In contrast to electronics, synthetic biology requires more than one signal carrier – five, as we have seen – to properly describe the possible interactions among parts, devices, and the cell environment. The flux of each signal carrier represents a different current that is to be measured in molars per second.

Finally, in spite of composability, not all the biological parts can be connected to each other in a synthetic gene circuit as they have to obey some ground biological rules. For instance, a coding region has to be always preceded by an RBS, whereas an sRNA always follows a promoter.

15.2.1.3 Pools and Fluxes

Free signal carriers can be thought to be stored inside *pools* [10, 11]. These pools are an abstract representation of the cellular (or extracellular, in the case of chemicals) environment where signal carriers hang about while not being involved in protein synthesis. There are several types of pools, depicted in Fig. 15.1.

In a circuit scheme (see Fig. 15.2 for an example), pools are connected to parts and appear to be the interfaces among different devices. So to speak, a regulatory factor that is produced inside a device will act on another device after passing through its corresponding pool. Furthermore, free signal carriers are a biological potential, in the sense that their amount determines circuit performance. In fact, as long as RNA polymerase and ribosome pools are disconnected from the circuit devices, there will be no protein synthesis (this corresponds to having all transfection plasmids outside the host cell). Then, as devices and pools are connected (insertion of the plasmids into a bacterium), RNA polymerases and ribosomes leave

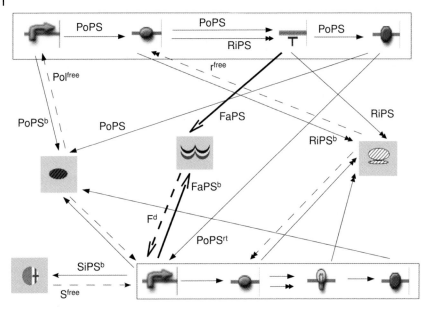

Fig. 15.2 A simple genetic circuit made of two transcription units. The first one produces a repressor that acts on the promoter of the second one. Here, chemicals can bind and inactivate the repressors, allowing the synthesis of a fluorescent protein. The exchange of signal carrier fluxes among parts, devices, and cell environment is here fully illustrated.

their pools and go inside promoters and RBSs in a quantity proportional to their reciprocal affinities. Differently from a more traditional model, where transcription and translation are described by Hill (switch-like) functions and both RNA polymerases and ribosomes are supposed to be infinite [12], this framework forces to consider a finite amount of molecules inside each pool. Therefore, it permits to estimate the circuit scalability, that is, how the circuit dynamics varies with respect to the total number of parts [10, 11].

Parts and pools are modeled independently by following mass-action kinetics. This framework permits to depict, in detail, the interactions between signal carrier molecules and DNA and mRNA chains. Fluxes of signal carriers are exchanged between parts and pools, and it is these biological currents that are responsible for the diffusion of information throughout the circuit. Every flux is generated into a unique type of parts (e.g., promoters give rise to polymerase flux, and RBSs to ribosome flux) or inside a pool (chemicals). Moreover, parts are not accessible to all signal carriers. For instance, ribosomes can never enter a promoter.

As we saw before, RNA polymerase, differently from the other signal carriers, goes through every standard part present in a transcription unit, in order to com-

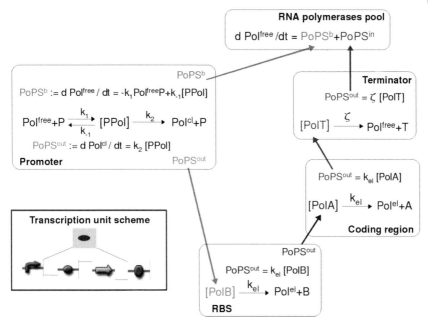

Fig. 15.3 The PoPS path along a transcription unit. In every standard part (and in the pool), all the reactions concerning RNA polymerases are described. Exchange of PoPS fluxes between parts is put in evidence.

plete mRNA transcription. As sketched in Fig. 15.3, free RNA polymerase (Pol^{free}) enters a promoter (P) and reacts with it following a Michaelis–Menten scheme

$$Pol^{free} + P \underset{k_{-1}}{\overset{k_1}{\rightleftharpoons}} [P\,Pol] \overset{k_2}{\to} P + Pol^{cl} \tag{15.1}$$

where k_1 and k_{-1} are, respectively, the association rate constant and the dissociation rate between promoter and RNA polymerase, k_2 is the transcription rate, Pol^{cl} represents the RNA polymerase in the action of leaving this part (promoter-clearance), and $[P\,Pol]$ is the so-called initiation complex. It should be noted that throughout this chapter we indicate biochemical complexes by means of square brackets.

From the initiation complex, RNA polymerase can either dissociate, becoming free and moving back to the pool, or start transcription by moving along the DNA chain. The former event contributes to a negative *balance* flux ($PoPS^b$) of free polymerases toward their pool,

$$PoPS^b = -k_1 Pol^{free} P + k_{-1}[P\,Pol] \tag{15.2}$$

the latter produces a positive balance flux directed to the RBS,

$$PoPS^{out} = k_2[P\,Pol] \tag{15.3}$$

where *PoPS* means polymerase per second. Inside the RBS, the incoming RNA polymerase is supposed to bind a site B and form a new complex [$PolB$] before starting the elongation phase (Pol^{el}), during which the mRNA is transcribed. The RNA polymerase flux from the RBS to the protein-coding region corresponds then to the product of elongation rate constant k_{el} and the [$PolB$] compound concentration. Entering the protein-coding region, the RNA polymerase flux is imagined to end up in a site A (in correspondence to the START codon) giving rise to the complex [$PolA$]. From this point, RNA polymerases can go on with the elongation phase and pass to the terminator. Here they bind a site T and produce the complex [$PolT$] before leaving the DNA and returning to the pool. The flux of free RNA polymerases directed to the pool is given by the product of [$PolT$] concentration and ζ, the dissociation rate constant. Inside the pool, the quantity of free RNA polymerase available to the circuit is calculated by integrating the sum of $PoPS^b$ (sent by the promoter) and $PoPS^{in}$ (incoming from the terminator). Fluxes associated with the path of the other signal carriers are calculated in a similar way, and it should be noted that they are fully responsible for the communication among the parts present in a circuit.

Since fluxes are measured in molar per second, it is straightforward to integrate them in a model based on ordinary differential equations (ODEs) that calculate the time derivatives of chemical species concentrations. Hence, each part is associated with the ODEs describing the dynamics of the embraced species and sends and receives one or more fluxes of signal carriers. As stated above, pools store free signal carriers whose amount is also calculated via ODEs that gather all the fluxes coming from the different devices.

15.2.2
Part Models

15.2.2.1 Promoters
In its simplest configuration (*constitutive*), a promoter contains only the RNA polymerase-binding site. Without any kind of regulation, RNA polymerases interact with the promoter via the reaction scheme in Eq. (15.1). More complex promoters host one or more operators, from now onward denoted as O. Promoters containing two operators are often used in synthetic gene circuits: they allow strong regulation and, thanks to *cooperativity* and *synergistic activation*, they can mimic Boolean gates [13]. Here we describe a model for this kind of regulatory element.

The *core* polymerase site is referred to as the $[-35\ldots-10]$ region, where the numbers indicate the location of DNA basis with respect to the transcriptional *start point* ($+1$). The relative positions of operators and core polymerase site are crucial for a proper promoter regulation. For instance, operators hosting repressors overlap, sometimes completely, the $[-35\ldots-10]$ region in order to prevent RNA polymerase binding. In contrast, binding sites for activators are placed upstream of the same region.

Two operators can be symmetric with respect to the transcription factors, that is, none is bound preferentially. Otherwise, the operator closest to the start point (O_1)

will have either the strongest affinity toward repressors or the weakest toward activators. These affinities are dictated by the values of the association rate constant (α) and of the dissociation rate (β) describing the interactions between transcription factors and operators.

Promoter negative regulation works efficiently even with a single operator, provided that it lies inside the core polymerase-binding site. A second operator, partially overlapping this region, generally improves the overall repression. Two repressors in the *active* configuration (R_1^a and R_2^a) can bind their respective operators either independently or cooperatively.

When acting cooperatively, and assuming that O_1 is stronger than O_2, R_1^a tends to reach the promoter first. The repressor binding O_1 causes a structural modification in the DNA chain (a rotation, for instance) that improves the affinity of O_2 toward R_2^a. In other words, O_2 can be bound more easily once O_1 is occupied by a repressor. On the contrary, the *state* of O_2 does not influence in any way the occupancy of O_1.

Since the two operators admit two possible states each (free and taken), there are four different states of a promoter, indicated as: $O_1^f O_2^f$ (both the operators are free), $O_1^f O_2^t$ and $O_1^t O_2^f$ (only one operator is occupied), and $O_1^t O_2^t$ (both the operators are occupied). Hence, the interactions among repressors and promoter are represented by four reversible reactions

$$R_1^a + O_1^f O_2^f \underset{\beta_{1f}}{\overset{\alpha_{1f}}{\rightleftharpoons}} O_1^t O_2^f \tag{15.4}$$

$$R_1^a + O_1^f O_2^t \underset{\beta_{1t}}{\overset{\alpha_{1t}}{\rightleftharpoons}} O_1^t O_2^t \tag{15.5}$$

$$R_2^a + O_1^f O_2^f \underset{\beta_{2f}}{\overset{\alpha_{2f}}{\rightleftharpoons}} O_1^f O_2^t \tag{15.6}$$

$$R_2^a + O_1^t O_2^f \underset{\beta_{2t}}{\overset{\alpha_{2t}}{\rightleftharpoons}} O_1^t O_2^t \tag{15.7}$$

where the first index of α and β refers to the operator where the repressor binds, and the second index indicates the state of the other operator. Therefore, to achieve cooperativity, one has to set $\alpha_{2t} > \alpha_{2f}$ and $\beta_{2t} < \beta_{2f}$ and, in case of symmetric operators, also $\alpha_{1t} > \alpha_{1f}$ and $\beta_{1t} < \beta_{1f}$. It should be noted that partial cooperativity can also be achieved. This is the case when only one inequality holds, either on α or on β. Furthermore, to enforce the first operator to have the strongest affinity among the two, one has to set $\alpha_1/\beta_1 > \alpha_2/\beta_2$.

Now, RNA polymerases only have access to the promoter when both operators are free. Hence, Eq. (15.1) becomes

$$Pol^{\text{free}} + O_1^f O_2^f \underset{k_{-1}}{\overset{k_1}{\rightleftharpoons}} [PolO_1^f O_2^f] \overset{k_2}{\rightarrow} O_1^f O_2^f + Pol^{\text{cl}} \tag{15.8}$$

In contrast, when a promoter is positively regulated, two activators are supposed to interact either through cooperativity or through synergistic activation. In the case of cooperativity, O_1 is considered the weakest operator when an activator (in its active configuration, A_1^a) can hardly bind without "being helped" by another activator (A_2^a) occupying O_2. When both the operators are occupied, RNA polymerases can bind to the promoter. In principle, the presence of the only A_1^a might be enough to let transcription start, but without the stabilizing action of having an A_2^a bound to O_2, A_1^a and O_1 will dissociate considerably fast. Therefore, the interaction between RNA polymerases and $O_1^t O_2^f$ can be neglected. In case of promoter activation, Eq. (15.1) is rewritten as

$$Pol^{\text{free}} + O_1^t O_2^t \underset{k_{-1}}{\overset{k_1}{\rightleftharpoons}} [PolO_1^t O_2^t] \overset{k_2}{\to} O_1^t O_2^t + Pol^{\text{cl}} \tag{15.9}$$

where the interaction between the activators and their operators – and the conditions about cooperativity – are symmetric to the those given for the promoter repression.

Synergistic activation presents a different scenario. In this case, activators bind to their operators without any reciprocal interaction. Furthermore, a hierarchy (in term of strength) between the two operators is not necessary. Both the activators can recruit RNA polymerases independently and the presence of only one of them on the promoter is sufficient to start transcription. When the promoter is fully occupied, RNA polymerases are recruited by the two activators together because they bind different sites on the RNA polymerase surface. This event results in a drastic increment of the promoter transcriptional activity. In contrast to the previous case, RNA polymerases are able to bind three promoter states ($O_1^t O_2^f$, $O_1^f O_2^t$ and $O_1^t O_2^t$) stably and Eq. (15.9) splits into three equations. In order to have synergistic activation, one has to assure that at least $k_{2sa} > k_2$, where k_{2sa} is the transcription rate at synergistic activation regime that corresponds to the promoter state $O_1^t O_2^t$.

Aside, and irrespective of the type of regulation, transcription factors must be active to regulate promoters. As we mentioned earlier, chemicals bind to transcription factors provoking structural changes that result in their activation or inactivation. Hence, these small molecules are a further control mechanism on protein synthesis. For instance, the action of inducers (I) on a free active repressor (R_1^a) is given by the reaction

$$n_1 I_1 + R_1^a \underset{\mu_1}{\overset{\lambda_1}{\rightleftharpoons}} R_1^i \tag{15.10}$$

where λ_1 and μ_1 are, respectively, the association rate constant and the dissociation rate between chemicals and transcription factors; R_1^i represents the inactive repressor configuration, and n_1 is the number of inducers required for the repressor inactivation (normally up to four). Equation (15.10) assumes *complete cooperativity*, namely, that the n_1 molecules must bind the transcription factor simultaneously. Furthermore, inducers can bind repressors hosted by their operators

$$n_1 I_1 + O_1^t O_2^f \xrightarrow{\gamma_1} O_1^f O_2^f + R_1^i \qquad (15.11)$$

$$n_1 I_1 + O_1^t O_2^t \xrightarrow{\gamma_1} O_1^f O_2^t + R_1^i \qquad (15.12)$$

where γ_1 is the association rate constant between inducers and bound active repressors. For simplicity, one can assume that $\gamma_1 = \lambda_1$.

To complete the description of the promoter dynamics, one has to consider the degradation of *nonfree* transcription factors, which are bound to DNA operators or to chemicals. Free transcription factors are supposed to decay into their pool. Following up on the example of a repressible promoter, R_1 is set to decay (with rate k_{D1}) according to the following reactions:

$$O_1^t O_2^f \xrightarrow{k_{D1}} O_1^f O_2^f \qquad (15.13)$$

$$O_1^t O_2^t \xrightarrow{k_{D1}} O_1^f O_2^t \qquad (15.14)$$

$$R_1^i \xrightarrow{k_{D1}} n_1 I_1 \qquad (15.15)$$

Equations (15.13) and (15.14) describe the decay of R_1^a when it is bound to O_1.

A promoter can be connected to three different kinds of pools, belonging to: RNA polymerase, transcription factors, and chemicals. Transcription factors and chemicals can be of different species and a promoter might also host, for instance, an activator (binding to O_2) and a repressor (binding to O_1) at the same time. Free signal carriers move from the pool to the promoter and backward, due to the reversibility of the binding reaction. The total effect is a balance flux from the promoter to each of these pools that updates the free signal carrier concentration. In Eq. (15.2), we had already introduced the RNA polymerase balance flux ($PoPS^b$). Inside a repressible promoter, it becomes

$$PoPS^b = -k_1 Pol^{free} O_1^f O_2^f + k_{-1}\left[PolO_1^f O_2^f\right] \qquad (15.16)$$

The balance fluxes of each transcription factor ($FaPS^b$, factors per second) and each chemical ($SiPS^b$, signals per second) are somewhat more elaborate:

$$FaPS_1^b = -\alpha_{1f} R_1^a O_1^f O_2^f + \beta_{1f} O_1^t O_2^f - \alpha_{1t} R_1^a O_1^f O_2^t$$
$$+ \beta_{1t} O_1^t O_2^t - \lambda_1 R_1^a I_1^{n_1} + \mu_1 R_1^i \qquad (15.17)$$

$$SiPS_1^b = -n_1\left(\gamma_1 I_1^{n_1} O_1^t O_2^f + \gamma_1 I_1^{n_1} O_1^t O_2^t + \lambda_1 R_1^a I_1^{n_1} - \mu_1 R_1^i - k_{D1} R_1^i\right)$$
$$\qquad (15.18)$$

Analogous equations hold for R_2^a ($FaPS_2^b$) and I_2 ($SiPS_2^b$).

Placed at the beginning of a transcription unit, a promoter is generally connected with an RBS, but it can be followed also either by another promoter or by an sRNA.

Furthermore, in the presence of *readthrough* (see below), a promoter can also be connected to the terminator at the end of the previous transcription unit along the DNA. With each of these parts, a promoter exchanges different types of *PoPS*.

As was described in Eq. (15.3), promoter output ($PoPS^{out}$) corresponds to the concentration of bound RNA polymerases multiplied by the transcription rate. In a repressible promoter, from Eq. (15.8), it follows that

$$PoPS^{out} = k_2 \left[PolO_1^f\, O_2^f \right] \tag{15.19}$$

Another output, present in all the promoters except for the constitutive one, is caused by the basal (*leakage*) production. This is proportional to the concentration of the promoter states that prevent RNA polymerase binding. The RNA polymerase flux due to the leakage in a repressible promoter is given by

$$PoPS^{lk} = k_2^{lk} \left(O_1^f\, O_2^t + O_1^t\, O_2^f + O_1^t\, O_2^t \right) \tag{15.20}$$

where k_2^{lk} is the leakage rate.

Just as promoter's flaws cause leakage, inefficiencies in the terminator are responsible for RNA polymerase readthrough. Namely, some molecules of RNA polymerases do not leave the DNA chain at the STOP codon but go into the next promoter carrying an mRNA tail that encodes for the abandoned transcription unit. By consequence, this transcription unit will be translated inside the next one, as long as the corresponding mRNA is not degraded. The $PoPS^{rt}$ (rt stands for readthrough) is generated by a terminator (see below) and sent forward to a promoter and backward to an RBS. For sake of simplicity, the "readthrough" RNA polymerases are not supposed to interact with the promoter, as they pass directly into the next RBS. As a result, the readthrough effect equally increases the mRNA transcription of two adjacent genes. However, this might influence some other transcription units, depending on several factors like mRNA half-life, RNA polymerase velocity, and gene length.

Finally, in the particular case where two promoters are placed in a row at the beginning of a transcription unit, the second promoter receives up to two RNA polymerase fluxes from the first promoter: $PoPS^{in}$, which corresponds to $PoPS^{out}$ in Eq. (15.19) and, potentially, $PoPS^{lk}$. For the sake of simplicity, the interactions between the leakage RNA polymerase and the second promoter are neglected (this is valid as long as $k_2^{lk} \ll k_2$ inside the first promoter). Moreover, $PoPS^{in}$ is assumed to flow into a new complex ($[PolQ]$) where RNA polymerases are in a *queue*, waiting for interaction with the promoter or their transfer back to the pool. Queued polymerases can bind to the promoter when the operators are free from repressors. In contrast, if they host activators, queued polymerase can become free or increase the leakage flux toward the next part.

15.2.2.2 Ribosome-Binding Sites

There are several analogies between the models representing a promoter and the RBS. RBS generates the ribosome current along the mRNA, measured in *RiPS* (ribosomes per second), and its activity can be regulated by proteins, sRNAs, and chemicals among others [9, 14, 15].

As was mentioned before, the RBS part sitting on the DNA receives $PoPS$ as an input from a previous part, and these RNA polymerases are supposed to flow entirely (symbol: $\Rightarrow$) into a new complex, $[PolB]$. This process is indicated as

$$PoPS^{in} \Rightarrow [PolB] \tag{15.21}$$

From $[PolB]$, RNA polymerases pass to the elongation phase (Pol^{el}):

$$[PolB] \xrightarrow{k_{el}} Pol^{el} + B + b \tag{15.22}$$

giving rise to the output polymerase flux ($PoPS^{out}$)

$$PoPS^{out} = k_{el}[PolB] \tag{15.23}$$

In Eq. (15.22), b represents the actual mRNA RBS, and it has two additional contributions: (1) from $PoPS^{lk}$, when the RBS is preceded by a nonconstitutive promoter and (2) from $PoPS^{rt}$, sent by the terminator at the end of the transcription unit the RBS belongs to. Namely,

$$PoPS^{lk} \Rightarrow b \tag{15.24}$$

$$PoPS^{rt} \Rightarrow b \tag{15.25}$$

Another $PoPS^{rt}$, coming from the promoter, is part of $PoPS^{in}$ in Eq. (15.21).

Upon the transcription of b, ribosomes can bind to it following a Michaelis–Menten scheme

$$r^{free} + b \underset{k_{-1r}}{\overset{k_{1r}}{\rightleftharpoons}} [rb] \xrightarrow{k_{2r}} b + r^{cl} \tag{15.26}$$

where r^{free} are free ribosomes and $[rb]$ depicts ribosomes bound to the mRNA; k_{1r} and k_{-1r} are the association rate constant and the dissociation rate between ribosomes and mRNA, respectively; k_{2r} is the translation rate, and r^{cl} represents ribosomes in the RBS clearance phase. Equation (15.26) is valid for constitutive RBSs, where ribosomes do not have to compete to have access to the mRNA.

Protein synthesis starts only in the next part, a coding region. For such condition, there are several control mechanisms generally used in synthetic biology, relying on sRNAs and on mRNA structures such as *riboswitches* and *ribozymes*. As mentioned before, sRNAs can bind their complementary sequence on the RBS repressing or activating translation. Following the nomenclature in [16] – where cis-repressing and trans-activating sRNA are described in detail – we refer to repressor-like sRNAs as *locks* and to activator-like sRNAs as *keys*. Additionally, riboswitches have two states: *off* that prevents ribosome binding and *on* that allows translation to start. In what follows, we do not consider ribozymes explicitly, but it should be noted that their dynamics are very close to those of riboswitches [17]. Moreover, in the cell, sRNAs and riboswitches are not necessarily confined to the RBS but can be found along the coding region as well. Without any loss of generality, we neglect this possibility in this representation.

Similarly to the promoter case, we consider RBS controlled by up to two regulatory factors, which can be either sRNAs or chemicals. These will bind the RBS corresponding to *b*-sites, the "operators" of the RBS. Furthermore, a *b*-site will be considered to be in two possible states: off (b^{off}) or on (b^{on}).

sRNAs do not show any cooperativity and cannot recruit ribosomes: they are able to interact only with the mRNA by forming or opening structures like *loops* that are hurdles on the ribosome path. Chemicals, on the contrary, can bind in a cooperative way to the aptamers of tandem riboswitches [18]. Mixed configurations where, by convention, the first *b*-site is occupied by a riboswitch and the second one is bound by an sRNA are allowed. However, the two sites have to share the same regulation type, meaning that they are either both repressible or both inducible.

A repressible RBS containing two *b*-sites in a mixed configuration is made of a riboswitch (*on* by default and turned *off* by a corepressor) followed by a lock-binding site. Hence, four states are possible: (1) $b_1^n b_2^n$, the riboswitch is not bound by any corepressor and the second *b*-site is free, on which translation can take place; (2) $b_1^n b_2^f$, the riboswitch is *on* but a lock has bound the second site, preventing ribosome binding; (3) $b_1^f b_2^n$, similar to the previous case, translation is prevented by the action of a corepressor on b_1; and (4) $b_1^f b_2^f$, the RBS is fully repressed. In this context, Eq. (15.26) becomes

$$r^{\text{free}} + b_1^n b_2^n \underset{k_{-1r}}{\overset{k_{1r}}{\rightleftharpoons}} [rb_1^n b_2^n] \overset{k_{2r}}{\rightarrow} b_1^n b_2^n + r^{\text{cl}} \tag{15.27}$$

and *b* in Eqs. (15.24) and (15.25) is replaced by $b_1^n b_2^n$.

The RBS regulation by means of a corepressor (C_1) and of a lock (l_2) binding to their corresponding sites is expressed by the following four reversible reactions:

$$C_1 + b_1^n b_2^n \underset{\xi_1}{\overset{\theta_1}{\rightleftharpoons}} b_1^f b_2^n \tag{15.28}$$

$$C_1 + b_1^n b_2^f \underset{\xi_1}{\overset{\theta_1}{\rightleftharpoons}} b_1^f b_2^f \tag{15.29}$$

$$l_2 + b_1^n b_2^n \underset{\xi_2}{\overset{\theta_2}{\rightleftharpoons}} b_1^n b_2^f \tag{15.30}$$

$$l_2 + b_1^f b_2^n \underset{\xi_2}{\overset{\theta_2}{\rightleftharpoons}} b_1^f b_2^f \tag{15.31}$$

where θ and ξ are, respectively, the association rate constant and the dissociation rate between chemicals or sRNAs and the corresponding *b*-sites.

The repressible RBS model is complete with the addition of mRNA degradation reactions:

$$b_1^n b_2^n \xrightarrow{k_d} \tag{15.32}$$

$$b_1^f b_2^n \xrightarrow{k_d} C_1 \tag{15.33}$$

$$b_1^n b_2^f \xrightarrow{k_d} \tag{15.34}$$

$$b_1^f b_2^f \xrightarrow{k_d} C_1 \tag{15.35}$$

$$[rb_1^n b_2^n] \xrightarrow{k_d} r^{\text{free}} \tag{15.36}$$

where k_d is the mRNA decay rate. From Eqs. (15.33) and (15.35), we imply that locks are degraded along with the mRNA when bound to the RBS. In contrast, free locks (and chemicals) are degraded into their pools, whereas ribosomes – like RNA polymerases – are not degraded.

An RBS can be connected to three different kinds of pools, belonging to: ribosomes, sRNAs, and chemicals. The exchange of information with the ribosome pool is similar to the one between a promoter and the RNA polymerase pool. Hence, according to Eq. (15.27), the balance flux of ribosomes sent to their pool by a repressible RBS is given by

$$RiPS^b = -k_{1r} r^{\text{free}} b_1^n b_2^n + (k_{-1r} + k_d)[rb_1^n b_2^n] \tag{15.37}$$

Equations (15.30) and (15.31) allow us to write the balance flux of locks to the sRNA pool as

$$RNAPS^b = -\theta_2 l_2 b_1^n b_2^n + \xi_2 b_1^n b_2^f - \theta_2 l_2 b_1^f b_2^n + \xi_2 b_1^f b_2^f \tag{15.38}$$

whereas $SiPS^b$ is determined by Eqs. (15.28) and (15.29) along with the degradation events in Eqs. (15.33) and (15.35):

$$SiPS^b = \xi_1 b_1^f b_2^n - \theta_1 C_1 b_1^n b_2^n + \xi_1 b_1^f b_2^f - \theta_1 C_1 b_1^n b_2^f + k_d \left(b_1^f b_2^n + b_1^f b_2^f \right) \tag{15.39}$$

On the DNA chain, the RBS is generally preceded by a promoter and followed by a coding region. Alternatively, an RBS can be preceded by an sRNA (working as a cis-repressing RNA) or a noncoding DNA. The output is of dual type: an RNA polymerase flux (see Eq. (15.23)) together with a ribosome one, derived from Eq. (15.27):

$$RiPS^{\text{out}} = k_{2r} [rb_1^n b_2^n] \tag{15.40}$$

Finally, nonconstitutive RBS are leaky. Hence, basal translation has to be taken into account and can be calculated as the product of a leakage translation rate (k_{2r}^{lk}) and the concentrations of the "off" RBS states. For the repressible RBS we have studied so far, the ribosome flux due to the RBS leakage is given as

$$RiPS^{lk} = k_{2r}^{lk}\left(b_1^n b_2^f + b_1^f b_2^n + b_1^f b_2^f\right) \tag{15.41}$$

This flux, as in the promoter case, is not summed to $RiPS^{out}$ but is treated as a separate output.

Inside polycistronic sequences, where several genes (cistrons) are separated by short DNA spacers, ribosomes also show a readthrough effect, similar to the one described for RNA polymerase on the DNA. In this case, instead of becoming free on reaching a spacer, ribosomes can continue translating the next protein-coding region. However, their final product has no biological meaning (which for simplicity we will term *garbage*) and the net result is a decrease in the synthesis of the proteins encoded by each cistron. We consider two ribosomal readthrough fluxes: the first captures the ribosomes at their first readthrough, whereas the second gathers all the ribosomes already equipped with a nucleotidic tail. Both fluxes contribute to a queued state entering an RBS part, upon which ribosomes can bind to the free b sites and continue garbage production.

15.2.2.3 Coding Regions

Depending on the information they store, coding regions can be of four types: general proteins, transcription factors, enzymes, and reporter (fluorescent) proteins. Essentially, they all share the same model except for transcription factors and enzymes that send their product to a pool.

A coding region gets from an RBS a dual input:

$$PoPS^{in} \Rightarrow [PolA] \tag{15.42}$$

$$RiPS^{in} \Rightarrow [ra] \tag{15.43}$$

where A and a represent the START codon on DNA and mRNA, respectively. Furthermore, if the RBS is not constitutive, a leakage flux of ribosomes is also present with the effect of increasing the concentration of the protein z synthesized by this part

$$RiPS^{lk} \Rightarrow z \tag{15.44}$$

From the complexes in Eqs. (15.42) and (15.43), RNA polymerases and ribosomes return to the elongation phase:

$$[PolA] \xrightarrow{k_{el}} Pol^{el} + A \tag{15.45}$$

$$[ra] \xrightarrow{k_{el}^r} r^{el} + a + z \tag{15.46}$$

where k^r_{el} indicates the ribosome elongation rate and r^{el} the ribosomes in the elongation phase. Equation (15.45) considers mRNA transcription implicitly, since total mRNA concentration corresponds to b in Eq. (15.22).

RNA polymerases scan the entire coding region part and finally flow into a terminator or a DNA spacer, if the coding region belongs to a polycistronic sequence

$$PoPS^{out} = k_{el}[PolA] \qquad (15.47)$$

Ribosomes, on the contrary, translate the coding region until they encounter the STOP codon (u). Here, they are supposed to form a new complex with the mRNA ($[ru]$) before unbinding and becoming free to go back to their pool. Considering a flux of ribosomes internal to the coding region ($RiPS^C$) and, from Eq. (15.46), equal to $k_{el}[ra]$, $[ru]$ can be seen as generated by the arrival of this flux at the STOP codon:

$$RiPS^C \Rightarrow [ru] \qquad (15.48)$$

Once u is reached, ribosomes dissociate from mRNA at a rate ζ_r

$$[ru] \xrightarrow{\zeta_r} r^{free} + u \qquad (15.49)$$

giving rise to an output flux directed to the ribosome pool

$$RiPS^{out} = \zeta_r[ru] \qquad (15.50)$$

We mentioned before that the only model difference lies in transcription factors and enzymes sending their product to a pool. Hence, if the coding region part corresponds to an enzyme or a transcription factor, z does not appear explicitly inside the coding region and a flux of proteins ($FaPS^{out}$, for instance) is then sent to the corresponding pool:

$$FaPS^{out} = RiPS^C + RiPS^{lk} \qquad (15.51)$$

In contrast, coding regions for general proteins and reporters additionally consider protein degradation.

Finally, in a polycistronic sequence, a protein-coding region receives from the RBS a readthrough flux of ribosomes. As described above, these ribosomes are subtracted from the synthesis of actual proteins, the concentration of which is naturally reduced.

15.2.2.4 Noncoding DNA
Noncoding DNA can only be found inside polycistronic sequences, at the end of each cistron. They are normally preceded by a coding region followed by an RBS, whereas only the spacer of the last cistron is connected to a terminator. Noticeably, when they are about 30 nucleotides long, they become responsible for the ribosome

readthrough, whereas longer spacers cause only a slight time delay in the protein synthesis.

In general, noncoding DNA can be seen as a coding region extension, where the STOP codon is supposed to be placed. Hence, in an operon, ribosomes no longer leave mRNA at the end of the coding region but only once they reach a spacer.

Together with the input fluxes of RNA polymerases and ribosomes, a non-coding DNA gets from a coding region one additional flux ($RiPS_{in}^{rt}$), due to the "readthrough" ribosomes (r^{rt}):

$$PoPS^{in} \Rightarrow [PolU] \tag{15.52}$$

$$RiPS^{in} \Rightarrow [ru] \tag{15.53}$$

$$RiPS_{in}^{rt} \Rightarrow [r^{rt}u] \tag{15.54}$$

where U indicates the STOP codon on the DNA.

Ribosomes and "readthrough" ribosomes can unbind from the STOP codon, u, releasing either a protein or some garbage (g), respectively:

$$[ru] \xrightarrow{\zeta_r} r^{free} + u + z \tag{15.55}$$

$$[r^{rt}u] \xrightarrow{\zeta_r} r^{free} + u + g \tag{15.56}$$

Alternatively, they can pass to the next RBS, giving rise to two different readthrough fluxes

$$[ru] \xrightarrow{\eta_r} r^{rt} + u \tag{15.57}$$

$$RiPS_{out1}^{rt} = \eta_r [ru] \tag{15.58}$$

$$[r^{rt}u] \xrightarrow{\eta_r} r^{rt} + u \tag{15.59}$$

$$RiPS_{out2}^{rt} = \eta_r [r^{rt}u] \tag{15.60}$$

where η_r is the ribosomal readthrough rate. It is worth noting that $RiPS_{out1}^{rt}$ is a current of ribosomes at their first readthrough, whereas $RiPS_{out2}^{rt}$ is made of "older" readthrough ribosomes. Both these fluxes, inside the RBS, end in the [$r^{rt}q$] complex, representing queued ribosomes that can subsequently bind to b and continue garbage synthesis.

Finally, a spacer is always connected to the ribosome pool and can be connected to either a transcription factor or an enzymatic pool, depending on the product of the associated coding region. If present, an $FaPS^{out}$ flux contains a term $\zeta_r[ru]$ due to Eq. (15.55) and a leakage term sent by the previous RBS, if this is a nonconstitutive one.

15.2.2.5 Small RNA

As mentioned before, sRNAs regulate translation (activation or repression) by binding to the RBS. In bacteria, they are generally between 7 and 20 nucleotides long [14]. An sRNA part is mostly found between a promoter and a terminator (active in *trans*), along the DNA chain. However, the sRNA can be cis-repressing when it is followed by an RBS.

An sRNA, when acting in trans, receives from the promoter an RNA polymerase flux that forms with the DNA a complex on a site called S:

$$PoPS^{in} \Rightarrow [PolS] \qquad (15.61)$$

From $[PolS]$, RNA polymerases dissociate restarting transcription at the elongation rate k_{el}

$$[PolS] \xrightarrow{k_{el}} Pol^{el} + S \qquad (15.62)$$

and the output RNA polymerase flux toward the terminator is

$$PoPS^{out} = k_{el}[PolS] \qquad (15.63)$$

In contrast to other parts, sRNA is not connected to any pool. In fact, free sRNAs ($sRNA^{free}$) are released by a terminator, as is explained in Section 15.2.2.6.

15.2.2.6 Terminator

A terminator marks the end of a transcription unit. It is here that the RNA polymerase leaves the DNA chain and detaches the mRNA tail, before returning to the pool. However, the terminator efficiency is always lower than one and this is the cause of RNA polymerase readthrough into the next promoter. Nevertheless, as we have seen above, only RBSs are sensitive to the readthrough flux.

A terminator can be preceded by a coding region, a spacer, and an sRNA. However, two terminators can be found or placed in a row. This configuration is often used in synthetic circuits to limit the readthrough effect. When readthrough is taken into account, in a circuit scheme, a terminator is connected "on the right" to a promoter (see Fig. 15.2).

The input to a terminator part is an RNA polymerase flux, which goes completely into a site T, forming the last complex with the DNA along a transcription unit:

$$PoPS^{in} \Rightarrow [PolT] \qquad (15.64)$$

From $[PolT]$, RNA polymerases either abandon the DNA at a dissociation rate ζ

$$[PolT] \xrightarrow{\zeta} Pol^{free} + T \qquad (15.65)$$

or pass into the next promoter at a readthrough rate η

$$[PolT] \xrightarrow{\eta} Pol^{rt} + T \tag{15.66}$$

where Pol^{rt} stands for "readthrough" polymerases.

A terminator is always connected to the RNA polymerase pool where, according to Eq. (15.65), it sends a flux

$$PoPS^{out} = \zeta[PolT] \tag{15.67}$$

whereas Eq. (15.66) determines the RNA polymerase readthrough flux to the next promoter

$$PoPS^{rt} = \eta[PolT] \tag{15.68}$$

Finally, when a terminator is preceded by an sRNA, it will also be connected to an sRNA pool. In such case, the outgoing sRNA flux ($RNAPS^{out}$) is equal to sum of the flux of the free RNA polymerases in Eq. (15.67) and a possible leakage contribution generated inside a promoter

$$RNAPS^{out} = \zeta[PolT] + PoPS^{lk} \tag{15.69}$$

15.2.3
Introducing Parts and Fluxes into Deterministic Equations

As will be seen in more detail in Section 15.3, under certain assumptions, the time evolution of synthetic circuits can be represented by differential equations, based on the laws of Mass Action. In such view, species concentrations will vary in time in accordance to only one factor: the chemical reactions that arise from the species interactions. Such a deterministic framework is only appropriate whenever all molecular species in a system are present in large numbers of molecules and, most importantly, when discreteness and internal noise have no noticeable effect. For the time being, let us assume we can adopt such a regime, in which case one could describe the time evolution as

$$\frac{dS}{dt} = Nv \tag{15.70}$$

where S indicates the vector of species concentrations, v the *flux* vector, and N the *stoichiometric* matrix.

When modeling synthetic gene circuits with parts, this framework becomes more complex due to the presence of fluxes of signal carriers. Even though they appear inside the time derivatives of some species, they are not included in the stoichiometric matrix. Furthermore, the reactions responsible for the signal carriers' production – as described in the previous sections – contain artificial variables, the dynamics of which need not be computed.

To better explain this concept, let us consider the dynamics of a constitutive promoter. According to Eq. (15.1), three reactions take place in this part. However, Pol^{cl} will be considered a "fictitious state" when modeling with parts. Namely, its time derivative will determine the flux $PoPS^{out}$, but it will no longer be tracked as a separate species. Furthermore, the variable Pol^{free} belongs to the RNA polymerase pool. Hence, we will only keep the time derivatives of P and $[P\,Pol]$, which can be calculated by multiplying the stoichiometric matrix

$$N = \begin{pmatrix} -1 & 1 & 1 \\ 1 & -1 & -1 \end{pmatrix} \qquad (15.71)$$

with the flux vector

$$v = \begin{pmatrix} k_1 P\,Pol^{free} \\ k_{-1}[P\,Pol] \\ k_2[P\,Pol] \end{pmatrix} \qquad (15.72)$$

It should be noted that in Eq. (15.71), rows are assigned, in the order, to P and $[P\,Pol]$; columns to k_1, k_{-1}, and k_2.

As a result, one obtains

$$\frac{dP}{dt} = -k_1 P\,Pol^{free} + k_{-1}[P\,Pol] + k_2[P\,Pol] \qquad (15.73)$$

$$\frac{d[P\,Pol]}{dt} = k_1 P\,Pol^{free} - (k_1 + k_2)[P\,Pol] \qquad (15.74)$$

Promoter dynamics is completed by the calculation of the two outgoing fluxes $PoPS^b$ and $PoPS^{out}$ as in Eqs. (15.2) and (15.3), respectively.

Now, let us suppose we can place a terminator after the promoter and connect both parts to the RNA polymerase pool. This small system is of no use in biology, but it can help understanding how signal carrier fluxes play the role of interfaces between parts and pools. Here, moreover, we wish to highlight how the fluxes algebraic expressions cancel out, yielding a set of elementary reactions to be input in a deterministic or stochastic solver. Equations (15.64) and (15.65) show that the compound $[PolT]$ is sufficient for a complete terminator description. Actually, the state T (free terminator) is fictitious: incoming RNA polymerases do not react with it but just flow into it, giving rise to $[PolT]$. Hence, this part contains only one variable and one reaction – associated with the rate ζ. The time derivative of $[PolT]$ contains a negative term $(-\zeta[PolT])$ to which the $PoPS^{in}$ flux – coming from the promoter – has to be summed

$$\frac{d[PolT]}{dt} = -\zeta[PolT] + PoPS^{in} \qquad (15.75)$$

Furthermore, the terminator dynamics requires the calculation of the $PoPS^{out}$ flux directed to the RNA polymerase pool (see Eq. (15.67)). Inside this pool, no reaction takes place. The only variable is Pol^{free} whose amount changes as a function of the fluxes $PoPS^b$ and $PoPS^{in}$, the former exchanged with the promoter, the latter received by the terminator. Therefore,

$$\frac{d\,Pol^{free}}{dt} = PoPS^b + PoPS^{in} \tag{15.76}$$

If we consider the system in its entirety, we can try to substitute the exchange of signal carrier fluxes with a "new" set of chemical reactions where the fictitious variables are also neglected:

$$Pol^{free} + P \underset{k_{-1}}{\overset{k_1}{\rightleftharpoons}} [P\,Pol] \tag{15.77}$$

$$[P\,Pol] \overset{k_2}{\to} [PolT] + P \tag{15.78}$$

$$[PolT] \overset{\zeta}{\to} Pol^{free} \tag{15.79}$$

Equations (15.77)–(15.79) give rise to a 4×4 stoichiometric matrix:

$$N = \begin{pmatrix} -1 & 1 & 1 & 0 \\ 1 & -1 & -1 & 0 \\ 0 & 0 & 1 & -1 \\ -1 & 1 & 0 & 1 \end{pmatrix} \tag{15.80}$$

where the rows correspond to P, $[P\,Pol]$, $[PolT]$ and Pol^{free} in the order and the columns to k_1, k_{-1}, k_2 and ζ, respectively. The flux vector contains the components of the three different $PoPS$ present in the system:

$$v = \begin{pmatrix} k_1 P\,Pol^{free} \\ k_{-1}[P\,Pol] \\ k_2[P\,Pol] \\ \zeta[PolT] \end{pmatrix} \tag{15.81}$$

By multiplying N in Eq. (15.80) with v in Eq. (15.81), one obtains the ODE system for this toy circuit. As expected, it contains Eqs. (15.73) and (15.74) for P and $[P\,Pol]$, respectively, together with

$$\frac{d[PolT]}{dt} = k_2[P\,Pol] - \zeta[PolT] \tag{15.82}$$

$$\frac{d\,Pol^{free}}{dt} = -k_1 P\,Pol^{free} + k_{-1}[P\,Pol] + \zeta[PolT] \tag{15.83}$$

that could be obtained directly from Eq. (15.75) and (15.76), respectively, by replacing every $PoPS$ term with its analytical expression. Hence, by closing the circuit, signal carriers fluxes and fictitious variable cancel out and the dynamics of the system is fully described again by an ODE system composed of elementary reactions that obeys to Eq. (15.70).

Now, pools are the places where free signal carriers are stored and from where they are distributed to the devices present in the circuit. Furthermore, pools can host biochemical reactions. For instance, the enzymatic reaction pool does not contain any of the five signal carriers, but it accommodates for the enzymes' substrate produced by some transcription units. Moreover, the generation, degradation, and dimerization of signal carriers can take place inside their pools.

The RNA polymerase pool is rather simple, despite the numerous connections it may have. Here – as we have seen in the example above – the total concentration of free RNA polymerases is modified by two kinds of incoming fluxes: $PoPS^b$ from promoters and $PoPS^{in}$ from terminators

$$PoPS^b \Rightarrow Pol^{free} \tag{15.84}$$

$$PoPS^{in} \Rightarrow Pol^{free} \tag{15.85}$$

As previously stated, RNA polymerases are not supposed to be degraded. If we denote the total number of promoters and terminators as N_P and N_T (they are, in general, different), the time variation of Pol^{free} (measured in molars per second) is simply given by the sum of the fluxes in Eqs. (15.84) and (15.85). Hence, we can formulate an ODE that describes the dynamics of free RNA polymerases as

$$\frac{dPol^{free}}{dt} = \sum_{i=1}^{N_P} PoPS^b_i + \sum_{j=1}^{N_T} PoPS^{in}_j \tag{15.86}$$

The model describing a ribosome pool is essentially identical to that of the RNA polymerase. Free ribosome concentration depends on the balance fluxes from RBSs and on the input fluxes from coding regions (or spacers). In addition, ribosome degradation is neglected. Hence, we can write

$$\frac{dr^{free}}{dt} = \sum_{i=1}^{N_R} RiPS^b_i + \sum_{j=1}^{N_C} RiPS^{in}_j \tag{15.87}$$

where N_R and N_C are the total number of RBSs and coding regions, respectively, present in the circuit. Note that spacers are not considered here.

The situation is different for transcription factor pools. When a repressor or an activator can only bind a promoter as a dimer, the transcription factor will enter its pool as a monomer (F^m) and will leave it as a dimer (F^d)

$$FaPS^{in} \Rightarrow F^m \tag{15.88}$$

$$FaPS^b \Rightarrow F^d \tag{15.89}$$

These two species interact according to the following scheme:

$$2F^m \underset{\varepsilon}{\overset{\delta}{\rightleftarrows}} F^d \tag{15.90}$$

where δ and ε are the association rate constant and the dissociation rate of the dimerization process, respectively. Additionally, both monomers and dimers are degraded inside the pool with the same rate (k_D)

$$F^m \overset{k_D}{\rightarrow} \tag{15.91}$$

$$F^d \overset{k_D}{\rightarrow} \tag{15.92}$$

Two coupled ODEs are necessary to depict the transcription factor dynamics properly. Both these differential equations contain a sum of fluxes and additional terms coming from Eqs. (15.88) to (15.90):

$$\frac{dF^m}{dt} = -2\delta F^{m^2} + 2\varepsilon F^d - k_D F^m + \sum_{i=1}^{N_{Cin}} FaPS_i^{in} \tag{15.93}$$

$$\frac{dF^d}{dt} = \delta F^{m^2} - \varepsilon F^d - k_D F^d + \sum_{i=1}^{N_{Pb}} FaPS_i^b \tag{15.94}$$

where N_{Cin} and N_{Pb} are the number of coding regions and promoters connected to the transcription factor pool, respectively.

On the contrary, sRNA pool only considers the chemical reaction for degradation, whose rate constant (k_{ds}) is generally lower than that of mRNA (k_d) [19]. Similar to our previously described pool models, the time derivative of the free sRNA concentration turns out to be

$$\frac{dsRNA^{free}}{dt} = -k_{ds} sRNA^{free} + \sum_{i=1}^{N_{Rb}} RNAPS_i^b + \sum_{j=1}^{N_{Tin}} RNAPS_j^{in} \tag{15.95}$$

where N_{Tin} and N_{Rb} are the number of connected terminators and RBSs, respectively.

The pool composed by chemicals presents a novelty. Here, free molecules of chemicals (S^{free}) are not only degraded but are also generally produced, whereas the pool itself represents the environment where the cell lives. In this configuration, only $SiPS^b$ fluxes coming from promoters or RBSs are present

$$SiPS^b \Rightarrow S^{free} \tag{15.96}$$

$$\overset{k_{sp}}{\rightarrow} S^{free} \tag{15.97}$$

$$S^{free} \overset{k_{sd}}{\rightarrow} \tag{15.98}$$

where k_{sp} and k_{sd} represent the signal production and degradation rates, respectively.

Moreover, an $SiPS^{in}$ reappears whenever the chemical production is performed by some other circuit devices that will be connected to the pool. Then, Eqs. (15.96)–(15.98) give rise to the following differential equation for the free chemicals dynamics:

$$\frac{dS^{\text{free}}}{dt} = -k_{sd}S^{\text{free}} + k_{sp} + \sum_{i=1}^{N_b} SiPS_i^b \tag{15.99}$$

where N_b indicates the number of promoters or RBSs connected to the pool.

Finally, we would like to describe the model of the enzymatic reaction pool, our last to consider. In its simplest configuration, this pool contains a definite amount of substrate and receives the enzymes as a flux ($EnPS^{in}$ – enzyme per second) from one or more coding parts. Alternatively, the substrate can also be provided by the rest of the circuit (as a flux $SuPS^{in}$ – substrate per second). Moreover, enzyme (E) and substrate (ζ) can be both proteins and sRNAs, provided that they interact following a Michaelis–Menten scheme such as Eq. (15.1).

The products (π) of the reaction can be any kind of molecule. If they are signal carriers, they are sent (as an outgoing flux) to their pool. Assuming that both enzymes and substrate are produced inside the circuit, into the pool they react, and are degraded as follows:

$$EnPS^{in} \Rightarrow E \tag{15.100}$$

$$SuPS^{in} \Rightarrow \zeta \tag{15.101}$$

$$E + \zeta \underset{k_{-1}}{\overset{k_1}{\rightleftharpoons}} [E\zeta] \overset{k_2}{\rightarrow} E + \pi \tag{15.102}$$

$$E \overset{k_{ED}}{\rightarrow} \tag{15.103}$$

$$\zeta \overset{k_{\zeta D}}{\rightarrow} \tag{15.104}$$

$$[E\zeta] \overset{k_{E\zeta D}}{\rightarrow} \tag{15.105}$$

where k_{ED}, $K_{\zeta D}$, and $k_{E\zeta D}$ are the enzyme, the substrate, and the complex $[E\zeta]$ degradation rate, respectively.

Four species are present in this pool. Their temporal evolution is fully described by three ODEs and an algebraic relation describing the outgoing product flux (as $ProPS^{out}$ – product per second):

$$\frac{dE}{dt} = -k_1 E\zeta + (k_{-1} + k_2)[E\zeta] - k_{ED}E + EnPS^{in} \qquad (15.106)$$

$$\frac{d\zeta}{dt} = -k_1 E\zeta + k_{-1}[E\zeta] - k_{\zeta D}\zeta + SuPS^{in} \qquad (15.107)$$

$$\frac{d[E\zeta]}{dt} = k_1 E\zeta - (k_{-1} + k_2 + k_{E\zeta D})[E\zeta] \qquad (15.108)$$

$$ProPS^{out} = k_2[E\zeta] \qquad (15.109)$$

15.3
Modeling Regimes and Simulation Techniques

It can be fairly naïve to think that once a system is characterized by an appropriate set of chemical reactions, the problem of providing a fair representation of it is basically solved. Many times, it is the choice of a modeling regime and solution technique that poses the bigger challenges. The correct selection of a particular modeling approach depends on several factors, such as molecular concentrations, distributions, the type of reactions and their timescales, whether discreteness and internal noise have noticeable macroscopic effects, and, finally, whether the model requires spatial resolution.

In the last section, we explained how a synthetic genetic circuit can be designed by juxtaposing parts as "building blocks," yielding a feasible set of chemical reactions representing our system of interest. The focus of this section is now on the selection of a simulation technique that best characterizes the particular features of our system, aiming for high accuracy within reasonable computation times.

Deterministic models assume a time evolution that is both continuous and predictable, and up to a couple of decades ago, they were fairly popular. Within these, one should distinguish between systems of ODEs, delay differential equations (DDEs), and partial differential equations (PDEs), the latter allowing for spatial resolution, inhomogeneous initial conditions, and/or molecular motion.

However, randomness is intrinsic to biological systems and a system's behavior is best described by the chemical master equation (CME). Intrinsic noise typically arises in a system with small to moderate numbers of key molecules and is due to the uncertainty of knowing when a reaction will occur and which type of reaction it might be. This source of randomness is entirely different to extrinsic noise, in which state changes are due to fluctuations in external conditions, such as temperature or fluctuations in the overall state of the cell. It is worth noting that different research groups use different definitions of noise sources [20] and careful attention should be paid to these subtleties when comparing results in the literature [21].

Many times, one is forced to account for intermediate biochemical steps or certain spatial effects, either explicitly or implicitly. By incorporating delays into the temporal model, one can capture essential information on a macroscopic level, each delay accounting for sets of biochemical processes and events on a microscopic timescale that would otherwise render us unable to compute cell dynamics in real time. For example, in eukaryotic gene expression, delays could be associated with transcription and translation, encompassing diffusion and translocation into and out of the nucleus, RNA polymerase activation, splicing, protein synthesis, and protein folding. The timescale of these processes can be considerably different to all other kinetic events and the effects of introducing such delays are very important, especially in the laying down of oscillating patterns of gene expression [22]. In general, one can expect more accurate and reliable predictions of cellular dynamics by incorporating delays into temporal models [23].

Now, it would all be perfect if solving a model's CME were an easy task. In most cases, obtaining a closed solution is simply impossible, whereas many times, a single simulation can be computationally expensive, if not unfeasible. When applying a stochastic simulation algorithm (SSA), the time steps in systems with large numbers of molecules and/or widely varying rate constants can become very small, thus limiting the feasible "real-time" span of the simulations. In order to overcome these limitations, coarse-graining techniques in time, space, or both can be applied. For instance, temporal coarse graining has been considered through the use of the so-called τ-leap methods [24–29], where a much larger time step can be used at the loss of a small amount of accuracy. Similar ideas have also been applied in the delay setting [30], thus rendering efficient algorithms that yield accurate simulations in time spans that are of actual interest to the experimentalists.

A further, yet natural, complication is that biological systems are in many cases characterized by complex spatial structure, low diffusion rates, or entail acute spatial dependencies, hence requiring spatially resolved simulations. Recent research suggests that molecular translocation processes can be well captured and modeled by means of time-delayed processes with specific delay distributions [31]. However, it is worth mentioning that spatially resolved algorithms are not replaceable in all cases. For instance, one must include spatial resolution whenever confronted with evidence of high spatial heterogeneity, anisotropies, molecular anomalous diffusion, or when single-particle tracking becomes strictly necessary, to name a few scenarios.

This section briefly explains the differences between a deterministic and a stochastic approach, pinpointing the limitations posed by adopting either regime. We then focus on stochastic systems available in the literature, portraying several exact SSAs and two coarse-grained methods incorporating time delays and spatial resolution.

15.3.1
Deterministic or Stochastic Modeling?

15.3.1.1 Deterministic Regime
A deterministic model is generally appropriate whenever all molecular species in a system are present in large numbers of molecules and, most importantly, when discreteness and internal noise have no noticeable effect in the time evolution of all relevant molecular species. If these conditions hold, the time evolution of synthetic

circuits (or any set of chemical reactions to that extent) can be represented by differential equations, based on the laws of Mass Action and assuming that reaction rates can be estimated on the basis of average values of the reactant densities.

In this regime, one can define $x = X(t)$ to be the vector containing the concentrations of the N species involved in the reactions at any time t, whereas any set of m chemical reactions can be characterized by two sets of quantities:

(1) The stoichiometric vectors $v_1, \ldots, v_m$, representing the update rules for each reaction.
(2) The propensity functions $a_1(X(t)), \ldots, a_m(X(t))$, representing the relative probabilities of each of the m reactions occurring.

The ODE that describes this chemical system is then given by

$$X(t)' = \sum_{j=1}^{m} v_j a_j(X(t)) \tag{15.110}$$

while the remainder of the problem lies in choosing an appropriate precompiled solver or, alternatively, to discretize the equations to a desired degree of accuracy without compromising stability. A detailed description of ODE solution methods does not lie within the scope of this chapter, but can easily be found in any Numerical Analysis textbook. However, natural phenomena are rarely deterministic and, more often than not, discrete effects and internal noise play a significant role in system's time evolution.

15.3.1.2 Stochastic Regime

Often the most important source of stochasticity stems from the fact that molecular reactions are random events, and it is impossible to say with certainty the specific type of reaction that will happen next, or when or where such event is to occur. Such intrinsic noise effects are captured by the CME, which describes the time evolution of the probability $P(X, t)$, for having x molecules in a system with m elementary reactions, N molecular species, and volume Ω at time t. It can be shown that for any state x, the probability distribution function satisfies the following discrete parabolic PDE, subject to appropriate initial and boundary conditions:

$$\frac{\partial P(x;t)}{\partial t} = \sum_{i=1}^{R} \alpha_j(x - v_j) P(x - v_j; t) - P(x;t) \sum_{i=1}^{R} \alpha_j(x) \tag{15.111}$$

The CME can be seen as a set of such PDEs, where each equation corresponds to each possible state of the system. The propensity functions are slightly different to those used in ODE models (cf. [32]), but depend only on the current state x. Hence, the process is memory less and considered Markovian.

Stochastic processes can be studied by trajectory-based approaches or by their underlying probability distribution function, which tracks how the probability of having a certain number of molecules changes over time. The first approach can be modeled, among some other options, through any of the variations of the SSA, first applied by [33] to simulate discrete chemical kinetics as the evolution of a discrete nonlinear Markov process.

The SSA is an exact procedure that describes a trajectory of a discrete nonlinear Markov process. It accounts for the inherent stochasticity of the m reacting channels and can only update the state vector by the addition or subtractions of integer numbers of molecules. There are several variations of the SSA without considering temporal coarse-graining that can be potentially more efficient on a number of different scenarios ([27, 34, 35] among many).

A very different approach is to note that the CME is a discrete parabolic PDE in which there is an equation for each configuration of the state space. When the state space is enumerated, the CME becomes a linear ODE and the probability density function is $p(t) = e^{At}p(0)$, where A is the state-space matrix. It should be noted that, even for relatively small systems, the dimension of A can be extremely large, rendering it a computationally unfeasible approach. However, in many cases, only a finite set of states are actually reachable and/or of interest to the particular biological question. A proposed finite state projection algorithm [36] reduces the size of the matrix A, while retaining a certificate on accuracy. Additionally, one can use Krylov subspace techniques [37] to efficiently compute the product of the exponential of a matrix and a vector.

Finally, it should be noted that there is a regime intermediate to the discrete stochastic and the continuous deterministic ODE ends of the spectrum, where some internal noise effects can still be captured while continuity arguments still apply. This leads to the so-called chemical Langevin equation that is an Itô stochastic ordinary differential equation, driven by a set of Wiener processes that describes the fluctuation in the concentrations of the molecular species:

$$dX = \sum_{j=1}^{m} v_j a_j(X(t)) + B(X(t))dW(t)$$

$$B(x) = \sqrt{C}$$

$$C = (v_1, \ldots, v_m) \operatorname{diag}(a_1(X), \ldots, a_m(X))(v_1, \ldots, v_m)^{\mathrm{T}} \qquad (15.112)$$

where $W(t) = (W_1(t), \ldots, W_N(t))$ is a vector of N independent Wiener processes. Noticeably, the Wiener increments $dW_j = W_j(t+h) - W_j(t)$ are normally distributed with a zero mean and a standard deviation equal to h, the discretization time step.

There are some numerical methods designed for the solution of SDEs (e.g., the Euler–Maruyama method) that can be used to simulate the chemical kinetics in this intermediate regime. However, it should be noted that implementing such a system may not be straightforward as some reaction propensities can become negative.

15.3.2
Stochastic Simulation Algorithms

It is only until recent years that discrete stochastic simulation techniques have been widely used to help understand the dynamic behavior of biochemical systems. However, SSAs have been around for several decades. In 1966, Young and Elcock [38] published the first basic features of the kinetic Monte Carlo (KMC) method, followed in 1975 by Bortz et al. [39], with the development of the *n-fold way* (also known as BKL), a KMC algorithm simulating the Ising model. In the following year, Gillespie coined the term SSA, stochastic simulation algorithm, for a KMC method describing chemical kinetic evolution in time. For simplicity, we will use the common terminology of SSA, but it should be noted that the algorithm and the time advancement scheme of the SSA are essentially the same as in BKL.

In the last decade, countless clever variations of the SSA have been published, dealing with acute separation of timescales [40–42], or simply more efficient computations [34, 35], among many. However, it is worth noting that there are two schemes in which natural extensions of the SSA are necessary: delays and spatial processes, both of which are explained later in the chapter. Additionally, we summarize some of the currently used exact and coarse-grained algorithms that can dramatically improve the computational performance.

15.3.2.1 Exact Algorithms
Stochastic Simulation Algorithm The SSA is a numerical Monte Carlo procedure that can be used to simulate the time evolution of a set of molecular species affected by a given set of reactions. In a nutshell, the SSA generates simulated trajectories of the system state, where each trajectory is an exact numerical realization of the CME. The SSA is exact with respect to the CME since it defines the probability, given any particular state, of only one reaction occurring within the next infinitesimal time internal.

Let us consider a well-stirred volume Ω kept at constant temperature, populated by molecules from N molecular species $\{S_1, \ldots, S_N\}$ interacting with each other through M chemical reaction channels $\{R_1, \ldots, R_M\}$. If we define $X(t)$ to be the vector containing the concentrations of the N species involved in the reactions at any time t, we can define a propensity function $a_j(x)$ in a given state $X(t) = x$ for each reaction R_j ($j = 1, \ldots, M$), such that $a_j(x)\,dt$ is the probability that one reaction R_j will occur somewhere inside Ω in the next infinitesimal time interval $[t, t + dt)$.

The missing ingredient is an update rule for the occurrence of reactions. For such, each reaction can be characterized by its stoichiometric vector v_j, defining the state change in the number of species due to the occurrence of one reaction R_j. Given a current state x, the probability of state change per unit of time is constant

$$a_0(x) = \sum_{k=1}^{M} a_k(x) \tag{15.113}$$

Algorithm 1: SSA

Data: reactions defined by reactant and product vectors, stoichiometry, reaction rates, initial state $X(0)$, simulation time T
Result: state dynamics
begin
    while $t < T$ **do**
        generate U_1 and U_2 as $U(0, 1)$ random variables
        $a_0(X(t)) = \sum_{j=1}^{m} a_j(X(t))$
        $\theta = \frac{1}{a_0(X(t))} \ln(1/U_1)$
        select j such that
            $\sum_{k=1}^{j-1} a_k(X(t)) < U_2 a_0(X(t)) \leq \sum_{k=1}^{j} a_k(X(t))$
        $X(t+\theta) = X(t) + \nu_j$
        $t = t + \theta$
end

Fig. 15.4 Algorithmic representation of the "direct method."

By consequence, the waiting time to the next reaction is an exponential random variable with mean $1/a_0(x)$. The reaction index j is an integer random variable with point probabilities $a_j(x)/a_0(x)$. These two random variables and their distributions are the basis of the SSA.

At each step, the SSA must (1) evaluate an exponential waiting time, τ, for the next reaction to occur and (2) indicate which reaction occurs. The state vector is updated after τ units of time by the addition of the jth stoichiometric vector to the previous value of the state vector, that is $X(t+\tau) = X(t) + \nu_j$, and this procedure is repeated to evolve the system through time. This update technique is quite straightforward, but what are the ways in which one can generate information about a type of reaction happening and its corresponding waiting time?

In [32], Gillespie introduced two basic mechanisms for implementing the Monte Carlo step. The first is the so-called direct method (Fig. 15.4), where two independent random numbers r_1 and r_2 are drawn from the uniform distribution in the unit interval $U(0, 1)$, to define τ and the reaction index j in the following manner:

$$\tau = \frac{1}{a_0(x)} \text{Ln}\left(\frac{1}{r_1}\right), \quad j \text{ st } \sum_{k=1}^{j-1} a_k(x) < r_2 \cdot a_0(x) < \sum_{k=j+1}^{m} a_k(x) \quad (15.114)$$

The second way is the so-called first reaction method, where a "tentative reaction time" for each reaction channel is calculated. Namely, one calculates $\tau_j = 1/a_0(x) \cdot \ln(1/r_j)$ for each $j \in [1, M]$, where each r_j is a random number from the uniform distribution in the unit interval. The next reaction happening in Ω is then defined to be of type R_ν, where $\tau_\nu = \min_{j \in [1...M]} \tau_j$.

Finally, a very clever modification of the "first reaction method" was proposed in [34]. In this algorithm (Fig. 15.5), all "tentative reaction times" and propensities are stored, recalculating only the affected ones at each time step, while using appropriate data structures to store these values and access them efficiently. Additionally,

Algorithm 2: Next reaction method

Data: stoichiometry, reaction rates, initial state $X(0)$, simulation time T, reaction dependency graph G, calculated propensity a_i and putative reaction time τ_i for each reaction R_i, the latter stored in an indexed priority queue Q
Result: state dynamics
begin
  while $t < T$ do
    let R_i be the next reaction, i.e., the reaction whose putative time τ_i stored in Q is least
    $X(t + \tau_i) = X(t) + \nu_i$
    $t = t + \tau_i$
    for each reaction R_j with edge (i, j) in G do
      Recalculate a_j
      if $i \neq j$ then
        set $\tau_j := (a_{j,old}/a_{j,new})(\tau_j - t) + t$
      if $i = j$ then
        generate tentative reaction time θ for R_i according to
        $\theta = \frac{1}{a_i(X(t))} \ln(1/r)$ with $r \in U(0,1)$
        set $\tau_i = t + \theta$
end

Fig. 15.5 Algorithmic representation of the "next reaction method."

reactions are organized in a dependency graph G, where reactions i and j are connected with a directed edge (i, j) if and only if any of the reactants and products of reaction i is a reactant of reaction j. Furthermore, due to its construction, this algorithm uses only one random number per simulation event and its computational time is no longer proportional to the number of reactions, but to their logarithm.

The Delayed Stochastic Simulation Algorithm (DSSA) If correctly posed, a time delay can account for sets of biochemical processes and events on a microscopic timescale that would otherwise render us unable to compute cell dynamics in real time. In such a perspective, key reactants can be processed and their corresponding products will not be present in the system until a certain future time point, that is, when time advances by the corresponding delay.

Mathematically, delays can be viewed as an additional parameter affecting the dynamic evolution of a biochemical system, where the DDE formulation is

$$X(t)' = \sum_{j=1}^{m} \nu_j a_j \big(X(t - \tau_j)\big) \tag{15.115}$$

Such a system of equations can be solved numerically, and there is an extensive literature on methods to do so (many of which are implemented in MATLAB).

However, if intrinsic noise plays a significant role, a generalization of the SSA to account for all possible delayed reactions is due.

Bratsun et al. [43] developed a delay SSA, but this algorithm does not consider waiting times for delayed reactions and only nonconsuming reactions can be specified as delayed. Independently, Barrio et al. [23] developed another delay SSA, taking proper account of waiting times for delayed reactions and including the possibility of specifying both consuming and nonconsuming reactions as delayed. Afterward, Cai [44] introduced a direct delay SSA method and showed that both the algorithm developed by Barrio et al. (from now on specified as delayed stochastic simulation algorithm (DSSA)) and the direct method are exact SSAs for chemical reaction systems with delays. A CME accounting for delays (DCME) has been derived from first principles, represented as a system of DDEs [23, 45]. Moreover, the probability density function of $X(t)$ in the delayed setting is completely determined by the DCME and, just as the trajectories simulated by SSA are exact realizations of the CME, the trajectories of the DSSA are shown to be exact trajectories of the DCME

The DSSA (Fig. 15.6) differs from the SSA by making a clear distinction between the reaction waiting time and reaction delay. The former is the time between two consecutive reactions whereas the latter is the time elapsed from the processing of the reactants to the appearance of the products. In a nutshell, simulation proceeds in the standard way (SSA) if nondelayed reactions take place. However, if the next reaction index points to a delayed reaction, then one has to necessarily distinguish between consuming and nonconsuming reactions. In the case of nonconsuming reactions, the corresponding reactants and products are not updated. Instead, the state update is scheduled for "present time + delay," which will be reached in a future simulation step. When that happens, the last drawn reaction is simply ignored and instead the state is updated according to the delayed reaction, whereas the simulation is continued at the delayed reaction time point. On the other hand, if the reaction is consuming, reactants and products of delayed consuming reactions must be updated separately: (1) reactant consumption updates the state when the delayed reaction is selected and (2) product generation is updated when the reaction is completed.

Spatially Resolved Stochastic Simulation Algorithms Temporal modeling of biological systems can many times fall short of describing experimental results, due to unforeseen or neglected spatial dependencies. For instance, a system can be embedded within complex spatial structures, or molecular motion is described by low diffusion rates, scenarios in which spatially resolved simulations may become mandatory.

The most straightforward spatial technique is the representation of chemical kinetics through reaction–diffusion PDEs. However, this deterministic approach is only valid when dealing with large molecular concentrations and when noise is not amplified throughout the system. If at least one of these conditions fails to hold, one must rely on spatial stochastic simulators, which can be discrete or continuous and have different levels of spatial resolution.

Algorithm 3: DSSA

Data: reactions defined by reactant and product vectors, consuming delayed reactions are marked, stoichiometry, reaction rates, initial state $X(0)$, simulation time T, delays

Result: state dynamics

begin
  while $t < T$ **do**
    generate U_1 and U_2 as $U(0, 1)$ random variables
    $a_0(X(t)) = \sum_{j=1}^{m} a_j(X(t))$
    $\theta = \frac{1}{a_0(X(t))} \ln(1/U_1)$
    select j such that
    $\sum_{k=1}^{j-1} a_k(X(t)) < U_2 a_0(X(t)) \leq \sum_{k=1}^{j} a_k(X(t))$
    if *delayed reactions are scheduled within* $(t, t + \theta]$ **then**
      let k be the delayed reaction scheduled next at time $t + \tau$
      if k *is a consuming delayed reaction* **then**
        $X(t + \tau) = X(t) + \nu_k^p$ (update products only)
      else
        $X(t + \tau) = X(t) + \nu_k$
      $t = t + \tau$
    else
      if j *is not a delayed reaction* **then**
        $X(t + \theta) = X(t) + \nu_j$
      else
        record time $t + \theta + \tau_j$ for delayed reaction j with delay τ_j
        if j *is a consuming delayed reaction* **then**
          $X(t + \theta) = X(t) + \nu_j^s$ (update reactants)
      $t = t + \theta$
end

Fig. 15.6 Pseudocode of the DSSA.

Spatially resolved simulations are, in general, very costly as compared with their solely temporal counterparts. As a result, one should always keep in mind the tradeoff between simulation time and level of resolution. The highly resolved end of the spectrum is represented by lattice and off-lattice particle methods, the first being commonly referred to as a KMC method. Particle methods can provide very detailed simulations of highly complex systems at the cost of exceedingly large amounts of computational time and, possibly, restrictions on the size of the simulation domain. Hence, such detailed simulations can often only yield short simulation time spans that may not be of interest to the experimentalists.

Lattice-based methods describe a computational mesh (generally 2D or 3D) that represents a membrane or the interior of some cellular compartment. The lattice is then "populated" with particles of different molecular species, either at random or at chosen spatial locations, depending on the theoretical question at hand. Each particle is able to diffuse throughout the simulation domain by jumping to empty neighboring sites and, depending on user-specified reaction rules, appropriate chemical reactions can take place with a certain probability. Some examples of lattice-based methods can be found in [46, 47]. In contrast, off-lattice methods only use a mesh to efficiently localize particles, but in this case, particles retain information on their specific spatial coordinates. For each particle, a reaction bin is drawn, whose size regularly depends on the particle's diffusion rate, among other possible factors. If one or more molecules happen to be inside such a bin, appropriate chemical reactions can take place with a certain probability, and if a reaction is readily performed, the reactant particles are flagged. An off-lattice simulator example can be found in [48].

An alternative to particle methods, albeit still computationally expensive in many scenarios, is the discretization of the reaction–diffusion chemical master equation (RDCME) into reactive neighboring subvolumes (SVs). There are a few algorithms in the literature extending discrete stochastic simulators to approximate solutions of the RDCME by introducing diffusion steps as first-order reactions with a reaction rate constant proportional to the diffusion coefficient. Baras and Mansour [49] provide the specific outline for extending discrete stochastic simulators to the RDCME regime, whereas the algorithms in [50, 51] provide clever extensions of the "next reaction method" [34]. Additionally, a great review on the construction of such methods can be found in [52].

15.3.2.2 Coarse-Grained Methods

All the above-described delayed and nondelayed, spatial and nonspatial SSAs have a common drawback: their high computational costs when dealing with large numbers of molecules or widely varying rate constants. In either case, time advances in exceedingly small simulation time steps, making the overall simulation computationally expensive or even infeasible. In order to overcome this limitation, one can coarse grain the simulation, accounting for many events in one single larger time step (coarse graining in time), averaging over spatial regions (coarse graining in space), or both. In what follows, we focus on temporal coarse-graining solely. Nevertheless, many examples of spatial coarse graining can be found throughout the literature, such as [53].

Poisson and Binomial τ-Leap Methods The general idea behind the so-called τ-leap methods is to account for many events in one single larger time step, where the accelerated time advancement comes at the loss of a small amount of accuracy. Initially, Gillespie [24] proposed the Poisson and the midpoint τ-leap methods, in which the number of reactions in each τ-leap are sampled from a Poisson distribution, and the τ step is controlled by a selection strategy that depends on a prespecified control parameter ε, such that $0 < \varepsilon \ll 1$. In the Poisson τ-leap method, the update procedure no longer depends on the regular propensity functions $a_j(X(t))$, but on samples from the Poisson distribution with mean $a_j(X(t))$. The update procedure can be written as

$$x(t+\tau) = x(t) + \sum_{j=1}^{m} K_j v_j$$

where $K_j = \text{Poiss}(a_j(X(t))\tau)$ for $j = 1, \ldots, M$ \hfill (15.116)

Further improvements were made by Gillespie and Petzold [54], Rathinam et al. [55], and Cao et al. [29, 56].

However, samples from a Poisson distribution are positive but unbounded and, by consequence, negative numbers of molecules can occur when updating the system if large enough step sizes are used. In order to avoid this, Tian and Burrage [25] and later Chatterjee et al. [57] proposed the binomial τ-leap method, in which the numbers of reactions in a leap are drawn from a binomial distribution. In other words, the various K_js take the form $K_j = B(N_j, P_j)$, where the variables N_j and P_j represent the sample size and probability of occurrence of reaction type j, respectively. There are some subtleties in the form of the N_j and P_j, for which we recommend the interested reader to refer to the original publications.

Finally, initial τ-leap algorithms (e.g. [25, 26]) were not able to capture accurately the dynamics of particular sets of chemical kinetics, for example, oscillating patterns of gene expression, due to insufficient numbers of reactions drawn in τ-leap steps. In [30], a new generalized τ-leap method (Bτ-DSSA) is presented that addresses the difficulties associated with complex chemical kinetics and, additionally, introduces delays into the τ-leap framework. We describe the full algorithm in the next subsection, where it should be noted that the pseudocode can be used for reaction sets containing reactions with or without delays.

Modified τ-Leap Method, for Delayed and Nondelayed Reactions (Bτ-DSSA) The devil lies in the details, and estimating a proper maximum number N_j of potential reaction events of type R_j for the binomial random variables $B(N_j, P_j)$ is essential for an accurate reproduction of system dynamics. In [30] the authors present a significant upgrade to temporal coarse-graining algorithms distinguishing between cases where (1) R_j is an isolated reaction that does not share reactants with any other reactions or (2) R_j is part of a large interacting reaction network where multiple reactions share the same reactants.

The algorithm (Fig. 15.7) contains many subtleties, for which the interested reader is encouraged to read the original paper. However, as a rough summary, the Bτ-DSSA samples reaction numbers from Binomial distributions $B(N_j'', P_j)$, where $N_j'' = N_j(x, \xi)$, with $\xi \equiv (\xi_1, \ldots, \xi_M)$ and $\xi_i \leq N_i(x)$. As mentioned before, $N_i(x)$ is the maximal number of potential reaction events of type R_i when $\xi_1, \ldots, \xi_M$ reactions of $R_1, \ldots, R_M$ occur in the τ-step. For $N_j(x, \xi)$, it is assumed that $\xi_j, \ldots, \xi_M = 0$ since only the already sampled reaction numbers $\xi_j, \ldots, \xi_{j-1}$ are considered. However, unlike the original binomial τ-leap method by Tian and Burrage [25], the N_j are calculated considering only those reactions R_i (and hence ξ_i) that share reactant species with R_j.

As a consequence, the maximal number of potential reaction events in the Bτ-DSSA is usually larger than in the original binomial τ-leap method, where num-

---
Algorithm 4: Bτ-DSSA

Data: reactions defined by reactant and product vectors, consuming
  delayed reactions are marked, stoichiometry: $\nu = -\rho + \pi$ (with ρ
  and π being update vectors for left-hand-side and right-hand-side
  of reaction, respectively), reaction rates, initial state $X(0)$,
  simulation time T, delays, pre-specified $K \in [1, 10]$

Result: state dynamics

begin

1. **Calculate** $a_1(x), \cdots, a_m(x)$, $a_0(x) = \sum_{j=1}^{m} a_j(x)$;
   and $N_1(x), \cdots, N_m(x)$

2. **Choose** a τ-selection procedure: update corresponding variables

3. **Check step-size conditions.** For each reaction R_j
   Calculate $N'_j = \min\{N_i(x), i \in I_j\}$ and $a'_j(x) = \sum_{i \in I_j} a_i(x)$
   If $(N'_j > 0)$ AND $(a'_j(x)\tau/N'_j > 1)$ then $\tau = N'_j/a'_j(x)$

4. If $\tau \leq K/a_0$ perform a normal (D)SSA step, otherwise go to (5)

5. **Sample.** Initialise $\xi \equiv (\xi_1, \cdots, \xi_m) = 0$
   For each reaction R_j
     Calculate $N''_j = N_j(x, \xi)$
     Generate a sample value ξ_j for the reaction number of type R_j
     If $N''_j > 0$ then $\xi_j = B(N''_j, P_j)$ with $P_j = a_j(x)\tau/N''_j$ else $\xi_j = 0$.

6. **Update non-delayed and delayed reactions.**
   The subscripts nd_j and d_j represent the jth nondelayed and
   delayed reaction, respectively.
   M' denotes the number of non-delayed reactions.

   6.1 Delayed R_{d_j} with delay δ_j:
     Record ξ_{d_j} random update time points $t + \delta_j + u_k\tau$
     with $u_k \in U(0, 1), k = 1, \cdot, \xi_{d_j}$
     If R_{d_j} is a consuming, delayed reaction,
       update $x(t + \tau) = x(t + \tau) + \rho_{d_j}$
   6.2 Nondelayed: $x(t + \tau) = x(t) + \sum_{j=1}^{M'} \xi_{nd_j} \nu_{nd_j}$
   6.3 Delayed (scheduled within $[t, t + \tau)$) :
     $x(t + \tau) = x(t + \tau) + \pi_{d_j}$ (consuming, delayed reactions)
     $x(t + \tau) = x(t + \tau) + \nu_{d_j}$ (remaining reactions)

end

Fig. 15.7 Pseudocode of the modified τ-leap method, accounting for delayed and nondelayed reactions.

bers of delayed reactions are sampled in the same way as numbers of nondelayed reactions. However, if a system includes delayed reactions, the update of the system state has to distinguish between delayed consuming and nonconsuming reactions scheduled within the τ-leap, and sample the update times of all delayed reactions drawn for the τ-leap.

Coarse-Grained Spatial Stochastic Simulation Algorithm: Bτ-SSSA The idea behind τ-leaping in spatial stochastic simulations is to account for several diffusion and reaction events in one larger time step, without compromising spatial or temporal accuracy. Marquez-Lago and Burrage [58] presented the binomial τ-leap spatial simulation algorithm, Bτ-SSSA, a coarse-grained version of an existing spatial SSA known as the next subvolume method (NSM, [50, 59, 60]).

In both the NSM and the Bτ-SSSA, the spatial domain is divided into separate subvolumes SVs that are small enough to be considered homogeneous by diffusion over the timescale of the reaction. At each step, the state of the system is updated by performing an appropriate reaction or by allowing a molecule to jump at to a random neighboring SV, where such "diffusion" is modeled as a unary reaction. In the beginning of the simulation, the expected time for the next event in each SV is calculated, including the reaction (a_j) and diffusion (d_j) propensities of all molecules contained in that particular SV at that particular time. However, time for next events will only be recalculated for those SVs that were involved in the current time step, on which the event queue is reordered.

In the Bτ-SSSA [58], the SV at the top of the time event queue (i.e., with the shortest reaction–diffusion τ-leap) is again selected at the beginning of each iteration. Then, within the chosen SV, all randomly chosen possible events are executed. However, such events might alter the concentrations of neighboring SVs, and so, a new τ-leap for all SVs that were involved in the current τ-leap is calculated. The time event queue in increasing time is reordered and, once more, the SV at the top of the time event queue is chosen, closing the loop. Of course, this all sounds simple, but many subtleties lie within, for which we refer the interested reader to the original publication. In a nutshell, the definitions of all matrices and vectors used in the Bτ-SSSA can be found in Table 15.1, whereas the pseudocode of the algorithm is presented in Fig. 15.8.

It is worth noting that the selection of a τ-leap for the selected SV implies several steps. Initially, step 3 of Algorithm 4 is obtained from matrix **F** and vectors **M**, **Σ**, and **T**, as defined in Table 15.1. The initial τ will be the minimum of the two entries in vector **T**. Subsequently, in step 4, a minimal τ will be chosen. Namely, for all molecular species inside the selected SV and for all reactions in which such species are involved, τ will necessarily comply with the restriction $P_j \leq 1$, where $P_j = M_j/N_j$, and depending on the type(s) of reaction(s):

(a) Heterodimer reaction (species k and m): $M_j = a_j + d_k + d_m$, $N_j = \min[C(SV, k), C(SV, m)]$.
(b) Homodimer reaction (species k): $M_j = a_j = 2d_k$, $N_j = \lfloor C(SV, k)/2 \rfloor$.
(c) Unary or Hill-type reaction (involving species k): $M_j = a_j = d_k$, $N = C(SV, k)$.
(d) A combination of any of the above. This is the case when any particular species involves more than one reaction, in which case, the restriction is $\tilde{P}_j \leq 1$, where $\tilde{P}_j = \tilde{M}_j/\tilde{N}_j$ and the $\tilde{M}_j$ is a sum of the corresponding M_j terms, as defined above, reflecting a reaction subnetwork within the system. On the other hand, $\tilde{N}_j$ is the minimum of all corresponding N_j terms.

Algorithm 5: Bτ-SSSA

Data: state $X(0)$, simulation time T, reaction and diffusion counter (set to zero), stoichiometric vectors, reaction and diffusion rates, the matrices C, R, N, and Q where Q is ordered according to the event time; ρ (lower limit on the number of events in a τ-step)
Result: state dynamics
begin

1. Choose the subvolume $SV = Q(1,1)$ with least event time
2. **If** total no. of molecules inside SV ($\sum_k C(SV,k)$) is ≤ 2:
   Let r_{SV} and s_{SV} be the reaction and diffusion propensity, respectively, in SV.
   2.1 If $r_{SV} = 0$ go to Step 7.
   2.2 Generate a random number ζ.
       If $\zeta < r_{SV}/(r_{SV} + s_{SV})$ go to Step 6.
       Else go to Step 7.
3. Calculate initial τ.
4. Check step-size conditions for τ (adjust τ if necessary).
5. **Single or multiple events?** If $r_{SV} > 0$:
   If $\tau < \rho Q(1,2)$ go to Step 2.2 else go to Step 8.
6. **Single chemical reaction event**
   6.1 According to the probability of occurence $P_j = a_j(SV)/r_{SV}$ of each reaction, sample the reaction j that occurred inside SV.
   6.2 Update the state (matrix C) in subvolume SV
   6.3 Increase the reaction counter by 1 and the time by $Q(1,2)$.
   6.4 Recalculate r_{SV} and s_{SV}. Determine a new next event time $Q(SV,2) = 1/(r_{SV} + s_{SV})\log(1/\zeta)$ for subvolume SV where ζ is a $U(0,1)$ random number.
   6.5 Reorder Q and go to Step 1.
7. **Single diffusion event**
   7.1 According to the probabilities $P_s = d_s(SV)/s_{SV}$ for each species, sample a species s in SV.
   7.2 Randomly select the neighbor SV_{new} of SV where a molecule of species s is diffusing into.
   7.3 Update the states (matrix C) in subvolume SV and SV_{new} according to the diffusion event
   7.4 Increase the diffusion counter by 1 and the time by $Q(1,2)$.
   7.5 Recalculate r_{SV}, s_{SV} and $r_{SV_{new}}$, $s_{SV_{new}}$
       Determine next event times $Q(SV,2)$ and $Q(SV_{new},2)$ for SV and SV_{new} (as in 6.4)
   7.6 Reorder Q and go to Step 1.
8. **τ-step**
   8.1 Sample $N_{j_{total}}$ from $B(N_i, P_j\tau)$ for each reaction R_j
       Sample the no. of reaction events ξ_j from $B(N_{j_{total}}, a_j/M_j)$
       Sample the no. of diffusion events ψ_s for species s from $B(N_{j_{total}} - \xi_j - \sum_{t=1}^{s-1} \psi_t, d_s/M_j)$
   8.2 Update reaction and diffusion counter; increase time by τ.
   8.3 Update C according to reaction events.
   8.4 Update C according to diffusion events.
   8.5 Recalculate r_{SV} and s_{SV}.
   8.6 Update Q, reorder Q and go to Step 1.

end

Fig. 15.8 Pseudocode of the Bτ-SSSA.

Table 15.1 Description of the matrices and vectors used in the NSM method and Bτ-SSSA.

N_R	Number of possible reactions within the system
N_S	Number of different molecular species
ε	Error approximation parameter
$\underline{D}$	$N_S \times 1$ vector containing the molecular diffusion rates of each species
$\underline{K}$	$N_R \times 1$ vector containing the reaction rate constants
$\underline{N}$	$N_V \times 6$ matrix that specifies, for each SV, the 6 possible neighboring SVs to which molecules can jump by diffusion (up, down, left, right, in front and behind)
$\underline{C}$	$N_V \times N_S$ matrix that specifies the number of molecules of each species in each SV
$\underline{R}$	$N_V \times 3$ matrix that contains, for each SV, the sums of reaction propensities $r_j = \sum_{i=1}^{N_V} a_{ij}$ (first column), the sum of diffusion propensities $s_j = \sum_{i=1}^{N_V} d_{ij}$ (second column), and $r_j + s_j$ (third column)
$\underline{Q}$	$N_V \times 2$ matrix that contains the SV's ID (first column) and the time in which the next set of events inside that SV will take place (second column). The matrix is ordered with respect to increasing times
$\underline{V}$	$N_S \times N_R$ matrix, containing the stoichiometric vectors v
$\underline{F}$	$(N_R + N_S) \times (N_R + N_S)$ matrix, with the following nonzero entries: $f_{jk} = \sum_{i=1}^{N_R} \frac{\partial a_j(x)}{\partial x_i} v_{ik}$ for $j, k = 1, \ldots, N_R$ and $g_{jk} = \sum_{i=1}^{N_S} \frac{\partial d_j(x)}{\partial x_i} e_{ik}$ for $j, k = 1, \ldots, N_S$
$\vec{M}$	$(N_R + N_S) \times 1$ vector, with entries $\mu_j = \sum_{i=1}^{N_R} f_{jk}(x) a_j(x)$ for values of $j, k = 1, \ldots, N_R$ and $\mu_j = \sum_{i=1}^{N_R} g_{jk}(x) d_j(x)$ for $j = N_{R+1}, \ldots, N_{R+S}$
$\vec{\Sigma}$	$(N_R + N_S) \times 1$ vector, with entries $\sigma_j = \sum_{i=1}^{N_R} f_{jk}^2(x) a_j(x)$ for values of $j = 1, \ldots, N_R$ and $\sigma_j = \sum_{i=1}^{N_R} g_{jk}^2(x) d_j(x)$ for $j = N_{R+1}, \ldots, N_{R+S}$
$\vec{T}$	$N_V \times 1$ vector, with entries $\tau_j = \min_k [\frac{\varepsilon(r_j+s_j)}{|\mu_k|}, \frac{\varepsilon^2(r_j+s_j)^2}{|\sigma_k|}], k \in [1, N_R + N_S], j \in [1, N_V]$

15.4
Application

Synthetic gene circuits can be constructed *in silico* as a simple composition of biological parts, devices, and pools. Process modeling tool (ProMoT) [61] is a computational tool for the graphical modeling of complex systems that provides the framework for this task. The software (written in Lisp) is equipped with a Java graphic user interface (GUI) that serves as canvas where parts and pools are placed and connected through wire-like lines following the "drag & drop" fashion commonly used in electronics tools (see, for instance, SPICE [62]).

The *de novo* design of a synthetic gene circuit requires, first, to generate all the necessary parts and pools. In our case, this is achieved by running the corresponding Perl scripts (now integrated into ProMoT) where parameter values and part

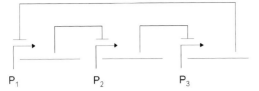

Fig. 15.9 The Repressilator scheme. Each transcription unit is represented by a promoter (P) followed by a straight line. The ring structure stems from the action of the third repressor on the first promoter.

structure, such as operator or b-site number, have to be specified as input. Every circuit component is encoded in model description language (MDL [63]) – the Lisp-based, object-oriented language of ProMoT. Inside an MDL file, parts and pools are structured in three blocks called *terminals*, *variables*, and *equations*. Terminals represent the interfaces between two parts and contain the names of the fluxes of signal carriers that have to be exchanged; variables can be either species concentrations or kinetic parameters, whereas equations are both the ODEs that account for the dynamics of the biochemical species and the algebraic relations that calculate the amount of the fluxes of signal carriers.

Once created, parts can be gathered into devices that are normally composed by one or more transcription units. Then, parts, pools, and devices are displayed on the ProMoT-GUI where matching terminals are connected to each other. When the circuit design is completed, ProMoT produces an MDL file that contains a description of the whole circuit. This file can then be exported into formats suitable for simulations, like Matlab or SBML (see http://www.sbml.org and [64]). Particularly, SBML is a widely used language in systems and synthetic biology and there is a large number of tools able to interpret it. Any SBML code describing a synthetic circuit will contain a list of all the biological species and the chemical reactions (with the corresponding rates) present in the system. Such information is required by any computational tool in order to simulate the circuit dynamics, either deterministically or stochastically.

In the following text, we illustrate how a synthetic gene circuit can be designed by means of composable parts and pools. Our sample system will be the Repressilator [7], a well-known synthetic gene circuit. We will then give a short description of the circuit ODE-based model that arises from its components and show results of deterministic simulations. Finally, we will perform exact stochastic simulations based on the elementary reactions composing the gene circuit, and discuss the differences between the time-dependent circuit behavior as a result of both deterministic and stochastic simulations.

15.4.1
The Repressilator

As sketched in Fig. 15.9, the Repressilator appears as a "ring oscillator" made of three genes. Every gene produces a different repressor that acts on the next tran-

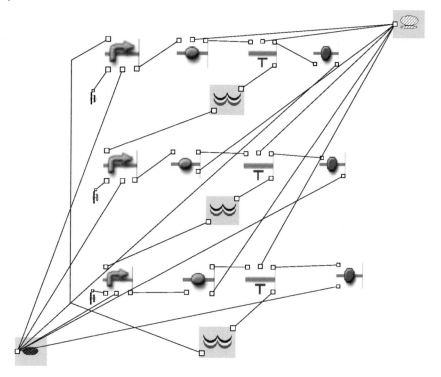

Fig. 15.10 Representation of the Repressilator using ProMoT. In this *in silico* implementation all the standard parts are shown and, beside each promoter, an "off" battery is placed to indicate that no chemicals are sent to this regulatory part.

scription unit, that is: the first gene represses the second one, the second gene represses the third one, and finally, the third gene closes the cyclic structure by repressing the first one.

In our *in silico* implementation (see Figs. 15.10 and 15.11), we consider three identical transcription units. This implies that we need only four different kinds of parts to build the whole circuit: a repressible promoter, a constitutive RBS, a coding region that synthesizes a repressor, and a terminator. To avoid cross talk among the transcription units, every repressor has to be stored into a different pool. Furthermore, the initial concentration of one of the three repressors has to be set to be greater than zero, as otherwise the oscillations damp in time and the network would reach a stable steady state. Finally, the RNA polymerase and ribosome pools complete the circuit design.

Looking in detail at each part, we observe that every promoter contains two operators to allow the binding of two molecules of the corresponding repressor (referred to as R_1^a), achieving in this way a strong repression. Following [7], we reproduce a partially cooperative behavior among repressors that is unveiled only in the transcription factors' dissociation from the DNA. Hence, we can make the following

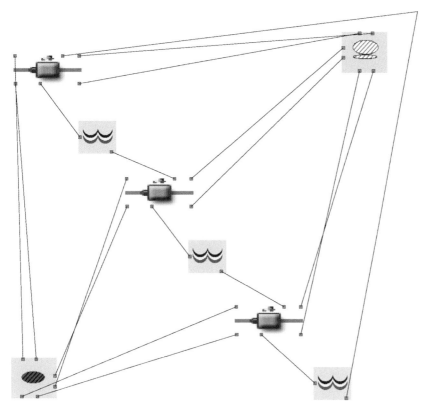

Fig. 15.11 Representation of the Repressilator using ProMoT. A more compact realization of the Repressilator requires to substitute every transcription unit with a device – a protein generator in this case.

simplifications on the α and β coefficients that account for the interaction between repressors and DNA (see Eq. (15.4)–(15.7)):

$$\alpha_{1f} = \alpha_{1t} = \alpha_1 \tag{15.117}$$

$$\alpha_{2f} = \alpha_{2t} = \alpha_2 \tag{15.118}$$

$$\beta_{1f} = \beta_{1t} = \beta_1 \tag{15.119}$$

Furthermore, repressors are supposed to be synthesized directly in their active configurations. Hence, no corepressors are needed in the system. The biochemical reactions present in this part reflect the ones outlined in Section 15.2.2.1. From these, we obtain the following set of ODEs, each describing the dynamics of a different promoter state:

$$\frac{dO_1^f O_2^f}{dt} = -\alpha_1 R_1^a O_1^f O_2^f + \beta_1 O_1^t O_2^f - \alpha_2 R_1^a O_1^f O_2^f + \beta_{2f} O_1^f O_2^t$$
$$+ k_{D1}(O_1^t O_2^f + O_1^f O_2^t) \quad (15.120)$$

$$\frac{dO_1^f O_2^t}{dt} = -\alpha_1 R_1^a O_1^f O_2^t + \beta_1 O_1^t O_2^t + \alpha_2 R_1^a O_1^f O_2^f$$
$$- \beta_{2f} O_1^f O_2^t + k_{D1}(O_1^t O_2^t - O_1^f O_2^t) \quad (15.121)$$

$$\frac{dO_1^t O_2^f}{dt} = \alpha_1 R_1^a O_1^f O_2^f - \beta_1 O_1^t O_2^f - \alpha_2 R_1^a O_1^t O_2^f + \beta_{2t} O_1^t O_2^t$$
$$+ k_{D1}(O_1^t O_2^t - O_1^t O_2^f) \quad (15.122)$$

$$\frac{dO_1^t O_2^t}{dt} = \alpha_1 R_1^a O_1^f O_2^t - \beta_1 O_1^t O_2^t + \alpha_2 R_1^a O_1^t O_2^f$$
$$- \beta_{2t} O_1^t O_2^t - 2k_{D1} O_1^t O_2^t \quad (15.123)$$

$$\frac{d[PolO_1^f O_2^f]}{dt} = k_1 Pol^{free} O_1^f O_2^f - (k_{-1} + k_2)[PolO_1^f O_2^f] \quad (15.124)$$

Aside, every promoter is connected to a single part, an RBS – toward which it sends two PoPS fluxes: $PoPS^{out}$ and $PoPS^{lk}$, as described in Eqs. (15.19) and (15.20) – and to two pools: the RNA polymerase and a transcription factor one. With the pools, the promoter exchanges two different balance fluxes: $PoPS^b$ as in Eq. (15.16) and $FaPS^b$, defined as follows:

$$FaPS^b = -\alpha_1 R_1^a (O_1^f O_2^f + O_1^f O_2^t) + \beta_1 (O_1^t O_2^f + O_1^t O_2^t)$$
$$- \alpha_2 R_1^a (O_1^f O_2^f + O_1^t O_2^f) + \beta_{2f} O_1^f O_2^t + \beta_{2t} O_1^t O_2^t \quad (15.125)$$

Now, as mentioned above, close to each promoter lies a constitutive RBS. Here, the RNA polymerase dynamics is described by the formation of the $[PolB]$ complex, described by Eq. (15.21), and by the successive start of the elongation phase (see Eq. (15.22)). On the other hand, the ribosomes interact with the mRNA in a Michaelis–Menten fashion, as shown in Eq. (15.26). Furthermore, we have to account for the degradation of the free and the occupied b site to complete the RBS picture:

$$b \xrightarrow{k_d} \quad (15.126)$$

$$[rb] \xrightarrow{k_d} r^{free} \quad (15.127)$$

15.4 Application

All together, three ODEs are required for the RBS dynamics, corresponding to the three states ($[PolB]$, b and $[rb]$) taken into account in our model:

$$\frac{d[PolB]}{dt} = PoPS^{in} - k_{el}[PolB] \tag{15.128}$$

$$\frac{db}{dt} = k_{el}[PolB] - k_{1r}r^{free}b + (k_{-1r} + k_{2r})[rb] - k_d b \tag{15.129}$$

$$\frac{d[rb]}{dt} = k_{1r}r^{free}b - (k_{-1r} + k_{2r} + k_d)[rb] \tag{15.130}$$

Additionally, an RBS exchanges with the ribosome pool a balance flux ($RiPS^b$)

$$RiPS^b = -k_{1r}r^{free}b + (k_{-1r} + k_d)[rb] \tag{15.131}$$

and it is connected to a coding region, toward which it sends both an RNA polymerase and a ribosome flux. The former ($PoPS^{out}$) is the same as in Eq. (15.23), the latter ($RiPS^{out}$) is equal to $k_{2r}[rb]$, according to Eq. (15.26).

Since a constitutive RBS cannot give rise to ribosome leakage, the protein production inside the next coding region is represented by a simplified version of Eq. (15.51), describing a flux of repressors toward their pool, given by

$$FaPS^{out} = RiPS^C \tag{15.132}$$

Furthermore, the RNA polymerases, coming from the RBS, flow into the complex $[PolA]$ (see Eq. (15.42)) before returning to the elongation phase. Analogously, incoming ribosomes form the complex $[ra]$ with the start codon, as described by Eq. (15.43). Upon this, they translate the mRNA until binding to a STOP codon, forming a complex $[ru]$ as shown in Eq. (15.48). Hence, three states characterize this part and yield the following ODEs:

$$\frac{d[PolA]}{dt} = PoPS^{in} - k_{el}[PolA] \tag{15.133}$$

$$\frac{d[ra]}{dt} = RiPS^{in} - k_{el}^r[ra] \tag{15.134}$$

$$\frac{d[ru]}{dt} = RiPS^C - \zeta_r[ru] \tag{15.135}$$

Aside, while ribosomes go back to their pool once they reach the end of the coding region, RNA polymerases pass into a terminator and form the complex $[PolT]$ (see Eq. (15.64)) before leaving the DNA, as described in Eq. (15.65). This part marks the end of each of the three transcription units involved in the Repressilator. For the sake of simplicity, the readthrough is here neglected and, by consequence, changes in the state [$PolT$] are simply given by

$$\frac{d[PolT]}{dt} = PoPS^{in} - \zeta[PolT] \qquad (15.136)$$

and the outgoing flux of RNA polymerases directed to their pool is identical to Eq. (15.67).

Finally, the time derivative of each free signal carriers is computed inside the corresponding pool as a sum of the incoming and the outgoing fluxes, as we explained before. It should be noted that, inside the RNA polymerase and ribosome pools, the constants N_P, N_R, N_C, and N_T have to be set equal to three. On the contrary, in every repressor pool, $N_{C\text{in}}$ and N_{Pb} are equal to one and a dimerization reaction, according to Eq. (15.90), is also taken into account.

On the whole, our model employs 44 ODEs and several kinetic parameters. However, this network contains a feedback loop and, as was reported in [7], oscillations are very sensitive to changes in the values of some kinetic parameters. In particular, sustained oscillations are strictly dependent on transcription and translation rates as well as on repressor and mRNA decay rates.

For our simulations, we relied on two different sets of data, with which we were able to reproduce the regime of sustained oscillations. Whenever possible, we considered the same parameter values provided in the original Repressilator [7]. However, as our model includes reactions portraying substeps "averaged out" in the original network, we were forced to search for additional values of kinetic parameters to account for these additional reactions. For such, we relied on other groups' published data, which had to be further tuned (see [10] for more details). In this way, we have shown the validity of our circuit design method, based on the composition of parts and pools.

In order to discuss simulation results depending on the modeling regime, we decided to portray the two simplest scenarios: deterministic and exact stochastic trajectories. For the sake of simplicity, we used the precompiled software COPASI [65], which provides the user with an ODE solver and an implementation of the Gibson–Bruck algorithm, allowing to run simulations directly from an SBML input file. However, we wish to stress that for other types of stochastic simulations (for instance, tailored coarse-graining, delayed, or spatial schemes), this software is not appropriate. In such scenarios, one should use other simulation tools or, as is often the case, one is simply forced to design and/or program an implementation from scratch.

Our first results (Figs. 15.12 and 15.13) show single deterministic and stochastic simulations of our "extended" Repressilator. As can be expected, the deterministic regime (Fig. 15.12) yields smooth and regular oscillations, without any variations in the period and amplitude on reaching its steady orbit. Furthermore, it should be kept in mind that rerunning the same set of reactions will always yield the exact same result, provided that the initial conditions are identical.

For the discrete stochastic case, it should be noted that one no longer considers molecular concentrations, but numbers of molecules. In our case, this implies that

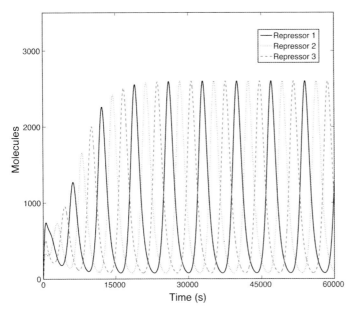

Fig. 15.12 Deterministic simulation of the Repressilator.

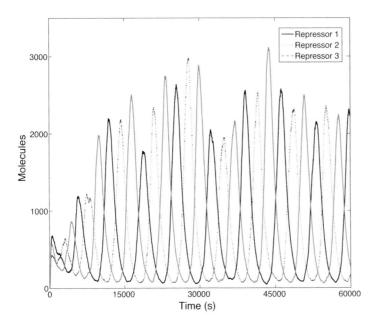

Fig. 15.13 Single stochastic simulation of the Repressilator.

the binary reactions' rates have to be transformed from inverse units of molarity time to inverse units of time, for which we considered a volume of 1.6 femtoliters. Contrary to the deterministic case, an exact stochastic simulation provides a time

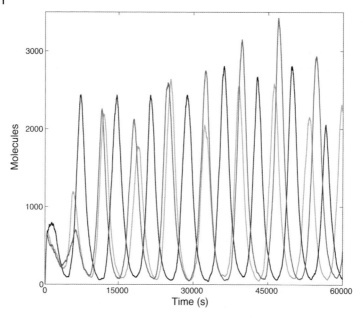

Fig. 15.14 Three independent time courses for the same repressor (first), under identical initial conditions.

course that, albeit is clearly oscillating, it no longer can be considered smooth (see Fig. 15.13). Moreover, the period and amplitude are no longer uniform and, in much contrast to the deterministic case, rerunning the same set of reactions with identical initial conditions will always yield different results. In order to illustrate this fact, Fig. 15.14 shows the time courses of the first repressor, obtained from three independent simulations with identical initial conditions. Here, one can readily observe that the oscillations from the same species are not in phase, and their amplitude varies considerably.

There are two elementary points to consider. The first is the averaging of oscillating time courses, resulting in a damped and biased oscillation, or a steady state in the limit of an infinitely large time-course sample. In order to avoid this damping bias, several techniques have been developed in which the single trajectories have to be either aligned (see for instance [66]) or studied separately.

The second point is: How can one determine the period and amplitude in a set of stochastic simulations? For such, and to the best of our knowledge, there is no clear-cut analytic answer. However, there are methodologies that can be used for this purpose. For instance, Tigges *et al.* [66] smooth the oscillation time courses and use a Fourier transform technique to extract driving period information from a power spectra. Once this is done, the time courses can be divided into windows corresponding to the driving period, in order to extract amplitude values, which in turn generate a distribution and not a single value.

It should be noted that in some scenarios, one could opt for a coarse-grained technique. Given the significantly shorter computational load, one is able to generate a larger time-course sample and, by consequence, derive "better statistics" of the driving period and amplitude of the oscillations. However, one should be cautious when doing so: choosing a high approximation tolerance can yield devastatingly different results or, if too small, it can create a computational overhead that in principle can result in larger computational times as compared with an exact algorithm. Finally, very different results can be obtained once delays (for instance, in transcription or translation) or spatial resolutions are included in a model, both of which are closer to the biological reality.

15.5 Conclusions

Synthetic biology is by no means fully an explored field. Even problems such as the right characterization of gene expression contain several open questions, where different assumptions may yield distinct results. This is exactly the reason why paying careful attention to biological detail is crucial, in order to successfully incorporate this knowledge into a model.

In this chapter, we have aimed at conveying this message throughout all sections. We first introduced a detailed modeling methodology based on gene circuit construction by means of composable parts. Aside from providing a rich analogy to electronic circuits and benefiting from their rich modeling literature, these parts can enhance a model by accounting for processes rarely introduced in models.

Afterward, we described many of the considerations to be accounted for when choosing "the right" simulation regime, pinpointing many of the shortcomings of using a deterministic framework. We introduced several SSAs, some of which account for delays in chemical reactions or are spatially resolved. We additionally summarized ways in which simulations can be coarse grained, in views of shortening the lengthy computational times needed for stochastic simulations.

Finally, we illustrated many of the topics discussed in the first two sections by means of a composable model of the Repressilator [7], outlining all modeling steps and highlighting some basic differences between choosing a deterministic or stochastic modeling regime for this case.

References

1 DRUBIN, D. A., WAY, J. C., SILVER, P. A., Designing biological systems, *Genes and Development* 21(3) (**2007**), pp. 242–254.
2 PURNICK, P. E., WEISS, R., The second wave of synthetic biology: from modules to systems, *Nature Reviews Molecular Cell Biology* 10(6) (**2009**), pp. 410–422.
3 ANDERSON, J. C., et al., Environmentally controlled invasion of cancer cells by engineered bacteria, *Journal of Molecular Biology* 355(4) (**2006**), pp. 619–627.
4 CHEN, M. T., WEISS, R., Artificial cell–cell communication in yeast *Saccharomyces cerevisiae* using signaling elements from *Arabidopsis thaliana*, *Nature Biotechnology* 23(12) (**2005**), pp. 1551–1555.
5 SAVAGE, D. F., WAY, J., SILVER, P. A., De-fossiling fuel: how synthetic biology can

transform biofuel production, *ACS Chemical Biology* 3(1) (**2008**), pp. 13–16.

6 Cases, I., de Lorenzo, V., Genetically modified organisms for the environment: stories of success and failure and what we have learned from them, *International Microbiology* 8(3) (**2005**), pp. 213–222.

7 Elowitz, M. B., Leibler, S., A synthetic oscillatory network of transcriptional regulators, *Nature* 403(6767) (**2000**), pp. 335–338.

8 Endy, D., Foundations for engineering biology, *Nature* 438(7067) (**2005**), pp. 449–453.

9 Lewin, B., *Gene VIII*, Pearson Prentice-Hall, Upper Saddle River, NJ, **2004**.

10 Marchisio, M. A., Stelling, J., Computational design of synthetic gene circuits with composable parts, *Bioinformatics* 24(17) (**2008**), pp. 1903–1910.

11 Marchisio, M. A., Stelling, J., Synthetic gene network computational design, in: *Proceedings of IEEE International Symposium on Circuits and Systems*, ISCAS 2009, **2009**, Taipei.

12 Kaern, M., Weiss, R., Synthetic gene regulatory systems, in: *System Modeling in Cellular Biology*, Szallasi, Z., Stelling, J., Periwal, V. (eds.), The MIT Press, Cambridge, MA, **2006**, pp. 269–295.

13 Bintu, L., et al., Transcriptional regulation by the numbers: applications, *Current Opinion in Genetic Development* 15(2) (**2005**), pp. 125–135.

14 Majdalani, N., Vanderpool, C. K., Gottesman, S., Bacterial small RNA regulators, *Critical Review in Biochemistry and Molecular Biology* 40(2) (**2005**), pp. 93–113.

15 Serganov, A., Patel, D. J., Ribozymes, riboswitches and beyond: regulation of gene expression without proteins, *Nature Reviews Genetics* 8(10) (**2007**), pp. 776–790.

16 Isaacs, F. J., et al., Engineered riboregulators enable post-transcriptional control of gene expression, *Nature Biotechnology* 22(7) (**2004**), pp. 841–847.

17 Isaacs, F. J., Dwyer, D. J., Collins, J. J., RNA synthetic biology, *Nature Biotechnology* 24(5) (**2006**), pp. 545–554.

18 Mandal, M., et al., A glycine-dependent riboswitch that uses cooperative binding to control gene expression, *Science* 306(5694) (**2004**), pp. 275–279.

19 Masse, E., Escorcia, F. E., Gottesman, S., Coupled degradation of a small regulatory RNA and its mRNA targets in *Escherichia coli*, *Genes and Development* 17(19) (**2003**), pp. 2374–2383.

20 Paulsson, J., Models of stochastic gene expression, *Physics of Life Reviews* 2(2) (**2005**), pp. 157–175.

21 Marquez-Lago, T. T., Stelling, J., Counter-intuitive stochastic behavior of simple gene circuits with negative feedback, *Biophysical Journal* 98(9) (**2010**), pp. 1742–1750.

22 Hirata, H., et al., Oscillatory expression of the bHLH factor Hes1 regulated by a negative feedback loop, *Science* 298(5594) (**2002**), pp. 840–843.

23 Barrio, M., et al., Oscillatory regulation of Hes1: Discrete stochastic delay modelling and simulation, *PLoS Computational Biology* 2(9) (**2006**), p. e117.

24 Gillespie, D. T., Approximate accelerated stochastic simulation of chemically reacting systems, *Journal of Chemical Physics* 115(4) (**2001**), pp. 1716–1733.

25 Tian, T., Burrage, K., Binomial leap methods for simulating stochastic chemical kinetics, *Journal of Chemical Physics* 121(21) (**2004**), pp. 10356–10364.

26 Peng, X., Zhou, W., Wang, Y., Efficient binomial leap method for simulating chemical kinetics, *Journal of Chemical Physics* 126 (**2007**), p. 224109.

27 Anderson, D. F., A modified next reaction method for simulating chemical systems with time dependent propensities and delays, *Journal of Chemical Physics* 127(21) (**2007**), p. 214107.

28 Anderson, D. F., Incorporating postleap checks in tau-leaping, *Journal of Chemical Physics* 128(5) (**2008**), p. 054103.

29 Cao, Y., Gillespie, D. T., Petzold, L. R., Efficient step size selection for the tau-leaping simulation method, *Journal of Chemical Physics* 124(4) (**2006**), p. 044109.

30 Leier, A., Marquez-Lago, T. T., Burrage, K., Generalized binomial tau-leap method for biochemical kinetics incorporating both delay and intrinsic noise, *Journal of Chemical Physics* 128(20) (**2008**), p. 205107.

31 Marquez-Lago, T. T., Leier, A., Burrage, K., Probability distributed times delays; integrating spatial effects into temporal models, *BMC Systems Biology* 4 (**2010**), p. 19.

32. GILLESPIE, D. T., General method for numerically simulating stochastic time evolution of coupled chemical reactions, *Journal of Computational Physics* 22(4) (**1976**), pp. 403–434.

33. GILLESPIE, D. T., Exact stochastic simulation of coupled chemical-reactions, *Journal of Physical Chemistry* 81(25) (**1977**), pp. 2340–2361.

34. GIBSON, M. A., BRUCK, J., Efficient exact stochastic simulation of chemical systems with many species and many channels, *Journal of Physical Chemistry A* 104(9) (**2000**), pp. 1876–1889.

35. SLEPOY, A., THOMPSON, A. P., PLIMPTON, S. J., A constant-time kinetic Monte Carlo algorithm for simulation of large biochemical reaction networks, *Journal of Chemical Physics* 128(20) (**2008**), p. 205101.

36. MUNSKY, B., KHAMMASH, M., The finite state projection algorithm for the solution of the chemical master equation, *Journal of Chemical Physics* 124(4) (**2006**), p. 044104.

37. BURRAGE, K., et al., A Krylov-based finite state projection algorithm for solving the chemical master equation arising in the discrete modelling of biological systems, in: *Proceedings of the Markov 150th Anniversary Conference*, Langville, A. N., Stewart, W. J. (eds.), Boson Books, Raleigh, NC, **2006**, pp. 21–38.

38. YOUNG, W. M., ELCOCK, E. W., Monte Carlo studies of vacancy migration in binary ordered alloys – I, *Proceedings of the Physical Society of London* 89(565P) (**1966**), p. 735.

39. BORTZ, A. B., KALOS, M. H., LEBOWITZ, J. L., New algorithm for Monte Carlo simulation of Ising spin systems, *Journal of Computational Physics* 17(1) (**1975**), pp. 10–18.

40. HASELTINE, E. L., RAWLINGS, J. B., Approximate simulation of coupled fast and slow reactions for stochastic chemical kinetics, *Journal of Chemical Physics* 117(15) (**2002**), pp. 6959–6969.

41. CAO, Y., GILLESPIE, D. T., PETZOLD, L. R., The slow-scale stochastic simulation algorithm, *Journal of Chemical Physics* 122(1) (**2005**), p. 14116.

42. BARIK, D., et al., Stochastic simulation of enzyme-catalyzed reactions with disparate timescales, *Biophys Journal* 95(8) (**2008**), pp. 3563–3574.

43. BRATSUN, D., et al., Delay-induced stochastic oscillations in gene regulation, *Proceedings of the National Academy of Sciences of United States of America* 102(41) (**2005**), pp. 14593–14598.

44. CAI, X., Exact stochastic simulation of coupled chemical reactions with delays, *Journal of Chemical Physics* 126(12) (**2007**), p. 124108.

45. TIAN, T., et al., Stochastic delay differential equations for genetic regulatory networks, *Journal of Computational and Applied Mathematics* 205(2) (**2007**), pp. 696–707.

46. TURNER, T. E., SCHNELL, S., BURRAGE, K., Stochastic approaches for modelling *in vivo* reactions, *Computers and Biological Chemistry* 28(3) (**2004**), pp. 165–178.

47. MORTON-FIRTH, C. J., BRAY, D., Predicting temporal fluctuations in an intracellular signalling pathway, *Journal of Theoretical Biology* 192(1) (**1998**), pp. 117–128.

48. PLIMPTON, S., SLEPOY, A., ChemCell: A particle-based model of protein chemistry and diffusion in microbial cells, Sandia National Laboratories, **2003**.

49. BARAS, F., MANSOUR, M. M., Reaction–diffusion master equation: a comparison with microscopic simulations, *Physical Review E: Statistical Physics Plasmas, Fluids, and Related Interdisciplinary Topics* 54(6) (**1996**), pp. 6139–6148.

50. ELF, J., DONCIC, A., EHRENBERG, M., Mesoscopic reaction–diffusion in intracellular signaling, *Proceedings of SPIE* (**2003**), pp. 114–125.

51. ANDER, M., et al., SmartCell, a framework to simulate cellular processes that combines stochastic approximation with diffusion and localisation: analysis of simple networks, *System Biology (Stevenage)* 1(1) (**2004**), pp. 129–138.

52. ERBAN, R., CHAPMAN, J., MAINI, P., Lecture notes, **2007**.

53. COLLINS, S. D., CHTTERJEE, A., VLACHOS, D. G., Coarse-grained kinetic Monte Carlo models: complex lattices, multicomponent systems, and homogenization at the stochastic level, *Journal of Chemical Physics* 129(18) (**2008**), p. 184101.

54. GILLESPIE, D. T., PETZOLD, L. R., Improved leap-size selection for accelerated stochas-

tic simulation, *Journal of Chemical Physics* 119(16) (**2003**), pp. 8229–8234.

55 Rathinam, M., et al., Stiffness in stochastic chemically reacting systems: the implicit tau-leaping method, *Journal of Chemical Physics* 119(24) (**2003**), pp. 12784–12794.

56 Cao, Y., Gillespie, D. T., Petzold, L. R., Accelerated stochastic simulation of the stiff enzyme-substrate reaction, *Journal of Chemical Physics* 123(14) (**2005**), p. 144917.

57 Chatterjee, A., Vlachos, D. G., Katsoulakis, M. A., Binomial distribution based tau-leap accelerated stochastic simulation, *Journal of Chemical Physics* 122(2) (**2005**), p. 024112.

58 Marquez-Lago, T. T., Burrage, K., Binomial tau-leap spatial stochastic simulation algorithm for applications in chemical kinetics, *Journal of Chemical Physics* 127(10) (**2007**), p. 104101.

59 Elf, J., Ehrenberg, M., Spontaneous separation of bi-stable biochemical systems into spatial domains of opposite phases, *System Biology (Stevenage)* 1(2) (**2004**), pp. 230–236.

60 Hattne, J., Fange, D., Elf, J., Stochastic reaction–diffusion simulation with MesoRD, *Bioinformatics* 21(12) (**2005**), pp. 2923–2924.

61 Mirschel, S., et al., PROMOT: modular modeling for systems biology, *Bioinformatics* 25(5) (**2009**), pp. 687–689.

62 Nagel, L. W., Pederson, D. O., SPICE (Simulation Program with Integrated Circuit Emphasis), in Memorandum No. ERL-M382, B. University of California, Editor, **1973**.

63 Ginkel, M., et al., Modular modeling of cellular systems with ProMoT/Diva, *Bioinformatics* 19(9) (**2003**), pp. 1169–1176.

64 Hucka, M., et al., The systems biology markup language (SBML): a medium for representation and exchange of biochemical network models, *Bioinformatics* 19(4) (**2003**), pp. 524–531.

65 Hoops, S., et al., COPASI – A COmplex PAthway SImulator, *Bioinformatics* 22(24) (**2006**), pp. 3067–3074.

66 Tigges, M., et al., A tunable synthetic mammalian oscillator, *Nature* 457(7227) (**2009**), pp. 309–312.

16
Identification of Physiological Models of Type 1 Diabetes Mellitus by Model-Based Design of Experiments

Federico Galvanin, Massimiliano Barolo, Sandro Macchietto, and Fabrizio Bezzo

Keywords

model-based design of experiments (MBDoE), physiological models, type 1 diabetes mellitus (T1DM), model identification, design of clinical tests, intravenous glucose tolerance tests (IVGTTs)

The quest for an artificial pancreas as a therapy for type 1 diabetes mellitus (T1DM) has long involved the process systems engineering (PSE) community. The importance of reliable physiological models as a support to define effective control algorithms, to mimic a subject's behavior in the development of clinical tests and therapies, to assess the performance of new devices and technologies, and more generally, to increase what is elsewhere called "process understanding" cannot be overestimated. Accordingly, model accuracy and reliability represent a critical issue: a physiological model needs to be identified through one or more clinical tests, which should be both safe and informative, and represent the specific physiological behavior of an individual subject. In this chapter, we will discuss how model-based design of experiments (MBDoE) techniques can be exploited to achieve such a goal for physiological models of T1DM.

The chapter is structured as follows. After introducing the general subject of T1DM, we will overview the issue of physiological model development and the need for a proper design of clinical tests for model parameter identification. Standard clinical tests used for model identification will then be described, and compartmental models will be briefly introduced. At this point, some general issues concerning model identifiability will be discussed. Successively, the MBDoE approach will be formally presented and its utilization for the identification of a model of glucose homeostasis will be shown. Some issues that may hinder the procedure will be illustrated and two advanced approaches that may help tackling them will be presented and demonstrated through simulated case studies.

16.1
Introduction

Diabetes is a disorder of the metabolism resulting in blood glucose concentration elevated beyond the normal levels (hyperglycemia). Persistent hyperglycemia conditions in diabetic subjects are associated with long-term complications and dysfunctions in various organs (especially eyes, kidneys, nerves, heart, and blood vessels). In the treatment of hyperglycemia, a diabetic subject may experience the opposite condition (hypoglycemia), which occurs when the glucose concentration in the blood falls to very low values. The effects of hypoglycemia are critical on a short-term basis, potentially leading to loss of consciousness and coma within just few hours.

According to the World Health Organization [1], there were 171 million people in the world suffering diabetes in the year 2000, and this number is projected to increase to 366 million by 2030. The national costs of diabetes in the USA for 2002 were US $132 billion, and are projected to increase to US $192 billion in 2020.

Type 1 diabetes mellitus (also called insulin-dependent diabetes) usually begins before the age of 40, and is characterized by a progressive destruction of the pancreatic islets and by an absolute deficiency of insulin secreted into the body. This determines the incapability to maintain the blood glucose concentration within a narrow range (normoglycemic levels). People with T1DM must, therefore, rely on exogenous insulin for survival. The most widespread treatment of T1DM is based on multiple daily self-injections (boluses) of insulin. However, the individual insulin requirement can be affected by many factors, such as a subject's body weight, carbohydrate (CHO) content of a meal, illness, degree of stress, and exercise. Thus, people affected by T1DM have to be instructed on how to regularly check their glycemia (usually several times a day, using fingersticks) and how frequently (and to which extent) perform insulin self-administration.

The most advanced research in T1DM management [2] seeks for the development of a wearable artificial pancreas (WAP), i.e., a device requiring an implantable glucose sensor, a control strategy to determine an appropriate insulin delivery rate, and a pump for the delivery of this rate of insulin. In fact, type 1 diabetes management closely resembles the problem of controlling a single-input, single-output chemical (or biochemical) process, where the manipulated input is the insulin delivery rate, and the controlled output is the blood glucose concentration. The main disturbances are represented by meal ingestion and by the variation of a subject's sensitivity to insulin (due to stress, exercise, circadian rhythm, and possibly other factors). This system is characterized by several challenges, which include nonlinear input/output responses, constraints in the input and output, unmeasured disturbances, and uncertainty (measurement noise).

16.1.1
Glucose Concentration Control Issues

In the few last decades, much research activity in the field of T1DM management has been dedicated to understand how the pancreas β-cells respond to control the glucose concentration in a healthy subject, and to subsequently use this information to determine how an artificial closed-loop algorithm for insulin delivery should behave [3]. The first attempts to develop a control algorithm to mimic the pancreatic activity came from the studies by Albisser [4], and the improved proportional plus derivative (PD) algorithm in [5] adopted in the Biostator, the first automated insulin delivery system. Nomura and coworkers [6] proposed a PD-based secretion model to reproduce the pancreatic islets activity. These first developed algorithms were patient-specific and they needed to be reprogrammed as the metabolic conditions of the subject changed in time.

More recently, hyperglycemic clamp tests have shown that the beta-cells respond with a biphasic insulin pattern to glucose challenges, while in the fasting state the response is characterized by a basal (steady) insulin level [7]. This response closely resembles the response of a standard proportional-integral-derivative (PID) controller, which is ubiquitously used in the chemical process industry (e.g., see Seborg et al. [8]), and therefore the PID control approach to glucose control received some attention in the literature (e.g., see [9]). It has been argued in [2] that a biphasic beta-cell response does not necessarily mean that the pancreas itself uses a PID algorithm to deliver the insulin in a healthy subject, and that the integral term in PID controllers can cause the overadministration of insulin, resulting in postprandial hypoglycemia. PD control is, therefore, also employed for glucose concentration control [10].

The major limitation of most of the published studies on PID control of diabetes is that simplified models are often used to represent the insulin/glucose system, and that in most cases neither noise, uncertainty in insulin sensitivity, nor unmeasured disturbances are taken into account. As discussed in [11], these issues may have a dramatic impact on the control performance and the results available so far in the field of standard PID control of glycemia may be questionable.

A PID switching control algorithm was successfully applied by Marchetti et al. [12] according to the idea that a time-varying glucose concentration setpoint is a more appropriate reference profile to be tracked. They also showed that a very mild integral action is useful to compensate for individual's changes in insulin sensitivity without leading to postprandial hypoglycemia. In a similar fashion, Percival et al. [13] proposed a practical approach to design and implement a PID control algorithm focusing on controller robustness when changes on insulin sensitivity, meal times, and meal sizes occur.

Advanced control strategies have also been proposed in the literature where the model itself is embedded in the control algorithm, as in model predictive control (MPC) [14, 15], adaptive control, optimal control [16, 17], neural networks, and H-infinity control [18]. As recognized in [19], the key issue is that the performance of model-based control systems is directly linked to model accuracy.

Some advanced control algorithms tested *in silico* have been also applied *in vivo*. El-Khatib *et al.* [20] have performed real-life trials in pigs using adaptive control with dual insulin and glucagon infusion, while Hovorka *et al.* [21] performed trials on subjects affected by T1DM adopting MPC strategies.

16.2
Introducing Physiological Models

It has long been recognized that a physiological model appropriately describing the insulin/glucose dynamics provides an invaluable aid to understand the complex metabolic mechanism controlling glucose homeostasis [22]. A model can be also regarded as the surrogate of a patient (a "virtual patient"), and in this respect it can be used to design a closed-loop control system and to evaluate its performance. Finally, a model itself can be embedded into a control algorithm [14].

Several physiological models have been developed in the last four decades to describe the dynamics of glucose/insulin system. The literature on this topic has been reviewed in survey articles [2, 19, 23]. Probably the most widely known and used model is the so-called minimal model by Bergman *et al.* [24], where the plasma glucose dynamics and plasma insulin dynamics are described using only three differential equations and few parameters. Despite its simplicity, and despite the fact that (strictly speaking) it was not derived to optimize insulin treatment in individuals with T1DM, this model is able to account for most of the physiological insulin–glucose relationships revealed by clinical evidence both in T1DM and T2DM [25]. Several modifications to the original minimal model have been reported (e.g., [16, 18, 26, 27]) to overcome its main limitations. These include, for example, the lack of representation of the dynamics of subcutaneous insulin infusion and the rate of glucose appearance following a meal. Recently, more detailed physiological models have been proposed that are better able to represent the overall glucoregulatory system, including the absorption of subcutaneously administered short-acting insulin, and glucose ingestion and absorption [28–31]. Additional details concerning a complex (compartmental) physiological model will be given in a later section.

16.3
Identifying a Physiological Model: The Need for Experiment Design

A critical issue for the use of a physiological model is the identification of model parameters for individual subjects. Typically, the identification of the minimal model has been achieved with frequently sampled intravenous glucose tolerance tests (IVGTTs), where a bolus of glucose is injected intravenously, and several samples of the glucose and insulin plasma concentrations are taken following the glucose injection [24]. This kind of test does not upset the subject excessively, but it is not guaranteed that the excitation pattern it generates is the most appropriate to estimate the model parameters with good precision. In fact, it is well known that the identifiability of a parametric model is strictly related to the structure of the model and to the level of excitation that the experimenter can realize during the experiments [32]. As early as 1981, Bergman *et al.* [24] recognized that, in order to estimate the metabolic parameters of the minimal model, the optimal input perturbation to be considered might well be different from that of an IVGTT, and different temporal patterns of glucose and/or insulin administration could lead to easier and more accurate parameter identification. Several modifications of the standard IVGTT have been proposed to improve parameter identification. For example, it has been shown that the infusion of insulin some time after the glucose injection in an IVGTT considerably improves parameter estimation [33, 34]. Some studies also aimed at defining the best input profile for the minimal model, using a measure of Fisher information matrix for a single test [35].

The availability of a model tailored to an individual subject would provide substantial benefits both to the clinician (who could devise a customized diabetes care solution for the subject) and to the engineer (who could design and test specifically tailored conventional and advanced glucose control techniques). However, this may be a very challenging task, particularly for detailed non-minimal physiological models, where identifying individual subject parameters from limited data may be extremely difficult. Therefore, while clinicians are interested in knowing the parameter values for individual subjects, researchers are more likely to analyze large response data sets from multiple subjects together and produce a "nominal" subject model, based on mean values [19]. A further complication arising when designing *ad hoc* a clinical test for individual parameter identification is that the test must be safe for the subject (i.e., the glycemic and insulinemic levels must always remain within the physiological bounds), sufficiently short, and easy to carry out.

This issue is very similar to the one that is faced by process engineers when they need to estimate the parameters of complex dynamic models by carrying out dynamic experiments in a chemical or biochemical process system. In this case, it has been shown [36, 37] that there exists a way (actually, an "optimal" way) to adjust the temporal profile of the inputs, the initial conditions, and the length of the experiment in such a way as to generate the maximum amount of information from the experiment itself, for the purpose of estimating the model parameters with an assigned precision. Optimal MBDoE procedures have been successfully applied in several applications of chemical and biological engineering (e.g., see [38–40]). The experiment design problem can be formulated as an optimal control problem, where the experiment decision variables are, for example, time varying and time invariant inputs, sampling times of response variables, experiment initial conditions, and duration. This leads to an optimal MBDoE problem for parameter identification in a dynamic system where constraints are present both in the inputs (manipulated quantities) and in the outputs (measured responses). A review about the state of the art on MBDoE can be found in [41].

Thus, MBDoE techniques can be adopted to optimally design an "experiment" (or a series of experiments) on a T1DM subject in order to develop an improved set of clinical tests from which the parameters of a glucose–insulin system dynamic model can be estimated with a higher degree of precision than possible so far. Most importantly, experiments can be designed so as to tailor a model to each individual patient [42].

16.4
Standard Clinical Tests

Standard clinical tests are used both to help diagnosing diabetes and to identify simple models of glucose homeostasis. Basically, an input pattern is used to excite the subject's glucoregulatory system in such a way as to subsequently extract some kind of information from the measured time-profiles of the plasma glucose and insulin concentrations. In most cases, the input excitation pattern is limited to the infusion or intake of glucose only, although the infusion of insulin is also possible. To provide a general overview of how these tests are carried out, the most widespread tests [1] are shortly recalled in the following:

1. *Oral glucose tolerance test (OGTT)*: This is the diagnostic test recommended by the World Health Organization [1]. The test is carried in the morning after about 3 days of unrestricted diet (greater than 150 g CHO daily) with the usual physical activity of the subject. A meal of 30–50 g CHO should be consumed the evening preceding the test. After collection of a fasting blood sample, the subject drinks a solution of 75 g of glucose in water over the course of 5 min. Blood is drawn at intervals for measurements of glucose, and sometimes insulin levels. The sampling frequency can vary according to the purpose of the test. For simple screening, one can take the samples at 0 and 2 h (only two samples collected), but in a research activity the sampling can be very frequent (e.g., a sample every 2 min).
2. *IVGTT*: This test is useful to evaluate the pancreatic activity *in vivo*, but it is mainly used in research activities, because it is much more invasive than the OGTT. Usually it consists in injecting 300 mg/kg of glucose over 60 s in an antecubital vein, and then measuring the plasma insulin and glucose concentrations. The sampling schedule of the standard IVGTT consists of taking three pretest samples and 23 additional 2-mL samples (the sampling frequency at the beginning of the test is of one sample every 2 min, and one sample every 20 min at the end of the experiment, lasting at least 3 h). In modified IVGTTs, insulin (30 mU/kg) is infused 20 min after the glucose ingestion [34].
3. *Postprandial glucose test*: This test is useful to screen for diabetes and to evaluate the effectiveness of treatment or dietary therapy for diabetic subjects. It is performed after the subject has eaten a balanced meal containing 100 g (or more) of glucose and then is fasted for 2 h before beginning the test (the sample policy can be variable).
4. *Euglycemic hyperinsulinemic clamp*: This test is used to quantify the insulin resistance of a subject by measuring the amount of glucose necessary to compensate for an increased insulin level without causing hypoglycemia. Through a peripheral vein, insulin is infused at 10–120 $mU/(m^2 \, min)$ [43]. At the same time, glu-

cose is infused to maintain blood sugar levels between 91 and 100 mg/dL. The blood sugar levels are controlled every 5–10 min and the rate of glucose infusion is adjusted accordingly. Different insulin doses can be used to discriminate between the different responses of peripheral tissues and the liver ones. The test takes about 2 h, and the rate of glucose infusion during the last 30 min of the test determines insulin sensitivity.

Tests 1 and 3 are oral tests and have the advantage of being physiological, that is, they are not invasive for the subject. An "ideal" test should yield the best compromise between level of stress for the subject (low), clinical effort (low), and the amount of information obtainable (high), while it is taken for granted that it should be safe (i.e., it should not drive the subject to either hyperglycemia or hypoglycemia).

"Normal" glucose levels are not easy to define [1] and not universally agreed upon. In this contribution, the safe upper and lower glucose concentration thresholds are set to 170 and 60 mg/dL, respectively.

16.5
A Compartmental Model of Glucose Homeostasis

A model of glucose homeostasis can be seen as a multiple-input single-output system usually described by a system of differential and algebraic equations (DAEs) where the output (measured) variable is the blood glucose concentration G and the input (manipulative) variables are the amount of CHO of the meal(s) and the subcutaneous insulin infusion rate.

The meal ingestion and the insulin infusion are modeled, respectively, by a system submodel of glucose absorption providing the rate of appearance of glucose in plasma (R_a), and by an insulin infusion submodel providing the rate of appearance of insulin in plasma (R_i). The connections between functional blocks for a model of glucose homeostasis are shown in Fig. 16.1. The relationships between the glucose/insulin systems, the endogenous glucose production, and the glucose utilization and elimination define the metabolic portrait of the individual and are inherently related to the mathematical structure of a specific model and the set of its parameters. Several submodels have been proposed to define the rate of appearance of glucose in plasma [44] and the kinetics of subcutaneous insulin absorption [45].

In this chapter, we will mainly refer to the model developed by Hovorka and coworkers [28], denoted as the Hovorka model (HM). Although the HM is known to suffer from some drawbacks [46], it has been applied in several studies on glucose control in T1DM [12, 14, 47]. This model has been shown to be identifiable when a proper reparameterization is realized [43]. Accordingly, it has been chosen as a suitable candidate to assess the performance of MBDoE techniques. The model and related parameters are presented in Appendix A. The purpose is to simulate the glucose response of a 56-year-old male subject with a body weight of 78 kg

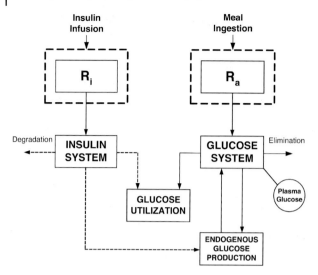

Fig. 16.1 Relationships between functional blocks for a model of glucose homeostasis. The insulin infusion submodel and the glucose absorption submodel are evidenced with dashed boxes.

affected by T1DM. In this chapter, simulated data only are considered. Therefore, the "real subject," too, is represented by a detailed physiological model.

16.6
Model Identifiability Issues

When considering a physiological model (in fact, any model), a critical aspect is its parametric identifiability. The goal of every model-building strategy is to tailor the model to the specificity of the phenomenon being studied, and therefore it is of great importance to assess whether or not the unknown parameters can be uniquely recovered from experimental data. The concept of identifiability can be introduced in several ways and several identifiability definitions have been proposed. First of all, as discussed in [48], a fundamental distinction has to be done between *a priori identifiability* (or structural identifiability testing) and *a posteriori identifiability*, based on collected experimental information.

A *priori* identifiability [49, 50] aims at verifying if, under ideal conditions of noise-free observations and absence of external disturbances, the unknown parameters of a postulated model can be estimated from a designed multi-input/multi-output experiment. Let us consider a physiological model with a given structure $M(\boldsymbol{\theta})$. The equality of the model inputs and outputs for two distinct sets of parameters $\boldsymbol{\theta}$ and $\boldsymbol{\theta}^*$ is denoted by

$$M(\boldsymbol{\theta}) \approx M(\boldsymbol{\theta}^*) \tag{16.1}$$

The parameter $\theta_i \in \boldsymbol{\theta}$ is *a priori* structurally globally identifiable (SGI) if for almost any $\boldsymbol{\theta}^*$

$$M(\boldsymbol{\theta}) \approx M(\boldsymbol{\theta}^*) \quad \Rightarrow \quad \theta_i = \theta_i^* \in \boldsymbol{\theta}^* \tag{16.2}$$

and it is structurally local identifiable (SLI) if for almost any $\boldsymbol{\theta}^*$ there exists a neighborhood $v(\boldsymbol{\theta}^*)$ such that (16.2) is still verified. Local identifiability is a necessary condition for global identifiability, and a model is said to be SGI if condition (16.2) is verified for the entire parametric set. A parameter that is not SLI is structurally nonidentifiable (SNI), and a model is said to be SNI if any of its parameters is SNI.

To test the identifiability of nonlinear parametric models, a local study may be misleading; a global identifiability test should be carried out instead. A method for testing the global identifiability is the one proposed in [51], based on the analysis of the series expansion of the output function, evaluated at time $t = 0$. Identifiability is assessed by determining the number of solutions for the given parametric set [52].

When the model is nonidentifiable, the identifiability analysis can be a very difficult task because of the computational effort required to solve the infinite set of equations of the exhaustive summary. The only way to solve the problem is to find a finite set of equations containing all the information of the exhaustive summary. Ljung and Glad [53] proposed a method and an explicit algorithm based on differential algebra, demonstrating how the testing of global structural identifiability can be reduced to the question of whether the given model structure can be rearranged as a linear regression. The authors also analyzed the condition of "persistent excitation" for the input that can be tested explicitly in a similar fashion, basically showing how identifiability and experiment design are highly correlated tasks. A new improved differential algebra algorithm based on the Buchberger algorithm [54] was proposed in [55].

As discussed in [56], *a priori* identifiability is a necessary condition to guarantee successful parameter estimation from real data (*a posteriori* identifiability) and, for structurally complex and large nonlinear dynamic models, the *a priori* identifiability testing could become an almost impossible task because of the computational complexity.

Global and local sensitivity analyses are widely used tools to assess *a posteriori* identifiability of large nonlinear dynamic models [57, 58]. Asprey and Macchietto [59] proposed an optimization-based approach to test identifiability where a distance measurement between two parameter vectors $\boldsymbol{\theta}$ and $\boldsymbol{\theta}^*$ providing the same model output is maximized. If this distance is arbitrarily small, the model can be deemed globally identifiable. The optimization algorithm can be constrained defining a validity domain for model parameters. Sidoli *et al.* [38] developed a perturbation algorithm coupling the previously mentioned optimization-based approach to test identifiability with a multilocal sensitivity analysis.

Soderstrom and Stoica [32] observed that the concept of model identifiability does not refer only to an intrinsic property of the model structure, but also to the identification procedure and the experimental conditions. The quality of the chosen parameter estimator (least squares, maximum likelihood, Bayesian estimator) can play a crucial role on estimating the model parameters from the data with acceptable statistical precision. Moreover, the information content of the experimental runs can by enriched by adding more measured variables, or planning the experiments through design of experiment techniques (either model-based or data-driven).

To summarize, the main advantage of assessing *a priori* structural identifiability adopting differential algebra or series expansion methods is that they provide a global identifiability test. The main drawback is that these methods are computationally expensive. On the other hand, *a posteriori* identifiability can be assessed by methods based on sensitivity analysis and perturbation studies that tend to be easier to carry out, even if they also may require a considerable computational effort when large systems are considered.

16.6.1
A Discussion on the Identifiability of the Hovorka Model

A simplified approach [42] will be illustrated in this section that allows to screen potential identifiability issues in the HM. A 30% difference in the model parameters is assumed between the physiological model representing the subject and the one that needs identifying. To assess the effect of the HM parameters on the glucose concentration response for the subject and for the model, a sensitivity analysis can be carried out considering the dynamic sensitivities in the following form:

$$q_i = \frac{\partial \Upsilon}{\partial \theta_i}, \quad i = 1, 2, \ldots, N_\theta \tag{16.3}$$

where here in particular Υ is the glucose response of the subject or of the model.

Figure 16.2 suggests that the dynamics of q_4 and q_5, that is, the sensitivities of glucose concentration to parameters θ_4 and θ_5, are very similar. Indeed, the two curves exhibit a symmetrical behavior, independently of the parameter values. This usually indicates a structural unidentifiability (at least within the constraints and assumptions considered here). It can be verified that a conventional experiment design approach produces unsatisfactory results when aiming at the identification of all five parameters of the HM, even if several experiments are carried out in sequence. In particular, θ_4 and θ_5 are always difficult to identify, and this occurs because of their strong correlation. To quantify the correlation between θ_4 and θ_5, one can analyze the correlation matrix **C**, whose elements (correlation coefficients) have the form

$$c_{ij} = \frac{v_{ij}}{\sqrt{v_{ii}}\sqrt{v_{jj}}} \tag{16.4}$$

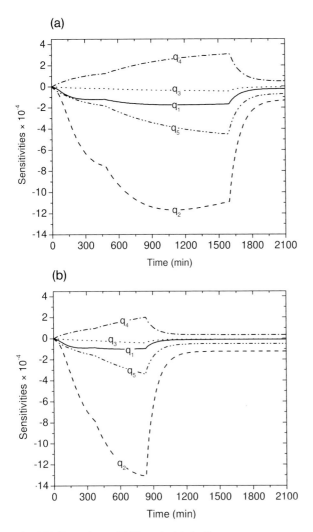

Fig. 16.2 Dynamic sensitivities calculated with $u_{basal} = 9.94$ mU/min and without insulin infusion for (a) the subject and (b) the model (after [42]).

where the v_{ij}'s are the elements of the variance–covariance matrix $\mathbf{V}_\theta$ of the HM parameters. As shown in [42], the analysis of correlation coefficients allows demonstrating that two parameters (θ_4 a θ_5) are structurally correlated. Thus, a new parameterization can be introduced: instead of θ_4 and θ_5, a new parameter θ'_4 is defined as the ratio between the two original parameters (i.e., $\theta'_4 = \theta_4/\theta_5$). With a small notation abuse, the set of model parameters initially available, after reparameterization into a modified set of four parameters, will be indicated with $\boldsymbol{\theta}^0$ in the remainder of the chapter.

16.7
Design of Experiments Under Constraints for Physiological Models

The HM belongs to the class of nonlinear dynamic models described by a set of DAEs. Standard model-based experiment design procedures aim at decreasing the model parameter uncertainty region predicted *a priori* by the model by acting on the experiment design vector $\varphi \in \Re^{n_\varphi}$ and solving the following set of equations:

$$\varphi^{\text{opt}} = \varphi\{\psi[V_\theta(\theta, \varphi)]\} = \varphi\{\psi[H_\theta^{-1}(\theta, \varphi)]\} \tag{16.5}$$

subject to

$$f(\dot{x}(t), x(t), u(t), w, \theta, t) = 0 \tag{16.6}$$

$$\hat{y}(t) = g(x(t)) \tag{16.7}$$

$$\tilde{C} = x(t) - \Gamma(t) \leq 0 \tag{16.8}$$

with the set of initial conditions $x(0) = x_0$. The symbol ^ is used to indicate the estimate of a variable (or a set of variables): thus, $y(t) \in \Re^{N_y}$ is the vector of measured values of the outputs, while $\hat{y}(t) \in \Re^{N_y}$ is the vector of the corresponding values estimated by the model. In these equations, V_θ and H_θ are the variance–covariance matrix of model parameters and the dynamic information matrix, respectively; $x(t) \in \Re^{N_x}$ is the vector of time-dependent state variables, $u(t) \in \Re^{N_u}$ and $w \in \Re^{N_w}$ are the time-dependent and time-invariant control variables (manipulated inputs), $\theta \in \Re^{N_\theta}$ is the set of unknown model parameters to be estimated, and t is the time. $\tilde{C}$ is an N_c-dimensional set of constraint functions expressed through the set $\Gamma(t) \in \Re^{N_c}$ of (possibly time-varying) active constraints on the state variables. The design vector

$$\varphi = \{y_0, u(t), w, t^{\text{sp}}, \tau\} \tag{16.9}$$

contains the N_y-dimensional set of initial conditions of the measured variables (y_0), the duration of the single experiment (τ), and the n_{sp}-set of time instants at which the output variables are sampled

$$t^{\text{sp}} = [t_1 \ \ldots \ t_{n_{\text{sp}}}]^T$$

The ψ function in (16.5) is an assigned measurement function of the variance–covariance matrix of model parameters and represents the design criterion. Different design criteria have been proposed (*D*-, *A*-, *E*-optimal criteria, considering the determinant, the trace, and the maximum eigenvalue of V_θ, respectively [60], SV-based [61], or P-based [62]).

In this particular case, the design vector is

$$\varphi = \{u(t), \mathbf{w}, \mathbf{t}^{sp}\} \tag{16.10}$$

where the time-dependent manipulated input $u(t)$ is the insulin subcutaneous infusion, while the glucose intake and the subcutaneous bolus administration are included in the time-invariant control vector $\mathbf{w}$. The time-dependent manipulated input $u(t)$ is approximated with a piecewise constant function (defined by n_z levels and n_{sw} switching times to be optimized). There is only one measurable output $y(t)$, which is the glucose concentration $G(t)$, and the blood sampling schedule is a design variable itself and is expressed through the vector $\mathbf{t}^{sp}$ of sampling times.

If we consider the design of the n_{exp}th experiment in a standard sequential approach, matrix $\mathbf{V}_\theta$ is the inverse of the $(N_\theta \times N_\theta)$ information matrix $\mathbf{H}_\theta$ defined as

$$\mathbf{H}_\theta(\boldsymbol{\theta}, \boldsymbol{\varphi}) = \sum_{k=0}^{n_{exp}-1} \mathbf{H}^*_{\theta|k.}(\boldsymbol{\theta}, \boldsymbol{\varphi}_k) + \mathbf{H}^*_\theta(\boldsymbol{\theta}, \boldsymbol{\varphi}) + (\boldsymbol{\Sigma}_\theta)^{-1}$$

$$= \mathbf{H}^*_\theta(\boldsymbol{\theta}, \boldsymbol{\varphi}) + \mathbf{K} \tag{16.11}$$

where $\mathbf{K}$ is a constant matrix comprising the information obtained from the previous $n_{exp} - 1$ experiments and from the $(N_\theta \times N_\theta)$ prior variance–covariance matrix of model parameters $\boldsymbol{\Sigma}_\theta$, and $\mathbf{H}^*_{\theta|k}$ is the dynamic information matrix (e.g., in the form proposed in [63]) of the kth experiment ($\mathbf{H}^*_{\theta|0}$ is the zero matrix, and superscript $*$ indicates that the information matrix refers to a single experiment). $\mathbf{H}^*_\theta(\boldsymbol{\theta}, \boldsymbol{\varphi})$ is defined as

$$\mathbf{H}^*_\theta(\boldsymbol{\theta}, \boldsymbol{\varphi}) = \sum_{i=1}^{N_y} \sum_{j=1}^{N_y} s_{ij} \mathbf{Q}_i^T \mathbf{Q}_j \tag{16.12}$$

where s_{ij} is the ijth element of the inverse of the $(N_y \times N_y)$ estimated variance–covariance matrix $\boldsymbol{\Sigma}$ of measurements errors, and $\mathbf{Q}_i$ is the matrix of the sensitivity coefficients for the ith estimated output at each of the n_{sp} sampling points:

$$\mathbf{Q}_i = \left[\frac{\partial \hat{y}_i(t_l)}{\partial \theta_m} \right], \quad l = 1, \ldots, n_{sp}, \ m = 1, \ldots, N_\theta \tag{16.13}$$

Prior information on the model parameter uncertainty region in terms of statistical distribution (for instance, a uniform or Gaussian distribution) can be included through matrix $\boldsymbol{\Sigma}_\theta$.

The set of constraints (16.8) on state variables aims at maintaining normoglycemia:

$$\tilde{C}_1 = y(\boldsymbol{\theta}, \boldsymbol{\varphi}, t) - \Gamma_1 \leq 0 \quad \text{and} \quad \tilde{C}_2 = \Gamma_2 - y(\boldsymbol{\theta}, \boldsymbol{\varphi}, t) \leq 0 \tag{16.14}$$

so that the glucose concentration is kept within an assigned range.

16.7.1
Design Procedure

A standard MBDoE procedure involves a sequential interaction between three key entities [59]:

1. design of the experiment (i.e., the clinical test);
2. execution of the experiment;
3. estimation of the parameters.

The procedure starts with the design of the first experiment, given an initial guess of parameters (say θ^0) and related statistics, and the (expected) features of the measurement noise of the glucose sensor. The identification test is executed afterward, with the designed experimental settings. Finally, the information embedded in the set of acquired data is exploited through a parameter estimation session. The procedure 1–3 can be iterated until a sufficiently precise parameter estimation is reached. The three design steps are structured as in Fig. 16.3.

The choice of the proper parameter estimation technique is crucial for MBDoE. Bayesian estimation techniques have been proved to be very efficient for physiological model identification [64], but the severe computational effort required and the lack of reliable *a priori* statistics often make them too challenging an approach. Thus, a maximum likelihood estimator is chosen in the following.

An additional important step in MBDoE is the evaluation of the quality of the estimates, and hence the setting of the stopping rule for the iterative scheme of the experiment design. In this study, the quality of the estimation is evaluated according to the following factors (with the assumption of Gaussian distribution of measurement errors):

1. *a posteriori* statistics of the estimates (in terms of t-test and confidence intervals);
2. goodness of fit (in terms of whiteness test and χ^2 test).

The confidence intervals κ_i at a given confidence level (95% in this case) of the estimates are evaluated from the output of the maximum likelihood estimator (from V_θ), while the t-values are calculated by

$$t_i = \frac{\theta_i}{\kappa_i}, \quad i = 1, \ldots, N_\theta \tag{16.15}$$

and compared with the tabulated reference t-value from the Student's t distribution with $(n_{sp} - N_\theta)$ degrees of freedom.

For numerical reasons, in this study all parameters have been normalized by dividing them by their true values; from now on, we will always refer to the parameters normalized values (indicated by symbol Θ). Therefore, note that the true (and unknown) value for each normalized parameter is 1.

Finally, it is important to stress that the optimization method also plays a critical role in the quality of the final results. Here, the gPROMS® modeling environment (by Process Systems Enterprise, Ltd.) [65] is used for modeling, simulation,

16.7 Design of Experiments Under Constraints for Physiological Models

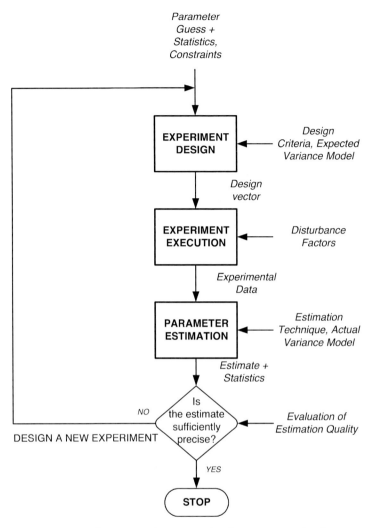

Fig. 16.3 Sequential experiment design procedure for parameter estimation.

and optimization purposes, as well as to design the experiments. The SRQPD optimization solver of gPROMS adopts an SQP sequential quadratic programming routine to solve the nonlinear optimization problem and, in principle, cannot guarantee the determination of global optima. We tried to mitigate the risk of incurring in local minima by a two-step multiple shooting technique [66]. In the first step, the optimal design problem is solved as a maximization of the trace of the dynamic information matrix over the experimental horizon. In the second step, the preliminary optimal design vector evaluated in the first step is randomized to provide different initial points for the subsequent multiple shooting optimization.

16.8
Design of Experimental Protocols

Once the designer starts the MBDoE exercise, the first design is based on an initial guess of the model parameters. From the point of view of the effectiveness of design, a large parametric mismatch can result in incorrect evaluation of the information content expected from an experiment, which may cause the experiment itself to be uninformative. In the HM perspective, unless a robust design approach is followed (potential methodologies are discussed in the following sections), the mismatch may lead to an erroneous evaluation of the insulin (if any) to be administered to the subject during execution of the designed experiment, which might drive her/him to hypoglycemic conditions. Therefore, it may be prudent to carry out a preliminary "reference" test, aimed at obtaining a rough (yet reasonable) estimation of the actual parameter values to be subsequently used to design the first experiment. An ideal reference test should be: (i) safe for the subject, independently of her/his pretest condition; and (ii) informative as well as (iii) quick and easy to perform. A possible reference test is discussed in [42]. Note that, in general, a reference test is not sufficient to estimate the parameters in a statistically sound way.

The design of an experimental protocol, following the reference test, is based on the following requirements:

1. Exclusion of nonphysiological tests (IVGTT, glucose clamp, or similar);
2. Possibility to manage a day-long test;
3. Possibility to manage multiple meal intakes and multiple insulin bolus administrations, or to modify the glucose intake policy of a standard OGTT;
4. Constraints on the glycemic curve: interior constraints to assure normoglycemia at all times (60–170 mg/dL); an end-point constraint on the glucose concentration (80 mg/dL) and an end-point constraint on the derivative of the glucose concentration to assure steady glycemia at the end of the test. The test formally ends with the last sampling point (which defines the duration of the experiment); however the above end point constraints are imposed to guarantee safe conditions for the subject also after the clinical test; the end point constraints must be fulfilled within a specified time interval;
5. Constraints on the insulin infusion rate: $u(t) = u_{\text{basal}}$ at the end of the test.

It must be also guaranteed that the subject returns to the basal settings after performing the day test.

The goal of the suggested protocol is to obtain sufficiently informative data so as to enable the estimation of the set of model parameters in a satisfactory manner with only one designed experiment being carried out after the reference test. A modified OGTT for parameter identification is discussed in the following section [42].

16.8.1
Modified OGTT (mOGTT)

A *D*-optimal experiment with multiple ingestions of a glucose solution and with insulin bolus intakes is designed. The optimization variables are as follows:

- the glucose content of the meals (glucose solution drink);
- the time interval between two consecutive meals (allowed to vary between 15 and 800 min);
- the amount of each insulin bolus;
- the sampling times (the number of samples n_{sp} is assigned *a priori* and the elapsed time between two consecutive samples cannot be shorter than 5 min).

The time schedule of insulin infusion is not optimized. The amount of insulin bolus per meal is modeled according to the following empirical relationship [42]:

$$u(t) = u_{\text{basal}} + \alpha \sum_{i=1}^{N_{\text{meals}}} \delta_i(t) k_i D_{g,i} \qquad (16.16)$$

where $D_{g,i}$ is the glucose content of the *i*th meal, u_{basal} is the basal insulin infusion rate, δ the Dirac impulse function, and $\alpha = 52.63$ mU/g. CHO represents the optimal insulin/CHO ratio for the subject considered in this study. Since parameter uncertainty can lead the design to constraint violation even if the optimal value of the insulin/CHO ratio is used, the "relaxing factors" k_i (also optimized in the design) are introduced to evaluate the possible discrepancy between the actual bolus release and the optimal ratio during a standard postprandial glucose test.

The end-point constraint on the glucose concentration and on the derivative of the glucose concentration must be fulfilled within 720 min (12 h) from the last meal. A long duration must be allowed for when designing an mOGTT test as it was verified that for the experiment to be informative, the three meals should be spaced over a sufficiently long period of time. Furthermore, as the designed ingestion of the last meal may occur toward the end of the experiment, a sufficiently long period may be needed before the interior points constraints can be fulfilled.

An experiment with 10 sampling points ($n_{sp} = 10$) is discussed here. The glucose concentration profiles are illustrated in Fig. 16.4. The four meals are taken in the first half of the test. The initial parameter mismatch leads to the glucose concentration profile in the designed experiment to be above the upper threshold during the actual execution of the test. As discussed before, this can be tolerated for diagnostic or identification purposes. The optimal design settings and the constraints are shown in Table 16.1. The test returns a satisfactory parameter estimation in statistical terms (Table 16.2).

The subject's glycemic stress can be reduced by increasing the number of meals [42]. The drawback is that a rather long experiment duration may be needed.

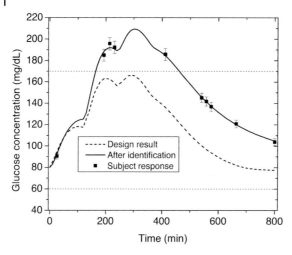

Fig. 16.4 Experiment mOGTT: glucose concentration profiles predicted by the model during the experiment design (broken line) and after the parameter identification (solid line); the subject's actual response to the designed experiment is indicated by the squares (after [42]).

Table 16.1 Experiment mOGTT: optimal settings and constraints [42].

Optimized design variables	Values
t^{sp} (min)	[25, 193, 214, 231, 412, 541, 557, 574, 663, 800]
t^{meals} (min)	[0, 120, 240, 360]
D_g (g)	[18.0, 36.0, 26.8, 10.0]
k	[0.45, 0.4, 0.5, 0.5]

Table 16.2 Parameter estimation after experiment mOGTT (the reference t-value is equal to 1.746) [42].

Model parameter	Final value	Initial guess	Confidence interval 95%	95% t-value	Standard deviation
Θ_1	0.98241	1.5924	0.1726	5.692	0.08141
Θ_2	0.82972	0.81871	0.315	2.634	0.1486
Θ_3	1.2115	1.3755	0.3655	3.315	0.1724
Θ_4	1.0689	1.2677	0.1386	7.715	0.06535

16.8.1.1 Effect of the Number of Samples

Equation (16.11) can be written as

$$\mathbf{H}_\theta(\boldsymbol{\theta}, \boldsymbol{\varphi}) = \sum_{l=1}^{n_{sp}} \sum_i^{N_y} \sum_j^{N_y} s_{ij} \left[\frac{\partial \hat{y}_i(t_l)}{\partial \theta_m} \right]^T \left[\frac{\partial \hat{y}_j(t_l)}{\partial \theta_m} \right] + \mathbf{K} = \sum_{l=1}^{n_{sp}} \mathbf{M}_l + \mathbf{K} \quad (16.17)$$

where the $\mathbf{M}_l$ matrix represents the contribution of the lth sample to the overall predicted information content. The information content is deeply affected by the frequency of sampling (in terms of the n_{sp}/τ ratio) and by the excitation pattern of the manipulated inputs; thus, the choice of the number of samples may have a strong impact on the effectiveness of the test in terms of information available at the end of the test itself. The possibility of increasing the number of samples is currently a feasible feature, in view of the fact that modern sampling techniques such as CGMs (continuous glucose monitoring systems) allow to increase the available information through a more precise and frequent measure of glycemia [67, 68].

An mOGTT with $n_{sp} = 5, 10, 20$ can be considered [69] to evaluate the impact of the number of samples on the quality of the final estimate. When $\tau = 840$ min (14 h), the estimate is statistically satisfactory only if more than five samples are taken (Table 16.3).

Figure 16.5 shows how increasing the number of samples may determine a faster and more effective increase of the information content available after the conclusion of the experiment. Note that five samples are insufficient to cross the threshold for a minimum desired information (estimated from $\mathbf{H}_\theta$ in the simplified assumption of uncorrelated parameters and 10% standard deviation).

16.9
Dealing with Uncertainty

Advanced approaches have been proposed to improve the effectiveness and applicability of MBDoE techniques. Adaptive optimal input design [70] and online model-based redesign of experiments (OMBRE) [71] allow exploiting the information acquired during an experiment thanks to intermediate parameter estimation sessions, thus performing an update of the optimally designed experimental conditions while an experiment is still running. The possibility to exploit the available information content during the experiment execution (rather than at the experiment completion) can help increasing the safety and feasibility of the clinical test since an online tuning of the model on the actual subject's response is made possible.

On a different perspective, backoff-based MBDoE techniques have proved to be very efficient on ensuring feasible and optimally informative tests under parametric uncertainty and have been successfully applied to a simple model of glucose homeostasis [72]. The possibility to incorporate a backoff strategy ensuring feasibility in the presence of uncertainty is also particularly important when, as usually is the case, the model is not a perfect representation of the real system. According to this approach, model uncertainty and parametric uncertainty are tackled by design in order to guarantee a safe and optimal clinical test. In the following, both advanced techniques will be introduced and assessed.

Table 16.3 Comparison of three mOGTT tests with different number of samples.

Design	Parameter values	Confidence interval 95%	95% t-value	t_{ref}
mOGTT-5	[1.096 0.800 1.231 1.024]^T	[±0.2437 ±0.4978 ±0.3028 ±0.1088]	[4.49 1.61[a] 4.06 9.41]	1.795
mOGTT-10	[0.902 1.007 1.089 1.022]^T	[±0.2139 ±0.1593 ±0.3054 ±0.1018]	[4.22 6.32 3.57 10.04]	1.745
mOGTT-20	[0.957 0.975 1.077 1.021]^T	[±0.1634 ±0.1176 ±0.2447 ±0.0817]	[5.86 8.29 4.40 12.49]	1.705

a) t-values failing the t-test (t_{ref} is the reference t-value) [69].

16.9.1
Online Model-Based Redesign of Experiments

Standard MBDoE techniques aim at solving Eqs. (16.5)–(16.8) optimization problem starting from a prior estimate of the model parameters. Nevertheless the prior parameter estimate might be considerably different from the parameter set describing the metabolic specificity of the subject and, as previously discussed, the experiment design is highly sensitive to parametric mismatch.

Equation (16.5) is sufficiently general to be extended for use within a strategy for online redesign of experiments. Through this strategy, one seeks to update the information available at a given updating time t^{up} by executing online, at the same time (either assigned or to be optimized), a parameter estimation session followed by a redesign of the remaining part of the experiment. In this way, the original trajectories of the control variables as well as the sampling schedule are adjusted for this remaining part. One or more updates can be attained in the redesign, each one adding a new component (in the form of (16.9)) to the design vector φ of the experiment, so that this vector can be rewritten as

$$\varphi = [\varphi_1, \varphi_2, \ldots, \varphi_j, \ldots, \varphi_{n_{up}+1}]^T \tag{16.18}$$

where n_{up} is the number of control updates, and φ_j is the design vector before the jth update. Note that φ_j contains the (sub-)lengths of the single updating intervals, whose actual duration may depend on the redesign strategy.

The amount of information that needs maximizing in the jth redesign can be expressed in terms of the dynamic information matrix:

$$\mathbf{H}^*_{\theta|j}(\boldsymbol{\theta}, \varphi_j) = \sum_{k=0}^{j-1} \tilde{\mathbf{H}}^*_{\theta|k}(\boldsymbol{\theta}, \varphi_k) + \tilde{\mathbf{H}}^*_{\theta}(\boldsymbol{\theta}, \varphi_j) + (\boldsymbol{\Sigma}_{\theta}^{-1})$$

$$= \tilde{\mathbf{H}}^*_{\theta}(\boldsymbol{\theta}, \varphi_j) + \mathbf{L} \tag{16.19}$$

where the sum between the prior information on model parameters ($\boldsymbol{\Sigma}_{\theta}^{-1}$) and the information acquired before the jth redesign can be expressed as a constant term $\mathbf{L}$. The symbol ($\sim$) indicates that the information matrix refers to a single updating interval, and $\tilde{\mathbf{H}}_{\theta|0}$ is the zero matrix. Note the similarity between information matrices (16.11) and (16.19): the main difference is that in (16.11) the vector to be optimized is φ, whereas in (16.19) the design vector is φ_j. It should be noted that if an OMBRE strategy is adopted, more degrees of freedom are available to the experiment designer for optimization and each of the optimization problems can be made less complex than the optimization required in standard sequential design.

A single experiment can be seen as a sequence of $(n_{up} + 1)$ subexperiments designed independently, each one of length $\tau_i = t_i^{up} - t_{i-1}^{up}$ (with $i = 1, 2, \ldots, n_{up} + 1$, $t_0^{up} = 0$), where τ_i is the length of the ith "updating interval." The designer can set the level of excitation during each updating interval, that is, the number

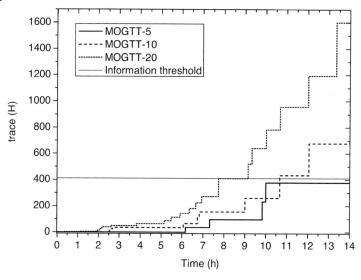

Fig. 16.5 Information profiles as given by the trace of the information matrix in Eq. (16.17) for different number of samples in mOGTT (after [69]).

of switching times per updating interval (e.g., if this number is same for all the updating intervals, then the subexperiments are said to be homogeneously excited). A further design variable is the number (and the time placement) of measurement samples taken per updating interval; this is a critical difference with respect to other adaptive methods proposed in the literature [70], where the sampling time is assigned *a priori*. For more details see [71].

An illustrative example of the application of an OMBRE strategy will be discussed at the end of the chapter, when discussing the effect of a structural difference between the model and a subject's response.

16.9.2
Model-Based Design of Experiment with Backoff (MBDoE-B)

The objective is to avoid unfeasible (unsafe) solutions "by design." Thus, a backoff from active constraints is introduced within the MBDoE framework [72], taking into account the parametric uncertainty in the design feasibility condition (16.8):

$$\mathbf{C} = \mathbf{x}(t) - \mathbf{\Gamma}(t) + \boldsymbol{\beta}\left(\tilde{\mathbf{x}}(t), \tilde{\mathbf{x}}(t), \mathbf{u}(t), \mathbf{w}, \tilde{\boldsymbol{\theta}}, t\right) \leq 0 \qquad (16.20)$$

where $\boldsymbol{\beta}$ is an N_c-dimensional backoff vector depending on the set $\tilde{\mathbf{x}}(t)$ of stochastic realizations of the state variables over a parameter uncertainty domain, defined by the set $\tilde{\boldsymbol{\theta}}$ of stochastic realizations of model parameters. The design optimization problem is the solution of the (16.5)–(16.7) equations with the (16.20) feasibility condition. A stochastic simulation procedure for the evaluation of the backoff vector evaluation has to be performed. The procedure consists of three key steps:

1. Characterization of the parametric uncertainty, that is, definition of the multidimensional uncertainty domain of model parameters and its sampling.
2. Mapping the uncertainty region of the state variables: a mean-variance approach [73] is used to provide a probabilistic description of the uncertainty region of state variables.
3. Backoff formulation and policy.

In the current study, the only variable being constrained is the glucose concentration, and the MBDoE optimization problem with backoff is the solution of (16.5)–(16.7) with the feasibility conditions

$$C_1 = y(\boldsymbol{\theta}, \varphi, t) + \beta(t) - \Gamma_1 \leq 0 \quad \text{and} \quad C_2 = \Gamma_2 - y(\boldsymbol{\theta}, \varphi, t) + \beta(t) \leq 0$$

(16.21)

Figure 16.6 clarifies the flux of information and the tasks needed in a backoff approach.

16.9.2.1 Backoff Application

To illustrate the effect of backoff, a simpler model of glucose homeostasis will be considered, that is, the model of glucose homeostasis in the form proposed in [27]. The model is represented by the following set of DAEs:

$$\frac{dG}{dt} = -\theta_1 G - X(G + G_b) + D(t) \tag{16.22}$$

$$\frac{dX}{dt} = -\theta_2 X + \theta_3 I \tag{16.23}$$

$$\frac{dI}{dt} = -n(I + I_b) + \frac{u(t)}{V_I} \tag{16.24}$$

where G is the blood glucose concentration (mg/dL), X the insulin concentration (mU/L) in the nonaccessible compartment, I the insulin concentration (mU/L), and $u(t)$ the rate of infusion of exogenous insulin (mU/min). Following [72], the meal disturbance model $D(t)$ is

$$D(t) = 2.5At \exp(-0.05t) \tag{16.25}$$

where A is the amount of CHO of the meal (fixed at 60 g). The basal parameters kept constants are: basal glucose concentration in the blood ($G_b = 81$ mg/dL), basal insulin concentration ($I_b = 15$ mU/L), insulin distribution volume ($V_I = 12$ L), and disappearance rate of insulin ($n = 5/54$ min^{-1}).

A constrained MBDoE with backoff is designed (the E-optimal criterion is chosen), where the optimized design variables are the profile of the insulin infusion rate $u(t)$ (approximated as piecewise constant with $n_{sw} = 7$ switching times and $n_z = 8$ switching levels) and the vector of sampling times ($n_{sp} = 12$ samples with

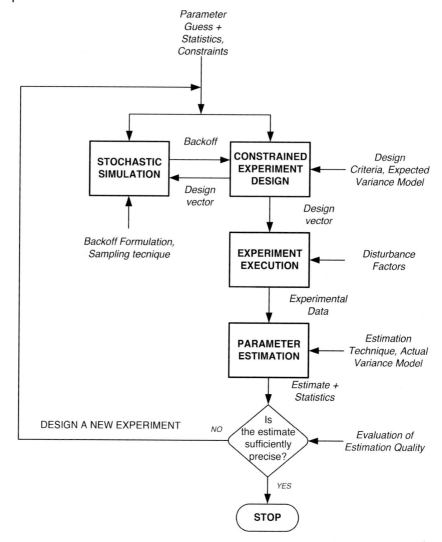

Fig. 16.6 Procedure for an MBDoE with backoff.

a minimum time between consecutive measurements of 10 min). The measured variable is G, with a 3% expected relative error on the measurements. The constraints on the system are related to normoglycemia attainment, and are the upper ($\Gamma_1 = 170$ mg/dL) and lower ($\Gamma_2 = 60$ mg/dL) thresholds on G. Strictly speaking, the lower bound only is a hard constraint not to be violated. However, here as a matter of example both constraints will be treated as hard ones. The constrained MBDoE is performed on the parametric set $\hat{\boldsymbol{\theta}} = [0.0287\ 0.0283\ 1.30\mathrm{E}{-}5]^\mathrm{T}$ describing a healthy subject, while the real subject is assumed to be affected by T1DM with $\boldsymbol{\theta} = [0.0155\ 0.0250\ 1.20\mathrm{E}{-}5]^\mathrm{T}$.

The backoff functions are evaluated through a stochastic simulation where the parameters belong to a family of independent normal distributions whose means and standard deviations are given by vectors $\hat{\boldsymbol{\theta}}$ and $\boldsymbol{\sigma} = [0.0065\ 0.0013\ 0.0033\text{E}{-}5]^\text{T}$.

The uncertainty region and the backoff vector are evaluated considering the maximum and the minimum profile of the measured variable over the expected uncertainty region of model parameters. Both MBDoE and MBDoE-B tests allow estimating the three parameters in a statistically sound way; however, Fig. 16.7 shows that only the backoff approach is capable of designing a safe test for the subject preserving at the same time the quality of the estimation.

16.9.3
Effect of a Structural Difference Between a Model and a Subject

It has been shown earlier that MBDoE can be used to identify complex models of T1DM in the presence of parametric mismatch. However, although constraints can be handled in the design formulation, the presence of a model (also called structural) mismatch (i.e., the subject's physiological behavior is different from the model representation) may severely affect the quality of the experiment. As discussed in [74], since the design methodology is model-based, both model mismatch and parametric mismatch can affect the consistency of the whole design procedure. The result could be a suboptimal design (scarcely informative) or, in the worst case, a dangerous or unfeasible test burdening on the subject's health (e.g., the actual subject's response may lead toward hyperglycemia or, even worse, hypoglycemia).

In order to mimic the response of a subject affected by T1DM, the Cobelli model (CM) [31] is adopted as the subject; within this model, the secretion compartment is substituted by a variation of a model described in [45] as presented in a recent simulation study in [75]. Conversely, the HM is used as the model during the MBDoE identification procedure.

A standard MBDoE approach may result unsafe and poorly informative because of the great uncertainty. In this case, a backoff strategy is often the most suitable one for MBDoE. However, here we will show that an OMBRE approach may prove successful, too. In fact, an OMBRE approach can be exploited to extract the information of the test by updating the optimally designed conditions as the test is running. Seven intermediate parameter estimations are carried out. Therefore, a clinical test can be seen as a sequence of eight separately planned subexperiments lasting 2 h each. During each subexperiment, five samples are taken and the insulin infusion profile, approximated by a piecewise constant function, is optimized by acting on two switching times and three switching levels.

In fact, the actual online redesign strategy is preceded by a preliminary first phase: during an overnight fast the subject, at first in slight hyperglycemic conditions [75], is kept under a continuous insulin infusion of $u = 11.7$ mU/min to normalize the glycemia. Glycemic levels are checked every 45 min and a parameter estimation is performed in order to reach a preliminary parameter estimation. The parameters estimated from this preliminary phase of the identification test are

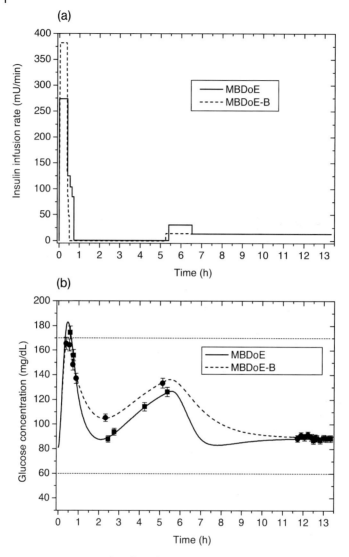

Fig. 16.7 (a) Optimized profiles of insulin infusion rate as given by MBDoE (solid line) and by MBDoE including backoff (broken line) and (b) glucose concentration profiles after identification. The subject's actual response is indicated by black circles (MBDoE) and squares (MBDoE with backoff).

only roughly, but allow for a more effective start-up of the design procedure. The OMBRE test considers four redesign updates by the end of the first phase and three additional (seven overall) updates by the end of the second phase.

The profile of insulin infusion rate and of the glucose concentration as dictated by OMBRE and as predicted by the model at the end of the test are shown in

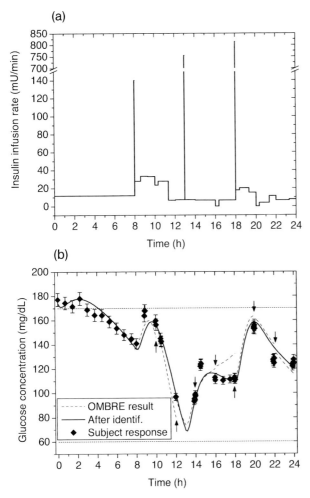

Fig. 16.8 OMBRE with model mismatch. (a) Optimized profile of insulin infusion rate and (b) profiles predicted by a redesign (broken line) and after identification with the HM (solid line); the subject's actual response (simulated by the CM) is indicated by diamonds. In (b), each arrow indicates the beginning of a new redesign.

Fig. 16.8. The parameter estimation is satisfactory considering that at the end of the test all the parameters except Θ_2 are estimated in a statistically sound manner (Table 16.4). It was noticed that parameter Θ_2 cannot be successfully estimated even by increasing the number of the clinical tests. In fact, this is not a surprising result as there is a structural difference between the model and the subject's response. Results cannot be compared to the ones discussed in the preceding sections where only a parametric mismatch was taken into account. As is usually the case, the model mimics reality but there may be some differences in the dynamic behavior

Table 16.4 Parameter estimation after an OMBRE designed test with seven updates (the reference t-value is equal to 1.676; asterisks denote t-values failing the t-test).

Model parameter	Final value	Initial guess	Confidence interval 95%	95% t-value	Standard deviation
Θ_1	0.1052	0.0807	0.0409	2.572	0.020
Θ_2	0.0010	0.0009	0.0745	0.013*	0.037
Θ_3	2.2589	2.2906	0.0380	59.450	0.019
Θ_4	0.8365	0.8248	0.0646	12.950	0.032

that cannot by fully explained by a model and thus may hinder a full identification. In such a case, it is important to assess whether the remaining uncertainty might affect the model reliability (at least within the expected domain of clinical conditions): here, it was verified that the uncertainty on parameter Θ_2 does not impede the model capability at representing the subject's response in a significant way. If it were not the case, then a better model should be sought.

The above results show that the possibility to update the parameters' value and to adjust the experimental plan in real time allow for an online tuning of the model according to the actual subject's response. As a result, the profile predicted by the design gets closer to the experimental data as the test approaches the completion.

The results clearly show that a redesign approach can offer several advantages and can be exploited to design a clinical test when the structural responses between the model and the subject are different. The test is safer for the subject in the presence of both model (structural) mismatch and parametric mismatch. Furthermore, the parameter estimation is more precise than the one provided by a standard MBDoE approach.

16.10
Conclusions

The quest for an artificial pancreas represents a highly ambitious multidisciplinary effort within which the contribution of PSE is of paramount importance. In this chapter, the role of a well-established PSE technique, namely optimal MBDoEs, has been reviewed with reference to the problem of individual parameter identification for complex physiological models of glucose homeostasis.

The parameter identification problem is a tradeoff between several issues: acquisition of a large information amount from a clinical test (thus enabling a reduction in the uncertainty region of the model parameters), compliance to a number of constraints in the system inputs and outputs (most importantly, safety for the subject in terms of glucose concentration levels during the test), practical applicability of the test (e.g., overall length and maximum number of blood samples).

It has been shown that MBDoE does allow designing effective and safe clinical tests, where the administration of CHO (i.e., glucose) and insulin is exploited to provide dynamic excitation to the body system, and a proper schedule of blood samples is used to collect the information generated during the test. The results show that it is possible to reach a statistically satisfactory parameter estimation using a single modified oral test by managing a proper meal intake and the insulin bolus administration. Improved results in the face of model (structural) mismatch can be obtained by redesigning online an experiment.

By using MBDoE, one can arrive at the identification of model parameters for individual subjects affected by T1DM. The availability of such a properly calibrated simulation model of the glucose–insulin system can be an invaluable tool for the development of customized solutions for diabetes treatment. We believe that the introduction of reliable continuous glucose monitoring systems will further enhance the potential of MBDoE techniques so as to deliver more design flexibility and the possibility to envisage faster, safer, and more effective clinical tests.

Appendix A: Hovorka Mathematical Model

The system of equations of the HM is here summarized. Note that while in the previous sections of this chapter the glucose concentration is expressed in mg/dL, here, as in the original paper [28], it is expressed in mmol/L. The insulin concentration is expressed in mU/L. The glucose accessible and nonaccessible compartments are represented by the following set of equations:

$$\frac{dQ_1(t)}{dt} = -\left[\frac{F_{01}^C}{V_G G(t)} + x_1(t)\right]Q_1(t)$$
$$+ k_{12}Q_2(t) - F_R + U_G(t) + EGP_0\left[1 - x_3(t)\right] \quad (A.1)$$

$$\frac{dQ_2(t)}{dt} = x_1(t)Q_1(t) - \left[k_{12} + x_2(t)\right]Q_2(t) \quad (A.2)$$

$$G(t) = Q_1(t)/V_G \quad (A.3)$$

where EGP_0 is the endogenous glucose production extrapolated to 0 insulin concentration, while F_{01}^C is the total noninsulin-dependent glucose flux (corrected for the ambient concentration):

$$F_{01}^C = \begin{cases} F_{01} & \text{if } G > 4.5 \text{ mmol/L} \\ F_{01}G/4.5 & \text{otherwise} \end{cases} \quad (A.4)$$

and F_R is the renal clearance rate above the glucose threshold of 9 mmol/L, given by

$$F_R = \begin{cases} 0.003(G-9)V_G & \text{if } G > 9 \text{ mmol/L} \\ 0 & \text{otherwise} \end{cases} \quad (A.5)$$

The insulin action subsystem is modeled through a three-compartment structure

$$\frac{dx_1}{dt} = -k_{a1}x_1(t) + k_{b1}I(t) \tag{A.6}$$

$$\frac{dx_2}{dt} = -k_{a2}x_2(t) + k_{b2}I(t) \tag{A.7}$$

$$\frac{dx_3}{dt} = -k_{a3}x_3(t) + k_{b3}I(t) \tag{A.8}$$

while the gut absorption rate is described by an exponential function of the following form:

$$U_G(t) = \frac{D_G A_G t e^{-\frac{t}{t_{max,G}}}}{t_{max,G}^2} \tag{A.9}$$

where $t_{max,G}$ is the time of maximum appearance rate of glucose (so that the $U_G(t)$ is modeled as a two-compartment chain with identical transfer rate $1/t_{max,G}$). The A_G parameter takes into account the fact that only a fraction of the whole CHO content of the meal (D_G) appears in plasma, while the remaining part is extracted by the liver. In the current study, the digesting dynamics was not modeled, and the focus is on the representation of insulin-dependent glucose flux. The Wilinska insulin absorption submodel is characterized by the following equations:

$$\frac{dS_{1a}}{dt} = ku - k_{a11}S_{1a} - LDA \tag{A.10}$$

$$\frac{dS_{1b}}{dt} = (1-k)u - k_{a22}S_{1b} - LDB \tag{A.11}$$

$$\frac{dS_2}{dt} = k_{a11}S_{1a} - k_{a11}S_2 \tag{A.12}$$

$$\frac{dS_3}{dt} = k_{a11}S_2 + k_{a22}S_{1b} - k_e I \tag{A.13}$$

$$I = \frac{S_3}{V_I} \tag{A.14}$$

$$LDA = V_{MAX,LD} \frac{S_{1a}}{(K_{m,LD} + S_{1a})} \tag{A.15}$$

$$LDB = V_{MAX,LD} \frac{S_{1b}}{(K_{m,LD} + S_{1b})} \tag{A.16}$$

where LDA is the local degradation at the injection site for continuous infusion, while LDB is the local degradation at the injection site for insulin bolus. For our purposes, the parameters of the Wilinska submodel are not estimated and are kept constants at the values reported in Table A.1 (which also includes a brief description of their physical meaning). The insulin and glucose distribution volumes, the

Table A.1 Parameters of the submodel for insulin absorption as reported in [12].

Constants	Description	Value
V_I	Insulin distribution volume	42.01×10^{-2} L/kg
k_{a11}	Slow channel transfer rate	0.011 1/min
k_{a22}	Fast channel transfer rate	0.021 1/min
k_e	Insulin elimination transfer rate	3.68×10^{-2} L/min
k	Proportion in slow channel	0.67 (unitless)
$V_{MAX,LD}$	Saturation level	1.93 mU/min
$K_{m,LD}$	Half-concentration constant	62.6 mU

insulin elimination rate constant, and the transfer rate contributions are also kept constants (Table A.2).

As suggested by the authors themselves, a new parameterization is considered allowing to split the insulin sensitivity into three terms (insulin sensitivity of distribution/transport, disposal, and endogenous glucose production, respectively):

$$S_{IT}^{f} = \frac{k_{b1}}{k_{a1}} \tag{A.17}$$

$$S_{ID}^{f} = \frac{k_{b2}}{k_{a2}} \tag{A.18}$$

$$S_{IE}^{f} = \frac{k_{b3}}{k_{a3}} \tag{A.19}$$

In this way, the k_{ai} parameters are also treated as constants. The values of the parameters are valid for a healthy male subject, and a body weight of 75 kg was considered in the simulations for this paper. With the alternative parameterization (A.17)–(A.19), the vector of model parameters is the set $\theta = [S_{IT}^{f} \ S_{IT}^{d} \ S_{IT}^{e} \ EGP_0 \ F_{01}]^T$, whose nominal values are shown in Table A.3.

Table A.2 Parameters of model that are kept constants, as reported in [42].

Constants	Description	Value
k_{12}	Transfer rate	0.066 min^{-1}
k_{a1}	Deactivation rate	0.006 min^{-1}
k_{a2}	Deactivation rate	0.06 min^{-1}
k_{a3}	Deactivation rate	0.03 min^{-1}
k_e	Insulin elimination from plasma	0.138 min^{-1}
V_I	Insulin distribution volume	0.12 L/kg
V_G	Glucose distribution volume	0.16 L/kg
A_G	CHO bioavailability	0.8 (unitless)
$T_{max,G}$	Time-to-maximum of CHO absorption	40 min

Table A.3 Nominal values of HM parameters for a healthy subject, as reported in [12].

Model parameters	Description	True values
S_{IT}^f	Insulin sensitivity of distribution/transport	51.2E−4
S_{ID}^f	Insulin sensitivity of disposal	8.2E−4
S_{IE}^f	Insulin sensitivity of EGP	520E−4
EGP_0	EGP extrapolated to zero insulin concentration	0.0161 mmol/kg min
F_{01}	Noninsulin dependent flux	0.0097 mmol/kg min

A normalization procedure was used because of the great difference in magnitude between parameters values, so as to avoid scale-dependent difficulties in the estimation and design procedures.

List of Symbols

General symbols

A – amount of carbohydrates in a meal
c_{ij} – ijth element of the correlation matrix for model parameters (**C**)
$\tilde{C}_i$ – ith element of the constraints vector ($\tilde{\mathbf{C}}$)
D – meal disturbance function
$D_{g,i}$ – glucose content of the ith meal
EGP_0 – endogenous glucose production extrapolated to zero insulin concentration
F_{01} – noninsulin dependent flux
F – differential and algebraic system implicit function
G – glucose concentration in the subject
$\hat{G}$ – glucose concentration in the model
G_b – basal blood glucose concentration
k_i – ith bolus release relaxing factor
I – insulin concentration
I_b – basal insulin concentration
N – disappearance rate of insulin
N_{meals} – number of meals
N_u – number of manipulated inputs
N_x – number of state variables
N_w – number of time invariant controls
N_y – number of measured variables
N_θ – number of model parameters
N_c – number of constraints
n_{exp} – number of experiments
n_{sp} – number of samples
n_{sw} – number of switching levels
n^{up} – number of control updates

n_φ – number of design variables
q_i – ith element of the dynamic sensitivity matrix (**Q**)
s_{ij} – ijth element of the inverse matrix of measurements errors
t – time
t_i – ith t-value
t_i^{up} – ith updating time
t^{sw} – switching time
U – insulin infusion rate
u_{basal} – time-invariant basal insulin infusion rate
u_{bol} – insulin bolus amount
v_{ij} – ijth element of the variance–covariance matrix $\mathbf{V}_\theta$
x – generic state variable
X – insulin concentration in the nonaccessible compartment
y – generic measured output

Vectors and Matrices [dimension]

C – correlation matrix $[N_\theta \times N_\theta]$
$\tilde{\mathbf{C}}$ – set of constraints $[N_c]$
C – vector of bounds on measured variables $[N_c]$
$\mathbf{D}_g$ – vector of the CHO content of the meals $[N_{meals}]$
g – measurements selection function
$\mathbf{H}_\theta$ – dynamic information matrix $[N_\theta \times N_\theta]$
$\mathbf{H}_\theta^0$ – preliminary information matrix $[N_\theta \times N_\theta]$
$\mathbf{H}_{\theta|k}$ – dynamic information matrix of the kth experiment $[N_\theta \times N_\theta]$
K – vector of relaxing factors for bolus release $[N_{meals}]$
K – constant dynamic information matrix in (16.11) $[N_\theta \times N_\theta]$
L – constant dynamic information matrix in (16.19) $[N_\theta \times N_\theta]$
$\mathbf{M}_l$ – dynamic information matrix of the lth sample $[N_\theta \times N_\theta]$
Q – sensitivity matrix $[n_{sp} \times N_\theta]$
$\mathbf{t}^{meals}$ – vector of meal ingestion times $[N_{meals}]$
$\mathbf{t}^{up}$ – vector of updating times $[n^{up}]$
$\mathbf{t}^{sw}$ – vector of switching times $[n_{sw}+1]$
u – vector of manipulated inputs $[N_u]$
$\mathbf{V}_\theta$ – variance–covariance matrix of model parameters $[N_\theta \times N_\theta]$
w – vector of time-invariant control $[N_w]$
x – vector of state variables $[N_x]$
$\mathbf{x}_0$ – vector of initial conditions $[N_x]$
y – vector of measured outputs $[N_y]$
$\hat{\mathbf{y}}$ – vector of estimated responses $[N_y]$
$\mathbf{y}_0$ – vector of initial conditions of measured outputs $[N_y]$
$\boldsymbol{\beta}$ – backoff vector $[N_c]$
$\boldsymbol{\varphi}$ – design vector $[n_\varphi]$
$\boldsymbol{\theta}$ – vector of values of model parameters for the subject $[N_\theta]$

$\hat{\boldsymbol{\theta}}$ – vector of estimated values of model parameters $[N_\theta]$
$\boldsymbol{\theta}^0$ – vector of initial guesses of model parameters $[N_\theta]$
$\boldsymbol{\Theta}$ – vector of normalized model parameters $[N_\theta]$
$\boldsymbol{\Sigma}$ – vector of standard deviations on model parameters $[N_\theta]$
$\boldsymbol{\Sigma}_\theta$ – prior variance–covariance of model parameters $[N_\theta \times N_\theta]$
$\boldsymbol{\Gamma}$ – set of active constraints on state variables $[N_c]$

Greek letters

α – unit conversion coefficient
δ – Dirac impulse function
Υ – response selection function
κ_i – ith confidence interval
θ_i – ith model parameter
Θ_i – ith normalized model parameter
τ – test duration
τ_i – ith updating interval
ψ – $\mathbf{V}_\theta$ measurement function

Acronyms

CGMs – continuous glucose monitoring systems
CHO – carbohydrates
HM – Hovorka model
IVGTT – intravenous glucose tolerance test
MBDoE – model-based design of experiments
mOGTT – modified oral glucose tolerance test
MPC – model predictive control
OGTT – oral glucose tolerance test
OMBRE – online model-based redesign of experiments
PSE – process systems engineering
SGI – structurally globally identifiable
SLI – structurally locally identifiable
SNI – structurally nonidentifiable
T1DM – type-1 diabetes mellitus

References

1 World Health Organisation, Department of Non-Communicable Disease Surveillance, Definition and diagnosis of diabetes mellitus and intermediate hyperglycaemia, **2006**.

2 BEQUETTE, B. W., *Diabetes Technology and Therapeutics* 7 (**2005**), p. 28.

3 DOYLE III, F. J., JOVANOVIC, L., SEBORG, D. E., PARKER, R. S., BEQUETTE, B. W., JEFFREY, A. M., XIA, X., CRAIG, I. K., MCAVOY,

T., *Journal of Process Control* 17 (**2007**), p. 571.
4 ALBISSER, A. M., LEIBEL, B. S., EWART, T. G., DAVIDOVAC, Z., BOTZ, C. K., ZINGG, W., *Diabetes* 23 (**1974**), p. 389.
5 CLEMENS, A. H., *Medical Program Technology* 6 (**1979**), p. 91.
6 NOMURA, M., SHICHIRI, M., KAWAMORI, R., YAMASAKI, Y., IWAMA, N., ABE, H. A., *Computers and Biomedical Research* 17 (**1984**), p. 570.
7 BELLAZZI, R., NUCCI, G., COBELLI, C., *IEEE Engineering in Medicine and Biology* 20 (**2001**), p. 54.
8 SEBORG, D. E., EDGAR, T. F., MELLICHAMP, D. A., DOYLE, F. J., *Process Dynamics and Control*, 3rd edn., Wiley, New York, **2011**.
9 STEIL, G. M., SAAD, M. F., *Current Opinion in Endocrinology and Diabetes* 13 (**2006**), p. 205.
10 DORAN, C. V., CHASE, J. G., SHAW, G. M., MOORHEAD, K. T., HUDSON, N. H., *Control Engineering Practice* 13 (**2005**), p. 1129.
11 FARMER, T. G., EDGAR, T. F., PEPPAS, N. A., *Industrial and Engineering Chemistry Research* 48 (**2009**), p. 4402.
12 MARCHETTI, G., BAROLO, M., JOVANOVIĈ, L., ZISSER, H., SEBORG, D. E., *IEEE Transactions on Biomedical Engineering* 55 (**2008**), p. 857.
13 PERCIVAL, M. W., DASSAU, E., ZISSER, H., JOVANOVIĈ, L., DOYLE III, F. J., *Industrial and Engineering Chemistry Research* 48 (**2009**), p. 6059.
14 HOVORKA, R., CANONICO, V., CHASSIN, L. J., HAUETER, U., MASSI-BENEDETTI, M., FEDERICO, M. O., PIEBER, T. R., SCHALLER, H. C., SCHAUPP, L., VERING, T., WILINSKA, M. E., *Physiological Measurement* 25 (**2004**), p. 905.
15 PARKER, R. S., DOYLE III, F. J., PEPPAS, N. A., *IEEE Transactions on Biomedical Engineering* 46 (**1999**), p. 148.
16 FISHER, M. E., *IEEE Transactions on Biomedical Engineering* 38 (**1991**), p. 57.
17 OLLERTON, R. L., *International Journal of Control* 50 (**1989**), p. 2503.
18 PARKER, R. S., DOYLE III, F. J., WARD, J. H., PEPPAS, N. A., *AIChE Journal* 46 (**2000**), p. 2537.
19 PARKER, R. S., DOYLE III, F. J., *Advanced Drug Delivery Reviews* 48 (**2001**), p. 211.
20 EL-KHATIB, F. H., JIANG, J., DAMIANO, E. R., *Journal of Diabetes Science and Technology* 1 (**2007**), p. 181
21 SCHALLER, H. C., SCHAUPP, L., BODENLENZ, M., WILINSKA, M. E., CHASSIN, L. J., WACH, P., VERING, T., HOVORKA, R., PIEBER, T. R., *Diabetic Medicine* 23 (**2006**), p. 90.
22 BERGMAN, R. N., *The Minimal Model Approach and Determination of Glucose Tolerance*, LSU Press, Baton Rouge, **1997**.
23 MAKROGLOU, A., LI, J., KUANG, Y., *Applied Numerical Mathematics* 56 (**2006**), p. 559.
24 BERGMAN, R. N., PHILLIPS, L. S., COBELLI, C., *Journal of Clinical Investigation* 68 (**1981**), p. 1456.
25 BERGMAN, R. N., *Diabetes* 56 (**2007**), p. 1489.
26 SORENSEN, J. T., *A Physiologic Model of Glucose Metabolism in Man and Its Use to Design and Assess Improved Insulin Therapies for Diabetes*, Massachusetts Institute of Technology, Cambridge, USA, **1985**.
27 LYNCH, S. M., BEQUETTE, B. W., Model predictive control of blood glucose in type I diabetics using subcutaneous glucose measurements, in: *Proceedings of the American Control Conference*, **2002**.
28 HOVORKA, R., SHOJAEE-MORADIE, F., CARROLL, P. V., CHASSIN, L. J., GOWRIE, I. J., JACKSON, N. C., TUDOR, R. S., UMPLEBY, A. M., JONES, R. H., *American Journal of Physiology Endocrinology and Metabolism* 282 (**2002**), p. E992.
29 WILINSKA, M. E., CHASSIN, L. J., SCHALLER, H. C., SCHAUPP, L., PIEBER, T. R., HOVORKA, R., *IEEE Transactions on Biomedical Engineering* 52 (**2005**), p. 3.
30 FABIETTI, P. G., CANONICO, V., ORSINI FEDERICI, M., MASSI BENEDETTI, M., SARTI, E., *Medical and Biological Computing* 44 (**2006**), p. 69.
31 DALLA MAN, C., RIZZA, R. A., COBELLI, C., *IEEE Transactions on Biomedical Engineering* 54 (**2007**), p. 1741.
32 SÖDERSTRÖM, T., STOICA, P., *System Identification*, Prentice-Hall, New York, **1989**.
33 YANG, Y. J., YOUN, J. H., BERGMAN, R. N., *American Journal of Physiology* 253 (**1987**), p. E595.
34 BOSTON, R., STEFANOVSKI, D., MOATE, P., SUMNER, A. E., WATANABE, R. M., BERGMAN, R. N., *Diabetes Technology and Therapeutics* 5 (**2003**), p. 1003.

35 Cobelli, C., Thomaseth, K., *Mathematical Bioscience* 83 (1986), p. 127.
36 Asprey, S. P., Macchietto, S., *Journal of Process Control* 12 (2002), p. 545.
37 Bauer, I., Bock, H. G., Körkel, S., Schlöder, J. P., *Journal of Computers and Applied Mathematics* 120 (2000), p. 1.
38 Sidoli, F. R., Manthalaris, A., Asprey, S. P., *Industrial and Engineering Chemistry Research* 44 (2005), p. 868.
39 Gadkar, K. G., Gunawan, R., Doyle, F. J., *BMC Bioinformatics* 6 (2005), p. 155.
40 Franceschini, G., Macchietto, S., *Industrial and Engineering Chemistry Research* 46 (2007), p. 220.
41 Franceschini, G., Macchietto, S., *Chemical Engineering Science* 63 (2008), p. 4846.
42 Galvanin, F., Barolo, M., Macchietto, S., Bezzo, F., *Industrial and Engineering Chemistry Research* 48 (2009), p. 1989.
43 De Fronzo, R. A., Jordan, D. T., Aundreb, A., *American Journal of Physiology* 237 (1979), p. E214.
44 Dalla Man, C., Cobelli, C., *IEEE Transactions on Biomedical Engineering* 53 (2006), p. 2472.
45 Nucci, G., Cobelli, C., *Computer Methods and Program in Biomedicine* 62 (2000), p. 249.
46 Finan, D. A., Zisser, H., Jovanovich, L., Bevier, W., Seborg, D. E., Identification of linear dynamic models for type 1 diabetes: a simulation study, in: *Proceedings of the International Symposium on Advanced Control of Chemical Processes*, 2006.
47 Bondia, J., Dassau, W., Zisser, H., Calm, R., Vehti, J., Jovanoviĉ, L., Doyle III, F. J., *Journal of Diabetes Science and Technology* 3 (2009), p. 89.
48 Davidescu, F. P., Jorgensen, S. B., *Chemical Engineering Science* 63 (2008), p. 4754.
49 Bellman, R., Aström, K. J., *Mathematical Bioscience* 7 (1970), p. 329.
50 Walter, E., Lecourtier, Y., *Mathematical Bioscience* 56 (1981), p. 1.
51 Pohjanpalo, H., *Mathematical Bioscience* 41 (1978), p. 21.
52 Walter, E., Pronzato, L., *Mathematics and Computers in Simulation* 42 (1996), p. 125.
53 Ljung, L., Glad, T., *Automatica* 30 (1994), p. 265.
54 Buchberger, B., *Aequationes Mathematicae* 4 (1998), p. 45.
55 Saccomani, M. P., Audoly, S., Bellu, G., D'Angiò, L., Cobelli, C., Global identifiability of nonlinear model parameters, in: *Proceedings of the SYSID '97 11th IFAC Symposium on System Identification*, 1997.
56 Saccomani, M. P., Audoly, S., D'Angiò, L., *Automatica* 39 (2003), p. 619.
57 Brun, R., Kühni, M., Siegrist, H., Gujer, W., Reichert, P., *Water Research* 36 (2002), p. 4113.
58 Kontoravdi, C., Asprey, S. P., Pistikopoulos, E. N., Mantalaris, A., *Biotechnology Progress* 21 (2005), p. 1128.
59 Asprey, S., Macchietto, S., *Computers and Chemical Engineering* 24 (2000), p. 1261.
60 Pukelsheim, F., *Optimal Design of Experiments*, John Wiley & Sons, New York, 1993.
61 Galvanin, F., Macchietto, S., Bezzo, F., *Industrial and Engineering Chemistry Research* 46 (2007), p. 871.
62 Zhang, Y., Edgar, T. F., *Industrial and Engineering Chemistry Research* 47 (2008), p. 7772.
63 Zullo, L., *Computer Aided Design of Experiments. An Engineering Approach*, The University of London, London, UK, 1991.
64 Pillonetto, G., Sparacino, G., Cobelli, C., *Mathematical Bioscience* 184 (2003), p. 53.
65 Process Systems Enterprise, *gPROMS Introductory User Guide*, Process Systems Enterprise Ltd., London, UK, 2004.
66 Bock, H., Kostina, E., Phu, H. X., Rannacher, R., *Modeling, Simulation and Optimization of Complex Processes*, Springer, Berlin, 2003.
67 Hirsch, I. B., Armstrong, D., Bergenstal, R. M., Buckingham, B., Childs, B. P., Clarke, W. L., Peters, A., Wolpert, H., *Diabetes Technology and Therapeutics* 10 (2008), p. 232.
68 Brauker, J., *Diabetes Technology and Therapeutics* 11 (2009), p. 25.
69 Galvanin, F., Barolo, M., Macchietto, S., Bezzo, F., On the optimal design of clinical tests for the identification of physiological models of type 1 diabetes mellitus, in: *Computer-Aided Chemical Engineering 27, 10th International Symposium on Process Systems Engineering*, 2009.

70 STIGTER, J. D., VRIES, D., KEESMAN, K. J., *AIChE Journal* 52 (2006), p. 3290.
71 GALVANIN, F., BAROLO, M., BEZZO, F., *Industrial and Engineering Chemistry Research* 48 (2009), p. 4415.
72 GALVANIN, F., BAROLO, M., BEZZO, F., MACCHIETTO, S., *AiChe Journal* 56 (2010), p. 2088.
73 APLEY, D. W., LIU, J., CHEN, W., *ASME Journal of Mechanical Design* 128 (2006), p. 945.
74 FORD, I., TITTERINGTON, D. M., KITSOS, C. P., *Technometrics* 31 (1989), p. 49.
75 DALLA MAN, C., RAIMONDO, D. M., RIZZA, R. A., COBELLI, C., *Journal of Diabetes Science and Technology* 3 (2007), p. 323.

Index

a

absolute supersaturation, crystallization processes 243
absorber and desorber performances, reactive separations 190–191
absorption
 – conventional and reactive 207–213
 – see also distillation
absorption column
 – ethyl acetate 226–227, 231
 – NO_x 222
abstraction hierarchy, synthetic biology 494
acetone–water–acetaminophen system 248–250
activity coefficient model, solubility 247
adaptive NEQ/OCFE 218–220
adenosine monophosphate, cyclic see cAMP
adsorbed solution theory, PSA 156
adsorbent bed models, PSA 144–145
adsorbent particle model, PSA 150–157
adsorbent particles
 – linear driving force 151–152
 – mass transfer in 151
adsorbents, carbon 163
adsorption layer model
 – axial dispersion 148–149
 – equation of state 148
 – PSA 146–150
 – thermophysical properties of gas mixture 148
adsorption processes, pressure swing 137–167
advection flux 473
agglomeration, crystallization processes 255
aggregation, crystallization processes 255
algorithm
 – delayed stochastic simulation 524–526, 528–529
 – design of experiments 431–432
 – stochastic simulation 521–532
ammonia, mass balance 417
analysis of variance (ANOVA) 422–423
anti-inflammatory response 343–344
antisolvent feedrate optimization 270–274
apoplast 472
apparent parameters 469
apple, void network 478
approximate lumped parameter model, desalination 294
aquaporins 472
area mean crystal size, definition 265
attrition, crystallization processes 255–256
autocatalyzed processes, reactive separations 175
"average" cells 405

b

Bτ-DSSA 528–529
Bτ-SSSA 529–532
backoff approach, model-based DOE 566–569
backward differentiation formulae based (BDF) methods 450
bacteria
 – biological parts 496
 – gramnegative 325
bacterial invasion, inflammatory response 323
baffle span, MTR 28
baffle window size, MTR 28
balance envelopes, error monitoring 219
batch cultures
 – monoclonal antibodies 413–434
 – static 409
batch drum granulation 56–58
batch free-radical solution polymerization, methyl methacrylate (MMA) 82–83

batch high-shear granulation, DEM-PBE Modeling 58
batch simulation, SAN copolymerization 95–97
BDF *see* backward differentiation formulae
binary interaction parameters, solubility models 248
binary multiscale integration frameworks 45
binomial τ-leap methods 527–528
biological materials, length scales 470–472
biological modeling, structure concept 406
biological models, classification 404
Biological Parts classification, Standard 494–496
biological significance grouping 425
biological systems
– definition 321–322
– mathematical modeling 404–405
– model development framework 411–412
– randomness 518
biological, bio-processing and biomedical systems
– dynamic model building using optimal identification strategies 441–463
– dynamic models of disease progression 321–360
– identification of physiological models of type 1 diabetes mellitus 545–572
– model development and analysis of mammalian cell culture systems 403–435
– multiscale modeling of transport phenomena in plant-based foods 469–485
– robust process control 369–397
– synthetic biology 493–541
biology, and systems theory 321–322
biomass concentration errors 380–381
bioprocess engineering, dynamic model building 441–463
bioprocesses
– model reduction techniques 377–383
– online measurements 377
– robust control 369–397
– state evolution 380–381
Blake–Kozeny correlation 148
blood glucose *see* type 1 diabetes mellitus
bottom–up modeling, granulation 43–44
brine flow rate, orifice model 306
brine heater fouling factor 301–302, 311
brine heater model
– dynamic modeling 307
– MSF desalination 296
bulk gas flow, PSA 146–148
bulk homopolymerization, methyl methacrylate (MMA) 85–86
butyl acetate, reactive multiphase distillation 231–234

c

CAD *see* computer aided design
cage effect, SAN bulk polymerization 93
cAMP 345
– inflammatory response modulation 354–355, 358–359
– neuroendocrine–immune system interactions 337
CAPM *see* computer-aided process modeling
carbohydrate (CHO) content, of meal 546, 550
carbon, as an adsorbent 163
catalyst packing, MTR 29
catalytic reaction model 26–27
CDH *see* corticotrophin-releasing hormone
cell culture systems, model development and analysis 403–435
cell cultures, feeding 413–434
cell growth
– Monod-type kinetics 409
– stoichiometric models 408
cell systems, dynamic modeling and construction 493–541
cells, "average" 405
cellular level, inflammation modeling 330–335
central nervous system (CNS) 325–326
CFD *see* computational fluid dynamics
CFX-4 181, 194
CFX-5 194–195
CFX-10 181, 197
chain length dependent rate coefficients 73–76
channels, in zero-gravity distillation 185–188
Chapman–Enskog equation 149
chemical master equation (CME) 518–521
chemical processing systems
– complex reactive and multiphase separation processes 203–234
– dynamic modeling for solving industrial problems 3–31

- dynamic multiscale modeling 35–61
- framework for the modeling of reactive separations 173–199
- modeling and control of proton exchange membrane fuel cells 105–132
- modeling multistage flash desalination process 287–315
- modeling of crystallization processes 239–276
- modeling of polymerization processes 67–97
- modeling of pressure swing adsorption processes 137–167

chemicals, interaction with RNA 497
CHO see carbohydrate (CHO) content
circuit interactions, granulation 39–40
circuit scale, granulation 50–51
circuits, synthetic biology 494
classification
- biological models 404
- biological parts 494–496
- modeling methods 176–178
- multiscale modeling 44–45

CM see Cobelli model
CME see chemical master equation
CNS see central nervous system
CO_2 transport model 483–487
coarse-grained methods 527–532
Cobelli model (CM) 569
coding regions 508–509
- standard biological parts 495
"collaborative" modeling 197–199
collocation, orthogonal 203–234
column dynamic model 219–220
compartmental model, of glucose homeostasis 551–552
complementary approach
- reactive separations 173, 196–199
complex fluid dynamics, hydrodynamic analogy approach 183–188
complex reactive separation processes, modeling 203–234
complexity pyramid 322
computational fluid dynamics, control applications 371
computational fluid dynamics (CFD) 22, 27–28
- complementary modeling 196–199
- fluid-dynamic approach 179–181
- granulation 44
- reactive separations 176
- structured packings 193–195

computed tomography (CT), plant material 477
computer aided design (CAD) tools 419
computer-aided process modeling (CAPM) 45
COMSOL Multiphysics 3.3a, CFD software 181
concurrent modeling, granulation 43–44
continuous cultures 409
continuous drum granulation
- DEM-PBE Modeling 59–61
- fault diagnosis 55–56
continuum models, plant-based foods 469
control design techniques, model reduction 383–397
control devices, key considerations 8
control vector parameterization (CVP) approach 450
controller synthesis, tubular reactors 389–392
convection–diffusion–reaction processes 371, 385, 397–398
coolant flowrate, MTR 28
cooling brine tube, dynamic modeling 305
cooling crystallization 241
- optimization 274–276
correlative models, solubility 244–245
corrugated-sheet structured packings 185–188, 193
- stream-line representation 197–198
corticotrophin-releasing hormone (CRH) 336
cortisol levels, plasma 348
crystal average size, population balance 260
crystal distributions, definition 265–266
crystal mean sizes 261–262
- definition 265
crystal shape 264
crystal size distribution (CSD) 257, 259, 261–262
- validation 273
crystal sizes, definition 265
crystallization processes
- agglomeration 255
- aggregation 255
- attrition 255–256
- crystal characterization 264–266
- crystallization methods 241–242
- dissolution 254–255
- driving force 242–243
- experimental design 267–268
- growth 254–255
- mass and energy balances 264

- mechanisms 251–256
- modeling 239–276
- nucleation 251–254
- optimization 270–276
- parameter estimation methods 268–269
- particle measurement 266
- population balance 256–263
- simulation environment 266–267
- solubility models 242–243
- supersaturation 242–243

CSD see crystal size distribution
CSS see cyclic steady state
cumulative mass distribution, definition 266
cumulative MWD 89–90
cumulative number distribution, definition 265
CVP see control vector parameterization
cybernetic modeling 406
cyclic 3′,5′-adenosine monophosphate see cAMP
cyclic steady state (CSS) 137, 141, 162–163
cytokine–cytokine receptor interactions 341

d

DAE see differential algebraic equations
Danckwert's boundary conditions 149
Darcy's law 473
data analysis, model-based 19
delayed stochastic simulation algorithm (DSSA) 524–526
- τ-leap methods 528–529

DEM see discrete element modeling
DEM-PBE modeling
- batch high-shear granulation 58
- continuous drum granulation 59–61

derivative based global sensitivity measures (DGSM) 425
desalination see multistage flash desalination
design
- computer aided design 419
- model-based see model-based DOE
- multitubular reactor see multitubular reactor design
- of controls 383–397
- optimal see optimal experimental design
- see also clinical tests; physiological models

design of experiments (DOE) 410–411
- algorithm 431–432
- and model validation 432–434
- see also model-based DOE

design procedure, physiological models 558–559
desorber and absorber performances, reactive separations 190–191
deterministic framework, part models 512–518
deterministic methods, global 452
deterministic modeling regimes 519–520
deterministic simulation, repressilator 539
devices
- electronics 494
- synthetic biology 494–496

DGM see dusty gas model
DGSM see derivative based global sensitivity measures
diabetes mellitus (type 1), model-based design of experiments 545–572
differential algebraic equations (DAE)
- cell culture systems 405
- in PSA modeling 162

differential MWD 89–90
diffusion, fundamentals 472–473
diffusion-controlled reactions, free-radical homopolymerization 69–71
diffusion mean model (DMM) 74
diffusion–convection–reaction processes 371, 385, 397–398
direct integration, MWD 85–86
discrete element modeling (DEM), granulation 48–49, 58–61
discretization
- population balance 258–259
- transport phenomena 478–479

disease progression dynamic modeling 321–360
- data collection 327–328
- human endotoxemia 325–327
- in-silico modeling of inflammation 323–325
- major response elements 340–343
- methods 328–340
- multilevel human inflammation model 328–340
- results 340–360
- transcriptional analysis 340–343
- see also inflammation modeling

dissipative processes, dynamic evolution 374
dissolution, crystallization processes 254–255

distillation
- conventional and reactive 207–213
- hydrodynamic analogy approach 184–188
- multiphase reactive 213–218
- *see also* absorption; multiphase separation processes; reactive separations

distributed process systems (DPSs)
- model reduction 372–377
- modeling 12

distributed processes and bioprocesses
- control design techniques 383–397
- robust control 369–397

DMM *see* diffusion mean model
DNA transcription 493
DOE *see* design of experiments
downscaling, multiscale modeling 476
DPS *see* distributed process system
driving force
- for crystallization 242–243
- in adsorbent particles 151–152

drowning out crystallization *see* salting out crystallization
drug–receptor complex 333–334
drum granulation, scale map for 46
dry pressure drop 193–194
DSSA *see* delayed stochastic simulation algorithm
dusty gas model (DGM) 153
dynamic brine heater fouling profile 301
dynamic model building 21
- identification strategies 441–463
- parameter estimation 443–447
- scheme 442

dynamic modeling
- background and basics 5–14
- basics 3–31
- combining models and experimental data 3–31
- disease progression 321–360
- distributed processes and bioprocesses 369–397
- distributed systems modeling 12
- equation-based modeling tools 11
- first-principles modeling 10
- high-fidelity predictive models 14–21
- hybrid modeling techniques 22
- incorporating hydrodynamics 22
- key considerations 7–9
- key modeling concepts 10–14
- mathematical formulation 4
- model-based engineering 14–23
- model-targeted experimentation 16
- modeling of operating procedures 9
- modeling software 5
- MSF desalination 303–307
- MTR design 23–31
- multiple activities from the same model 13
- multiscale modeling 10–11, 35–61
- predictive process models 7–9
- reduced order 203–234
- synthetic biology 493–541

dynamic models, biological systems 405
dynamic observers
- validation 379–383
- *see also* state observers

dynamic optimization 13
dynamic simulation 13
- basics 3–31
- distributed processes and bioprocesses 369–397
- future challenges 313
- MSF desalination 311–312
- reactive absorption of NO_x 220–225
- reactive distillation of ethyl acetate 225–231
- reactive multiphase distillation of butyl acetate 231–234
- reduced order modeling 220–233

dysregulation, inflammatory response 349–353

e

electrocardiogram (ECG), disease progression dynamic modeling 327
electronics, devices 494
emergence, definition 321
empirical solubility model 243–244
endoplasmic reticulum (ER) 408, 414, 417–418
endotoxemia, multiscale modeling 325–327
endotoxin *see also* lipopolysaccharides
endotoxin injury, modeling the effect on heart rate variability 338–340
endotoxin tolerance 353
energetic response 343–344
energy balance 204, 215, 234
- adaptive NEQ/OCFE 218–220
- at collocation point 207, 212–213
- crystallization processes 264
- error monitoring 219

enthalpy balance, MSF desalination 295

environmental impact
- MSF desalination modeling 302–303, 313–314
enzyme kinetics, mathematical expressions 407
EPI *see* epinephrine
epinephrine (EPI) 345, 354
- neuroendocrine–immune system interactions 337
equation-based modeling tools 11
equation of state, adsorption layer model 148
equations, in MDL 533
equilibrium, key considerations 8
equipment geometry, key considerations 8
ER *see* endoplasmic reticulum
Ergun correlation 148
error convergence
- using FEM 380–381
- using ROM 381–383
error monitoring
- balance envelopes 219, 230
ethyl acetate, reactive distillation 225–231
euglycemic hyperinsulinemic clamp 550
eukaryotic cell modeling 406
- *see also* mammalian cell culture systems
evaporative crystallization 241
exact algorithms, SSA 522–527
experimental design
- model-based *see* model-based DOE
- optimal *see* optimal experimental design
experimental protocols, design 560–563
extended Langmuir isotherms, PSA 154–156

f

factor per second (FaPS), genetic circuit 498
failure mode effects analysis (FMEA) 55–56
FAST *see* Fourier amplitude sensitivity test
FBRM *see* focused beam reflectance measurement (FBRM)
fed-batch cultures 409
- monoclonal antibodies 413–434
fed-batch reactor, production of gluconic acid 457–463
feeding
- cell cultures 413–434
- optimal 435

FEM *see* finite element method
Fick's laws, of diffusion 472–473
Fickian description, of reactant diffusion 71–72
film-like-flow problems, reactive separations 178–179
film model, rate-based approach 188–190
FIM *see* Fisher information matrix
finite element method (FEM) 374–376
- transport phenomena 478–479
- orthogonal collocation 86–87, 203–234
finite state machine (FSM), PSA 160
"first-layer" analysis, monoclonal antibody modeling 424–427
first-principles models 10
- constructing 17
- monoclonal antibodies 415–421
Fischer–Tropsch tubular reaction process 12
Fisher information matrix (FIM) 448
- gluconic acid production 460–462
fixed-bed adsorber
- PSA 137, 144–145
fixed-bed reactor model 12
- multitubular reactor 26–27
fixed pivot technique (FPT) 88
flashing brine model, dynamic modeling 305
flow regimes, cocurrent and countercurrent 178–179
flow reversal, key considerations 8
fluid dynamics, hydrodynamic analogy approach 183–188
fluid-dynamic approach (FDA), reactive separations 176–183
fluid dynamics complexity, and modeling rigor 176
flux vector 512
fluxes, synthetic biology 497–500
FMEA *see* failure mode effects analysis
focused beam reflectance measurement (FBRM), particle measurement 266
fouling factor, brine heater 301–302, 311
Fourier amplitude sensitivity test (FAST) 423
FPT *see* fixed pivot technique
frameworks
- binary multiscale integration 45
- model development 411–412
- modeling of reactive separations 173–199

free-radical homopolymerization 68–77
 – chain length dependent rate coefficients 73–76
 – diffusion-controlled reactions 69–71
 – Fickian description 71–72
 – free-volume theory 72–73, 75–76
 – fully empirical models 76–77
 – kinetic modeling 68–69
free-radical multicomponent polymerization 77–80
 – polymer composition 80
 – pseudo-homopolymerization approximation 78–79
free-volume theory, free-radical homopolymerization 72–73, 75–76
freshwater demand, worldwide 287–288
front-capturing methods 179–180
front tracking methods 179
fruit
 – multiscale gas exchange 480–487
 – multiscale structure 470–472
fuel cells, proton exchange membrane 105–132
full-order model representation, NEQ/OCFE models 213
full reactor model, construction of 28

g
GA see gluconic acid
gained output ratio see GOR
 – MSF process 291
gas antisolvent (GAS) crystallization 242
gas-phase turbulence, complementary modeling 197–198
gas valve model, PSA 157
gas–solid phase equilibrium isotherms, PSA 154–157
Gauss–Newton method 451
gel effect, SAN bulk polymerization 94
gene circuits
 – repressilator 533–541
 – synthetic 494, 532
gene expression data, multilevel human inflammation model 329
general balance equations
 – adsorbent particle model 150–151
 – adsorption layer model 146
generalized simulated annealing (GSA) 422–432
 – "first-layer" analysis 424–427
 – "second-layer" analysis 427
genetic circuits 493
 – polymerases per second 498–499
 – simple 498
genetically structured model 406
geometric mean model (GMM) 73–74
geometrical model, macroscale 476–478
glass effect, SAN bulk polymerization 92–93
global optimization methods
 – gluconic acid production 458–461
 – NLP 452
global sensitivity analysis 448
gluconic acid (GA)
 – concentration errors 380–381
 – production 378–380, 457–463
glucose, mass balance 416
glucose concentration control issues, T1DM 547
glucose homeostasis, compartmental model 551–552
glutamine, mass balance 416
GMM see geometric mean model
Golgi apparatus 414, 418
GOR (gained output ratio) 291
 – steady-state simulation 310
gPROMS®
 – ethyl acetate 227–230
 – monoclonal antibodies modeling 419
 – reactive absorption of NO_x 223–224
 – simulation environment 267
gramnegative bacteria 325
granulation 35–61
 – basic mechanisms 39
 – batch drum 56–58
 – batch high-shear 58
 – bottom–up modeling 43–44
 – circuit interactions 39–40
 – circuit scale 50–51
 – computational fluid dynamics (CFD) 44
 – concurrent modeling 43–44
 – continuous drum 55–56, 59–61
 – discrete element modeling (DEM) 48–49
 – equipment, phenomena, and mechanisms 37–39
 – information flows in multiscale modeling 61
 – macroscale issues 41
 – mesoscale challenges 40
 – microscale models 40
 – middle-out modeling 43–44
 – multiscale modeling 42–45, 52–61

- need for and challenges of modeling 39–41
- operation and its significance 36
- scales of observation 45–52
- top–down modeling 43–44

granulation flowsheet simulation, commercial 51–52
granule agglomeration 47
granule scale 48
granule structure formation 47
grid dependency analysis
- interval effect on computational time 263
- population balance 260–263

grouping parameters 424–432
- biological significance 425
- GSA 424–425

growth
- cell 408–409
- crystallization processes 254–255

growth models 453–457
GSA *see* generalized simulated annealing

h

harmonic mean model (HMM) 74
hazard and operability studies (HAZOPs) 55–56
HAZOPs *see* hazard and operability studies
heart rate variability (HRV) 326–328
- effect of endotoxin injury 338–340

Heat and Temperature Management, PEM modeling 112
heat balance, adsorption layer model 147
heat transfer
- PSA 144–148, 163
- rate-based approach 188–190

heat transfer coefficient, effect of NCGs 303
heat transfer equation, MSF desalination 296
Helal–Rosso models
- brine heater model 296
- extension to dynamic models 304
- makeup mixers model 297–299
- splitters model 297
- stage model 292–296

heterogeneously catalyzed processes, reactive separations 175
high fidelity models 411
high-fidelity predictive models 9, 14–15
- applying 22–23
- constructing 16–21

high order SVD (HSVD) 371
Hill equation 407
HMM *see* harmonic mean model
homogeneously catalyzed processes, reactive separations 175
homogenization, multiscale modeling 476
homopolymerization
- free-radical 68–77
- methyl methacrylate (MMA) 85–86

Hovorka model (HM) 551, 573–576
- identifiability 554–555

HPA *see* hypothalamic-pituitary-adrenal
HRV *see* heart rate variability
HSVD *see* high order SVD
human endotoxemia, multiscale modeling 325–327
hybrid modeling techniques 22
hybridoma cultures 420–421, 434
- modeling 408–409

hydrodynamic analogy approach (HA)
- reactive separations 177, 183–188

hydrogen separation, from SMR 163–166
hyperglycemia 546
hypoglycemia 546
hypothalamic-pituitary-adrenal (HPA) axis 325, 336

i

ideal adsorbed solution (IAS) theory 157
identifiability
- of parameters 447–449
- physiological models 552–555

identifiability analysis
- Monte Carlo based 460–462
- robust 456–457

identification, of transcriptional responses 328–330
identification strategies 545–576
- dynamic model building 441–463

IDR *see* indirect response
IκB kinase (IKK) complex, activity 332–335
IMC *see* internal model control
immune system–neuroendocrine interactions, modeling 336–337
in silico modeling of inflammation 323–325
in silico gene circuits 534
in silico synthetic biology 494
indirect response (IDR) modeling 328, 330–331, 336
industrial problems, solved by dynamic processing modelling 3–31
infections
- responses to 323–324

– surgical site 322
Inflammation and Host Response to Injury Large Scale Collaborative Project 327
inflammation modeling
 – at cellular level 330–335
 – at systemic level 335–336
 – effect of endotoxin injury on HRV 338–340
 – estimation of relevant parameters 345–347
 – identification of transcriptional responses 328–330
 – *in-silico* 323–325
 – multilevel 328–340, 343–345
 – neuroendocrine–immune system interactions 336–337
 – qualitative assessment 347–360
 – systemic 321–360
 – *see also* disease progression; mathematical models
inflammatory response
 – memory effects 353–354
 – modes of dysregulation 349–353
 – modulation 354–360
 – self-limited 346, 348
 – stress hormone infusion 354–360
 – to bacterial invasion 323
information flows, multiscale modeling 61
information matrix, Fisher 448, 460–462
initial value problem (IVP) 446
injury models *see* inflammation modeling
INSERT (Integrating Separation and Reaction Technologies) 191–192
insulin-dependent diabetes *see* type 1 diabetes mellitus
integral, partial differential and algebraic equations *see* IPDAE systems
interfaces, multiphase reactive distillation 215–217
internal model control (IMC) 370
interval effect, grid dependency analysis 263
INTINT (Intelligent Column Internals for Reactive Separations) 193
intravenous glucose tolerance tests (IVGTTs) 548–551
 – experimental protocols 560
inverters, standard biological parts 495–496
IPDAE systems 4
isotherms, gas–solid phase equilibrium 154–157
IVGTTs *see* intravenous glucose tolerance tests

IVP *see* initial value problem

j
JAK-STAT signaling pathway 341
Jouyban–Acree model, solubility 245

k
kinetic modeling
 – free-radical homopolymerization 68–69
 – transport phenomena 475
kinetic models 405
 – mechanistic structured 409
Knudsen diffusion 153–154, 473
Kumar model 323–324

l
Lagrange interpolation polynomials, NEQ/OCFE models 207–210
Langmuir isotherms, extended 154–156
Laplacian spectral decomposition (LSD) 373, 377, 386–387
Latin hypercube sampling (LHS) 448
Lauffenburger model 323
LDF *see* linear driving force
leakage, synthetic biology 504
τ-leap methods 527–532
 – DSSA 528–529
length mean crystal size, definition 265
length scales, biological materials 470–472
LEQ *see* local equilibrium
leukocyte transcriptional elements 343–344
leukocytes, multilevel human inflammation model 328–329
level set (LS) method 180–181
Levenberg–Marquardt method 451
LHS *see* Latin hypercube sampling
limit cycle dynamics 385–387, 389–392, 394
linear driving force (LDF), in adsorbent particles 151–152
linear isotherms, PSA 154–155
lipopolysaccharides (LPS) 325–327
 – critical level 348–349
 – inflammation at the cellular level 331–332
 – *see also* endotoxin
liquid-film pertraction 184
liquid–liquid (LL) flash calculation 217–218
liquid–liquid extraction processes 179–180
literature review, proton exchange membrane (PEM) fuel cells 108–109

"live" and "dead" polymer chains 84–88
LL flash *see* liquid–liquid flash
loading ratio correlations (LRC)
– PSA 154, 156
local equilibrium (LEQ), in adsorbent particles 151–152
local methods, nonlinear programming (NLP) solvers 451
localization, multiscale modeling 476
low-dimensional subspaces 371
LPS *see* lipopolysaccharides
LRC *see* loading ratio correlations
LSD *see* Laplacian spectral decomposition

m
MAb *see* monoclonal antibodies
macromolecular species mass balances, MWD 85–86
macroscale geometrical model 476–478
macroscale modeling
– multiscale gas exchange 480–482
– transport phenomena 474
makeup mixers model, MSF desalination 297–299
mammalian cell culture systems
– model development and analysis 403–435
– *see also* eukaryotic cell modeling
MAP-DAP *see* mono- and di-ammonium phosphate
MAPK signaling pathway 341
Marangoni convection 180
mass balance 234
– adsorbent particle model 150–151
– adsorption layer model 147
– at collocation point 207, 209–210, 213
– crystallization processes 264
– error monitoring 219, 230
– MMA bulk homopolymerization 85–86
– monoclonal antibodies modeling 416–418
– multiphase reactive distillation 214–215
– styrene–acrylonitrile (SAN) bulk polymerization 91–92
mass transfer
– in adsorbent particles 151
– rate-based approach 188–190
mass transport, fundamentals 472–473
master equation, chemical 518–521
material balance *see* mass balance

material transport, key considerations 8
mathematical modeling
– of biological systems 404–405
– parameter estimation 444
mathematical models
– dynamic 4
– inflammation 323–324, 334, 336, 339
– mammalian cell culture systems 406–410
Matlab, simulation environment 266–267
maximum mean size optimization 276
Maxwell–Stefan (MS) diffusion 189
Maxwell–Stefan (MS) equations 153
– NEQ/OCFE models 211
MBDoE *see* model-based DOE
MBDoE-B *see* model-based DOE with backoff
MBE *see* model-based engineering
MDL *see* model description language
meal
– carbohydrate (CHO) content 546, 550
mechanistic structured kinetic model 409
membrane fuel cells, proton exchange 105–132
memory effects, inflammatory response 353–354
metabolic models 405
metaheuristics 453
metal sheet packing 185–188
method of moments, population balance 257
methyl methacrylate (MMA)
– batch free-radical solution polymerization 82–83
– bulk homopolymerization 85–86
Michaelis–Menten kinetics 407, 475–476
– respiration of fruit 481, 483–484
microbial growth
– kinetic mathematical expressions 407
– modeling 453–457
microbial systems, and mathematical modeling 406
microscale modeling
– multiscale gas exchange 482, 485
– transport phenomena 474–475
middle-out modeling, granulation 43–44
minimum variance optimization 276–277
MMA *see* methyl methacrylate
MMPC *see* multimodel predictive control
model analysis techniques 421–432
model-based data analysis 19

model-based DOE   20
- experimental protocols   560–563
- model–subject difference   569–572
- monoclonal antibody modeling   424–425
- physiological constraints   556–559
- procedure   558–559
- type 1 diabetes mellitus   545–572
- uncertainty   563–572

model-based DOE with backoff (MBDoE-B)   566–569

model-based engineering (MBE)   4, 14–23
- schematic overview   21

model-based experiment design   see model-based DOE

model-based redesign of experiments   563, 565–572

model building loop   442

model description language (MDL)   533

model development framework, for biological systems   411–412

model identification   441–463, 545–576

model parameters, estimation   18–19

model-predictive control (MPC)   370, 547
- PEM   106–107
- real-time optimization   394–398

model reduction   371
- bioprocesses   377–383
- control design   383–397
- distributed process systems   372–377
- partial differential equations   372

model-targeted experimentation   16

model–subject difference, model-based DOE   569–572

modeling
- dynamic   see dynamic modeling
- multiscale   see multiscale modeling
- steady-state   3
- vs. simulation   13

modeling regimes
- and simulation techniques   518–532
- deterministic   519–520
- stochastic   520–521

modeling rigor, against fluid dynamics complexity   176

modeling software   5

modeling tools, equation-based   11

modified OGTT (mOGTT)   561–563
- number of samples   562–566

mOGTT   see modified OGTT

molecular diffusion   472–473

molecular weight distribution (MWD)
- comparison of model predictions and experimental measurements   89
- cumulative and differential   89–90
- polymerization processes   80–90

moments
- of polymers   84–86
- population balance   257

momentum balance, adsorption layer model   148

mono- and di-ammonium phosphate (MAP-DAP) production   51

monoclonal antibodies (MAb)   403, 408
- batch and fed-batch cultures   413–434
- concentration   428–430
- feeding   413–434
- "first-layer" analysis   424–427
- first principles model derivation   415–421
- gPROMS®   419
- model analysis   421–432
- model-based DOE   424–425
- "second-layer" analysis   427
- structured models   414

Monod equation   407

Monod-type kinetics, cell growth   409

monomer conversion profiles   83

monomer mass balances, MWD   85–86

Monte Carlo based approach, parameter identifiability   448–449

Monte Carlo based practical identifiability analysis   460–462

MOSCED model
- solubility   245–246, 248

mp-MPC   see multiparametric model-predictive control

MPC   see model predictive control

mRNA
- mass balance   417
- translation of   493

MS   see Maxwell–Stefan

MSF   see multistage flash desalination process

MTR   see multitubular reactor

multibed PSA model   158–159

multicistronic sequences   496

multicomponent polymerization, free-radical   77–80

multidimensional population balance models (PBMs)   56–58

multilevel inflammation modeling   328–340, 343–345

multimodel predictive control (MMPC) 394–398
multiparametric model-predictive control (mp-MPC) 106–107
multiphase reactive distillation, NEQ/OCFE models 213–218
multiphase separation processes, orthogonal collocation on finite elements (OCFE) 203–234
multiple shooting, parameter estimation 446
multiscale gas exchange
 – CO_2 transport model 483–487
 – in fruit 480–487
 – macroscale model 480–482
 – microscale model 482
 – O_2 transport model 482–483
multiscale modeling 10–11
 – classification of 44–45
 – disease progression 321–360
 – granulation 42–45, 52–61
 – human endotoxemia 325–327
 – information flows 61
 – ontology of 44–45
 – plant-based foods 469–485
 – three-dimensional 56–58
 – transport phenomena see transport phenomena
multistage flash (MSF) desalination process 287–315
 – dedicated journals 292–293
 – dynamic modeling 303–307
 – dynamic simulation 311–312
 – environmental impact 302–303, 313–314
 – future challenges in modeling 312–315
 – issues of 289–291
 – scale formation modeling 299–303
 – stages 289–291
 – steady-state modeling 292–303
 – steady-state simulation 308–311
multitubular reactor (MTR) design 23–31
 – detailed equipment design 28–30
 – fixed-bed performance 26–27
 – tube center temperature distribution 30–31
 – see also tubular reactors
MWD see molecular weight distribution

n
NCG see noncondensable gases
NEQ/FULL model

 – ethyl acetate 227–230
 – reactive absorption of NO_x 223–225
 – technical specifications 229
NEQ/OCFE models see nonequilibrium (rate-based) models
networks, definition 321
neuroendocrine–immune system interactions, modeling 336–337
next subvolume method (NSM) 530–532
NF-κB
 – activity 350–352
 – concentration 331–335
NLP see nonlinear programming
NN see normal-to-normal intervals
NO_x, reactive absorption 220–225
noncoding DNA 509–510
 – standard biological parts 495–496
noncondensable gases (NCGs) 292
 – effect on performance of MSF process 309–310
 – effect on scale formation modeling 301–303
nonequilibrium (NEQ) (rate-based) models 203, 205–218
 – adaptive 218–220
 – butyl acetate 232–233
 – conventional and reactive absorption and distillation 207–213
 – ethyl acetate 227–230
 – Lagrange interpolation polynomials 207–210
 – multiphase reactive distillation 213–218
 – reactive absorption of NO_x 223–225
 – technical specifications 229
nonlinear dynamic models 441–443
nonlinear programming (NLP) 446
 – global methods 452
 – local methods 451
 – population-based strategies 453
 – solvers 450–453
normal-to-normal (NN) intervals 327
NRTL model, solubility 245
NRTL-SAC model, solubility 246–247
NRTL-SAC parameters, solubility models 248
NSM see next subvolume method
nuclear concentration, and nuclear activity 332
nucleation
 – crystallization processes 251–254
 – of granules 47

nucleation models, thermodynamic 252–253
nucleotide sequence, biological modeling 406
number mean crystal size, definition 265
number of samples, modified OGTT (mOGTT) 562–566
number percent distribution, definition 266
numerical solutions
 – optimal experimental design 450
 – PSA modeling 162–163
 – transport phenomena 476–480

o

O_2 transport model 482–483
observers see dynamic observers; state observers
OCFE see orthogonal collocation on finite elements
OED see optimal experimental design
OGTT see oral glucose tolerance test
online measurements, bioprocesses 377
online model-based redesign of experiments (OMBRE) 563, 565–572
 – model–subject difference 569–572
online strategy, real-time optimization 396
optimal experimental design (OED) 441–463
 – gluconic acid production 457–463
 – microbial growth 453–457
 – minimizing a cost function 449–450
 – nonlinear programming solvers 450–453
 – numerical methods 450
 – see also dynamic model building
optimal identification strategies, dynamic model building 441–463
optimization, real-time 394–397
optimization methods, crystallization processes 270–276
oral glucose tolerance test (OGTT) 550
 – experimental protocols 560–563
 – modified 561–563
orifice model, dynamic modeling 306
orthogonal collocation on finite elements (OCFE) 86–87, 203–234
 – see also nonequilibrium (rate-based) models

p

packing, structured 185–188
paracetamol, solubility 244
paracetamol–ethanol system, solubility models 247
parameter estimation 13, 18–19, 412, 419–420, 443–447
 – cost function 445
 – crystallization processes 268–269
 – experimental scheme 444–445
 – gluconic acid production 458–463
 – mathematical model formulation 444
 – numerical methods 446
 – physiological models 558–559, 562, 572
 – reactive separations 193–196
parameter identifiability 447–449
parameter validation, crystallization processes 269–273
parameter vector 411–412
parameters
 – apparent 469
 – grouping 424–432
 – "refined" model 433–434
PARSIVAL, simulation environment 266–267
part composability 494
part models 500–512
 – coding regions 508–509
 – deterministic framework 512–518
 – noncoding DNA 509–510
 – promoters 500–504
 – ribosome-binding sites 504–508
 – small RNA 511
 – terminators 511–512
part standardization 494
partial differential and algebraic equations (PDAEs), in PSA modeling 139, 162
partial differential equations (PDEs), model reduction 372
particle scale, primary 47
particle size distribution (PSD) 39
parts, synthetic biology 494–496
PBM see population balance modeling
PD see pore diffusion
PD/PK models 333–334
PDE see partial differential equation
pear
 – geometrical model 477
 – O_2 and CO_2 diffusivities 487
 – respiratory gas concentration 486
 – void network 478
PEM see proton exchange membrane
penetration/surface renewal model, rate-based approach 188
persistent inflammatory response 350

phase disappearance, key considerations 8
phases see multiphase
phenotype (response), of a biological system 321–322
physiological models
– constraints 556–559
– design procedure 558–559
– experimental protocols 560–563
– identifiability 552–555
– identification 545–576
– need for experiment design 548–550
– standard clinical tests 550–551
PI see polydispersity index
Picard iteration, in PSA modeling 163
PID see proportional-integral-derivative
plant-based foods
– continuum models 469
– length scales 470–472
– multiscale modeling of transport phenomena 469–485
plasma cortisol levels 348
plasmodesmata 472
POD see proper orthogonal decomposition
Poisson τ-leap methods 527–528
Pol (RNA polymerase) 498–499
– pools 513, 515
polydispersity index (PI) 81
polymer chains, "live" and "dead" 84–88
polymer composition, free-radical multicomponent polymerization 80
polymer molecular properties, modeling 80–90
polymer moments 84–86
polymerases per second (PoPS), genetic circuit 498–499
polymerization processes
– free-radical homopolymerization 68–77
– free-radical multicomponent polymerization 77–80
– modeling 67–97
– molecular properties 80–90
– SAN 90–97
polyolefins, MWD 81
pools, synthetic biology 497–500
PoPS see polymerases per second
population balance
– crystal average size 260
– crystal size distribution (CSD) 257, 261–262
– discretization methods 258–259
– grid dependency analysis 260–263
– method of moments 257
– supersaturation 260
– volume mean size 261–262
population balance equation (PBE)
– batch high-shear granulation 58
– continuous drum granulation 59–61
– general 50
population balance modeling (PBM) 406, 410
population balance models (PBMs) 49–50
– multidimensional 56–58
population-based strategies, NLP solvers 453
pore diffusion (PD), in adsorbent particles 151, 153–154
postprandial glucose test 550
postprocessing method, residence time distribution 195
PPol (promoter–RNA polymerase complex) 499
practical identifiability analysis 447–449
– Monte Carlo based 460–462
predictive microbiology 453–457
predictive process models 6
predictive thermodynamic models 244–247
– activity coefficient model 247
– Jouyban–Acree model 245
– MOSCED model 245–246
– NRTL-SAC model 246–247
– UNIFAC model 247
pressure drop studies 193–197
pressure swing adsorption (PSA) processes 137–167
– adsorbent bed models 144–145
– adsorbent particle model 150–157
– adsorption layer model 146–150
– case-study applications 144–163
– configurations 163–164
– DAEs 162
– gas valve model 157
– heat transfer 144–148, 163
– model formulation 163–166
– multibed PSA model 158
– numerical solution of model 162–163
– PDAEs 139, 162
– single-bed adsorber 145–146
– state transition network approach 158–161
– successive substitution method 163
– transport properties 149
primary particle scale 47

pro-inflammatory response 343–344
process control and control devices, key considerations 8
process modeling, future challenges 312–313
Process Modeling Tool (ProMoT) 532
process optimization, future challenges 314–315
process systems engineering (PSE) 545, 572
processing systems
 – biological 321–572
 – chemical 1–315
proinflammatory mediators 324, 354
prokaryotic cells, and mathematical modeling 406
ProMoT 532–533
promoter dynamics 513
promoters 500–504
 – standard biological parts 495
proper orthogonal decomposition (POD) 373, 376, 381
 – real-time optimization 396
 – robust reactor control 386–389
proportional-integral-derivative (PID) controller 370
 – diabetes 547
protein synthesis, scheme 414
proton exchange membrane (PEM) fuel cells
 – heat and temperature management 112
 – literature review 108–109
 – modeling and control 105–132
 – motivation 109–113
 – MPC 106–107
 – reactant flow management 112
 – water management 113
PSA see pressure swing adsorption
PSE see process systems engineering
pseudo-homopolymerization approximation 78–79, 94–95

q

quench crystallization see salting out crystallization

r

randomness, in biological systems 518
rate-based approach (RBA) 177–178, 188–193
rate coefficients, chain length dependent 73–76
RBS see ribosome-binding sites

reactant diffusion, Fickian description 71–72
reactant flow management, PEM modeling 112
reactive absorption, of NO_x 220–225
reactive and multiphase separation processes 203–234
reactive distillation
 – butyl acetate 231
 – of ethyl acetate 225–231
reactive dividing wall column 190–192
reactive multiphase distillation, butyl acetate production 231–234
reactive separations (RS)
 – absorber and desorber performances 190–191
 – autocatalyzed processes 175
 – classification of modeling methods 176–178
 – complementary approach 173, 196–199
 – computational fluid dynamics 176
 – film-like-flow problems 178–179
 – fluid-dynamic approach 178–183
 – hydrodynamic analogy approach 183–188
 – modeling of 173–199
 – orthogonal collocation on finite elements 203–234
 – parameter estimation and virtual experiments 193–196
 – rate-based approach 177–178, 188–193
 – structured packings 174–175
reactor diameter, MTR 28
reactor tube model, construction of 27
reactors see multitubular reactor design; tubular reactors
real-time optimization 394–397
 – online strategy 396
reboiler heat duty, ethyl acetate reactive distillation 228
recrystallization methods 241–242
recursive projection methods (RPM) 371
reduced order dynamic modeling, reactive and multiphase separation processes 203–234
reduced order model (ROM) 381–383
 – NEQ/OCFE 213
reference trajectory 389–390, 392
"refined" model 433–434
relative supersaturation, crystallization processes 242–243

repressilator   533–541
- deterministic simulation   539
- ribosome-binding sites (RBS)   537
- scheme   533
- stochastic simulation   539
repressor
- biological parts   496
- time courses   540
residence time distribution   195
response
- energetic   343–344
- indirect   see indirect response
- of a biological system   321–322
- persistent inflammatory   350
- pro-inflammatory   343–344
- systemic inflammatory   322, 325–326
- transcriptional   328–330
- see also inflammatory
ribosome-binding sites (RBS)   504–508
- repressilator   537
- standard biological parts   495
ribosome pools, and RNA polymerases   497–498
ribosomes per second (RiPS), genetic circuit   498
riboswitches   497, 505
ribozymes   497, 505
RiPS   see ribosomes per second
RNA
- interaction with chemicals   497
- see also mRNA; small RNA
RNA polymerases   495–496
- and ribosome pools   497–498
robust control
- distributed processes and bioprocesses   369–397
- proper orthogonal decomposition (POD)   386–389
- tubular reactors   386–394
robust identifiability analysis   456–457
ROM   see reduced order model
RPM   see recursive projection methods

S

SA   see sensitivity analysis
salting out crystallization   241
SAN   see styrene acrylonitrile
Sauter crystal size   see area mean crystalsize
SAX method   329
SBML   533
scale formation modeling   299–303
- effect of NCGs   301–302
- environmental impact   302–303
- estimation of dynamic brine heater fouling profile   301
scales of interest, in granulation   45–52
Schulz distribution   81–82
SCM   see single cell model
SD   see surface diffusion
SDNN   see standard deviation of normal interbeat intervals
seawater
- specific heat capacity/enthalpy   298
- see also multistage flash (MSF) desalination process
"second layer" analysis, monoclonal antibody modeling   427
secretion process, modeling   410
seeded cooling crystallization optimization   274–276
seeded crystallization   242
segregated models   405
self-limited inflammatory response   346, 348
semibatch simulation, SAN copolymerization   96–97
sensitivity analysis (SA)   408, 411–412
- monoclonal antibodies   421–432
sensitivity indices (SI)   411–412
- analysis   423
- high   427–431
- viable cell concentration   428–429
sensitivity test, Fourier amplitude   423
separation processes   see multiphase separation processes; reactive separations
sequential quadratic approach (SQP)   451
SI   see sensitivity indices
signal carriers
- synthetic biology   494, 496–497
signal per second (SiPS)
- deterministic fluxes   516–517
- genetic circuit   498
simulation
- and dynamic modeling   3–31
- and modeling regimes   518–532
- batch and semibatch   95–97
- see also stochastic simulation algorithm
simulation environment, crystallization processes   266–267
single-bed adsorber, PSA   139–140, 145–146, 158
single cell model (SCM)   405–406
- coupled with PBM   410
single shooting, parameter estimation   446

single-tube experiment setup, MTR design 27
singular value decomposition (SVD) 371
SiPS   *see* signal per second
SIRS   *see* systemic inflammatory response
Skarstrom cycle, two-bed 138
small RNA (sRNA) 506, 511
– standard biological parts 495
SMR   *see* steam-methane reforming off gas
SNS   *see* sympathetic nervous system
Sobol' global SI 423, 425, 427
solubility models 243–250
– correlative thermodynamic 244–245
– empirical 243–244
– examples 247–250
– parameters 248
– predictive thermodynamic   *see* predictive thermodynamic models
– solution concentration measurement 250
solution theory, IAS 157
solution concentration
– measurement 250
– validation 271–272
spatially resolved stochastic simulation algorithms (SSSA) 525–527
– coarse-grained 529–532
species concentrations vector 512
specific enthalpy, of water 298
specific heat capacity, of water 298
splitters model, MSF desalination 297
sRNA   *see* small RNA
SSA   *see* stochastic simulation algorithm
SSSA   *see* spatially resolved stochastic simulation algorithms
stage model
– dynamic modeling 305
– MSF desalination 292–296
stage to stage model, desalination 294
standard basic parts 494
Standard Biological Parts classification 494–496
standard clinical tests, physiological models 550–551
standard deviation of normal interbeat intervals (SDNN) 327–328
state evolution, bioprocesses 380–381
state observers
– general considerations 377–378
– *see also* dynamic observers
state transition network (STN) approach, PSA 158–161

static batch cultures 409
steady-state design 13
steady-state modeling 3
– MSF desalination 292–303
steady-state optimization 13
steady-state simulation 13
– future challenges 313
– MSF desalination 308–311
steam-methane reforming off gas (SMR) 137, 163
stem cell modeling 406
STN   *see* state transition network
stochastic methods, global 452
stochastic modeling regimes 520–521
stochastic simulation, repressilator 539
stochastic simulation algorithm (SSA) 519, 521–532
– coarse-grained methods 527–532
– delayed 524–526
– exact algorithms 522–527
– spatially resolved 525–527
stoichiometric matrix 512
stoichiometric models 405
– animal cell growth 408
STOP codon 510
stream-line representation
– corrugated sheet packing 197–198
– pressure drop studies 195–196
stress hormone infusion, inflammatory response modulation 354–360
structural difference, between a model and a subject 566, 569, 571
structure concept, biological modeling 406
structured models
– advantages 410
– monoclonal antibodies 414
structured packings (SP) 174–175
– computational fluid dynamics 193–195
– hydrodynamic analogy approach 185–188
styrene–acrylonitrile (SAN) polymerization 90–97
– batch and semibatch simulation 95–97
– diffusion limitations 92–94
– kinetic diagram 90–91
– mass balances 91–92
– modeling results 95–97
– pseudo-homopolymerization approximation 94–95
subject–model difference, model-based DOE 569–572

successive substitution method, in PSA modeling 163
supersaturation
– crystallization driving force 242–243
– grid dependency analysis 260
surface diffusion (SD) 151–152
surgical site infections 322
surrogate models 325
SVD *see* singular value decomposition
sympathetic division 325
sympathetic nervous system (SNS) 336–337
symplast 472
synthetic biology
– abstraction hierarchy 494
– devices 494–496
– dynamic modeling and construction of cell systems 493–541
– fluxes 497–500
– general nomenclature 494–500
– *in silico* 494
– part models 500–512
– parts 494–496
– pools 497–500
– signal carriers 494, 496–497
– standard basic parts 494
synthetic gene circuits 494
system identification, dynamic model building 441–463
systemic inflammation in humans, multiscale model 321–360
systemic inflammatory response (SIRS) 322, 325–326
systemic level inflammation modeling 335–336
systems biology 493
systems theory, and biology 321–322

t

T1DM *see* type 1 diabetes mellitus
TBT *see* top brine temperature
temperature elevation (TE) 292
– empirical correlations for 299–300
terephthaldehyde (TPAL), production of 25
terminals, in MDL 533
termination rate distribution (TRD) 84–85
terminator dynamics 513
terminators 511–512
– standard biological parts 495–496
thermal death time (TDT) model 454
thermal swing adsorption (TSA) 138–139
thermodynamic nucleation models 252–253

thermodynamic solubility models 244–247
thin-film model representation, NEQ/OCFE models 205–206
three-dimensional multiscale modeling, granulation 56–58
three-stage polymerization model (TSPM) 76
time courses, repressilator 540
TLR4 receptor, dynamics 331
TNF-a 354
top brine temperature (TBT), MSF process 291, 308
top–down modeling, granulation 43–44
total sensitivity index (TSI) 423
TPAL *see* terephthaldehyde
tracer method, residence time distribution 195
trajectory
– dissipative processes 374
– reference 389–390, 392
transcription, DNA 493
transcription factor pools 515–516
transcription factors, involved in inflammation 331
transcription unit 495
transcriptional analysis, and major response elements 340–343
transcriptional responses, identification 328–330
transcriptional state (TS) 329–330
translation, RNA 493
transport phenomena
– in plant-based foods 469–485
– kinetic modeling 475
– macroscale approach 474
– mass transport fundamentals 472–473
– microscale approach 474–475
– molecular diffusion 472–473
– multiscale modeling 472–476
– numerical solution 476–480
transport properties, PSA 149
tray-by-tray model structure, NEQ/OCFE models 207
TRD *see* termination rate distribution
TS *see* transcriptional state
TSA *see* thermal swing adsorption
TSI *see* total sensitivity index
TSPM *see* three-stage polymerization model
tube arrangement, MTR 29
tube limit, MTR 28
tube size, MTR 29
tubular reactors 377

- control applications 383–398
- controller synthesis 389–392
- gluconic acid production 378–380
- multimodel predictive control 394–397
- robust control of 386–394
- *see also* multitubular reactor design

two-film model, rate-based approach 188–190

type 1 diabetes mellitus (T1DM), model-based DOE 545–572

u

ultrasound crystallization 242
uncertainty, in model-based DOE 563–572
UNIFAC model 247–248
UNIQUAC model 245, 247
unseeded crystallization 242
unsegregated models 405
upscaling, multiscale modeling 476

v

van Laar model 245
variables, in MDL 533
vector
- control 450
- flux 512
- of species concentrations 512
- parameter 411–412

vessel scale, granulation 49
viable cell concentration, sensitivity indices 428–429
virtual experiments, reactive separations 193–196
void network, apple 478

volume mean crystal size, definition 265
volume mean size
- optimization 275, 277
- population balance 261–262
- validation 271–272

volume of fluid (VOF) method 180–181
volume percent crystal size distributions, validation 273
volume percent distribution, definition 266

w

WAP *see* wearable artificial pancreas
water, specific heat capacity/enthalpy 298
water management, PEM modeling 113
water supply, worldwide 287–288
wave evolution studies, reactive separations 181–183
WCLD *see* weight chain length distribution
wearable artificial pancreas (WAP) 546
weight chain length distribution (WCLD) 82–83
weight molecular weight distribution (WMWD) 86
Wilson model 245
WMWD *see* weight molecular weight distribution
world population, and freshwater demand 288

x

p-xylene, oxidation to TPAL 25

z

zero-gravity distillation 184–188